Lecture Notes in Physics

For information about Vols. 1–151, please contact your bookseller or Springer-Verlag.

Vol. 152: Physics of Narrow Gap Semiconductors. Proceedings, 1981. Edited by E. Gornik, H. Heinrich and L. Palmetshofer. XIII, 485 pages. 1982.

Vol. 153: Mathematical Problems in Theoretical Physics. Proceedings, 1981. Edited by R. Schrader, R. Seiler, and D.A. Uhlenbrock. XII, 429 pages. 1982.

Vol. 154: Macroscopic Properties of Disordered Media. Proceedings, 1981. Edited by R. Burridge, S. Childress, and G. Papanicolaou. VII, 307 pages. 1982.

Vol. 155: Quantum Optics. Proceedings, 1981. Edited by C.A. Engelbrecht. VIII, 329 pages. 1982.

Vol. 156: Resonances in Heavy Ion Reactions. Proceedings, 1981. Edited by K.A. Eberhard. XII, 448 pages. 1982.

Vol. 157: P. Niyogi, Integral Equation Method in Transonic Flow. XI, 189 pages. 1982.

Vol. 158: Dynamics of Nuclear Fission and Related Collective Phenomena. Proceedings, 1981. Edited by P. David, T. Mayer-Kuckuk, and A. van der Woude. X, 462 pages. 1982.

Vol. 159: E. Seiler, Gauge Theories as a Problem of Constructive Quantum Field Theory and Statistical Mechanics. V, 192 pages. 1982.

Vol. 160: Unified Theories of Elementary Particles. Critical Assessment and Prospects. Proceedings, 1981. Edited by P. Breitenlohner and H.P. Dürr. VI, 217 pages. 1982.

Vol. 161: Interacting Bosons in Nuclei. Proceedings, 1981. Edited by J.S. Dehesa, J.M.G. Gomez, and J. Ros. V, 209 pages. 1982.

Vol. 162: Relativistic Action at a Distance: Classical and Quantum Aspects. Proceedings, 1981. Edited by J. Llosa. X, 263 pages. 1982.

Vol. 163: J.S. Darrozes, C. Francois, Mécanique des Fluides Incompressibles. XIX, 459 pages. 1982.

Vol. 164: Stability of Thermodynamic Systems. Proceedings, 1981. Edited by J. Casas-Vázquez and G. Lebon. VII, 321 pages. 1982.

Vol. 165: N. Mukunda, H. van Dam, L.C. Biedenharn, Relativistic Models of Extended Hadrons Obeying a Mass-Spin Trajectory Constraint. Edited by A. Böhm and J.D. Dollard. VI, 163 pages. 1982.

Vol. 166: Computer Simulation of Solids. Edited by C.R.A. Catlow and W.C. Mackrodt. XII, 320 pages. 1982.

Vol. 167: G. Fieck, Symmetry of Polycentric Systems. VI, 137 pages. 1982.

Vol. 168: Heavy-Ion Collisions. Proceedings, 1982. Edited by G. Madurga and M. Lozano. VI, 429 pages. 1982.

Vol. 169: K. Sundermeyer, Constrained Dynamics. IV, 318 pages. 1982.

Vol. 170: Eighth International Conference on Numerical Methods in Fluid Dynamics. Proceedings, 1982. Edited by E. Krause. X, 569 pages. 1982.

Vol. 171: Time-Dependent Hartree-Fock and Beyond. Proceedings, 1982. Edited by K. Goeke and P.-G. Reinhard. VIII, 426 pages. 1982.

Vol. 172: Ionic Liquids, Molten Salts and Polyelectrolytes. Proceedings, 1982. Edited by K.-H. Bennemann, F. Brouers, and D. Quitmann. VII, 253 pages. 1982.

Vol. 173: Stochastic Processes in Quantum Theory and Statistical Physics. Proceedings, 1981. Edited by S. Albeverio, Ph. Combe, and M. Sirugue-Collin. VIII, 337 pages. 1982.

Vol. 174: A. Kadić, D.G.B. Edelen, A Gauge Theory of Dislocations and Disclinations. VII, 290 pages. 1983.

Vol. 175: Defect Complexes in Semiconductor Structures. Proceedings, 1982. Edited by J. Giber, F. Beleznay, J.C. Szép, and J. László. VI, 308 pages. 1983.

Vol. 176: Gauge Theory and Gravitation. Proceedings, 1982. Edited by K. Kikkawa, N. Nakanishi, and H. Nariai. X, 316 pages. 1983.

Vol. 177: Application of High Magnetic Fields in Semiconductor Physics. Proceedings, 1982. Edited by G. Landwehr. XII, 552 pages. 1983.

Vol. 178: Detectors in Heavy-Ion Reactions. Proceedings, 1982. Edited by W. von Oertzen. VIII, 258 pages. 1983.

Vol. 179: Dynamical Systems and Chaos. Proceedings, 1982. Edited by L. Garrido. XIV, 298 pages. 1983.

Vol. 180: Group Theoretical Methods in Physics. Proceedings, 1982. Edited by M. Serdaroğlu and E. İnönü. XI, 569 pages. 1983.

Vol. 181: Gauge Theories of the Eighties. Proceedings, 1982. Edited by R. Raitio and J. Lindfors. V, 644 pages. 1983.

Vol. 182: Laser Physics. Proceedings, 1983. Edited by J.D. Harvey and D.F. Walls. V, 263 pages. 1983.

Vol. 183: J.D. Gunton, M. Droz, Introduction to the Theory of Metastable and Unstable States. VI, 140 pages. 1983.

Vol. 184: Stochastic Processes – Formalism and Applications. Proceedings, 1982. Edited by G.S. Agarwal and S. Dattagupta. VI, 324 pages. 1983.

Vol. 185: H.N. Shirer, R. Wells, Mathematical Structure of the Singularities at the Transitions between Steady States in Hydrodynamic Systems. XI, 276 pages. 1983.

Vol. 186: Critical Phenomena. Proceedings, 1982. Edited by F.J.W. Hahne. VII, 353 pages. 1983.

Vol. 187: Density Functional Theory. Edited by J. Keller and J.L. Gázquez. V, 301 pages. 1983.

Vol. 188: A.P. Balachandran, G. Marmo, B.-S. Skagerstam, A. Stern, Gauge Symmetries and Fibre Bundles. IV, 140 pages. 1983.

Vol. 189: Nonlinear Phenomena. Proceedings, 1982. Edited by K.B. Wolf. XII, 453 pages. 1983.

Vol. 190: K. Kraus, States, Effects, and Operations. Edited by A. Böhm, J.W. Dollard and W.H. Wootters. IX, 151 pages. 1983.

Vol. 191: Photon Photon Collisions. Proceedings, 1983. Edited by Ch. Berger. V, 417 pages. 1983.

Vol. 192: Heidelberg Colloquium on Spin Glasses. Proceedings, 1983. Edited by J.L. van Hemmen and I. Morgenstern. VII, 356 pages. 1983.

Vol. 193: Cool Stars, Stellar Systems, and the Sun. Proceedings, 1983. Edited by S.L. Balliunas and L. Hartmann. VII, 364 pages. 1984.

Vol. 194: P. Pascual, R. Tarrach, QCD: Renormalization for the Practitioner. V, 277 pages. 1984.

Lecture Notes in Physics

Edited by H. Araki, Kyoto, J. Ehlers, München, K. Hepp, Zürich
R. Kippenhahn, München, H. A. Weidenmüller, Heidelberg
and J. Zittartz, Köln

215

Computing in Accelerator Design and Operation

Proceedings of the Europhysics Conference
Held at the Hahn-Meitner-Institut
für Kernforschung Berlin GmbH
Berlin, Germany, September 20–23, 1983

Edited by W. Busse and R. Zelazny

Springer-Verlag
Berlin Heidelberg GmbH 1984

Editors

Winfried Busse
Hahn-Meitner-Institut für Kernforschung Berlin GmbH
Bereich Kern- und Strahlenphysik
Glienickerstr. 100, D-1000 Berlin 39

Roman Zelazny
RCC CYFRONET, IAE
PL-05-400 Otwock-Swierk, Poland

ISBN 978-3-540-13909-6   ISBN 978-3-540-39130-2 (eBook)
DOI 10.1007/978-3-540-39130-2

This work is subject to copyright. All rights are reserved, whether the whole or part of the material is concerned, specifically those of translation, reprinting, re-use of illustrations, broadcasting, reproduction by photocopying machine or similar means, and storage in data banks. Under § 54 of the German Copyright Law where copies are made for other than private use, a fee is payable to "Verwertungsgesellschaft Wort", Munich.

© Springer-Verlag Berlin Heidelberg 1984
Originally published by Springer-Verlag Berlin Heidelberg New York in 1984
2153/3140-543210

PREFACE

Accelerators became long ago a very important research tool in nuclear physics and its industrial and medical applications.

Recently we have observed their impressive development, so well illustrated by the design and construction of very large accelerator systems like the CERN-SPS in Geneva and Fermilab in Batavia. New and still larger accelerators are under study and will be built with international support, for example LEP, once again on the CERN site in Geneva.

During feasibility studies, design and construction, and also during operation, computing plays a very essential role in enabling designers and operators to perform their duties properly. In all these phases of the accelerator life-cycle computers are used very extensively.

It is difficult to state in which of these phases the application of computers is most important. Some people claim that without digital control the usage of accelerators in research would not - at present - be possible. Additionally, physical experiments with particle-accelerator beams cannot be conceived without the decisive role of computers in acquisition and processing of experimental data.

All this means that computers and accelerators are tightly affined to each other. This symbiosis is the essence of progress in both fields.

This explanation makes it obvious that a conference on computing in accelerator design and operation had to be organized. Due to the initiative of one of us (R.Z.) while a member of the Computational Physics Group of the European Physical Society, the board of the CPG decided to convene such a conference under the Europhysics Conference label. The European Physical Society supported this idea vigorously. The sponsoring organizations are acknowledged with gratitude. Without their support and assistance the idea would not have materialized.

The conference was organized around three topical subjects: computing for design applications, for digital control of accelerators and for operational aspects. The subjects of invited lectures as well as the lecturers were carefully selected by the Scientific Advisory Committee. The invited lectures were the only oral presentations in plenary sessions. Each subject was introduced by a 15-20 minutes talk by a leading prominent personality in the field. Invited lectures were given 45 minutes for presentation and discussion. All contributed papers were presented at poster sessions, a format which was positively accepted by the participants. In the framework of the conference two workshops were organized on request. The first was devoted to lattice calculations of accelerator structures, the second to local area network concepts in the field of digital control of accelerators.

It was felt that this conference was necessary to bring together accelerator designers, builders and users, because a common understanding between them is still to be created. Therefore, as an important corollary, both the Scientific Advisory Committee and the participants of the conference endorsed unanimously the idea of organizing such a conference each third year. The Computational Physics Group Board has been approached with this suggestion. Let us hope that the year 1986 will be the year of the next Europhysics Conference in Accelerator Design and Operation.

Roman Zelazny
Regional Computing Centre CYFRONET
Otwock-Swierk, Poland

Winfried Busse
Hahn-Meitner Institute
Berlin, Germany

International Scientific Advisory Committee

R. Zelazny	RCCA, Otwock-Swierk	Poland	(chairman)
F. Beck	FNAL, Batavia	USA	
W. Busse	HMI, Berlin	Germany	
M. C. Crowley-Milling	CERN, Geneva	Switzerland	
M. Edwards	RAL, Didcot	United Kingdom	
G. N. Florov	JINR, Dubna	USSR	
O. Houssin	CGR MeV, Buc	France	
E. Keil	CERN, Geneva	Switzerland	
W. Klotz	BESSY, Berlin	Germany	
S. Kulinski	INST, Otwock-Swierk	Poland	
F. Peters	DESY, Hamburg	Germany	
M. Promé	GANIL, Caen	France	
J. Schwabe	INP, Krakow	Poland	
H. Sherman	Daresbury Lab., Warrington	United Kingdom	
A. N. Skrynski	INP, Novosybirsk	USSR	
P. Strehl	GSI, Darmstadt	Germany	

Local Organizing Committee

W. Buchholz	BESSY	
W. Busse	HMI P - VICKSI	(chairman)
K. H. Degenhardt	HMI D/M	
H. Kluge	HMI P	
W.-D. Klotz	BESSY	
G. Liar de Martin	HMI P - VICKSI	(conf. secretariat)
K. H. Maier	HMI P	
R. Maier	BESSY	
B. Martin	HMI P - VICKSI	
R. Michaelsen	HMI P - VICKSI	
B. Spellmeyer	HMI P - VICKSI	
K. Ziegler	HMI P - VICKSI	

Sponsors

European Physical Society
Deutsche Physikalische Gesellschaft
Regional Computation Centre of Atomic Energy, Otwock-Swierk, Poland
Hahn-Meitner-Institut für Kernforschung Berlin GmbH

Commercial Sponsors

DANFYSIK - Jyllinge, Denmark
HEINZINGER Regel- und Meßtechnik - Rosenheim, Germany
INCAA Special Systems for Industry and Science - Apeldoorn, Holland
KINETIC SYSTEMS INTERNATIONAL SA - Geneva, Switzerland
KNÜRR AG - München, Germany
LEYBOLD-HERAEUS GmbH - Köln, Germany
SILENA Wissenschaftliche Instrumente GmbH - Hasselroth, Germany

Welcome Address by   Prof. K.H. Lindenberger, Scientific Director of the
                                Hahn-Meitner Institute, Berlin

Dear Colleagues,

On behalf of the Hahn-Meitner Institute I welcome you heartily to our town. We are pleased and feel honoured that you have chosen Berlin as the place of your conference. By helping to organize this meeting we can, in a certain way, pay back some of the debt which we owe to the community of accelerator builders.

When, more than ten years ago, we started to convert our small Van de Graaff installation into a heavy-ion facility, we had little experience in accelerator technology, especially in how to run such a system with the help of a computer. In this situation we thought it best to ask the professionals for help and we got this help in a really generous way. This talk is not the right opportunity to give a full record of this story, but I would like to mention as an example two outstanding members of your community whose skill, ambition and enthusiasm had great impact on the project and who were essential for its success.

Prof. Hagedoorn from Eindhoven contributed much to the understanding of the orbit-dynamics in our cyclotron. One direct result of this is a programme by which the control computer can center and isochronize the beam. Dr. Susini from CERN was in charge of the design and the construction of the RF-systems of our cyclotron and he did it in such a way that they are really computer-compatible. In the meantime, our accelerator crew has joined your community, and that you meet at our place may be a hint that they now are passing as professionals. But to say it once more: without your assistance we would not have been able to get such a system running with the very good performance that, as we think, we now have at our disposal. I am glad that I can use this opportunity to express our gratitude.

CERN was an especially important source of information and also of very practical help. So I am very pleased that the opening honorary lecture will be given by Dr. Adams, who was twice responsible for the construction of the large accelerators in Geneva. We all know how brilliantly this job was done, what high technical standards have been achieved and what important experiments can be done with those machines. Here I would like to make a remark on the sideline: I was always very impressed how effective and smooth is the international cooperation in the field of particle physics and accelerator building. I think it would be of great value for all of us if, in other technical and political matters of world-wide impact, the same efficiency of cooperation could be achieved as in particle physics. Once more my special welcome to you, Dr. Adams, here in Berlin.

It is according to the hopes I just mentioned that the chairman of your conference, Prof. Zelazny, comes from Poland, from the Nuclear Research Centre at Swierk. The Hahn-Meitner Institute has a number of scientific contacts with this institute and we are happy that we can cooperate with you, Prof. Zelazny, in running this conference. But, as I mentioned, we have also other contacts to Swierk: the volleyball team of Swierk beat the Hahn-Meitner crew 2:1 when they met at Swierk in 75. Best welcome also to you Prof. Zelazny. I hope the local staff will make life easy for you in your job as a chairman.

I would like to thank the Free University that we can hold the conference in this place. We had to do so because, at our institute, we have no facilities to handle a meeting of this size.

Now nothing is left but to wish you a lively and interesting conference which gives you new ideas for your work at home. But I also hope that beside the work here you will find some time to stroll around Kurfürstendamm, to meet Nefretiti at the Egyptian Museum or to find out what else is going on in our town. After the afternoon session today, however, we would be very pleased if you could visit us at the Hahn-Meitner Institute to have a look at our installations and to join us for a cocktail. Once more, welcome to Berlin and good luck for your conference.

Thank you.

Welcome **Address** by   the chairman of the conference,
                                 Prof. Roman Zelazny, CYFRONET
                                 Otwock-Swierk, Poland

Ladies and Gentlemen, Dear Colleagues,

All of you may observe the enormous development of activities in the field of accelerator construction and their application in research, industry and medicine. New accelerators are proposed, are under design and construction, start their operation.

Computers play a very important role in the design, in feasibility studies and in operation of accelerators. They are applied in many interesting and innovative ways: for computer-assisted design, for digital control, for beam administration and particularly in experiments with beams of accelerator particles.

The conference organized under the auspices of the European Physical Society and its Computational Physics Group is devoted to various aspects of computing in accelerator design and operation.

Originally scheduled to be organized in Poland in September 1982, it has been postponed and moved to Berlin. I wish to thank very much the Hahn-Meitner Institute authorities, particularly Prof. Lindenberger and Dr. W. Busse for their willingness to take over the organization of the conference. In a short period of time the local organizers performed a very good job, enabling us to meet today to open this, as I do hope, interesting and important meeting.

Using this chance, I wish to thank not only the Local Organizing Committee headed by Dr. Busse but also other sponsoring organizations: the Deutsche Physikalische Gesellschaft and the Regional Computing Centre of Atomic Energy CYFRONET. It is my special pleasure to thank all members of the Scientific Advisory Committee for their effort concerning the scientific programme of the conference and all invited lecturers for accepting the invitation to deliver the invited talks. Their contributions make the conference an important and interesting event.

Particular thanks are due to Sir John Adams for his acceptance to deliver the honorary lecture at this conference. It seems to me that the community of European physicists shall consider that this homage to his activities in the field of accelerator development is well deserved and that they join the Scientific Advisory Committee's opinion with applause.

I welcome sincerely all the participants. Without you all the conceptual and organizational efforts would be empty. You are the salt of the earth. It is done for you, you will make it finally a success by the contribution of your papers, you make it vivid by discussion and exchange of views and experience. All the organizers have done their duty. The critical mass for a chain reaction necessary to create a peaceful explosion of ideas, concepts, interactions among interesting people has been prepared. Let it go. I declare the Europhysics Conference on "Computing in Accelerator Design and Operation" open.

Thank you very much for your attention.

CONTENTS

Future High Energy Accelerators (Honorary Invited Lecture)   1
J. Adams

A: COMPUTING for ACCELERATOR DESIGN

Beam Optics and Dynamics   11
E. J. N. Wilson

Design of R.F. Cavities   21
T. Weiland

Computer Aided Magnet Design   33
C. W. Trowbridge

Beam Instabilities and Computer Simulations   50
A. Piwinski

Calculation of Polarization Effects   59
A. W. Chao

Particle Tracking in Accelerators with Higher Order Multipole Fields   75
A. Wrulich

Programs for Designing the Accelerating Cavities for Linear Accelerators   86
S. Kulinski, L. Sawlewicz, J. Sekutowicz

The MAGMI Program for Double Pass Electron Linear Accelerators   92
T. Czosnyka, K. Deutschman, S. Kulinski, S. Zaremba

A FORTRAN Program (RELAX3D) to Solve the 3 Dimensional Poisson (Laplace) Equation   98
H. Houtman, C. J. Kost

Calculation of Three Dimensional Electric Fields by Successive Over-Relaxation in the Central Region of a Cyclotron   104
S. Oh, R. Pogson, M. Yoon

The Design of the Accelerating Cavity for SUSE with the Aid of the Three-Dimensional Cavity Calculation Program CAV3D   110
W. Wilhelm

The Further Development of the Calculation of the Three Dimensional Electric Field in the Central Region of the INR Cyclotron   116
Mao-bai Chen, Wen-bin Sen

Particle Tracking Using Lie Algebraic Methods   122
A. J. Dragt, D. R. Douglas

Numerical Investigation of Bunch-Merging in a Heavy-Ion-Synchrotron   128
I. Bozsik, I. Hofmann, A. Jahnke, R. W. Müller

Nonlinear Aspects of Landau Damping in Computer Simulation of   134
the Microwave Instability
I. Hofmann

The Transport Theory of Particle Beam-Congregation in   140
Six-Dimensional Phase Space
Cao Qing-xi, Guan Xia-ling

The MAD Program (Methodical Accelerator Design)   146
F. Ch. Iselin

Analogue Computer Display of Accelerator Beam Optics   152
K. Brand

A Monte Carlo Beam Transport Program, REVMOC   158
C. Kost, P. A. Reeve

Multiparticle Codes Developed at GANIL   164
J. Sauret, A. Chabert, M. Promé

MIRKO - An Interactive Program for Beam Lines and Synchrotrons   170
B. Franczak

Aperture Studies of the BNL Colliding Beam Accelerator with   176
Reduced Superperiodicity
G. F. Dell

The Study of Misalignment Characteristics of Beam Optical   182
Components of HI-13 Tandem
Guan Xia-ling, Cao Qing-xi

Calculations for the Design and Modification of the 2 Cyclotrons   188
of S.A.R.A.
P. S. Albrand, J. L. Belmont, F. Ripouteau

Magnetic Field Optimization and Beam Dynamics Calculations for   193
SUSE
W. Schott, E. Zech, N. Rösch

'DFLKTR' The Code for Designing the Electrostatic Extraction   199
System for Cyclotrons
R. C. Sethi, A. S. Divatia

RFQ Design Considerations   206
P. Junior, H. Deitinghoff, K. D. Halfmann, A. Schempp, N. Zoubek

Effects of Higher Order Multipole Fields on High Current RFQ   212
Accelerator Design
G. E. McMichael, B. G. Chidley

Versatile Codes and Effective Method for Orbit Programming with   218
Actually Existing First Harmonics in Cyclotron
Mao-bai Chen, Sen-lin Xu, Wen-bin Sen

Calculations of the Heavy Ion Saclay Tandem Post Accelerator   224
Beams
S. Valero, B. Cauvin, J. P. Fouan, P. M. Lapostolle

Electron Injector Computer Simulations 231
D. Tronc

Numerical Simulations of Orbit Correction in Large Electron 237
Rings
G. Guignard, Y. Marti

Simulation of Polarization Correction Schemes in e^+e^- Storage 243
Rings
D. P. Barber, H. D. Bremer, J. Kewisch, H. C. Lewin, T. Limberg,
H. Mais, G. Ripken, R. Rossmanith, R. Schmidt

Computation of Electron Spin Polarisation in Storage Rings 249
J. Kewisch

ARCHSIM: A Proton Synchrotron Tracking Program Including 255
Longitudinal Space Charge
H. A. Thiessen, J. L. Warren

A Method for Distinguishing Chaotic from Quasi-Periodic Motions 261
in Orbit Tracking Programs
J. M. Jowett

PATH - A Lumped-Element Beam-Transport Simulation Program with 267
Space Charge
J. A. Farrell

WORKSHOP No.1: Computer Programs for Lattice Calculations 273

B: DIGITAL CONTROL OF ACCELERATORS

Digital Control of Accelerators - The First Ten Years 275
H. Frese

Distributed Digital Control of Accelerators 278
M. Crowley-Milling

Centralized Digital Control of Accelerators 289
R. E. Melen

Concurrent Control of Interacting Accelerators with Particle 300
Beams of Varying Format and Kind
P. P. Heymans, B. Kuiper for the PS Controls Group

Integrated Control and Data Acquisition of Experimental 311
Facilities
F. Bombi

Software Engineering Tools 316
R. Zelazny

Centralization and Decentralization in the TRIUMF Control System 332
D. A. Dohan, D. P. Gurd

The Fermilab Accelerator Controls System 338
D. Bogert, S. Segler

The Control System for the Daresbury Synchrotron Radiation Source D. E. Poole, W. R. Rawlinson, V. R. Atkins	344
The Microprocessor-Based Control System for the Milan Superconducting Cyclotron F. Aghion, S. Diquattro, A. Paccalini, E. Panzeri, G. Rivoltella	351
The ELSA Control System Hardware Ch. Nietzel, M. Schillo, H. J. Welt, C. Wermelskirchen	355
Computer Control System of Polarized Ion Source and Beam Transport Line at KEK J. Kishiro, Z. Igarashi, K. Ikegami, K. Ishii, T. Kubota, A. Takagi, E. Takasaki, Y. Mori, S. Hukumoto	361
Computer Control System of TRISTAN A. Akiyama, K. Ishii, E. Kadokura, T. Katoh, E. Kikutani, Y. Kimura, I. Komada, K. Kudo, S. Kurokawa, K. Oide, S. Takeda, K. Uchino	367
The System for Process Control and Data Analysis Based on Microcomputer and CAMAC Equipment in the LAE 13/9 Linear Electron Accelerator Z. Zimek, J. R. Zablotny	372
Some Features of the Computer Control System for the Spallation Neutron Source (SNS) of the Rutherford Appleton Laboratory T. R. M. Edwards	377
Design Criteria for the Operation of Accelerators Under Computer Control P. D. Eversheim, P. von Rossen	386
Computer Aided Control of the Bonn Penning Polarized Ion Source N. W. He, P. von Rossen, P. D. Eversheim, R. Büsch	391
Treatment and Display of Transient Signals in the CERN Antiproton Accumulator T. Dorenbos	398
Fast CAMAC-Based Sampling Digitizers and Digital Filters for Beam Diagnostics and Control in the CERN PS Complex V. Chohan, C. Johnson, J. P. Potier, M. Miller	405
Automated Cyclotron Magnetic Field Measurement at the University of Manitoba V. Derenchuk, J. Bruckshaw, I. Gusdal, J. Lancaster, A. McIlwain, S. Oh, R. Pogson, J. S. C. McKee	411
On the Problem of Magnet Ramping E. Bozoki	416
High Level Control Programs at NSLS E. Bozoki	420
The Minicomputer Network for Control of the Dedicated Synchrotron Radiation Storage Ring BESSY G.v.Egan-Krieger, W.-D. Klotz, R. Maier	425

The Electronic Interface for Control of the Dedicated 436
Synchrotron Radiation Storage Ring BESSY
G.v.Egan-Krieger, W.-D. Klotz, R. Maier

WORKSHOP No.2: Which LAN to Use for Accelerator Control 445

C: COMPUTING IN ACCELERATOR OPERATION

Introduction to Computing for Accelerator Operation 446
W. Joho

Man-Machine Interface Versus Full Automation 455
V. Hatton

Models and Simulations 465
M. J. Lee, J. C. Sheppard, M. Sullenberger, M. D. Woodley

Operations and Communications Within the Daresbury Nuclear 473
Structure Facility Control System
S. V. Davis, C. W. Horrabin, W. T. Johnstone, K. Spurling

Consoles and Displays for Accelerator Operation 481
G. Shering

Operator Interface to the Oric Control System 491
C. A. Ludemann, B. J. Casstevens

Computer Aided Setting Up of VICKSI 497
W. Busse, B. Martin, R. Michaelsen, W. Pelzer, B. Spellmeyer,
K. Ziegler

GANIL Beam Setting Methods Using On-Line Computer Codes 503
GANIL Operation Group and Computer Control Groups

A Multi-Processor, Multi-Task Control Structure for the CERN SPS 509
C. Saltmarsh

Computer Codes for Automatic Tuning of the Beam Transport at the 518
UNILAC
L. Dahl, A. Ehrich

Interactive Testprogram for Ion Optics 524
V. Schaa, G. Fliss, P. Strehl, J. Struckmeier

Numerical Orbit Calculation for a LINAC and Improvement of Its 530
Transmission Efficiency of a Beam
A. Goto, M. Kase, Y. Yano, Y. Miyazawa, M. Odera

The Computerized Beam Phase Measurement System at GANIL - Its 536
Applications to the Automatic Isochronization in the Seperated
Sector Cyclotrons (SSC) and Other Main Tuning Procedures
J. M. Loyant, F. Loyer, J. Sauret

On-Line Optimization Code Used at SATURNE 542
J. M. Lagniel, J. L. Lemaire

Automatic Supervision for SATURNE 553
C. Fougeron, J. Gontier, J. M. Lagniel, P. Mattêi

A Local Computer Network for the Experimental Data Acquisition 557
at BESSY
W. Buchholz

Closing Remarks 561
M. C. Crowley-Milling

Conference Attendees 563

Author Index 573

FUTURE HIGH ENERGY ACCELERATORS

John ADAMS

European Organization for Nuclear Research (CERN)
1211 Geneva 23 - Switzerland

Introduction

I feel very honoured to be asked to give the opening talk at this conference on computing in accelerator design and operation. The subject of my talk is future high-energy particle accelerators and colliders, that is to say machines that may be built after those that are now in operation or under construction. The reason for this choice of subject is that I believe that these future machines will make even heavier demands on computing than the present ones, especially on computer control systems. I realise that this is not a very original thought since it only follows the trend which has been evident in recent years. Nevertheless, we are a still a long way from the cybernetic machine proposed many years ago by scientists at the Radio Technical Institute in Moscow but I believe future machines will push us much further in this direction.

The needs of the research

In presenting this subject to you I thought that I should start by saying something about the needs of the research in the years ahead, since it is these needs which should determine what kind of accelerators and colliders are built in the future. These research needs are usually determined by theoretical predictions so one may ask the question, what does theory predict?

The most depressing prediction is that nothing much will happen after the W and Z particle energy range of about 0.1 TeV until one reaches energies which are well beyond any accelerator or collider that can be conceived today. In other words, there stretches before us a featureless desert whose further boundary is way beyond our reach. This view is based on two assumptions. Firstly that there are no new gauge forces except the know SU(3), SU(2) and U(1) operating between the presently accessible energy range and some very much higher energy level and secondly that no new particles will occur in this energy range which upset the value of the Weinberg angle, $\sin^2 \theta_W = 3/8$. With these two assumptions and the known quarks and leptons, the renormalization group extrapolation shows that the effective couplings of the three gauge forces converge to the same value at the same upper energy level and that this level has the very high value of 10^{11} TeV. Beyond that energy level there is another at about 10^{16} TeV which comes from the supergravity ideas and the possibility of unifying gravity with the other forces. Thus the desert stretches from 0.1 TeV out to at least 10^{11} TeV. This prediction is not, of course, very encouraging to experimentalists and machine builders nor is it very much use in fixing the characteristics of future machines, at least not until we know how to get to 10^{11} TeV. In fact, it led Abdus Salam to entitle a talk which he gave last year on this subject at the International Particle Physics Conference "The impending demise of high energy accelerators" and I am indebted to that talk for the explanation of the desert syndrome which I have just given you. Incidently, he concluded his talk with the advice – "Do not ask theorists which energy to aim at for future machines. Aim at the highest possible".

A more useful prediction, at least for machine builders, is that there may be flowers blooming this side of the desert which with a great effort we might be able to reach. This view is based on the observation that the present so called standard model based on the unified electroweak theory of Glashow, Weinberg and Salam and the QCD theory of strong interactions cannot be the final answer. For example, it does not predict the numbers of families of quarks and leptons or their masses or their mass ratios. Also there is the all important symmetry breaking which causes the gauge bosons that mediate between the weak interactions, the W particles, to acquire very large masses whereas those that mediate between the electromagnetic interactions, the photons, are massless. The "deus ex machina" is said to be the Higgs mechanism with its scalar Higgs particle. Unfortunately, nothing seems more elusive than the Higgs particle. Is it a particle or a set of particles or – and I quote – "an approximation of dynamical effects which manifest themselves at energies a few times the inverse square root of the Fermi weak interaction coupling constant", i.e. roughly 1 TeV ? Clearly the search for and study of the Higgs particle or its equivalent is of the highest interest and priority and fortunately the energy range in this case could conceivably be reached by particle colliders in the forseeable future.

The machine energies required to explore the Higgs sector depend on whether the machine is a proton collider or an electron collider. A rule of thumb is that an electron-positron collider of one sixth to one tenth the centre of mass energy of a hadron collider will explore the same general domain of hard processes or heavy particle production. So, if the Higgs sector has a mass scale of about 1 TeV, the future electron-positron collider should give about 2 TeV in the centre of mass system and a proton-proton collider about 20 TeV.

This seems for the moment the best guide we can get from theory concerning future machines but before turning to these machines, I should point out that past predictions about the future needs of the research show a marked lack of correlation between the reasons given for building the machines and the important discoveries they made.

T. D. Lee in a recent talk at Brookhaven listed the twenty most outstanding discoveries made using accelerators and colliders during the last 35 years starting with pion production at the 184 inch cyclotron at Berkeley in the late 1940's and ending with the intermediate vector bosons at the SPS collider at CERN this year. He pointed out that only two of these twenty landmark discoveries, the anti-nucleons at the Berkeley Bevatron and the intermediate vector bosons at the CERN collider were anticipated at the time the relevant machines were approved. Another remarkable feature he found was that the major discoveries arrived very regularly over the 35 years - almost one every two years. It seems that Nature reveals her secrets unexpectedly but rather regularly, but she does not read machine prospectuses.

After these cautionary remarks I will now pass on to the machines themselves.

Future accelerators and colliders

Two machines have emerged in recent years as possible candidates for the accelerators and colliders of the future. The first is a hadron collider, either proton-proton or proton-antiproton, which might also be used as a fixed target machine and the second is a linear electron-positron collider. Both of these machines were studied in some depth at two workshops organized by the International Committee for Future Accelerators (ICFA) in 1978 and 1979. More recently, the hadron collider, under the name of the Desertron, or Superconducting Super Collider (SSC), has been taken up enthusiastically in the USA. There was a Summer Study on particle physics and future facilities held at Snowmass in Colorado in July 1982. This was followed by a Technical Workshop on a 20 TeV hadron collider held at Cornell University in March 1983 and by a Workshop on hadron collider detectors held at Berkeley in April 1983.

From all these studies and workshops, the general conclusion emerges that a hadron collider with a centre of mass energy of about 20 to 40 TeV is technically feasible, that it would cost about 2 billion dollars and maybe more, and that detectors could be designed to measure the events produced in the collisions at these high energies and extract the relevant data. To reduce the machine cost down to 2 billion dollars it is thought that 3 or 4 years of design and development work will be needed on its components before construction can start. There is less agreement on how long all this would take assuming, of course, that the U.S. government is willing to agree to such a large and expensive project. Estimates range from 9 years to 15 years. In other words, if approval is given to this venture in 1984, the collider might be operating at the earliest in 1993 or at the latest in 1998; let us say sometime in the second half of the 1990's.

Let me now say something about this machine. Since it is assumed that new particles or sets of particles beyond the W will have smaller production cross-sections following roughly the $s^{1/2}$ law, the highest machine luminosity seems to be desirable and this can best be achieved by a proton-proton collider, i.e. an intersecting ring machine like the ISR or CBA in which luminosities approaching 10^{33} per cm^2 per second are thought possible.

The magnet system for this machine will have to use superconducting coils to reduce its electrical power consumption to an acceptable level. Three magnet systems were studied at the Cornell Workshop, the first used bending magnetic fields of 2-3 Tesla, the second 5-6 Tesla and the third 8-10 Tesla. The first would use iron to shape the field and superconducting coils to save power. The second would be based on Tevatron or CBA technology using NiTi conductor at 4.5° K. For the third, 8 Tesla could be reach with NiTi conductor by operating at 2°K, but 10 Tesla would need Nb$_3$Sn conductor. A 3 Tesla machine for 20 TeV beam energy would be about 160 km in circumference, a 5 Tesla machine 100 km and an 8 Tesla machine about 60 km. For comparison, the largest machine now under construction is LEP which is 27 km in circumference. This future machine is therefore several times the size of LEP. Curiously enough rough estimates of the total cost of the machine made at Cornell showed little difference whichever bending field level is chosen as can be seen in Table 1.

Table 1

Estimated Machine Costs (Million Dollars) [20 + 20 TeV.pp]

	3 Tesla	5 Tesla	8 Tesla
Fixed costs [1]	540 ± 80	540 ± 80	540 ± 80
Enclosure, etc. [2]	300 ± 120	190 ± 70	130 ± 50
Magnets [3]	450 ± 150	710 ± 240	780 ± 260
Accelerator components [4]	350 ± 80	280 ± 60	230 ± 50
TOTAL	1640 ± 320	1720 ± 360	1680 ± 360

[1] Fixed costs include the site infrastructure (but not the cost of the site) the injector machines, experimental areas and the magnet factory.

[2] Enclosure costs include the machine tunnel, access roads, service buildings and power distribution.

[3] Magnet costs include all magnet elements and their cryogenic systems.

[4] Accelerator components costs include the refrigerators, vacuum, RF, controls, injection and abort systems, power suppliers, robots, etc., and installation costs.

One of the tasks during the initial development period of this machine will be to chose between these three magnet systems. Another even more important task is to see whether the present cost estimates are realistic since if one compares them with superconducting magnet machines like the Tevatron or CBA, one sees that large reductions have been made in the unit costs of the components to get the total machine cost down below 2 billion dollars.

The size of these reductions can be seen from estimates made by R.B. Palmer for a 5 Tesla 20 + 20 TeV pp collider based on the latest Fermilab cost data for Tevatron magnets. He arrived at total cost of 6.6 billion dollars compared with the 1.7 of Table 1 and cost reduction factors for magnets and tunnels ranging between 4 to 6. Achieving such large cost reduction factors will not be easy. Even the alternating gradient focusing principle, when it was introduced in 1953, only reduced total machine costs by a factor of 2.

Different ways are proposed to make these cost reductions, for example, using a very small magnet aperture giving a good field region of about 20 - 30 mm diameter and getting the beam once round the machine by coaxing it sector by sector round the 100 km circumference; installing the bending magnets of each ring side by side in the same cryostat to reduce heat losses and save refrigerator power; using very small cross section tunnels, in the limit only sufficient for the machine but not for human beings; and using modern techniques to reduce production costs of machine components.

Let me now turn to the other future machine, the linear electron-positron collider. At the ICFA Workshops of 1978 and 1979, a tentative design was made for such a machine to give 700 GeV in the centre of mass system. A linear machine was chosen since limiting synchrotron radiation losses in circular electron machines with beam energies above about 250 GeV gives machine circumferences which are prohibitively large. Two solutions were studied, a linear collider using room temperature RF cavities and one using superconducting RF cavities. On balance the room temperature solution looked more feasible although it required a peak RF power of the order of 10^6 MW for driving the cavities. Since then the Novosibirsk and SLAC laboratories have continued with these studies. At SLAC the construction is going ahead of a machine called SLC using the existing 30 GeV electron linac upgraded to give 50 GeV. Both electrons and positrons will be accelerated in this linac and at the end of it the electrons will travel round one semi-circular arc to meet head on the positrons which travel around a second arc. To reach the desired luminosity of 6×10^{30} per cm^2 per second the two beams, or rather bunches of particles, have to be focused down to a diameter of about 1 or 2 microns. High precision and stability are necessary in space and time to ensure that such small diameter bunches actually hit each other at the collision point. This machine, which is planned to come into operation in 1987 will give the first opportunity to study experimentally the problems likely to be encountered in linear colliders particularly the disruptive effect which one bunch has on the other when they collide.

In addition to this experimental machine a tentative study has recently been made of a linear electron-positron collider to give 2 TeV in the centre of mass system based on existing technology. Some of the parameters of such a machine are given in the next table.

Table 2

1 + 1 TeV electron-positron linear collider

RF frequency	2856 MHz (S band)
Length	2 x 50 km
RF gradient	20 MV/m
Repetition rate	185 Hz
Bunches per pulse	12
Bunch length	2 mm
No of particles per bunch	$1.4 \cdot 10^{10}$
No of klystrons	2 x 3500
Peak klystron power	330 MW
Average klystron power	23 kW
Total peak RF power	$2.4 \cdot 10^6$ MW
Total average RF power	160 MW

The total length of this machine, 100 km, is about the same as the circumferential length of the 20 TeV hadron collider and a very rough estimate of its cost suggests a figure about twice as large.

One of the problems of linear colliders is that there is only one region where the two beams meet head on. To run as many experiments as with a circular hadron collider which has several interaction areas around its circumference, the experiments of a linear collider have to be placed side by side and the linac beams switched to each experiment in turn on a pulse to pulse basis. Since the beams consist of bunches 2 mm long and a few tenths of microns in diameter, colliding the bunches at the correct place inside each experiment does not seem so easy.

I can hardly leave this part of my talk without mentioning the pressing need for new ideas for accelerating particles in order to reduce the size and cost of the accelerators and colliders. The two future machines which I have just described really are monsters; 100 km in circumference or length and costing several billion dollars each. It is by no means certain that governments or even groups of governments will be willing to finance such machines and we may be forced to find cheaper solutions or stop building very high energy accelerators. Let me try to explain what these new ideas should be aiming to achieve.

To reduce the size of future machines higher accelerating gradients are needed, the accelerating gradient being defined as the maximum particle energy of the machine divided by its length or circumference. Electron linacs are now approaching gradients of 20 MeV/m although short test cavities have reached voltage gradients of up to 150 MV/m. In the case of proton synchrotrons the accelerating gradient as I have defined it is set by the maximum bending magnetic field. A 5 Tesla machine achieves 150 MeV/m and a 10 Tesla machine would achieve 300 MeV/m. Therefore the new ideas of accelerating particles if they are to enable smaller machines to be built should aim at accelerating gradients of several 100 MeV/m and preferably at a few GeV/m. Since at a few 100 MeV/m it becomes impossible to maintain the necessary voltage gradients between metal surfaces, the accelerating field has then to be established in a medium such as a plasma column or an intense electron beam. Some of the new ideas aim in this direction, for example the beat wave accelerator in which two laser beams running along a plasma column have a frequency difference equal to the plasma frequency and by beating together set up intense localised charge concentrations and hence very high field gradients which can then be used to accelerate particles. Several GeV/m are promised by this scheme at least theoretically. However, these new ideas are still in their infancy and even if they are found promising experimentally, it will take many years to develop them into an accelerating system for a machine to give several TeV beam energy.

Also they must at least hold out the promise of less cost per GeV. A shorter but more expensive machine is not a solution to this problem.

Future machines and computers

I would like to end now with a few remarks about the implications of future machines to computing and so try to link my talk with the subject of this conference.

There are, as everyone knows, five distinct but overlapping phases of machine building. These are the design phase, the construction phase, the installation phase, the commissioning phase and finally the operating phase. I notice that this conference only covers computing for the design and operating phases. I would like to suggest that for future machines the other three phases will also need a great deal of computing.

Let me illustrate this point by taking first the construction phase. Until bright new ideas actually succeed in reducing the size and cost of future particle colliders, we are faced with machines which will be of the order of 100 km in circumference or length. These machines will contain thousands of components of a limited number of types - magnets, RF cavities and power units, vacuum pumps and so on. These components will have to be cheaply mass produced to very tight tolerances. Up to now machine builders and industrial firms have used manufacturing technologies which, although achieving the desired products to the required tolerances, are relatively primitive compared with the methods now employed in the most advanced mass production industry. In the case of superconducting magnet production, Fermilab and Brookhaven have set up their own factories on site but the Tevatron and the CBA are very small machines compared with a future 20 + 20 TeV hadron collider. To manufacture the magnets of the latter in the same time as it took for the Tevatron one would need 20 Fermilab factories working in parallel. It seems therefore that much more automation in production perhaps using robotic systems under computer control will be needed for future machines. This will also allow a closer quality control which can then be integrated into the production process rather than be superimposed as periodic inspection as has been the case up to now. This same technology will be required for all the other machine components which will be needed in their thousands.

Turning now to the assembly stage, the problem is to install thousands of components in the correct order via a limited number of access points in a tunnel 100 km in circumference or length and to align them to a very high accuracy. If all this is to be done in a reasonable time, like two or three years, it will require superb logistic organization and a great deal of automation. There is first the storage of finished components on the surface in sufficient number to ensure a smooth supply for installation and their distribution to the access points. There is then the transport of these components into and around the tunnel in the correct sequence, since, for economic reasons, its size will not allow vehicles to overtake each other in the tunnel. Each component must then be installed at the correct place and finally there is the alignment of the components and their connection to preinstalled electrical and other supplies. For the SPS machine at CERN, a computer system was used for keeping track of all the components and their installation in the correct place, and a fleet of free moving vehicles was used for transporting them in the tunnel. For LEP a more elaborate data base system will be used for marshalling the components and a relatively fast monorail for their transport in the tunnel. For a future machine it may be necessary to use robots under computer control both for marshalling components and transporting them to their allotted position in the tunnel in the right order. And if the tunnel is so small that human beings cannot work in it, then, in addition, the robots will have to put the components in place, align them and connect them to their supplies. I leave the experts to imagine the computer control system necessary for this kind of operation.

I come now to the commissioning stage. As I have already mentioned, cost saving in future machines will require that the vacuum chamber and the good field region of the magnet be as small as possible. If no allowance is made for initial closed orbit deviations the machine will have to be aligned using the beam section by section round the machine. Non-linear effects particulary of the dipole fields on the beam dynamics will have to be studied in advance with elaborate tracking programmes and more feed-back systems used to control the beam in the machine. The multi-TeV hadron collider presents an additional problem. Circulating beams of several amperes inside a small bore superconducting magnet for hours on end without letting a very small fraction of the beam, a few milliamperes, reach a magnet element and quench it will call for very precise and reliable beam control. Scraping the beam, as regularly done in the ISR machine at CERN in order to prepare it for experiments, without the scraped-off part hitting a magnet is another problem. And finally, dumping the beam safely outside the superconducting magnet at the end of each run or in an emergency without spraying magnet units with secondary particles is yet another. Hopefully, solutions to these problems will be found and tried out with machines like the Tevatron before the large hadron collider design is finalized but whatever the solutions that are found, I am sure more computers and computing will be required.

Finally, there is the problem of the maintainance of a machine 100 km in circumference or length so that it achieves a high operating efficiency in terms of hours per year for physics research. The planning of how the machine will be subsequently maintained has, of course, to form part of its initial design. To a large extent, the operating efficiency will depend on how quickly faults can be detected and localised and then corrected either by adjustments or component replacements. Given the size of the machine and the time needed to reach a component, maintainance could take a very long time unless it is carried out by fast robots backed up by computer systems.

Conclusion

I hope that these remarks about future machines and computing convince you that more computer systems will be needed in the future and that they will be used not only in the design and operating phases of future machines but also in their construction, installation, commissioning and maintainance. This is the rather cheerful message which I would like to pass on to you at the beginning of this conference.

BEAM OPTICS AND DYNAMICS

E.J.N. Wilson
European Organisation for Nuclear Research (CERN)

1211 Geneva 23, Switzerland

Abstract

After introducing the fundamental equations which determine the optics of beams in circular machines and how computers find their solutions, the paper describes how a typical matching problem of designing a low-beta region might be tackled. The limitations of existing optics programs are also discussed.

1. Introduction

I shall not attempt to describe in depth the art of designing synchrotrons or to discuss the frontiers of the theory of particle dynamics. Later papers in this conference will provide material to tax the intellect. The aim of this paper is to explain to the computational specialist who knows more about programming than about accelerators, an outline of accelerator theory[1], the way in which computers apply this theory to help the designer and to suggest a few directions which might be explored to improve the tools available.

As we shall see, the analysis of the optics of synchrotrons is largely a matter of multiplying together a large number of matrices, each describing the transport of a particle through a magnet. Regular patterns of magnets, or lattices, can be calculated in closed form but computers can be used to graft in special regions of the machine, where other components such as rf cavities or extraction magnets replace bending magnets or where the beam is focused to a narrow waist to collide with another beam.

2. The Lattice Structure

A typical pattern of magnets or lattice of a modern accelerator, the SPS, is shown in Fig. 1. The first element in a cell is a horizontally defocusing magnet, a quadrupole, characterised by its normalised gradient.

$$k = (1/B\rho)\partial B_y/\partial x$$

where x, y are transverse to the beam direction. Half a cell later is a defocusing quadrupole of opposite sign. There are four bending magnets between each quadrupole and the 64 m long cell is repeated 108 times around the circumference of the accelerator. In some places the bending magnets are omitted leaving space for equipment and in others, special patterns of magnets called insertions interrupt the regular lattice to make the beam very narrow.

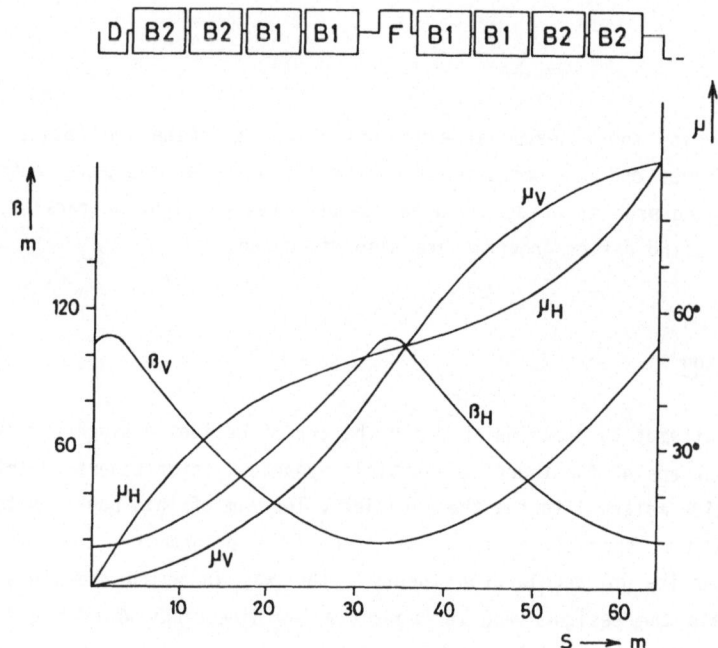

Figure 1 - Lattice Functions in a Cell.

In the naive approximation used in the days before computers, quadrupoles, length, ℓ, were treated as thin lenses of focal length $f = 1/k\ell$. But, to a higher degree of exactitude the circulating particles obey Hill's equation[2]:

$$\frac{d^2x}{ds^2} + k(s)x = 0 .$$

This equation and its solution:

$$x = \varepsilon^{1/2}\beta^{1/2}(s) \cos[\psi(s) + \lambda]$$

are reminiscent of harmonic motion with a phase advance ψ and an amplitude $\beta^{1/2}(s)$, which is envelope of the motion. The quantity ε, the emittance, is constant depending only on the size of the injected beam and its subsequent history. The difference is that the amplitude like the restoring force both vary with s, the distance round the ring.

In solving the equation of motion we wish to calculate $\beta(s)$, from the lattice pattern $k(s)$ and to do so we use the fact that the solution of a linear differential equation can be written as a transport matrix from a point s_1 on the circumference to a point s_2:

$$\begin{pmatrix} x(s_2) \\ x'(s_2) \end{pmatrix} = \begin{pmatrix} a & b \\ c & d \end{pmatrix} \begin{pmatrix} x(s_1) \\ x'(s_2) \end{pmatrix}$$

operating on the displacement from the central orbit x and its derivative $x' = dx/ds$.

The transport matrix is simply the product of a number of matrices, one for each element either a quadrupole or the drift length between lenses.

drift length   quadrupole

$$\begin{pmatrix} 1 & \ell \\ 0 & 1 \end{pmatrix} \qquad \begin{pmatrix} \cos\sqrt{k}\,\ell & 1/\sqrt{k}\,\sin\sqrt{k}\,\ell \\ -\sqrt{k}\,\sin\sqrt{k}\,\ell & \cos\sqrt{k}\,\ell \end{pmatrix}$$

Dipole magnets are slightly different from drift lengths in that their ends have a focusing effect. One complication is that a quadrupole which is focusing in the horizontal plane is defocusing vertically and vice-versa. The matrix multiplication must be carried out independently for x and y motion. The defocusing matrix comprises hyperbolic terms and is obtained by substituting -k for k in the matrix above.

The transport matrix from a point around one complete turn can be computed numerically by matrix multiplication of the several thousand elements or it can be expressed analytically in terms of the betatron amplitudes at the start/finish point:

$$M = \begin{pmatrix} \cos\mu + \alpha\sin\mu & \beta\sin\mu \\ -\gamma\sin\mu & \cos\mu - \alpha\sin\mu \end{pmatrix}$$

where $\alpha = -\beta'/2$,
        $\gamma = (1 + \alpha^2)/\beta$,
        μ = phase advance/turn = $2\pi Q$.

The four numerical elements of the computed matrix can be compared with the algebraic expression and solved to find:

$$\mu = \cos^{-1}\{TrM/2\} ,$$

$$\beta = M_{12}/\sin\mu ,$$

$$\alpha = (M_{11} - M_{12})/(2 \sin\mu) .$$

It is the task of the lattice program to perform this calculation starting at each point in the ring and for both planes and forming a table of μ, β and α around the ring. Fig. 2 shows a typical output for one cell.

LENGTH	ANGLE	K(V)	ALPHA(P)	BETA(H)	ALPHA(H)	MUH/2PI	BETA(V)	ALPHA(V)	MUV/2PI	AH/2	AV/2
3,085000	0,000000	-,015063	1,386440	104,884855	2,452160	,004571	19,011703	-,520345	,026571	65,715663	9,917560
,360000	0,000000	0,000000	1,374653	103,127965	2,428089	,005122	19,395014	-,544408	,029555	64,547513	10,017039
6,260000	,008445	0,000000	1,196124	75,348859	2,009521	,016433	28,828710	-,962519	,072198	64,004371	12,212911
,400000	0,000000	0,000000	1,186405	73,751941	1,982775	,017287	29,609417	-,989248	,074377	54,751341	12,376825
6,260000	,008445	0,000000	1,060742	51,548094	1,564207	,033474	44,610910	-1,407071	,101928	54,174091	15,192432
,390000	0,000000	0,000000	1,054559	50,338182	1,538130	,034692	45,718585	-1,433122	,103302	45,428681	15,379447
6,260000	,008445	0,000000	,981762	33,701223	1,119563	,058975	66,274961	-1,850527	,121441	44,905056	18,517478
,380000	0,000000	0,000000	,978948	32,860011	1,094154	,060793	67,691002	-1,875896	,122344	36,980337	18,713705
6,260000	,008445	0,000000	,959017	21,781569	,675586	,098381	93,787676	-2,292753	,134861	36,534921	22,028267
2,342700	0,000000	0,000000	,961450	18,983146	,518942	,116758	104,896272	-2,449038	,138621	30,069327	23,295624
3,085000	0,000000	,015037	1,034354	18,983068	-,518916	,143368	104,901620	2,447388	,143191	28,349412	23,716525
,350000	0,000000	0,000000	1,050730	19,354500	-,542318	,146275	103,196611	2,424067	,143726	28,638028	23,296218
6,260000	,008445	0,000000	1,370047	28,764399	-,960879	,189011	75,452122	2,007802	,155027	35,089639	23,106121
,380000	0,000000	0,000000	1,391035	29,504322	-,988287	,191088	73,935822	1,982463	,155836	35,546047	19,757412
6,260000	,008445	0,000000	1,763219	44,472640	-1,404847	,218731	51,724094	1,565610	,171975	43,750575	19,557880
,390000	0,000000	0,000000	1,788053	45,578591	-1,430924	,220109	50,513067	1,539589	,173189	44,298587	16,358398
6,260000	,008445	0,000000	2,213103	68,113699	-1,849484	,238298	33,849177	1,122280	,197377	53,470174	16,165762
,400000	0,000000	0,000000	2,241952	67,603985	-1,876229	,239251	32,962034	1,095579	,199283	54,079136	13,233307
6,260000	,008445	0,000000	2,719868	93,714254	-2,294790	,251780	21,859390	,677943	,236745	63,830251	13,058741
2,352700	0,000000	0,000000	2,909420	104,882261	-2,452099	,255558	19,038995	,520847	,255140	67,592709	10,634409
3,085000	0,000000	-,015063	2,946010	104,882266	2,452098	,260129	19,038106	-,520546	,281673	68,853088	9,924676
,360000	0,000000	0,000000	2,925443	103,125421	2,428027	,260680	19,421551	-,544579	,284653	67,665889	10,023890
6,260000	,008445	0,000000	2,594240	75,347037	2,009487	,271992	28,854181	-,962177	,327246	67,105194	12,218305
,400000	0,000000	0,000000	2,574765	73,750162	1,982722	,272846	29,634602	-,988874	,329424	57,546939	12,382087
6,260000	,008445	0,000000	2,296428	51,546933	1,564162	,289032	44,628208	-1,406185	,356957	56,950187	15,195377
,390000	0,000000	0,000000	2,280734	50,337057	1,538085	,290251	45,735180	-1,432204	,358331	47,899567	15,382238
6,260000	,008445	0,000000	2,055264	33,700612	1,119525	,314534	66,276862	-1,849098	,376466	47,358928	18,517744
,380000	0,000000	0,000000	2,043182	32,859428	1,094117	,316352	67,691805	-1,874435	,377369	39,127022	18,713817
6,260000	,008445	0,000000	1,870577	21,781395	,675557	,353941	93,766993	-2,290782	,389888	38,663082	22,025838
2,342700	0,000000	0,000000	1,815875	18,983101	,518917	,372318	104,865902	-2,446875	,393648	31,892336	23,292251
3,085000	0,000000	,015037	1,873603	18,983178	-,518943	,398928	104,862544	2,447912	,398220	30,027986	23,712598

Figure 2 - Lattice Program Output for One Cell.

Another quantity of great importance is the dispersion, α_p, which is the function which describes the horizontal displacement per unit error in beam momentum, p. Beam width is in fact the sum of a betatron term and a dispersion term:

$$w(s) = 2 \left[\sqrt{\varepsilon\beta(s)} + \alpha_p(s) \frac{\Delta p}{p} \right] .$$

Dispersion arises because each time a particle with a momentum error, Δp, is bent on radius ρ, it gets an additional kink in its orbit which modifies Hill's equation:

$$\frac{d^2x}{ds} + k(s)x = \frac{1}{\rho(s)} \frac{\Delta p}{p} .$$

In terms of the computed matrix elements:

$$\alpha_p' = d\alpha_p/ds = \frac{M_{13}M_{21} + (1 - M_{11})M_{23}}{(1 - M_{11})/(1 - M_{22}) - M_{21}M_{12}} ,$$

$$\alpha_p = (M_{12} \alpha_p' + M_{13})/(1 - M_{11}) .$$

3. Matching

A computation of the β and α values around the ring takes only a few seconds of CP time even for a large synchrotron and so it is natural to ask the computer to vary the focusing strength k_F and k_D and after some iterations to produce the desired va- values of phase advance or wave number per turn, $Q = \mu/2\pi$, in both x and y direc- tions. This Q value or tune is of paramount importance in designing the machine for if it is related to an integer or close to a vulgar fraction one sequence of turns tends to be repeated indefinitely and any errors and perturbations build up into nasty resonances which eject the beam. Fig. 3 shows a typical Q diagram and the lines where numerological relations between Q_H, Q_V and integers cause resonances.

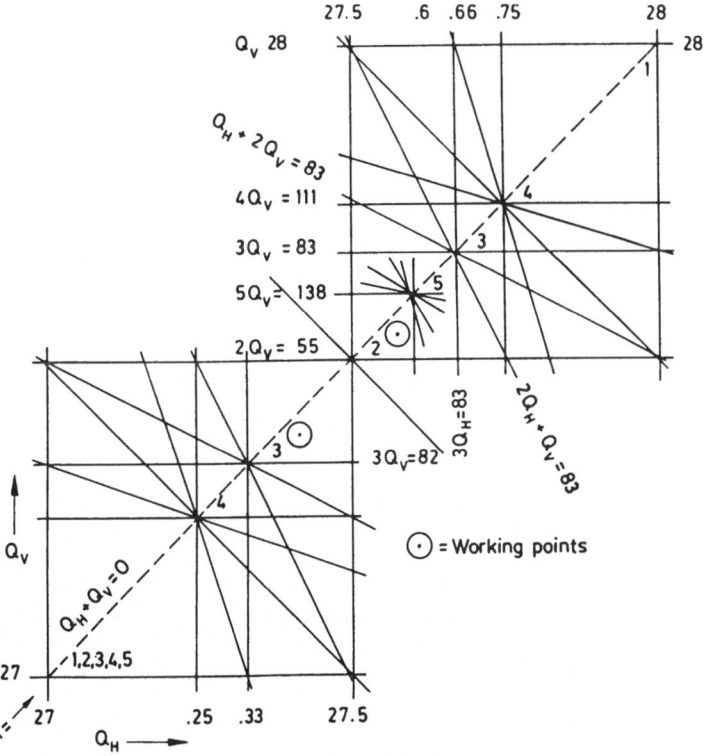

Figure 3 - Example of a Q Diagram Showing Resonant Conditions to be avoided as Lines.

Adjusting Q is a relatively trivial task for a lattice program. A more demanding task is to graft in an insertion such as that shown in Fig. 4. This is to make a narrow waist where, for example in a p$\bar{\text{p}}$ storage ring counter rotating beams collide with greater probability. If we start at the centre of the insertion where α_H, α_V are zero, α_p is usually required to be zero and β_H and β_V some small desired values, the lattice program multiplies matrices of the quadrupoles to form a matrix which gives the values of β_H, β_V, α_H, α_V, α_p and α'_p at the point where it will match up with the normal lattice period. It must then vary at least six parameters, usually quadrupole strengths until the six functions match the input values for the normal period. An imperfect match will cause beating or sausaging of the beta functions elsewhere in the ring.

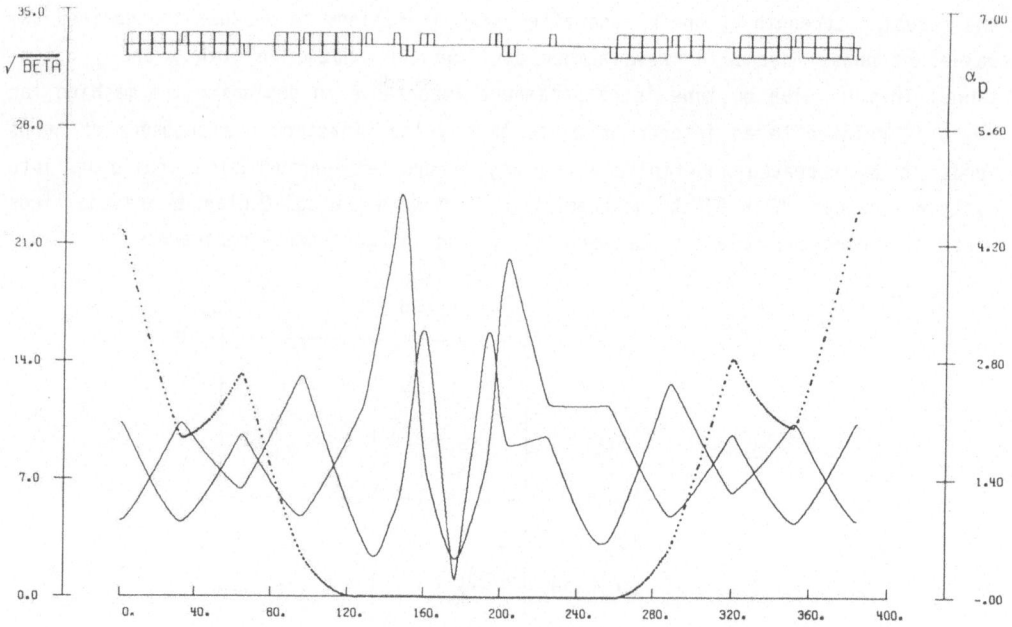

Figure 4 - Low Beta Insertion for the SPS.

Some computer programs will match up to a dozen functions with as many variables though this can easily take half an hour of CP time. Such multivariable minimisation routines are notorious for getting bogged down in "mountain lakes" and never finding the real solution. Splitting the matching into a number of almost orthogonal steps helps to avoid this.

Let us take as an example how one might tackle the design of a regular machine with six low-beta insertions to illustrate this. Suppose the machine has six-fold symmetry and reflection symmetry about the centre of each superperiod. It is only necessary to consider one twelfth of the ring i.e. a series of regular cells (Fig. 1) followed by a sequence of quadrupoles which match into a waist at the end of th sequence which is the centre of the long straight section (Fig. 4).

The first step is usually to compute the characteristics of a machine consisting of normal periods matched to some nominal phase advance 60 or 90° per cell with cell length chosen to give the desired number of periods, total bending angle and adequate space for hardware in the final ring.

The aim of the matching is then to find a pattern of 8 quadrupoles to transform the β_H, β_V, α_H, α_V, α_p and α'_p at the exit of the normal cell into the values desired in the centre of the straight section and to make up the phase advance in each plane to give a safe value of Q. Altogether there are eight conditions to satisfy.

In most cases we require α_p and α'_p to be zero throughout the long straight section and this is best achieved by inserting a dispersion suppressor as the first special element in the sequence following the normal periods. The dispersion suppressor must contain at least one bending magnet and two variables either quadrupole strengths or drift lengths, to match from the normal α_p and α'_p to zero at the exit of the bending magnet. If the cell phase advance is 90° or 60° dispersion suppressors are just normal cells with some magnets omitted[3].

We are now left with six variables and six conditions. One may rapidly arrive at a minimum beta value in both planes at the centre of the long straight section by asking for α_H and α_V to be zero there. These are the slopes of β and, if zero, will automatically ensure a minimum. Only two quadrupoles F and D are needed to achieve this. Stepwise adjustment of their spacing and position can then be combined with a little experience to arrive at the actual values of beta needed at the low beta point thus satisfying a further four conditions.

Finally, one is left with the task of ensuring that the total phase advance of the ring gives safe Q values. Here one must choose between adding additional variables to the last procedure or returning to the beginning and adjusting the number of periods and/or period length as necessary. Either procedure usually converges rather rapidly.

Final polishing of the match once we are confidently in the correct valley, may consist of adjustments to lengths and positions of quadrupoles to ensure that they have the same strength and may be powered in series. It may also be necessary to introduce extra variables to restrain the excursion of the β functions within reasonable limits. In this context not only geometrical aperture but sensitivity to chromaticity argues for modest beta values.

Other more sophisticated tasks include shaping the dispersion function to arrive at a desired value of $\eta = (1/\gamma^2 - 1/\gamma_{tr}^2)$ and designing special insertions for the extraction of beams from synchrotrons to fixed target experiments.

4. Limitations of Linear Programs

So far we have ignored that off-momentum particles see either more or less focusing strength than the $\Delta p/p = 0$ particle. The β and μ for these particles will differ from the reference particle. The first effect of this is to modify the Q and the first derivative of Q with respect to $\Delta p/p$ is known as the chromaticity. This must be corrected if the beam is not to be an extended line in the Q diagram which cuts across dangerous resonances. The remedy is a "quadrupole" whose strength varies as horizontal position i.e. as $\alpha_p(\Delta p/p)$. A sextupole has such a field and modern computer programs handle such higher order non-linear lenses and match them to make Q independent of momentum.

The chromaticity may be written as the sum of two terms integrated around the ring. The first term due to the normal quadrupoles is the natural chromaticity, the second may be due to sextupole errors in dipole magnets or may be the sextupole pattern used to make $\xi = 0$.

$$\xi = (\Delta Q/Q)/(\Delta p/p) = [-1/4\pi Q(B\rho)]\int[B'(s) + B''(s)\alpha_p(s)]\beta(s)ds$$

where $B\rho$ is the magnetic rigidity,
      B' is the gradient in a quadrupole,
      B'' is the second derivation or strength of a sextupole.

Of course, there are higher order derivatives in the expansion of α_p, Q and β as a function of β and much attention has been given to minimising these higher order terms by subdividing sextupole correction into families and by careful choice of the position of these elements. However, here most orbit programs which are linear in concept break down and one must turn to others like HARMON[4].

Orbit programs developed with approximations for large rings give notoriously bad results in small rings which have large angles of bend and in which, like the Antiproton Accumulator, the fringe fields of magnets play an important role. In particular the fact that beam envelopes change within the fringe field produce extra non-linear effects[5-6]. A new small ring program ORBIT[7] is under development specifically to provide accurate results for these small ring machines.

5. Computer Codes

There are a number of general purpose lattice programs which do all of the above things. Table 1[8] lists a few. They mainly differ in the way in which the magnet data is read in and assembled into a ring and in their numerical techniques for minimisation and matching. MAGIC and TRANSPORT solve a set of n non-linear equations for m > n unknowns x_i:

$$0 = f_i(x_1, \ldots x_m) \quad i = 1, \ldots, n$$

while AGS and SYNCH construct a mismatch function which is minimised using a general purpose package MINUIT[4].

$$F = \prod_{i=1}^{n} (1 + f_i^2 w_i^2) - 1$$

where w_i are weights which can be chosen by the user.

Name	Ref.	Contact	Address
MAGIC	9	M.J. Lee	SLAC
SYNCH	10	A.A. Garren	LBL
TRANSPORT	11	K.L. Brown	SLAC
		D. Carry	FNAL
		F.C. Iselin	CERN
AGS	12	E. Keil	CERN
MAD	13	F.C. Iselin	CERN

Table 1 - General-Purpose Optics Programs.

Input data format is often different for the various programs, some of which economise input by automatically associating gaps between magnets with the magnets. They all provide graphical output to TV screens or plotters and some produce binary tapes to pass tables of lattice functions to subsequent programs. Most of the programs were written long before computers were used interactively with graphics screens and light pens, and there is some scope for imaginative improvement here, especially since much of the work not done automatically involves stepwise trials of moving magnets around and inspecting the change in the functions.

Conclusions

Lattice programs are a standard tool of the designer of a modern accelerator or storage ring. They solve the linear optics problem efficiently and new developments are directed towards including the end effects in small rings. Their limitation to essentially linear solutions have spawned other programs which track by simulation[14-18] which are described elsewhere in these proceedings.

References

1. E.J.N. Wilson, CERN 77-07 (1977).
2. E.D. Courant and H.S. Snyder, Ann. Phys. 3, 1 (1958).
3. E. Keil, CERN 77-13, p. 29 (1977).
4. M.H.R. Donald, PEP Note 311 (1979).
5. G. Wüsterfeld, ANL/AAD-N-26 (1982).
6. S.X. Fang, CERN, PS/AA/LT/Note 26, Part. 9 (1982).
7. B. Autin, M. Bell, Private communication.
8. E. Keil, CERN Academic Training Course (1983).
9. A.S. King, M.J. Lee and W.W. Lee, SLAC-183 (1975).
10. A.A. Garren and J.W. Eusebio, UCID-10153 (1975).
11. K.L. Brown, D.C. Carey, Ch. Iselin and F. Rothacker, CERN 80-04 (1984).
12. E. Keil, Y. Marti, B.W. Montague and A. Sudboe, CERN 75-13 (1975).
13. F.C. Iselin, Private communication (1983).
14. F. James and M. Roos, CERN Library Program D506 (1967).
15. K.L. Brown and F.C. Iselin, CERN 74-2 (1974).
16. H. Wiedemann, PEP Note 220 (1976).
17. K. Steffen and J. Kewisch, DESY PET 76/09 (1976).
18. E. Close et al., PEP Note 271 (1978).

DESIGN OF R.F. CAVITIES

T. Weiland
Deutsches Elektronen-Synchrotron DESY
Notkestraße 85, 2000 Hamburg 52

In linear accelerators and electron storage rings the r.f. accelerating system represents a major part of investment and operating cost. For many years r.f. cavities have been designed with the aim of maximising shunt impedance so as to minimise the power input for a given gradient. Many parasitic collective effects are caused by the cavities such as beam loading, instabilities, bunch lengthening, head tail turbulence and beam break-up. In recent years these effects have been found to cause severe performance limitations in many high energy physics facilities. As a consequence, the design goal for cavities has to be redefined in a much broader perspective. With recently developed computer codes the overall effects of accelerating cavities can now be studied ranging from shunt impedance considerations to the most complicated beam dynamic aspects.

Introduction

The physics of charged particle acceleration deals with two kinds of forces which form together the driving term is Newton's law:

$$q (\vec{E} + \vec{v} \times \vec{B}) = \vec{F} = m \vec{a} . \qquad (1)$$

We distinguish external forces (such as bending and focusing forces of magnets) and self forces (such as space charge). The external forces are dominant in the limit where the charge of the accelerated particles vanishes. These forces are directly under our control and consist mainly of bending- focusing- and accelerating forces. A major part of an accelerator physicist's work is to design apparatus for bending and focusing magnets or static deflectors. Subsequently the particle's motion under the influence of these forces is studied. For both of these tasks many computational tools have been developed and are the subject of papers in this volume /1/2/3/.

In this paper we deal primarily with the external forces which act parallel to the particle velocities and serve as means for acceleration. With few exceptions these forces are applied by means of r.f. electromagnetic fields in resonators, the so-called accelerating cavities (or often just cavities). However, as a result of practical experience with cavities, in many existing accelerators it is found that such cavities produce many parasitic effects which severly limit the performance (e.g. beam break-up, emittance growth, head tail turbulence and all kind of instabilities). The strength of these effects is proportional to the accelerated charge (collective effects). Since the maximum charge or the maximum current that can be accelerated is probably the most important design and performance criterion, the design of a cavity must take these effects into account. Thus we will have to deal with the combined action of external driven and self excited forces.

For over 30 years work has been published on how to design a cavity in order to accelerate particles efficiently. For over 20 years computer codes have been used to optimize realistically shaped structures. Quantitative prediction about collective effects however became possible only very recently (last five years) with new computational tools and theories about the complicated collective interaction between charged moving particles and surrounding cavities (or other structures).

After a brief review of the long history of "conventional" cavity design we will present "unconventional" design procedures that optimize the over-all effect of accelerating structures. All modern design tools are large computer codes needing of the order of one megaword of core and an enormous amount of cpu time. However, the striking results from these codes in recent years underline that this is the way to go.

Some Definitions

A very simple r.f. accelerating cavity device is shown in figure 1: a cylindrically symmetric "pill-box" with small side tubes. The tubes are large enough to fit the beam dimensions and small enough so as not to perturb the field pattern of the driven mode very much. The TM010 resonance is driven by an external power supply. This mode has a longitudinal electric field and thus can accelerate charged particles. When a charged particle with charge q traverses this structure it experiences a force yielding a net change in energy after the particle has left. For constant offset a and constant velocity $v = \beta c \hat{z}$ we find for the energy change ΔU:

$$\Delta U = q \, \text{Re}\{\underline{V}\} \;,\; \underline{V} = \int_{-L}^{+L} E_z(r=a, \varphi=\varphi_1, z=\beta ct) e^{i\omega z/\beta c} e^{i\psi} dz \;. \quad (2)$$

For small beam ports, L does not extend very far and can basically taken as the gap length. ψ is an arbitrary phase between the cavity mode and the particle's arriving time. In order to obtain a large net effect one must make sure that the oscillating term in eq.2 does not change sign within the significant range of integration. Obviously any cavity design must take the particle's speed into account and cavities will be very different for different β. The first step is to adjust the cavity shape so that a high ΔU is obtained for a given amount of the externally supplied power. Several amplitude independent quantities are defined, the r/Q ("r over Q"), the loss parameter k and the shunt impedance R_s as:

$$k = \underline{V} \, \underline{V}^*/4W \quad (W = \text{stored energy}) \;, \quad (3)$$

$$r/Q = 4k/\omega \;, \quad (4)$$

$$R_s = (r/Q) \cdot Q \quad (\text{or } R'_s = (r/Q) \cdot Q/\text{unit length}) \;. \quad (5)$$

Note that all these quantities assume a constant speed of the particle and that they all depend on β in a complicated way. k and r/Q are purely geometric quantities that do not invoke the conductivity of the cavity material. For a given geometry the shunt impedance then depends on the quality factor Q. Unfortunately the shunt impedance (and the r/Q) is then also defined by:

$$P = \underline{V} \, \underline{V}^*/R_s \;. \quad (6)$$

This power law misses a factor of two in the denominator compared with a.c.-circuit theory. (The shunt impedance R in a RLC model is then $R = R_s/2$). Given a power P and a structure of shunt impedance R_s the maximum energy gain of a particle becomes:

$$U_{max} = q \sqrt{P \, R_s} \quad (7)$$

The actual energy gain varies as $\cos\psi$, see eq.2.

Eq.7 indicates that R_s is the figure of merit as are r/Q and Q. The r/Q can be optimized by changing the geometry, Q is influenced by the choice of material (copper, aluminum or superconducting materials).

In large accelerators cavity cells are grouped together in mechanical units and used as travelling wave or standing wave modules. Just to give the two most outstanding examples in linac and circular accelerator technology: The SLAC linac is an arrangement of over 80 000 cavity cells with more than 3.000 m length; the world largest e^+e^- storage ring PETRA now has over 800 cavity cells with over 200 m length.

History

The history of optimization of accelerating cavities by means of computers dates back to the 50's. Starting from a closed pill-box cavity as shown in figure 2 (which can easily be solved analytically) chains of pill-boxes with connecting beam tubes were analysed. See figure 3. The method uses eigenmode expansions in simply shaped subregions and matches the expansion coefficients at interface areas. By this method both cylindrically symmetric modes (monopole) and modes with variation in azimuthal direction (dipole, quadrupole, etc.) can be analysed. Among many other authors I give here only a few references to Bell, Gluckstern, Hahn, Helm, Hereward, Nakamura and Walkinshaw /4-9/.

In the 60's the first mesh codes (MESSYMESH by Edwards /10/ and LALA by Hoyt and Simmonds /11/) were used to calculate arbitrarily shaped cavities of cylindrical symmetry. Apart from a few rare objects such as r.f. kickers or r.f. quadrupoles most cavities were built cylindrically symmetric and driven in the TM010 mode. Thus there was no need to develop mesh codes for deflecting modes and the existing codes (which were restricted to cylindrical symmetry for the fields and the geometry) could cover the necessary work for optimizing accelerating structures.

In the same decade accelerating cavities attracked the machine physicist's interest from a different point of view. Any cavity with small beam ports has - among TM010 - many other resonances which couple to a beam. While gaining energy from the driven TM010 mode, particles also loose energy into all higher order modes (and into the TM010 mode reducing the effective voltage seen).

The total parasitic energy loss is given by the sum of the partial energies lost into the individual modes. In order to obtain this number the existing methods had to be extended. SUPERFISH by Halbach and Holsinger /12/ could compute some ten's of cylindrically symmetric modes in otherwise arbitrary cavities. KN7C by Keil /13/ and TRANSVRS by Bane and Zotter /14/ were able to compute hundreds of modes in chains of pill-box with beam tubes for cylindrically symmetric fields and for fields with azimuthal variation.

Anomalous bunch lengthening in storage rings indicated that apart from loosing energy into a cavity there is a serious effect of the potential well inside the beam originating from the beam cavity interaction. Bane and Wilson /15/ could construct this potential well from the calculated modes (by KN7C and later TRANSVRS). Although this approach was limited to disk loaded waveguides it could be used as a model for many existing structures.

In recent years the problem of the potential well has been studied by a quite different approach in Novosibirsk /16,17/ and CERN /18/ (BCI). Instead of calculating the many modes that are needed one can solve Maxwell's equations directly in time domain and obtain all fields and forces. Furthermore it can be shown that frequency and time domain approaches give the same results /19/.

Some additional contributions have come to the subject of field computation in the last two years: The mesh codes were extended to calculate now modes with azimuthal variation. Similarly the PRUDE-code by Daikovskii, Portugalov and Riabov /20/ and ULTRAFISH by Gluckstern, Halbach, Holsinger and Minerbo /21/ both solve for dipole and higher resonances. Another code named URMEL /22/ was developed last year and has already been used in many laboratories for studying deflecting modes. The time domain code BCI was generalized to TBCI /42/ and now calculates transient fields of off-axis beams. Finally, the history of calculating fields in rotationally symmetric cavities comes to end and only minor improvements are still being worked on.

A somewhat similar history could be outlined for cavities of translational symmetry (or constant cross-section cavities). Since such a cavity can be considered as a waveguide section we have to refer to many r.f. studies outside the accelerator field. It should be only briefly mentioned here that this chapter is also well closed and covered by rather sophisticated programs /23-26/.

Although 3D calculations with mesh codes have been made for over ten years (e.g. Albani and Bernardi /27/) they are not yet incorporated in standard design procedures due to their inherent problems with accuracy and computational effort. Recently two more 3D resonator codes were described by Wilhelm /28/ and Hara, Wada, Fukasara and Kikuchi /29/ and a 3D code using the FIT-theory /30/ is being worked on (Furthermore, this code will also solve the 3D time domain problem).

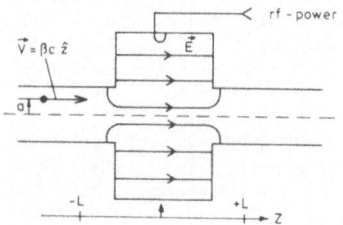

Figure 1:
Accelerating cavity of cylindrical symmetry with beam ports ($\vec{E}$ indicates the electric field of the TM010 accelerating mode)

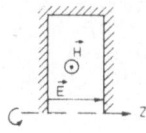

Figure 2:
Simple pill-box cavity and direction of the electric field of TM010 mode

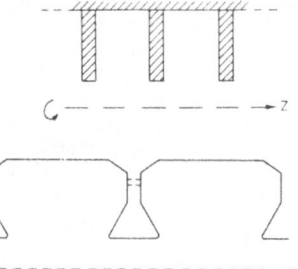

Figure 3:
Chain of cylindrically symmetric cavities with beam hole (disk loaded waveguide)

Figures 4, 5, 6:
Typical accelerating structures of cylindrical symmetry

4:   nose cone slot coupled π-mode structure

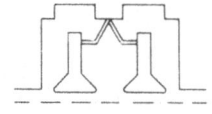

5:   disk and washer cavity

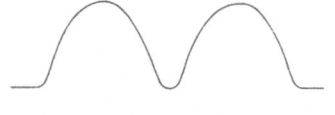

6:   elliptical-elliptical superconducting cavity

Conventional Design

In a "conventional cavity design" all complicated beam cavity interaction is neglected. The optimization is mainly concerned with the shape of the cavity. Starting from a simple pill-box cavity with side tubes "nose cones" were introduced which shorten the significant path length of integration in eq.2 and consequently weaken the oscillating term /31-33/, see figure 4. This cavity type is now standard in storage rings. A further increase in shunt impedance was promised by the "disk and washer" cavity /34-35/ and at many places this structure is under investigation, see figure 5. Higher order mode loss calculations have been taken into account in the recent LEP design /36/ but did not influence the final shape.

A different design procedure is applied to superconducting cavities where R_s is less important than other effects resulting from the high surface field strength /37/. However, this procedure is still conventional and the task is to have a high field in the TM010 and damping antennas for higher order modes. Considerations other than those connected with r.f. have not been taken into account. A typical cavity of elliptical-elliptical shape is sketched in figure 6 /38-39/.

The conventional design procedure makes use of resonator programs results (see e.g. figure 7) and is by now well established.

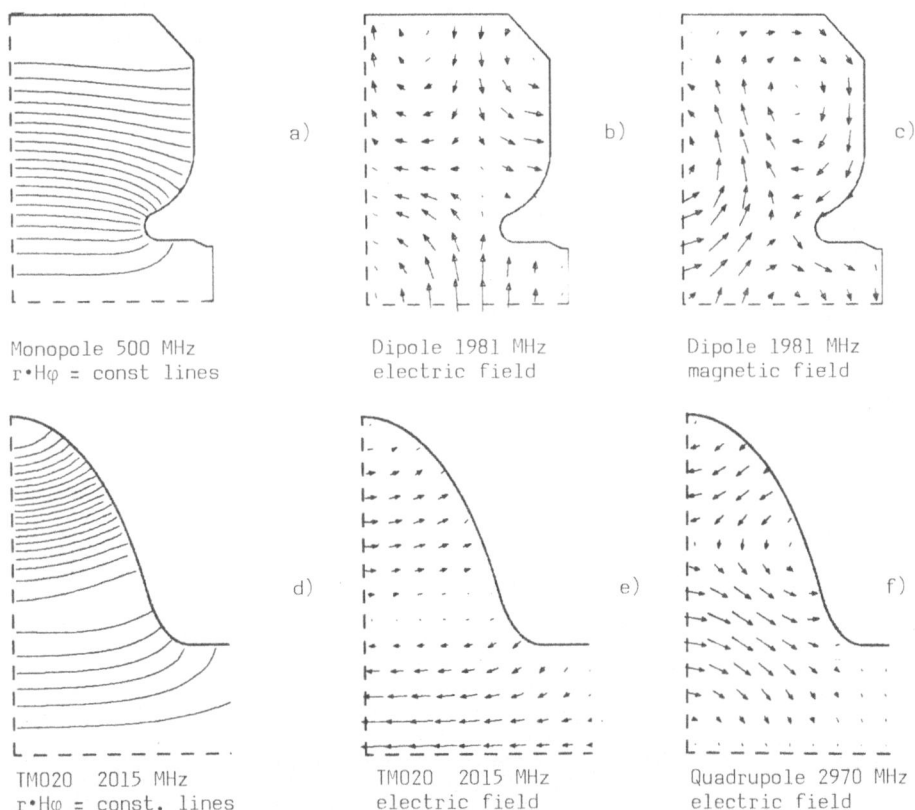

Monopole 500 MHz
r·Hφ = const lines

Dipole 1981 MHz
electric field

Dipole 1981 MHz
magnetic field

TM020 2015 MHz
r·Hφ = const. lines

TM020 2015 MHz
electric field

Quadrupole 2970 MHz
electric field

Figure 7: Typical computer output for the new PETRA cavity (a, b, c) and the DESY superconducting cavity prototyp (half cell) (d, e, f)

Unconventional Design

The fact that the maximum beam current in PETRA was limited by a transverse instability /40/ which was found to be mainly due to parasitic cavity effects /41/ was one of the first indications that the parasitic effects caused by accelerating cavities are much stronger than originally expected.

In order to understand this phenomenon we forget for a while that cavities are used for acceleration and consider just undriven cavities and their interaction with particle beam. The results in figure 8 showing transient fields excited by a Gaussian bunch traversing three PETRA cells have been calculated with the previously mentioned time domain code TBCI /18,42/. These self excited fields act back on the particles inside the bunch. After the passage of the structure each individual particle will have experienced a certain deceleration and deflection due to the fields. The wake potentials inside a Gaussian bunch are shown in figure 9.

Each accelerator component generates wake potentials but it has been found experimentally and theoretically that the cavity contributions are dominant. This statement applies to all cavities, i.e. also bellows, pumps and separator tanks. As a consequence one preceeds as follows: Compute wake potential due to the unavoidable accelerating cavities, compute wake potentials due to other objects and try to remove all objects that contribute significantly in comparison with the accelerating structures. This recipe was qualitatively applied years before one could calculate such effects and PETRA has been built with careful avoidance of any unneccessary cavities. A typical design procedure using the computational means now available is being undertaken along with the LEP design /43/.

Handy quantities for the wake potential effects are the total loss and the total kick:

$$k_{tot} = \int \rho(s) w_{\parallel}(s) \, ds / (\int \rho(s) ds)^2 \qquad (8)$$

$$k_{\perp} = \int \rho(s) w_{\perp}(s) \, ds / (\int \rho(s) ds)^2 \qquad (9)$$

$w_{\parallel}(s)$ decelerating wake potential ($\Delta p_{\parallel}$)
$w_{\perp}(s)$ deflecting wake potential ($\Delta p_{\perp}$)
$\rho(s)$ charge density along the bunch

Since that portion of the total energy loss that goes into the fundamental mode is not lost, one usually defines the parasitic loss parameter as the total loss minus the fundamental loss:

$$k_{par} = k_{tot} - k_{fund} \qquad (10)$$

Since the cavity is used to accelerate particles it seems suitable to compare the parasitic effects with the desired effects by defining the "goodness"-functions:

$$g_0(\lambda/\sigma) = k_{par}/k_{fund} \qquad (11)$$

$$g_1(\lambda/\sigma) = k_{\perp}/k_{fund}$$

λ = r.f. wave length of accelerating mode

σ = rms bunch length (12)

k_{par}, $k_{\perp}$ for a Gaussian charge distribution

These functions basically give the "loss per shunt impedance" and "deflection per shunt impedance" (Higher order g_m functions for parasitic quadrupole, sextupole field are defined analogously). Since the natural bunch length is somewhat proportional to the rf wavelength as a result of some beam dynamics equations it seems to use λ/σ as parameter ratner than the bunch length σ itself. A typical range in storage ring operation is $20 < \lambda/\sigma < 100$.

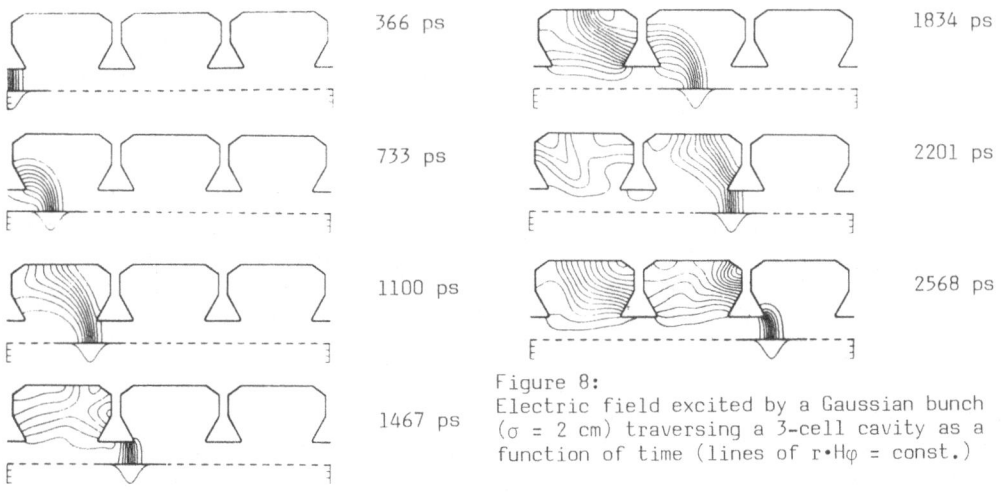

Figure 8:
Electric field excited by a Gaussian bunch (σ = 2 cm) traversing a 3-cell cavity as a function of time (lines of $r \cdot H\varphi$ = const.)

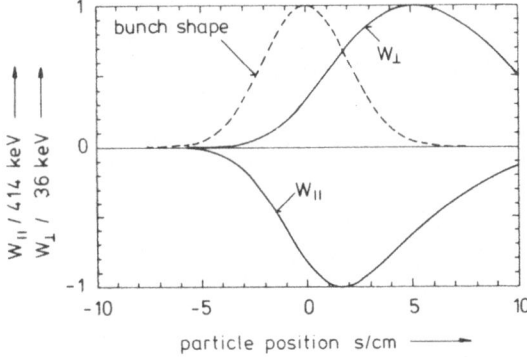

Figure 9:
Wake potentials inside a Gaussian bunch (charge 1µC, σ =2cm) after the passage of a single PETRA cavity cell at a distance 0.5 cm off axis as a function of particle position (leading particles on the left)
$w_{\parallel}(s)$ = energy loss due to axis-symmetric (monopole) fields
$w_{\perp}(s)$ = transverse change in momentum due to dipole fields

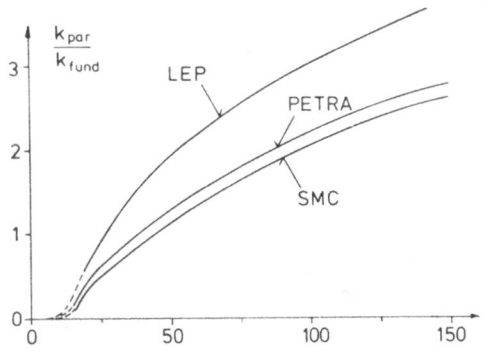

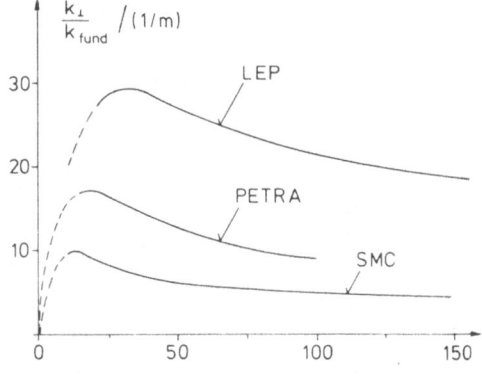

Figure 10:
Ratio of parasitic mode loss to fundamental loss parameter (TM010 shunt impedance) as a function of λ/σ (r.f. TM010 wave length divided by rms bunch length) for LEP, PETRA and SMC structure

Figure 11:
Ratio of average transverse change in momentum and fundamental loss parameter as a function of λ/σ.

Figures 10 and 11 show g_0 and g_1 for the LEP cavity /36/, the original PETRA five cell cavity /33/ and a recently developed SMC-cavity /44/. Surprising result: the non-computer-optimized PETRA cavity has better g_0 and g_1 values when compared to the fairly well optimized LEP cavity. The SMC cavity (Single Mode Cavity) beats both. In particular g_1 is about five times lower than the LEP value.

The shape of the SMC cavity is shown in figure 12. Although this structure was developed for superconducting rf, it could in principle be made out of copper as well. In this case R'_s would be 16 MΩ/m (compared to 24 MΩ/m for PETRA and 26 MΩ/m for LEP).

Since PETRA is (and LEP is believed to become) limited by transverse wake potential effects such a cavity would allow for twice (PETRA) or five (LEP) times the maximum beam current. Note that a) the comparison takes into account the different bunch length at different frequencies and b) a factor of five in beam current means 25 in luminosity in the present design and c) that the power per integrated luminosity is reduced by more than a factor of ten, although the instantaneous power in increased by about 50 percent).

Compared to the superconducting cavity being developed at DESY /39/ this SMC type would result in an increase of a factor of three in single bunch current. There is an important byproduct: this cavity was designed - by means of URMEL - so that it has only one monopole mode, the TM010. There is no higher order resonance of TM0 type and only two weakened dipole modes. Thus multiturn effects carried by resonant modes are eliminated. Figure 13 shows the standard PETRA superconducting nine cell unit in the TM010-π mode and the corresponding SMC structure.

Although g_0 and g_1 are rough figures giving only integrated averaged quantities, they can serve as a preliminary design aid. It takes only a few short TBCI runs to obtain these functions and the cpu time consumption is about the same as the time going into the mode calculations. Once a $\pm$ 10 % optimized cavity is found, one has to go one step further and investigate the full beam dynamics taking into account all details of the wake potentials.

Just recently it became possible to study beam dynamics with wake potentials for realistic accelerators (The ingredient missing for long time were the wake potentials, the theory was known). These tracking codes simulate the particle motion by some thousands of "super particles" which are tracked through an accelerator lattice. When the particles pass a cavity, wake potential effects are added according to the charge distribution and by this one obtains a self consistent solution of the combined equations of motion and electromagnetic field equations.

About two years ago measured and simulated bunch lengthening in PETRA were found to agree very well /45/. The full equations of motion (including betatron oscillations) are meanwhile being combined with the wake effects in storage ring codes /45,46,47,48/ and in linac tracking codes /49/.

The only drawback so far is that these codes - expecially for storage ring application - need an immense amount of cpu time even in the most primitive versions. Several ten hours on an IBM 3081 may be necessary to obtain results such as maximum beam current versus tune. Just as a typical example (in figures 14 and 15) we give a result of WAKERACE /48/ for PETRA at injection energy. The theoretically predicted mode coupling /41/ is verified and it yields a beam blow up.

So far computer tools for an electromagnetic analysis of accelerator structures are quite advanced but computer tools for beam dynamics with collective effects are still subject to research. Current tracking codes are too complicated and too unreliable to become widely used.

Approaching the ultimate design procedure

The unconventional design procedures described in the previous section invoke various computer codes for quite different tasks. In order to make the entire package useful and easily usable the user should not need to know all details. The ideal situation would be that the input data contain the lattice, the cavity geometry and some beam data. The output would be the particle motion versus time displayed just like real experiments are in control rooms.

To some extent the above requirements are met by the computer codes TBCI-URMEL-WAKERACE-WAKETRAC /18,22,48,49/. From a single input deck for the cavity geometry, TBCI and URMEL calculate wake potentials, decelerating and deflecting modes. Results are fed via storage disk to the two tracking codes for linacs (WAKETRAC) and storage rings (WAKERACE). Lattice parameters are input to the tracking codes from existing data sets used for beam optics calculations. The storage ring tracking code is an extension of RACETRACK /50/ which simulates higher order multipoles in lattice elements.

Present work focusses on extending the beam tracking codes in order to include as many effects as possible. Furthermore 3D versions of TBCI and URMEL are being developed in order to complete the wake potentials which so far include only the contributions from the accelerating structures and not the contributions from e.g. separator tanks. The ultimate goal is to have one single code that can be used for computer experiments in the same way that one now uses accelerators.

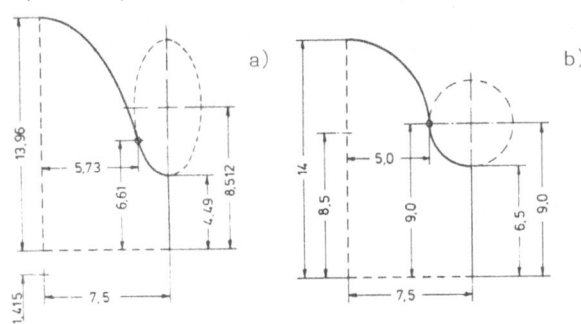

Figure 12:
Half cell of superconducting multicell cavity

a) DESY prototyp of elliptical-elliptical shape

b) Single Mode Cavity (SMC) structure with reduced parasitic effects

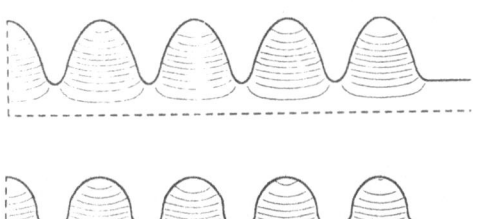

Figure 13:
Accelerating π-mode in 9-cell cavities

a) DESY prototyp of elliptical-elliptical shape

b) Single Mode Cavity (SMC) structure with reduced parasitic effects

Future Development

a) Electromagnetic codes for cylindrically symmetric structures

Existing computer codes are still being improved with respect to accuracy and speed /52,53,54,55/ and extended to codes solving for the r.f. heating problem /56/. An ultimate version of URMEL /22/ including triangular meshes and arbitrarily shaped blocks of anisotropic magnetic or electric material will be available soon. TBCI has recently been improved enormously /51/ and can now (in principle) solve for "infinetely" long structures using a mesh with "unlimited" number of nodes. Probably next year or soon after, the chapter on calculating electromagnetic properties of cavities of cylindrical symmetry can be closed. Another few years later these codes will be a standard tool for any rf cavity design.

b) Cavities with translational symmetry

Cavities of an arbitrary shape in a plane and no change of shape perpendicular to the plane occur less often in accelerator technology. Such cavities can be considered as just a section of a waveguide. Thus all the work done for waveguide theory can be applied and results in sophisticated computer codes solving for arbitrarily shaped anisotropic magnetic, dielectric and conducting materials (e.g. the LW-code, now in the CERN computer program library T201/202).

c) 3D electromagnetic codes

The real challenges are still fully 3D codes. For the last ten years papers have been published on this subject /25,26,27,28,29/ but none of these computer codes is without problems when applied to realistic structures. There is of course the problem of the enormous cpu time and need of storage locations but there is also a theoretical problem hidden in the equations. For time harmonic fields in vacuum, Maxwell's equations can be written as (with $k = \omega/c$):

$$\operatorname{curl} \vec{H} = k \vec{E} , \qquad (13)$$

$$\operatorname{curl} \vec{E} = k \vec{H} , \qquad (14)$$

or as

$$\operatorname{curl} \operatorname{curl} \vec{E} = k^2 \vec{E} \qquad (15)$$

The eigenvalue spectrum obviously contains zeros since any static electric field obeying

$$\operatorname{curl} \vec{E} = 0$$
$$\operatorname{div} \vec{E} = \rho$$

is also solution of eq.(15) with $k = 0$. In order to "repair" this situation (which yields to a matrix problem with many unwanted solutions) one has to impose the condition

$$\operatorname{div} \vec{E} = 0 \qquad (16)$$

on the solution. However there is a free parameter left and it is necessary to study the dependence of k on this parameter in order to find out whether a solution is valid or not /29/.

Another approach is ti implement the divergence condition right away and to solve:

$$(\nabla^2 + k^2) \vec{E} = 0 \qquad (17)$$

Any correct solution of the above equation has no static contents. However, numerical solutions seem in some case to have such spurious results /28/ and the reason is probably hidden in the fact the method of numerical solution is not consistent with all four Maxwell's equations. The FIT ansatz /30/ - which is also some kind of finite difference method - does solve all four Maxwell equations and very recently a new 3D-code has been written by the author. This code is not yet completed; figure 16 shows an output example of the mesh generator.

In order to economize the coding this new program will become a complete system including 3D-BCI, 3D-URMEL and the existing 3D codes for magneto- and electrostatic problems /57/ and 3D eddy currents /58-60/. However, the everyday use of these 3D codes will not take place before the next generation of computers is available.

d) Particle dynamics codes with collective effects

Probably the most difficult area of near future research is that of particle dynamics codes. Present codes must make crude approximations for many complicated effects and as a consequence results are not more reliable than within some factor of five (say). The problem is solely in the speed of present computers - the necessary theoretical components are well understood (basically F = ma and beam optics theory).

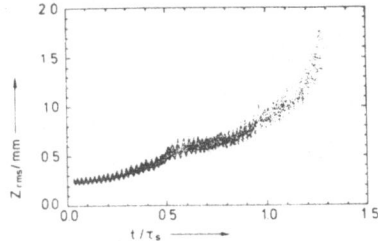
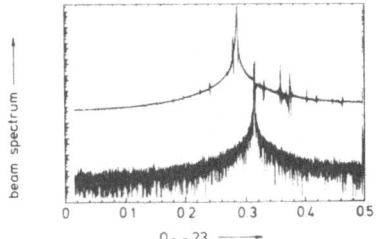

Figure 14:
Beam height in PETRA versus time as obtained from computer simulation including transverse wake field effects (τ_s=synchr. damping time)

Figure 15:
Beam spectrum in PETRA obtained from computer simulation
a) zero current (no synchrotron side bands)
b) above threshold current of the transverse instability (many synchrotron side bands)

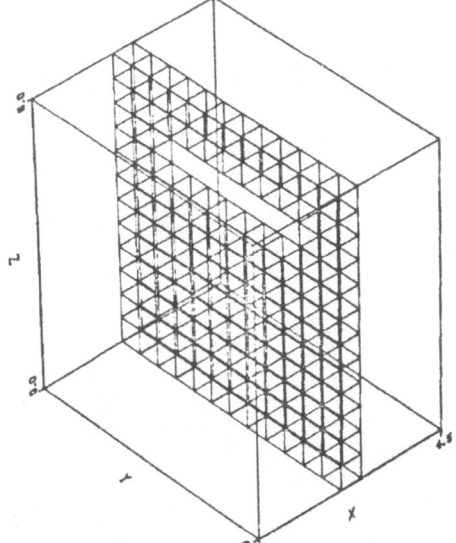

Figure 16:

3D view of two slot coupled square box cavities

Acknowledgements
The author wishes to thank D. P. Barber for careful reading of the manuscript.

References

/1/ C. W. Trowbridge, this conference
/2/ E. Wilson, this conference
/3/ A. Wrulich, this conference
/4/ W. Walkinshaw et al., AERE report TM/104 (1954)
/5/ M. Bell, H. Hereward, CERN/63-33/1963
/6/ H. Hahn, Rev. Sci. Inst. $\underline{34}$ (1963), p. 1094
/7/ R. L. Gluckstern, 1961 Internat. Conf. on High Energy Accel., p.129
/8/ M. Nakamura, Japanese Journal of Applied Physics $\underline{7}$(1968), p.257
/9/ R. H. Helm, SLAC-PUB-813, 1970 and references therein
/10/ T. Edwards, MURA rep. 622 (1961)
/11/ H. C. Hoyt, D. D. Simmonds, W. F. Rich, Rev. Sci. Instr. $\underline{37}$(1966), p.755
/12/ K. Halbach, R. F. Holsinger, Part.Accel. $\underline{7}$(1976), p. 213
/13/ E. Keil, NIM $\underline{100}$(1972), p.419
/14/ K. Bane, B. Zotter, XI-th Int.Conf. on High Energy Accel., Geneva 1980, p.581
/15/ K. Bane, P.B. Wilson, SLAC, PEP-226A (1977)
/16/ V. E. Balakin et al., SLAC-TRANS-188(1978)
/17/ A. V. Novokhatskij, Inst. of Nucl. Phys., Siberian Div. of the USSR, Preprint 82-157 (in Russian) and references therein
/18/ T. Weiland, XI-th Internat.Confl. on High Energy Accel., Geneva 1980, p.575
/19/ K. Bane, T. Weiland, SLAC-AP1,(1983) and 12th Conf.on High Energy, 1983
/20/ A. G. Daikovskii, In. I. Portugolov, A. D. Riabov, Part. Accel. $\underline{12}$, p.59/1982)
/21/ R. L. Gluckstern et al., 1981 Linear Accel. Conf., Santa Fe, p.102
/22/ T. Weiland, DESY $\underline{83}$-005, 1983 and DESY M-82-24(1982)
/23/ Z. J. Csendes, R. M. J. Silvester, IEEE MTT $\underline{18}$(1970), p.1124
/24/ J. S. Hornsby, A. Gopinath, IEEE MTT $\underline{17}$(1969), p.684
/25/ S. Akhtarazad, P. B. Johns, IEEE Vol.$\underline{122}$(1975), p.1134
/26/ T. Weiland, AEÜ $\underline{33}$(1979), p.170
/27/ M. Albani, M. Bernardi, IEEE MTT $\underline{22}$(1974), p.446
/28/ W. Wilhelm, Part. Accel. $\underline{12}$(1982), p.139
/29/ M. Hara et al., IEEE-NS $\underline{30}$, (1983), p.3639
/30/ T. Weiland, AEÜ $\underline{31}$(1977), p.116
/31/ M. A. Allen and P. B. Wilson, Proc. of the 1974 Internat. Conf. on High Energy Accel., Stanford, p.92 and references therein
/32/ M. A. Allen et al., IEEE, NS-$\underline{24}$(1977), p.1780
/33/ H. Gerke et al., DESY PET-77/$\underline{08}$(1977)
/34/ S. O. Schriber, LASL report, LA-UR 79-463(1979)
/35/ J. M. Potter et al., 1979 Particle Accelerator Conference, San Franzisko(1979)
/36/ LEP design report, CERN/ISR-LEP/79-13
/37/ e.g. discussion and references in: M. Tigner, H. Padamsee, Cornell, CLNS82/553
/38/ P. Kneisl et al., NIM $\underline{188}$(1981), p.665
/39/ W. Ebeling et al., IEEE NS-$\underline{30}$(1983), p.3357
/40/ D. Degèle et al., XI Internat. Conf. on High Energy Accel., Geneva 1980, p.16
/41/ R. D. Kohaupt, XI Internat. Conf. on High Energy Accel. Geneva 1980, p.562
/42/ T. Weiland, DESY $\underline{82}$-015 (1982)
/43/ H. Henke, LEP note 454 (1983)
/44/ T. Weiland, DESY $\underline{83}$-073(1983)
/45/ T. Weiland, DESY $\underline{81}$-088(1981)
/46/ R. H. Siemann, CBN 82-27, 1982
/47/ D. Brandt, LEP note 444, 1983
/48/ T. Weiland, A. Wrulich, WAKERACE-Code, to be published
/49/ T. Weiland, F. Willeke, 12th Int. Conf. Accel., 1983
/50/ A. Wrulich, RACETRACK-Code, being published
/51/ K. Bane, T. Weiland, 12-th Internat. Conf. Accel., 1983
/52/ R. L. Gluckstern, Proc. of the IV-th Compumag Conf., Geneva 1983, being publ.
/53/ B. B. Fomel et al., Part. Accel. $\underline{11}$,(1981), p.173
/54/ P. Fernandes, R. Parodi, Part.Accel.$\underline{12}$(1982),p.131
/55/ J. Tückmantel, CERN/Ef-RF/83-5
/56/ R. F. Holsinger, S. O. Schriber, IEEE NS-$\underline{30}$(1983),p.3545
/57/ H. Euler et al. AfE $\underline{65}$(1982), p.299
/58/ T. Weiland, AfE $\underline{60}$(1978), p.345
/59/ H. Euler, T. Weiland, AfE $\underline{61}$(1979), p.103
/60/ T. Weiland, Archiv der etz $\underline{1}$(1979), p.263

Computer Aided Magnet Design

C W Trowbridge
Rutherford Appleton Laboratory
Chilton, Didcot, Oxon, OX11 OQX, UK

1. Introduction

The purpose of this paper is to review the status, and highlight some of the difficulties, in the computer aided design of electromagnetic devices for particle accelerators, beam lines and detectors. Magnet design is a complex activity involving many techniques and so the material presented here will be limited to those aspects on the computer modelling of magnets which are used to predict the fields, forces etc, ie to those techniques concerned with the numerical solution of the field equations. In 1972 the author carried out a review of this subject which was presented at the 4th international magnet technology conference at Brookhaven[1]; in this report an attempt was made to review the then state-of-the-art in magnet computation and concluded that idealised two dimensional time independent models, could be analysed to high accuracy (< 0.1%) and that a number of viable computer codes were available for routine use. It was also a period of vigorous activity in constructing reliable algorithms for solving the corresponding three-dimensional problem and several promising developments were described.

Furthermore the period between 1965 and 1972 was very productive with many active groups of researchers involved in developing magnet computation throughout the world and it was also an exciting period because of the advent of the large digital computer. At the Brookhaven conference it was predicted that the decade following would see numerous achievements in algorithm development and a rapid expansion in the use of interactive graphics techniques. This has happened to some extent; the evolution of the multiuser mini-computer (MUMs) in the mid-seventies provides a more democratic use of high-speed graphics terminals with some real-time interactive capability, and more recently the introduction of new single user mini-computers (SUMs) with raster operations is changing methods of working.

A prediction that has not been realised was the wide-spread use of automatic optimisation, ie it was confidently expected that the 'inverse problem' would be routinely solvable by 1980, in which a designer specifies a field shape and a computer program is used to compute the model shape. At present this procedure is only practical for extremely simple cases which is disappointing since the essence of design is optimisation[2]. A major effort needs to be made in order to take up the advanced numerical techniques in optimisation because the difficulty is not only a lack in available computing power.

In this paper the elements of a computer aided magnet design system will be outlined then the methods of solving the field equations will be briefly reviewed followed by a survey of available codes; next some particular examples of the use of the RAL codes in solving 3D magnetic design problems will be given. In the final section some of the likely future developments will be discussed.

2. The Elements of a Computer Aided Magnet Design System

The flow diagrams in Figure 1 illustrate the software components of a typical CAMD system. An idealisation of the design process is shown in Figure 1(a) in which the designer usually iterates toward a satisfactory solution by a heuristic approach. There are three conceptual stages involved, thus: (a) Data input or pre-processor for defining geometric and material data, (b) Solution Processor for solving the equations numerically, and (c) Post-processor for examining and extracting the results, ie fields, gradients, integrals and forces, etc. All three processors should be file driven and access a common data base. The pre- and post-processors will be interactive utilising interactive graphics techniques and reside in a multi user machine environment. Except for small problems the solver processors will reside on

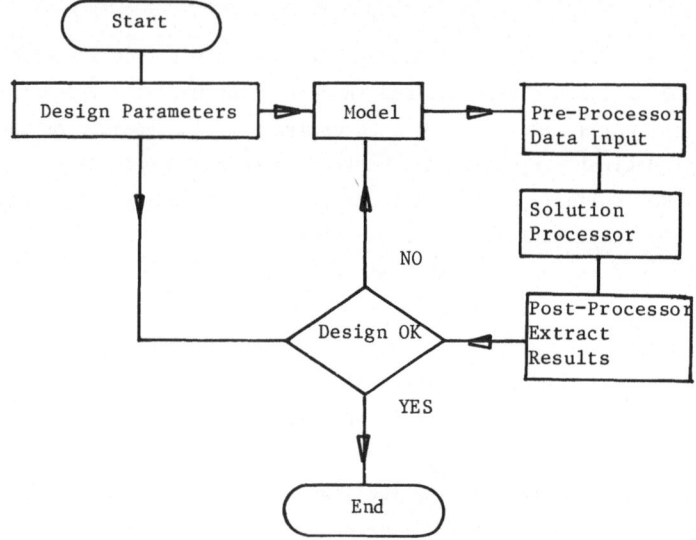

Figure 1(a) Heuristic Design

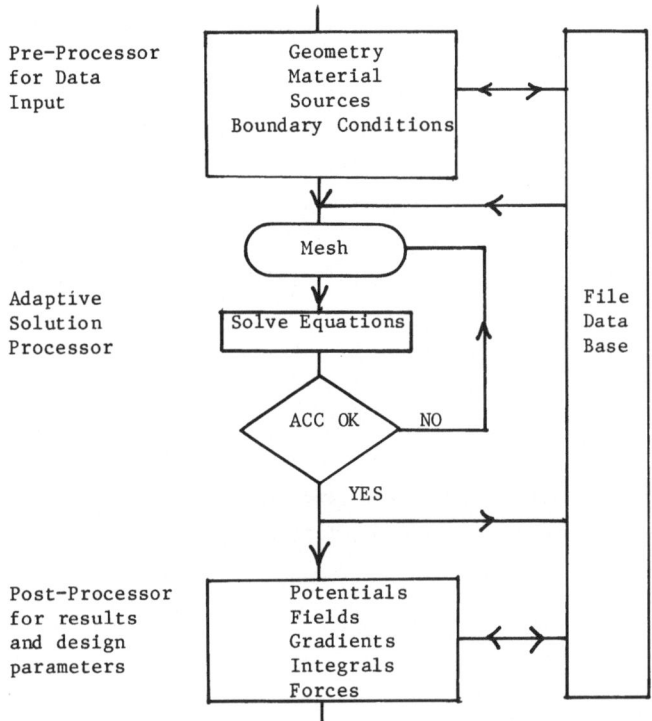

Figure 1(b)   Components of the Ideal Numerical
Modelling Software System

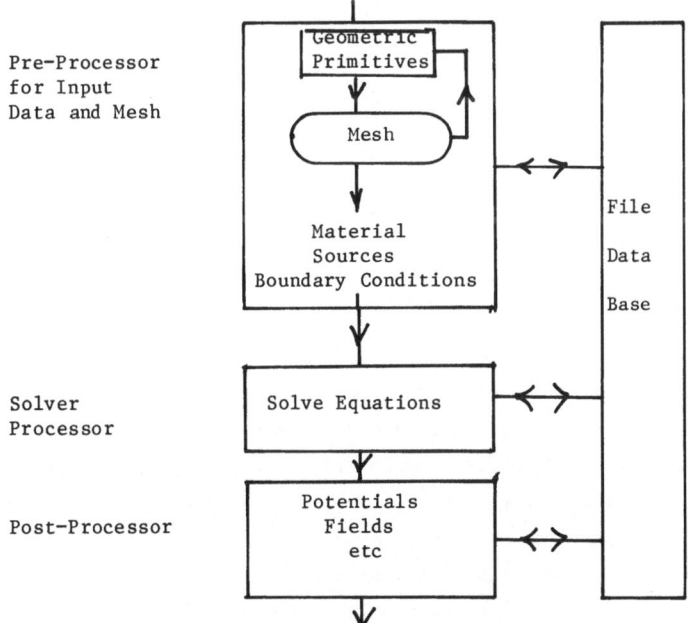

a mainframe with file transfer between machines locally networked or perhaps widely distributed. In Figure 1(b) and 1(c) the components are specified in a little more detail; the user can be imagined as controlling the pre- and post-processing phases at a graphics terminal using heuristic techniques to optimise his design. Because the problem is to be solved numerically the model has to be discretised, ie meshes of elements have to be generated to conform to the data-model and the accuracy of the result will depend upon the level of discretisation. In the ideal system shown in Figure 1(b) this discretisation process is taking place inside a solver generating sufficient mesh to achieve a specified accuracy. It is a current limitation that this ideal has not yet been satisfactorily met, at least for three dimensional systems despite considerable research under way. The present systems usually follow the scheme outlined in Figure 1(c) in which the pre-processor includes a mesh generation stage. There are some software packages that do not have interactive pre- and post-processing facilities at all and have to be run in a 'batch mode'.

3. Solving the Field Equations

3.1 Differential and Integral

There is no need to give a detailed review of the various field formulations and numerical algorithms here since this subject has been exhaustively covered elsewhere[3,4], it will suffice to give some general remarks on the main methods that have been successfully used. The first point to make is that only statics and eddy current effects will be considered; this at once 'rules out' all considerations of high frequency devices from the scope of this paper.

The customary starting point is the low frequency sub set of Maxwell's equations[5], these equations plus the non-linear constitutive laws define the electromagnetic field for slowly-varying time-dependent and static fields. Those equations are differential and, in many cases, the normal approach to their solution is to eliminate the field vectors by introducing scalar and vector potentials to produce a set of partial differential equations. Provided the essential boundary conditions for the problem are prescribed these equations are then solved numerically by a more or less standard technique, eg Finite Differences[6], or Finite Elements[7]. In order to do this the problem space must be discretised into points or elements to allow the non-linear continuum problem to be transformed into a linear algebraic problem. Hence the heart of solution process is one of linear

algebra characterised by a system matrix of coefficients. For the
differential equation formulation the following two points apply:

(a) The system matrix is sparse: since the differential nature implies only
    local coupling of equations.
(b) The whole problem space, including the external world, has to be
    discretised.

There is an alternative formulation based on the integral forms of the field
equations, ie Gauss 'theorem' is the integral form of Poisson's equation. If
the field integral forms are used instead of the differential formulation,
the discretised equations are algebraic with a system matrix of a very
different character, ie

(a) The system matrix is fully populated: since every point is coupled to
    every other point.
(b) Only the material regions of the problem space have to be discretised.
    This is because the integrals are only defined for the active regions
    which implies that the boundary conditions are intrinsically satisfied,
    and that the far field problem has been automatically taken into
    account. Another important consequence is that for linear problems only
    the surfaces of the active regions have to be discretised.

At first sight it may appear that the integral approach will be superior to
the differential approach since the mesh generation problem is far far
simpler - the empty space does not have to be filled with elements and the
far field boundary conditions are automatically satisfied. However, there
are very significant difficulties because the matrix, although smaller, is
fully populated and the solution times will vary as the cube of the number of
degrees of freedom. Where as for the differential case, although the matrix
is larger it is sparse and possibly symmetric and the solution times will
vary, at most, as the square of the number of degrees of freedom.

Extensive comparisons have been made[8,9] between differential and
integral methods and some guidelines have been evolved governing the
choice[10]. One general remark is appropriate here; if the problem is
relatively simple in that it can be discretised into a modest number of
elements, depending upon the power of the computer available, then the
integral method offers an accurate technique involving a minimum of data
preparation and yields smooth solutions.

3.2 Solution Potential

There is also considerable latitude of choice of solution potential; for the two dimensional case the simplest and probably the most effective selection is the single component vector potential but in three dimensions there is no single choice. In this case it is usual to separate out the statics case, for which it is possible to define a scalar potential, from the time-dependent case which by its nature must be vector field problem at least in the eddy current regions. The type of scalar potential for the statics 3D case also is not self-evident since there is more than one possibility. The essential consideration here is that if prescribed currents are to be included the total scalar potential ($H = -\nabla \psi$) may be multivalued and it is usual to introduce a reduced scalar potential by subtracting the source fields for the currents[1], however exclusive use of the reduced potential will introduce serious numerical cancellation errors[11] due to subtraction of large quantities so it is better to use a combination of both potentials, the total potential everywhere except in conductor regions when the reduced potential should be used[12].

The more complex 3D eddy case is receiving considerable research attention with only test and research codes available, see table 4. Some of these developments are showing considerable promise and in the next few years viable packages should become available[13]. However the full non-linear transient case, will require enormous computer power.

3.3 Available Codes

Tables 1 to 3 attempt to list the codes that, in the author's opinion, should be of practical use to the magnet designer. The list is not exhaustive and there could well be omissions due to ignorance or oversight. The references should be consulted for technical details and for information on validation and the accuracies achievable, however, some results for the RAL codes[14] will be given in section 4.

4. CASE STUDIES

4.1 SNS Trim Quadrupole Type QT137

The GFUN integral code[15] was used in the design stage of this magnet which had the design parameters shown in table 5.

Table 1. 2D Differential Electromagnetics Codes

Date	Name/Centre	Authors	Formulation	Numerical Method and Discretisation	Remarks	Environment
1964	POISSON(Trim)/ Lawrence Livermore Berkeley USA	Winslow Colonias Halbach Holsinger etc	Vector potential Variational and finite difference Plane XY Axisymmetry	First Order triangular mesh with fixed topology	Non-Linear Magnetostatics	Batch
1974	MAGGY/ NV Philips Eindhoven Holland	Munro Polak Wachters Van Welij	Vector Potential Finite Element Plane XY Axisymmetry	First Order quadrilaterals mesh with fixed topology	Non-Linear Magnetostatics Permanent Magnets	Batch
1976	PE2D/RAL Oxford UK	Simkin Trowbridge Armstrong Biddlecombe Diserens	Vector Potential Scalar Potential Finite Element Plane XY Axisymmetry	First Order and Second Order triangle elements with variable topology	Non-Linear Magnetostatics Permanent Magnets Eddy Currents Steady State Eddy Currents Transients Electrostatics	Batch plus interactive Pre- and Post-Processor on a MUM
1977	FLUX/ Laboratoire d'Electro-technique de Grenoble	Sabonnadiere Meunier Morel	Vector Potential Scalar Potential Finite Element Plane XY Axisymmetry	First and Second Order Isoparametric Elements with Variable Topology	Non-Linear Magnetostatics Eddy Currents Steady State Electrostatics	Batch plus interactive Pre- and Post-Processors on a MUM
1978	MAGNET-11 McGill Montreal Canada Imperial College London UK	Lowther Silvester Freeman	Vector Potential Finite Element Plane XY	First Order Triangle Elements with variable topology.	Non-Linear Magnetostatics	Interactive Pre- and Post-Processor and Solver on a MUM
1979	AOS/Magnetic AOS Data Systems Milwaukee USA	Brauer Bodine Larkin	Vector Potential Finite Element Plane XY	First Order	Non-Linear Magnetostatics Eddy Currents Steady State Skin effect	Batch
1982	APPLE RAL Oxford UK	Simkin Trowbridge	Vector Potential Scalar Potential Plane XY Axisymmetry	First Order Triangle Elements with fixed topology	Linear Magnetostatics Electrostatics Flow	Interactive Pre- and Post-Processor with Rubber Banding on a SUM

Table 2. 3D Magnetostatic Codes

Date	Name/Centre	Authors	Formulation	Numerical Method and Discretisation	Remarks	Environment
1972	PROFI – Program for calculation of Fields Technische Hockschule Darmstadt W Germany	Müller Krueger Wolf	Reduced Scalar Potential Differential	Finite Difference	Non-Linear Magnetostatics Electrostatics	Batch
1977	TOSCA/RAL Oxford UK	Simkin Trowbridge	Mixed Total and Reduced Scalar Potential Differential Formulation	Isoparametric Finite Elements First and Second Order Hexahedra Variable Topology	Non-Linear, Permanent Magnets Electrostatics Distributed Conductors Sources. Avoids cancellation errors associated with exclusive use of the reduced scalar potential	Batch plus Interactive Pre- and Post-Processor
1981	PADDY/NV Philips Eindhoven Holland	de Beer Polak Wachters van Welij	Mixed Total and Reduced Scalar Potential Differential	Finite Element First Order hexahedra fixed topology	Non-Linear, Permanent Magnets. Filament Sources See Tosca above	Batch

Table 3. Integral Operator Codes for Magnetostatics

Date	Name/Centre	Authors	Formulation	Discretisation	Remarks	Environment
1970	GFUN/RAL Oxford UK	Newman Trowbridge Turner	Magnetisation Integral Equation	Moments Point Matching Constant Magnetisation Prisms	Non-Linear Magnetostatics 2D and 3D	Batch with interactive Pre- and Post-Processor on Main Frame
1973		Collie Diserens		Tetrahedra		
1975		Armstrong Simkin				
1976	BIM/RAL Oxford UK	Armstrong Collie Diserens Simkin Trowbridge	Scalar Potential	Point Matching Linear Quadratic and Cubic Triangles	Linear Magnetostatics 2D 3D	Interactive Batch
1981		Simkin		Galerkin		

Table 4. 3D Eddy Current Research Codes[22]

Project/Centre	Authors	Formulation	Numerical Method	Remarks
GEC Stafford UK	Preston Reece Riley	(T, Ω) Curl $T = J$ $H = T + \nabla\Omega$	Finite Element	Normal component of T discontinuous at conducting interface
RAL Oxford UK	Biddlecombe Heighway Simkin Trowbridge	(R, S) Curl $R = J$ $H = R + S$ $R_s \equiv 0$	Finite Element	Field Vector R zero over interface; cancellations may be problem
RAL Oxford UK	Emson Simkin	(A, Φ) Curl $A = B$ div $\sigma A = 0$	Finite Element	Maximum economy for potentials achieved Avoids electric scalar potential in conducting regions. Magnetic Scalar Potential in air
GE/Schenectady	Chari Konrad Palno Angelo	(A, V) div $A = 0$	Finite Element	Vector potential throughout space
TRIFOU/ Electricite de France	Bossavit Verite	(H)	Finite Element Coupled Boundary Integrals	Maximum economy for Fields achieved, 3 unknown per mesh point in conducting region. Infinite Exterior region handled by Boundary elements
Bath University UK	Balchin Davidson	(H, E)	Circuit analogue method (Finite Differences)	Coupled Electric and Magnetic Circuits
Bath University UK	Rogers Eastham	(E, Φ)	Finite Element	Achieves maximum economy and avoids electric scalar potential in conducting regions
EDDYNET/ Argonne USA	Turner	(J)	Integral Circuit Analogue (Wire Grid)	Conductors only but includes transients

Table 5   QT137 Design and Computed Parameters

Aperture dia	0.274 m	Peak Gradient Integral G	0.1832 T
Iron Length	0.2034 m		(mean of 23 magnets)
Peak Current	300 A	Effective Length measured	
Peak Gradient	0.6026 T/m	at 300A	0.304 m
Turns/pole	15		(mean of 23 magnets)
		Effective Length calculated	
		by GFUN at 290A	0.307 m
		GFUN Error in Length L_o	+ 1%

Twenty-three magnets were built and measured. The average change in the gradient integral, $\Delta G/Go$, is shown for the $\theta = 0$ plane, compared with GFUN at 300A. See Figure 2. The asymmetry in measured values between left (power) and right (non-power) sides can be seen to be significantly less than the error in the GFUN values. The effective length predicted by GFUN is 1% more than that measured. The change in effective length with current from 50A to 300A is about 1.5%.

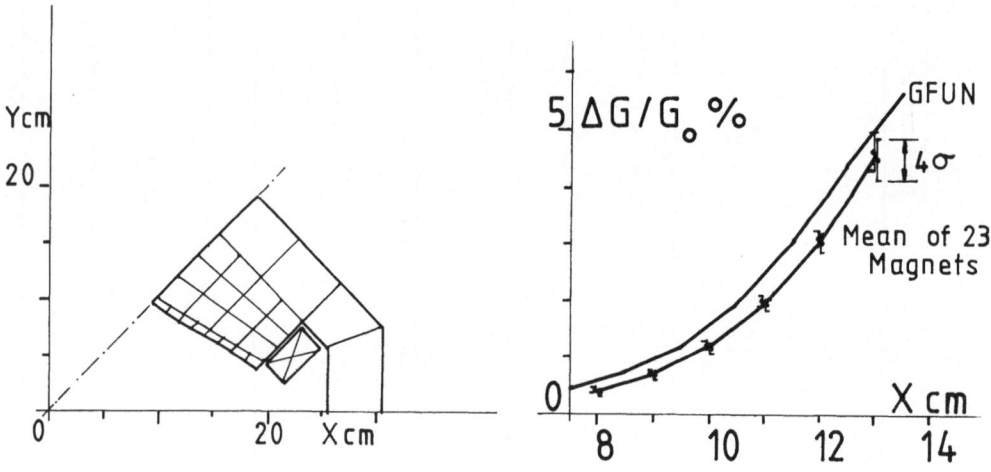

Figure 2   Model and Field Gradient Integrals

Although the measurements are self-consistent and the GFUN prediction is outside the 90% confidence limits the search coils used for the measurement were not properly compensated for the higher harmonics of field and this may have contributed to the discrepancy between the calculation and measurement.

The figure on the left shows the elements used in GFUN to represent one octant
in X-Y. With 3 layers in the z direction there were 360 tetrahedra in 1/16th
of the entire magnet. The coils are racetracks. This example is near the
limit of applicability of GFUN, with 360 elements, the computing time (½ hour
IBM 195) will be dominated by the cube law and any further refinement of the
model, say factor of 2, will require hours of computing. However the
designers were able to construct a prototype according to the calculated
dimensions which was sufficently close to the prediction that no modifications
were required before going into production.

4.2 EVB2 Magnet[16]

The Spallation Neutron Source synchrotron, has a vertical extraction system
for the 800 MeV proton beam. After the fast-pulsed vertical-kick magnets the
extracted beam is bent by a d.c. septum magnet through an angle of some
22° to clear the following main-ring dipole. Above the main dipole is
placed the EVB2 magnet which restores the extracted beam to the horizontal
plane.

EVB2 is required to produce a sector field of 1.9 T-m on a radius of 3.56 m,
with an aperture of 0.24 m radially x 0.14 m axially. The original design
consisted of a C-yoke, bent through 22° with the open side down towards
the top of the main dipole. The effect of the main dipole on the leakage
field from the C was not allowed for until the magnet was out to tender; the
tenderers asked for some design modifications to be investigated, including
the effects of changing various coil parameters and estimates of the forces on
the C, when it was realised that there were several defects in the concept.

The design was then changed to an asymmetric straight window-frame yoke with
extra radial aperture to accommodate the beam sagitta and wedge-shaped
pole-ends to produce the sector field. Although this is no cheaper, the
leakage field towards the main dipole is quite negligible, so there is no
field distortion. The straight coil and yoke are easier to make, and the yoke
is much stronger than the C. The wedge shim is clearly visible near the coil,
and is also shown on the computer model Figure 3. The wedge angle is
12°, which is about one degree more than the angle required in the
effective field boundary, (e.f.b.); this offset was established for the
Oxford Spectrometor, and is approximately correct for EVB2.

The TOSCA[12,17] differential code was used for the calculations, including
tracking some protons through the end-field to estimate how far the sector

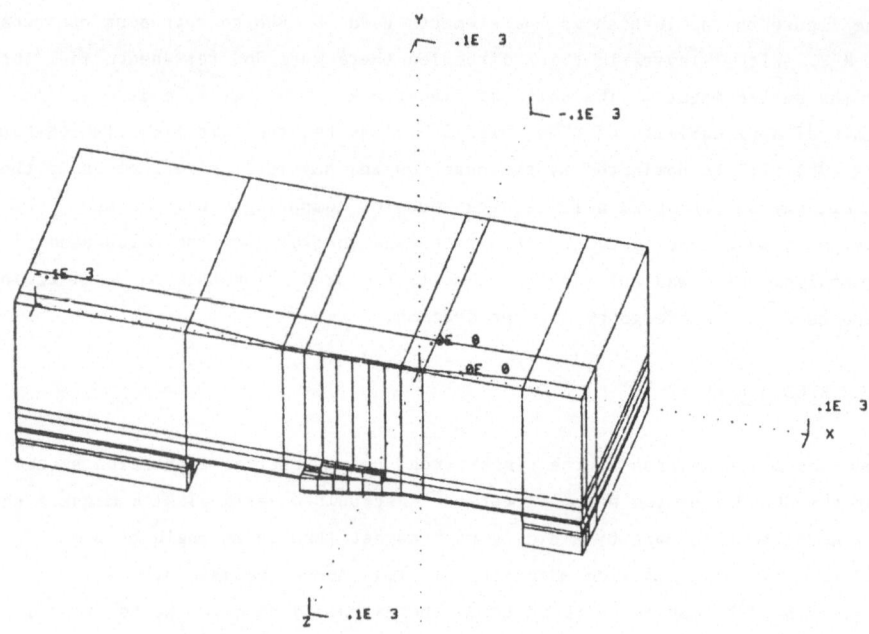

Figure 3 Computer Model of EVB2

field requirement had been met. It was too difficult to measure the field along the notional proton tracks; instead the field was measured along straight lines in the median plane, and re-calculated as nearly as possible at the measurement positions. Figure 4 shows that the e.f.b. from the tracking (T) is very close to that from straight-line integrals (15), which confirms the value of the latter. The e.f.b. (17) was calculated from field values at the same position, current, gap and, as near as possible, B-H data as the measurements (M) at the non-powered end of the magnet. The discrepancy of 2mm (0.3% of the effective half-length) arises from the difference in the calculated field at the position under the coil where the field gradient is highest.

Contours of the difference between TOSCA and measurement at the non-powered end are shown in Figure 5. It is not possible at this stage to know how much of the difference comes from manufacturing and measurement errors, and how much from calculation errors: the contours show some correlation with the TOSCA mesh, but they are typically of the same height as those obtained from a comparison between the two ends of the magnet (Figure 6). The latter are substantially the same at the lower current of 1445A, showing that the differences between the actual magnet and the TOSCA model are probably significant.

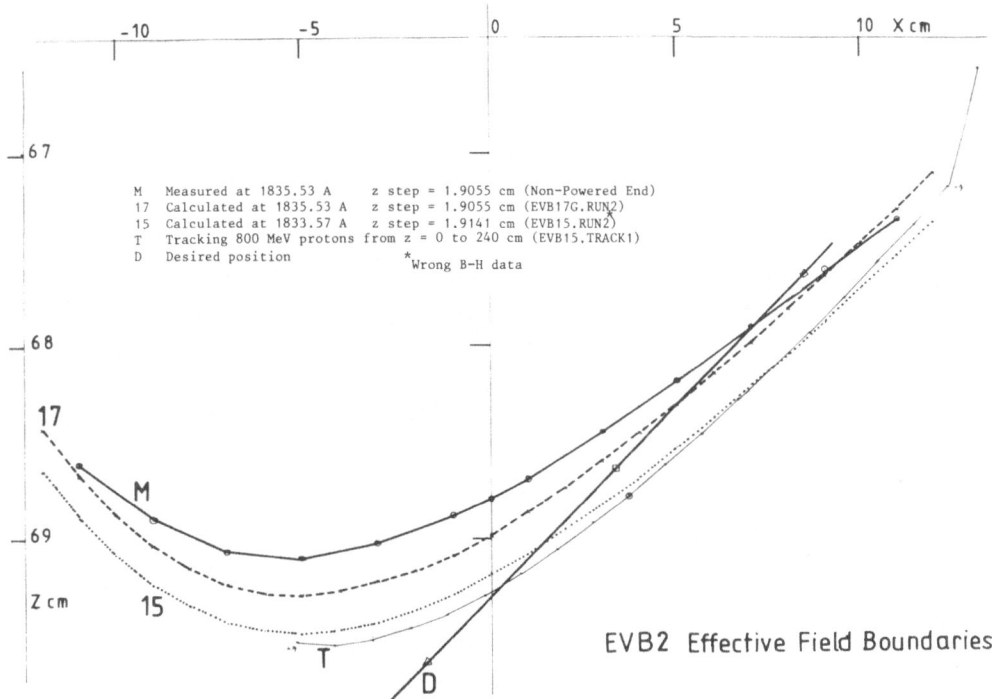

Figure 4   EVB2 Effective Field Boundaries

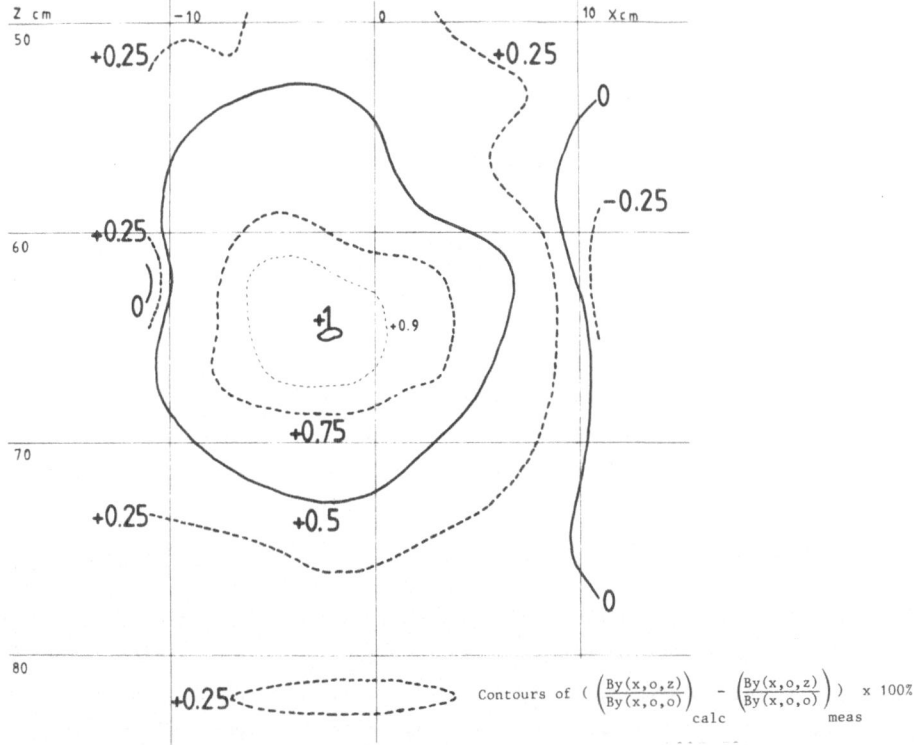

Figure 5   EVB2 Calculation cf Measurement at 1835.53 A

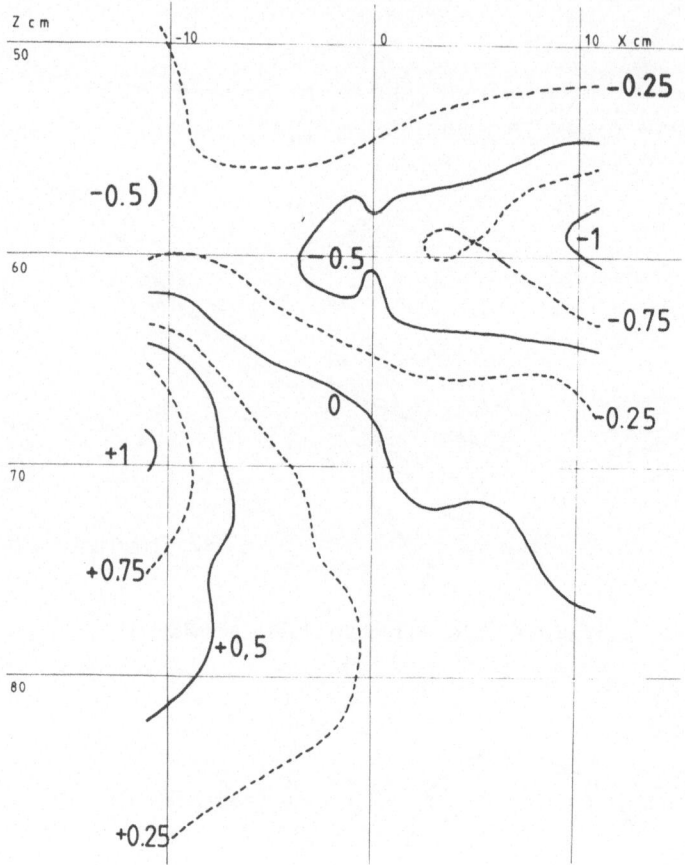

Figure 6   EVB   Measurement at 1835.53 A

Two possible sources of error which remain to be explored are the mesh distortion used to model the wedge-shaped pole ends, and errors in either the coil field or its integration to find the scalar potential (r.h.s. in the equation). A full analysis of these results will be given in reference 16.

5.   CONCLSUIONS AND FUTURE DIRECTIONS

Clearly good results are obtainable with computer codes for predicting fields; the accuracy achievable will depend on the degree of discretisation and on having a precise knowledge of the properties of the materials involved. The degree of discretisation depends critically on the complexity of the design

and on the computing power available. It is difficult to estimate a priori
the discretistion errors although some significant work has been carried out
and should make possible the adaptive mesh schemes already mentioned[18,19].
At present the answer to the question 'How accurate is your code?' still is
'that depends on ....' and perhaps more importantly on 'who is to carry out
the work?'. There is no doubt that the users with considerable expertise in
both electro- magnetics and numerical modelling are prerequisite to achieve
the best results.

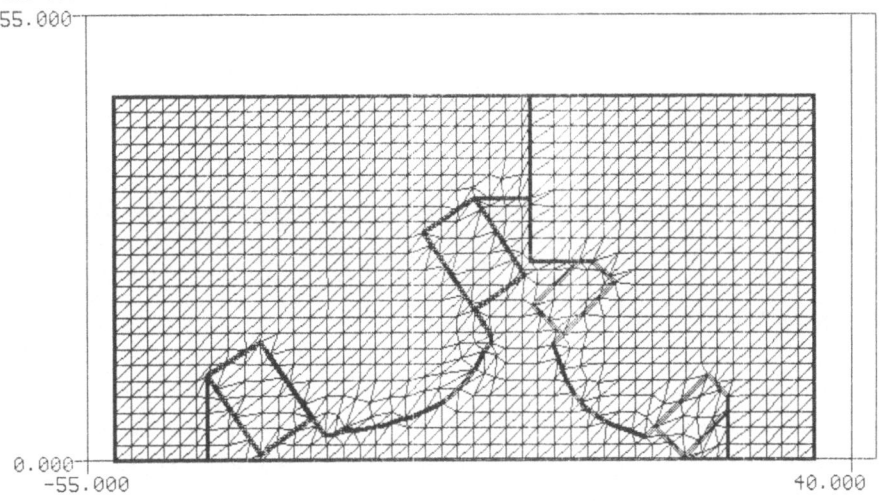

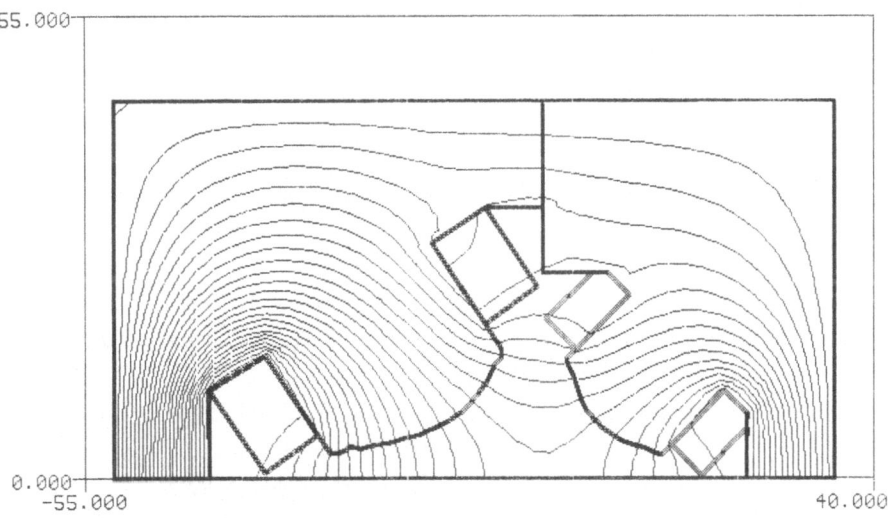

Figure 7   2D Model, Mesh and Solution of an Asymmetric Quadrupole
using the ICL PERQ Single User System. The Software is
written in Pascal and Allows Real Time Interaction of
Geometry and Mesh[20].

What does the future hold in store? Apart from the three dimensional eddy current codes using differential operator finite elements there should be a considerable revitalising interest in integral methods when the impact of parallel processing hardware is assimilated; also to be optimistic single user machines networked together sharing file servers and accessing larger machines as required, should create a truly democratic interactive environment[20]. See Figure 7. Computing is moving into the 5th generation era[21] in which far greater emphasis will be placed on features like man-machine interfaces, data flow architectures, logic programming and knowledge-handling capabilities.

These developments will have a considerable impact on engineering design and at some stage, the sooner the better, detailed thought must be given to use of these techniques in numerical modelling and hence computer aided magnet design.

6. ACKNOWLEDGEMENTS

The author is indebted to Alan Armstrong of CAG Group RAL for providing the information on the EBV2 magnet and for his skill and patience in solving field problems. He is also grateful to Mrs Pam Peisley for her skill and patience with word-processing in preparing the typescript.

REFERENCES

1. Trowbridge, C W, "Progress in Magnet Design by Computer", Proc 4th Intern Conf on Magnet Technology, Brookhaven Laboratory, 1972.
2. Armstrong, A G A M, Fan, M W, Simkin, J, Trowbridge, C W, "Automated Optimisation of Magnet Design using Boundary Integral Method", IEEE Trans Mag, vol. Mag-18, No.2, March 1982.
3. Trowbridge, C W, "Numerical Solution of electromagnetic Field Problems in Two and Three Dimensions", Contributed chapter in "Numerical Methods in Coupled Systems", pub: John Wiley and Sons, 1983.
4. Compumag Conference Proceedings: Oxford 1976, published, Rutherford Appleton Laboratory, Ed: J Simkin.
   Grenoble 1979, published, Laboratoire d'Electrotechnique de Grenoble, ERA 524 CNRS, Ed: J C Sabonnadiere.
   Chicago 1981, IEEE Transactions in Magnetics, vol. Mag-18, No.2, March 1982.
   Genoa 1983, IEEE Transactions in Magnetics, November 1983.

5. Stratton, J A, "Electromagnetic Theory", McGraw Hill, New York, 1941.
6. Smith, G D, "Numerical Solution of Partial Differential Equations", OUP 1971.
7. Zienkiewicz, O C, "The Finite Element Method", 3rd Edition, McGraw Hill 1977.
8. Trowbridge, C W, "Three Dimensional Field Computation", IEEE Transactions in Magnetics, vol. Mag-18, No.1, January 1982.
9. Simkin, J, "A Comparison of Integral and Differential Equation Solutions for Field Problems", IEEE Transactions in Magnetics, vol.Mag-18, No.2, March 1982.
10. Simkin, J, Magnet Technology Conference, Invited Paper on Field Computation, Grenoble, September 1983.
11. Simkin, J, and Trowbridge, C W, "On the Use of the Total Scalar Potential in the Numerical Solution of Field Problems in Electromagnets", Int J Numer Methods Eng, vol.14, pp423-440.
12. Simkin, J, and Trowbridge, C W, "Three-Dimensional Non-Linear Electromagnetic Field Computations, Using Scalar Potentials", Proc Inst Elec Eng, vol.127, part B, no.6, Nov 1980.
13. Emson, C R I, Simkin, J, "An Optimal Method for 3-D Eddy Currents", to appear IEEE Trans Mag, November 1983 (COMPUMAG Conference).
14. "Computing Aids for Engineers", Computing Applications Group, Rutherford Appleton Laboratory, Chilton, Didcot, Oxon OX11 OQX.
15. Newman, M J, Trowbridge, C W, and Turner, L R, "GFUN: An Interactive Program as an Aid to Magnet Design", Proc 4th Conf Magn Technol, Brookhaven National Laboratory, Brookhaven, NY, pp.617-626, 1972.
16. Armstrong, A G A M, "Design Calculations Compared with Measurements on the EVB2 (SNS) Magnet". To be published. Rutherford Appleton Laboratory.
17. Armstrong, A G A M, Riley, C P, Simkin, J, "Tosca User Guide - Version 3.1", Rutherford Appleton Laboratory Report, RL-81-070 (1982).
18. Polak, S J, den Heijer, C, Bielen, J, "A Future of Mesh Adaptive Solving", IEEE Trans in Magnetics, vol. Mag-18, No.2, March 1982.
19. Penman, J, Lees, P, Fraser, J R, Smith, J R, "Complementary Energy Methods in the Computation of Electrostatic Fields", to appear IEEE Trans Mag, November 1983 (COMPUMAG Conference).
20. Simkin, J, Trowbridge, C W, "Electromagnetics CAD Using a Single User Machine (SUM)", to appear IEEE Trans mag, November 1983 (COMPUMAG Conference).
21. Simons, G L, "Towards Fifth-Generation Computers", published by NCC Publications 1983.
22. Lari, R J, "Comparison of eddy current programs", to appear IEEE Trans Mag, November 1983 (COMPUMAG Conference).

BEAM INSTABILITIES AND COMPUTER SIMULATIONS

A. Piwinski

Deutsches Elektronen-Synchrotron DESY
Notkestraße 85, D-2000 Hamburg 52, W. Germany

1. Introduction

The most important instability which limits the luminosity in existing electron-positron storage rings is caused by beam-beam interaction. But also synchro-betatron resonances are a serious problem and can limit the luminosity. The beam-beam interaction leads to a blow-up of the beams which reduces the luminosity and the life time of the colliding bunches. Synchro-betatron resonances or satellite resonances occur if the relation

$$k Q_x + \ell Q_z + m Q_s = n \tag{1}$$

is satisfied where k, ℓ, m and n are integers and Q_x, Q_z and Q_s are the betatron and synchrotron frequencies in units of the revolution frequency. With increasing bunch current the number and the width of the satellite resonances increase such that finally the currents are limited.

These satellite resonances are mainly excited by dispersions in the accelerating cavities and by transverse fields with a longitudinal gradient in the cavities. Satellite resonances due to the chromaticity are usually prevented by sextupoles. Dispersions in the cavities cannot be supressed completely. Especially during energy ramping when the optics is changed and the orbit is shifted by small amounts, spurious dispersions are produced. Transverse fields with a longitudinal gradient are generated when passing the cavities off-axis which is difficult to avoid during energy ramping. An analytical treatment of satellite resonances is difficult. Especially the increase or decrease of the oscillation amplitudes during many revolutions cannot be calculated. Computer simulations, however, can show the general behaviour of the particles on such a resonance.

Satellite resonances can also be excited by the beam-beam interaction when the beams cross at an angle or when the dispersion at the interaction point is not zero. In that case many satellites of nonlinear betatron resonances appear which were found by computer simulations and seen in the storage ring DORIS I, where the beams crossed at an angle. This mechanism might also become important for the new ep-project HERA.

The blow-up of the beams at head-on collision is caused by the strong nonlinearities of the space charge forces. Many resonances can be excited, especially coupling resonances of horizontal and vertical betatron oscillations such that the oscillation energies between the horizontal and vertical plane are exchanged. The beam height, which is usually much smaller than the beam width, is then increased drastically whereas the width remains nearly constant. An analytical treatment of this effect is, at the moment, not yet possible. Computer simulations are, therefore, the best way to investigate the dependence of the blow-up on various machine parameters and to find cures to suppress the blow-up as far as possible.

2. Synchro-betatron resonances

2.1 Excitation caused by dispersions in a cavity

For each particle with an arbitrary energy deviation exists a closed orbit that can be described as the product of the dispersion D times the relative momentum deviation or approximately, the relative energy deviation:

$$x_{c.o.}(\ell) = D_x(\ell) \frac{\Delta p}{p} \approx D_x(\ell) \frac{\Delta E}{E} \qquad (2)$$

Around this closed orbit the particles perform betatron oscillations. The dispersion can be horizontal or vertical and we consider horizontal or vertical satellite resonances. In the cavity the energy is changed and therefore the closed orbit is shifted. Since the total coordinates x and x' cannot be changed in such a short interval the betatron coordinates are changed:

$$\delta x = - D_x \frac{\delta E}{E} \qquad (3)$$

$$\delta x' = - D_x' \frac{\delta E}{E} \qquad (4)$$

$$\delta E = eU(\sin(\psi + 2\pi s/\lambda) - \sin\psi) \qquad (5)$$

with e = elementary charge, U = cavity voltage, ψ = synchronous phase, s = longitudinal position, λ = wave length of the voltage.

We take into account only that part of the energy change which varies with the synchrotron frequency. The other constant part which replaces the radiation losses does not play a role for a resonance. The longitudinal position is changed in the curved sections of the machine. The change of s per revolution is given by

$$\delta s = \oint \frac{1}{\rho} (x + D_x \frac{\Delta E}{E}) \, d\ell = A_1 x + A_2 x' - \alpha_M C \frac{\Delta E}{E} \qquad (6)$$

with
$$A_1 = - \frac{1}{\beta_x} (D_x \sin\mu_x - F_x(1 - \cos\mu_x)) - A_2 \alpha_x$$

$$A_2 = - D_x(1 - \cos\mu_x) - F_x \sin\mu_x$$

$$F_x = \alpha_x D_x + \beta_x D_x' \qquad \alpha_x = - \beta_x'/2$$

α_M is the momentum compaction factor and C is the circumference, ρ is the radius of curvature and μ_x is the betatron phase advance per revolution.
x and x' are the betatron coordinates at the beginning of the revolution.

A simulation of the coupled synchrotron and betatron oscillations on a digital computer has been done [1] for small currents taking into account the natural nonlinearity of the synchrotron oscillation (Eq. (5)). The simulation has the following steps: In the cavity the longitudinal coordinate s of the synchrotron oscillation remains constant whereas the energy is changed according to Eq. (5). The change of the betatron coordinates is given by Eqs. (3) and (4). In the arc the change of s is given by Eq. (6) whereas the energy deviation is constant. The betatron coordinates x and x' are transformed with

$$M = \begin{pmatrix} \cos\mu_x + \alpha_x \sin\mu_x & \beta_x \sin\mu_x \\ -(1 + \alpha_x^2)\sin\mu_x/\beta_x & \cos\mu_x - \alpha_x \sin\mu_x \end{pmatrix} \qquad (7)$$

Fig. 1 shows the typical behaviour of the amplitudes during 1200 revolutions. Fig. 1a shows the betatron and synchrotron amplitudes for a sum resonance where always one amplitude increases while the other one decreases. Fig. 1b shows that on a difference resonance both amplitudes increase or decrease at the same time. In this case the beat is larger than in the first case.

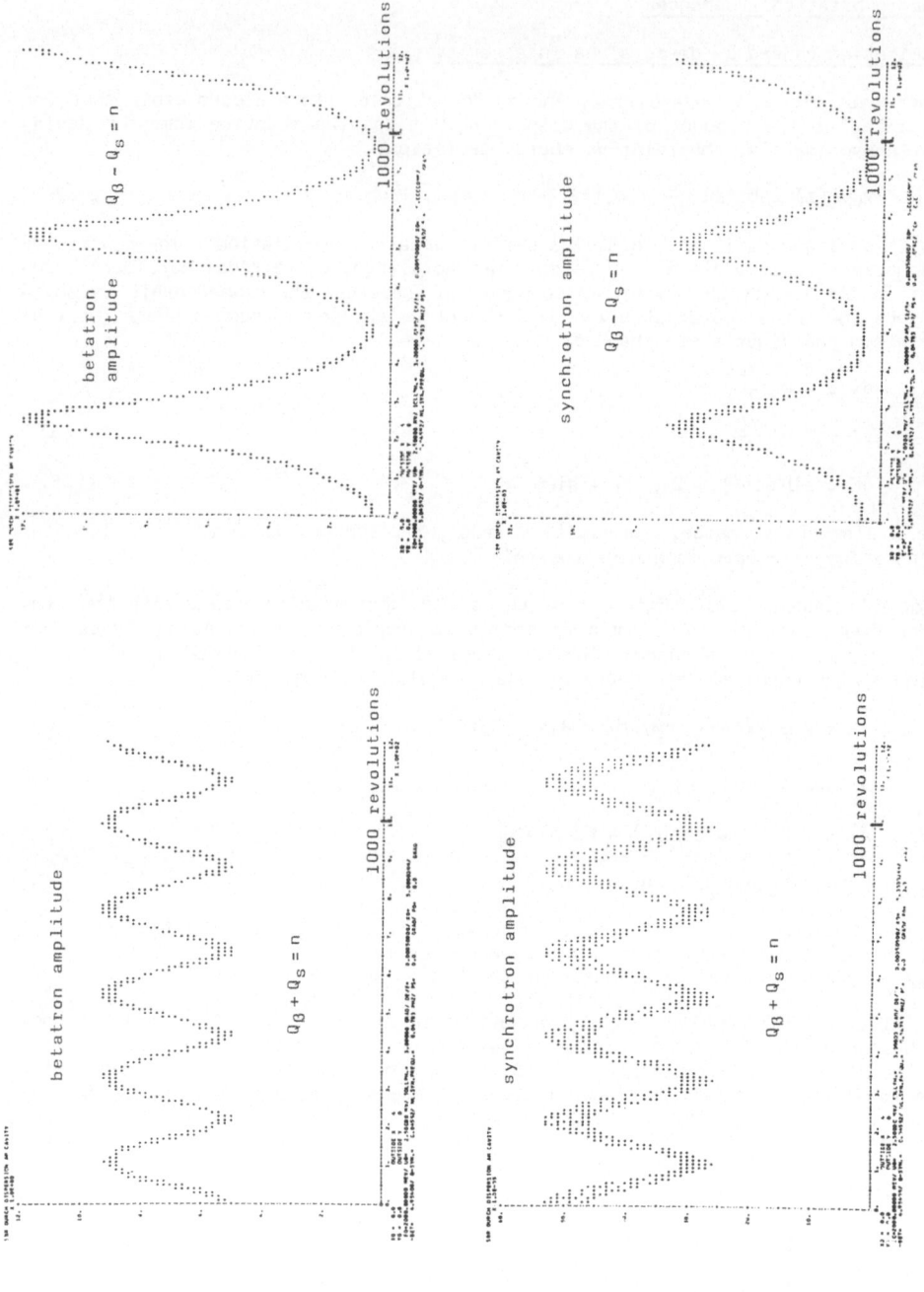

Fig. 1: Betatron and synchrotron amplitudes in arbitrary units as a function of time
a) on a sum resonance
b) on a difference resonance

2.2 Excitation caused by transverse fields with a longitudinal gradient

Transverse fields which vary with the longitudinal position of a particle in a bunch can be caused by the accelerating voltage due to an angle between the cavity axis and the closed orbit or due to asymmetries in the cavities[2] and by the bunch itself due to off-center passage or due to cavity asymmetries[3]. The vertical kick of a particle produced by transverse fields is given by

$$\delta z' = \frac{e}{p} \int (E_z + v B_x) \, dt \tag{8}$$

where E_z and B_x are the electric and magnetic fields and v is the particle velocity. The integral is taken along the path of the particle between two limits where the fields vanish. The region of the fields is assumed to be short as compared to the wave length of the betatron oscillation.

We consider that part of $\delta z'$ which varies linearly with the longitudinal position s of a particle in the bunch and obtain with Maxwell's equations

$$\delta z' = s \frac{e}{p} \int \left(\frac{\partial E_z}{\partial s} + v \frac{\partial B_x}{\partial s} \right) dt = s \frac{e}{p} \int \frac{\partial E_s}{\partial z} \, dt = A s \tag{9}$$

If the transverse fields have a longitudinal gradient the longitudinal field must have a transverse gradient, and that means that the betatron oscillation influences the synchrotron oscillation. That part of the energy change which varies linearly with the transverse position of a particle in the bunch is given by[4]

$$\frac{\delta E}{E} = z \frac{e}{E} \int \frac{\partial E_s}{\partial z} v \, dt = \frac{v^2}{c^2} A z \approx A z \tag{10}$$

A comparison of Eqs. (3), (4) and (9) shows that the increase of betatron oscillation depends in the same way on the synchrotron oscillation, namely on s. The two mechanisms, caused by a dispersion and by transverse fields, can at least for slowly changing synchrotron amplitude compensate each other. This is also possible for higher harmonics of the synchrotron frequency produced by nonlinearities. Then the compensation depends on the amplitude of the synchrotron oscillation and on the bunch current.

2.3 Excitation caused by the beam-beam interaction at a crossing angle

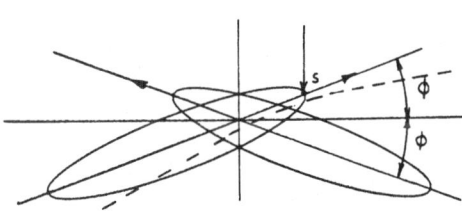

Fig. 2: Orbit distortion due to a crossing angle

We consider a particle which has a distance s from the center of its own bunch. This particle does not pass through the center of the other bunch and hence gets a vertical kick. The kick has the direction indicated in Fig. 2 if we assume particles with different signs as electrons and positrons. Because of this kick the axis of the bunch cannot be the closed orbit for the considered particle, but the closed orbit must look like the dotted line.

The kick, i.e. the change of the vertical betatron angle z' can be written in the form

$$\delta z' = f(z + s\phi) \tag{11}$$

where f describes the space charge forces as a function of the vertical position z and the longitudinal position s.

The synchrotron oscillation is also influenced by the betatron oscillation since the energy is changed for a particle that crosses the bunch at an angle. The energy change is understandable, if one takes into account that the kick or momentum change is vertical. One can then decompose the change of the momentum and obtains a longitudinal component p_s which is given by Φp_z. The energy change is then

$$\frac{\delta E}{E} = \frac{\delta p}{p} = \frac{\delta p_s}{p} = \Phi \frac{\delta p_z}{p} = \Phi \delta z' = \Phi f(z + s\Phi) \tag{12}$$

If the effective cross section is circular as in DORIS I ($\sigma_{zef} \approx \sigma_x = \sigma$) an exact expression for the function f (Eq. (11)) can be obtained:

$$f(u) = \frac{2r_e N_b}{\gamma u} \left(e^{-\frac{u^2}{2\sigma^2}} - 1 \right) \tag{13}$$

with r_e = electron radius, N_b = number of particles per bunch, γ = particle energy divided by its rest energy, σ = standard deviation of the Gaussian-like particle distribution.

A simulation of the betatron and synchrotron oscillation including the nonlinear coupling (Eq. (13)) has been done on a digital computer. Between the interaction points the oscillations were transformed linearly. Fig. 3 shows the ratio of the maximum to the minimum betatron amplitude during 2000 revolutions for Φ = 12 mrad, $N_b = 8 \cdot 10^9$, γ = 3500, σ = .23 mm, Q_s = .034, β_z^* = 1 m. More than 1000 Q_β-values between 6.02 and 6.48 are investigated but only those resonances are shown which reach an increase of more than 50 % of their initial amplitude. Measurements with the storage ring DORIS I have proved that exactly at these frequencies the life time of the bunch dropped to a few seconds[5]).

Analytical investigations and the latest measurements with DORIS II have shown that also a dispersion at the interaction point can excite similar synchro-betatron resonances[5,6]).

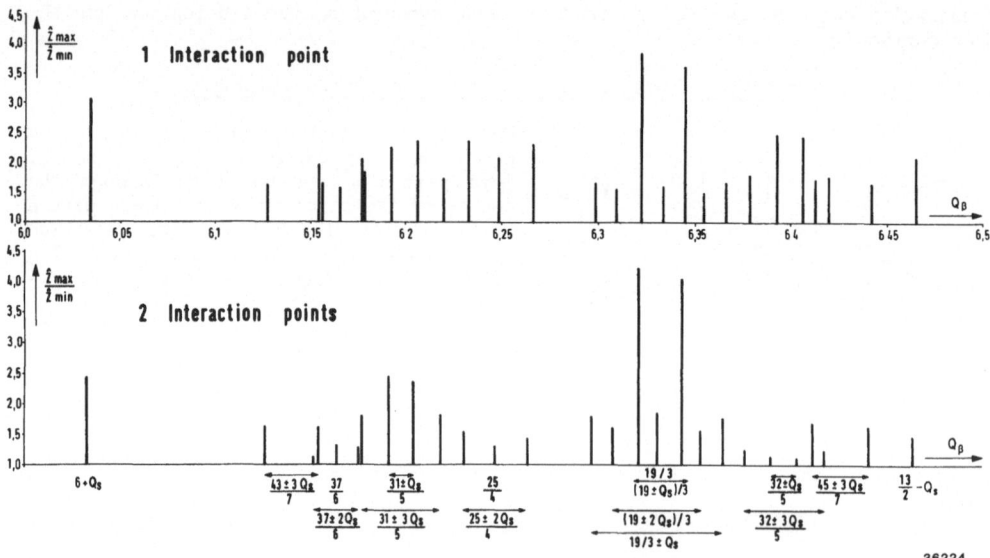

Fig. 3: Maximum betatron amplitude as a function of the betatron frequency

3. Beam-beam interaction

In simulating the beam-beam interaction on a digital computer, significant progress was made in 1980[7,8,9,10]. Since that time, many simulations were done by several authors, and a better understanding of the mechanism of the blow-up was achieved. In the following the simulations for PETRA are briefly described.

The space charge forces are exactly calculated for a relativistic bunch with a Gaussian particle distribution. The change of the betatron angles x' and z' at each interaction point is then given by

$$\left| \begin{array}{c} \Delta x' \\ \Delta z' \end{array} \right| = -\frac{2\pi}{1-V} \frac{1-V^2}{\int_0^{\cdot}} \exp\{-as - \frac{bs}{1-s}\} \left| \begin{array}{c} x\xi_x/\beta_{xo} \\ \frac{z\xi_z/\beta_{zo}}{1-s} \end{array} \right| \frac{ds}{\sqrt{1-s}} \quad (14)$$

with $\quad V = \sigma_z/\sigma_x; \; a = \frac{1}{2}\frac{x^2}{\sigma_x^2 - \sigma_z^2}; \; b = \frac{1}{2}\frac{z^2}{\sigma_x^2 - \sigma_z^2}; \; \xi_{x,z} = \frac{r_e N_b \beta^*_{x,z}}{2\pi\gamma\sigma_{x,z}(\sigma_x + \sigma_z)}$

The kicks were calculated for a two-dimensional grid of points and then interpolated quadratically for calculating the transverse kick of a particle at each passage. The longitudinal motion of the interaction point seen by a particle due to its synchrotron oscillation is always taken into account. Between the interaction points the horizontal and vertical betatron oscillation and the synchrotron oscillation are transformed linearly. The radiation damping is included. The quantum fluctuation is simulated by applying random kicks on all three modes of oscillation. The motion of particles is observed over several damping times, i.e. over a large number of revolutions. Both cases, "weak-strong" and "strong-strong" were investigated.

As an example Fig. 4 shows the vertical motion of a single particle in the phase diagram. The horizontal axis gives the position and the vertical axis the angle of the vertical betatron oscillation at a symmetry point of the machine for each revolution. The vertical amplitude of the particle starting with zero amplitudes remains within one or two standard deviations of the Gaussian distribution of the opposing bunch during the first 8000 revolutions (a). Then its amplitude increases rapidly due to quantum fluctuation and the nonlinearity of the space charge forces, and it moves into a third order resonance where the phase advance is about $2\pi/3 + 2\pi \times$ integer (b). After about three quarters of a damping time it comes out of resonance and leaves the three fixed points, but is immediately captured by other three fixed points of the third order resonance (c) which are a mirror image and equivalent to the first three points (without optical asymmetries). After about 600 revolutions the particle leaves the third order resonance (d).

A similar behaviour can be observed for several resonances. More often, however, coupling resonances between horizontal and vertical betatron oscillations appear. These resonances can be found by counting the betatron oscillations and by observing the variation of the amplitudes. In all cases the particles usually do not stay longer than a damping time on a resonance. Due to quantum fluctuation and damping they can leave the resonance and can then be captured by another resonance.

The computer simulations have shown that small disturbances of the ideal machine increase the number and strength of the resonances which can be excited. Those disturbances are small differences in betatron phase advance between the interaction points and spurious dispersions at the interaction points. Thus machine imperfections enlarge the blow-up of the beams and become more important with increasing number of interaction points.

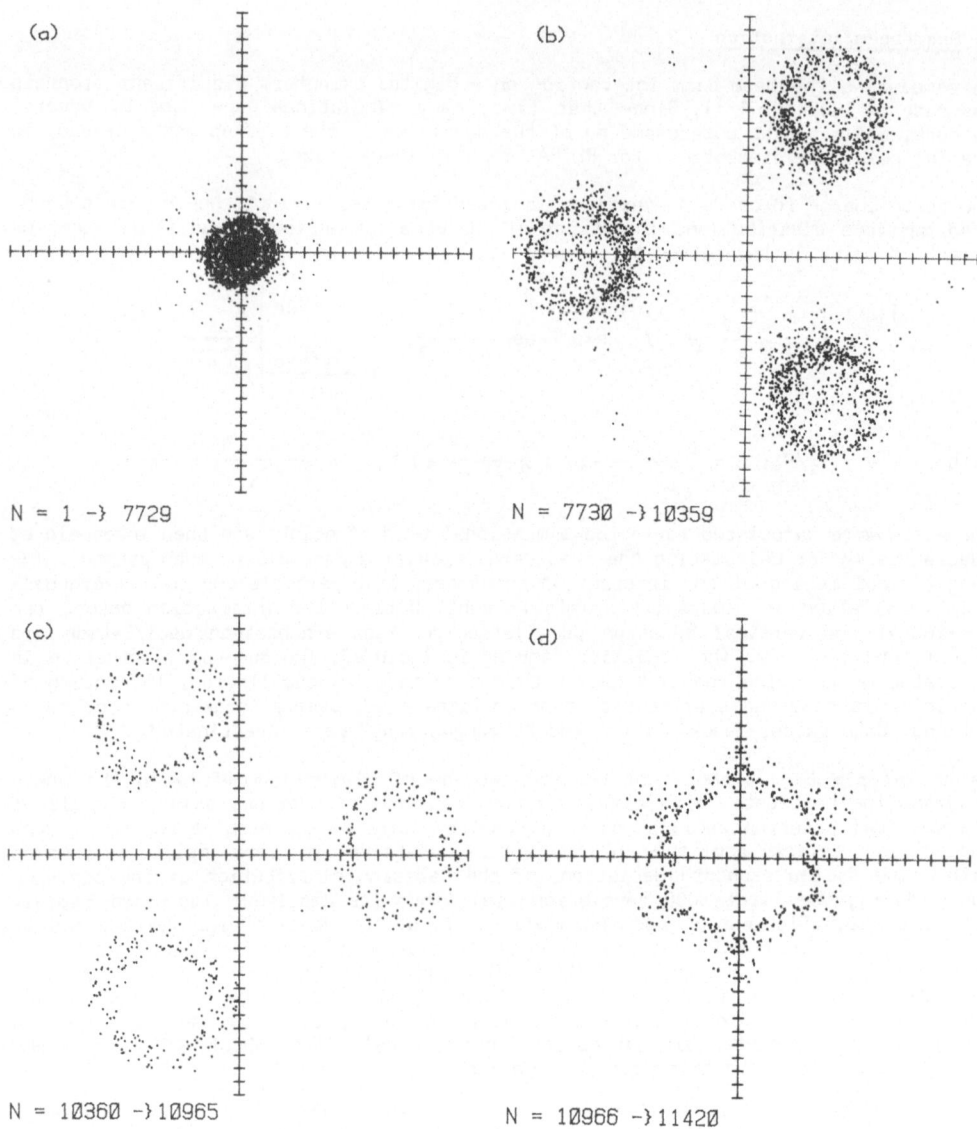

Fig. 4: Particle motion in the phase diagram z-z'
(Q_x = 25.2, Q_z = 23.32, Q_s = .07, $\xi_x = \xi_z$ = .04, σ_x^*/σ_z^* = 15, 4 interaction points, 1 damping time = 3000 revolutions)

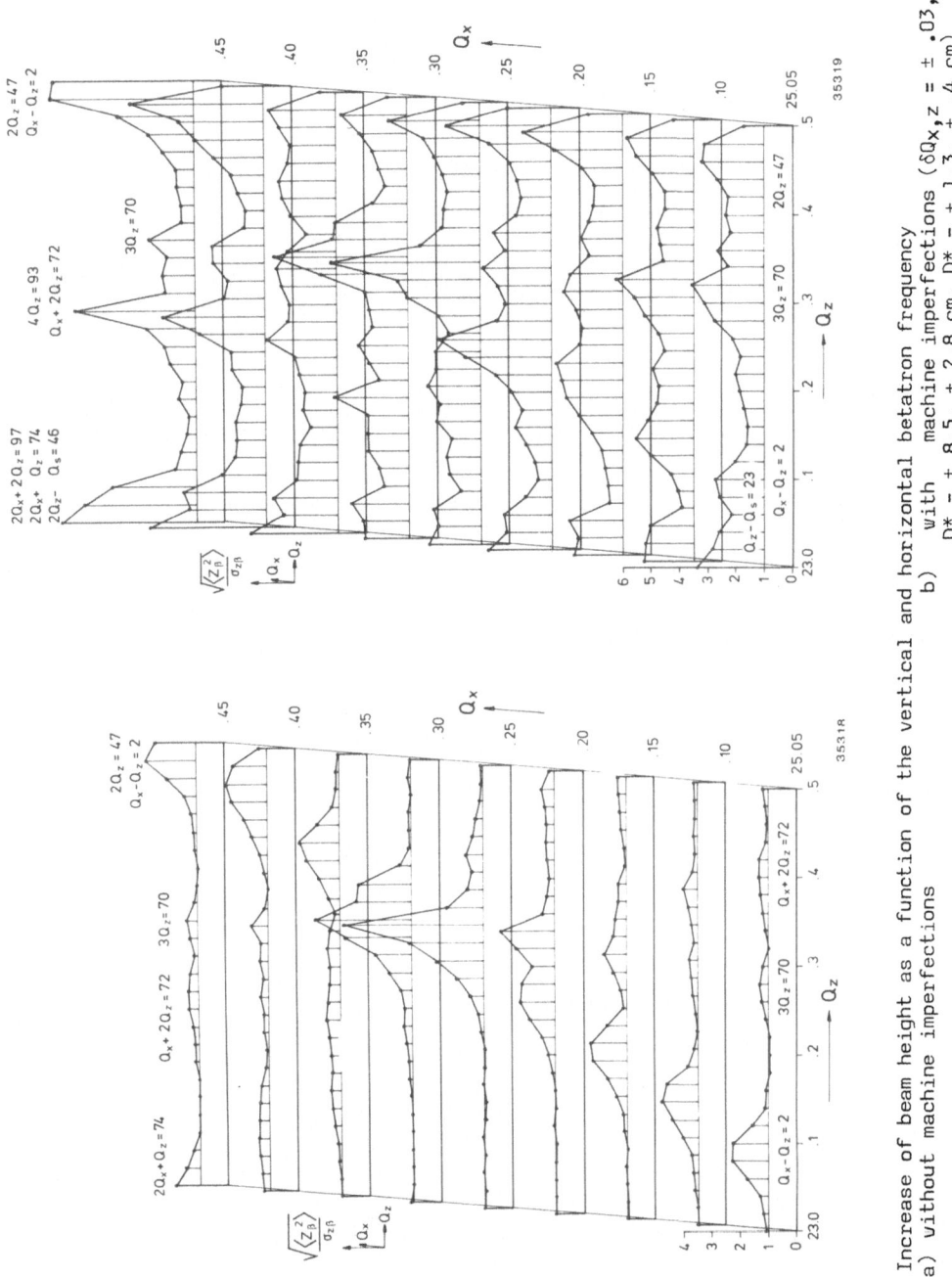

Fig. 5: Increase of beam height as a function of the vertical and horizontal betatron frequency
a) without machine imperfections
b) with machine imperfections ($\delta Q_{x,z} = \pm .03, \pm .01$, $D_x^* = \pm 8.5, \pm 2.8$ cm, $D_z^* = \pm 1.3, \pm .4$ cm)

The increase in beam height is given by the root mean square of the vertical betatron coordinates of many particles over many revolutions:

$$\sqrt{\langle z_\beta^2 \rangle} = \sqrt{\frac{1}{N} \sum_{i=1}^{N} z_{\beta i}^2} \tag{15}$$

N includes all particles at all interaction points and all revolutions after 4 damping times.

Fig. 5 shows the influence of machine imperfections on the blow-up. To reduce the computer time these simulations were done only for the case "weak-strong", however, simulations of the case "strong-strong" have shown that the dependence on the working point is very similar in both cases. The assumed phase asymmetries can be produced by the usually observed orbit displacements in the sextupoles, and the magnitude of the spurious dispersions at the interaction points is scaled from measurements of the dispersions in the straight sections outside the mini beta insertion.

References

1. A. Piwinski, A. Wrulich: Excitation of Betatron-Synchrotron Resonances by a Dispersion in the Cavities, DESY 76/07 (1976)

2. N.A. Vinoburow et al.: Synchrobetatron Resonances at Zero Value of Chromaticity, 10th Intern. Conf. on High Energy Accel., Protvino, 254 (1977)

3. R.M. Sundelin: Synchrobetatron Oscillation Driving Mechanism, IEEE Trans. Nucl. Sci. NS-26, 3604 (1979)

4. A. Piwinski: Synchro-Betatron Resonances, 11th Intern. Conf. on High Energy Accel., Geneva, 562 (1980)

5. A. Piwinski: Limitation of the Luminosity by Satellite Resonances, DESY 77/18 (1977)

6. H. Nesemann, K. Wille: Operational Experience with DORIS II, 12th Intern. Conf. on High Energy Accel., Fermilab (1983)

7. J. Tennyson, Univ. of Calif., Berkeley (1980) (unpublished)

8. S. Peggs, R. Talman: Observations at CESR and Theory of the Beam-Beam Luminosity Limitation, 11th Intern. Conf. on High Energy Accel., Geneva, 754 (1980)

9. A. Piwinski: Computer Simulations of the Beam-Beam Interaction, 11th Intern. Conf. on High Energy Accel., Geneva, 751 (1980)

10. A. Piwinski: Computer Simulation of the Beam-Beam Interaction, DESY 80/131 (1980)

CALCULATION OF POLARIZATION EFFECTS

ALEXANDER W. CHAO*
Stanford Linear Accelerator Center
Stanford University, Stanford, California 94305

1. Introduction

Basically there are two areas of accelerator applications that involve beam polarization. One is the acceleration of a polarized beam (most likely a proton beam) in a synchrotron. Another concerns polarized beams in an electron storage ring. In both areas, numerical techniques have been very useful.

2. Accelerating Polarized Beams in a Synchrotron

In a proton synchrotron, a polarized beam is injected and then accelerated with its polarization parallel to the guiding magnetic field. If particles see only the guiding field, then there will be no depolarization effects. But perturbing magnetic fields, such as the quadrupole magnetic fields seen by an off-axis particle, will cause the particle spin to deviate from the vertical direction $\hat{y}$. This spin deviation will then precess around $\hat{y}$ with a precession frequency $a\gamma f_{rev}$, where f_{rev} is the revolution frequency, γ is the Lorentz energy factor and a is a fundamental constant given by

$$a = \begin{cases} 1.793 & \text{proton} \\ 0.001160 & \text{electron} \end{cases} \quad (1)$$

The quantity

$$\nu = a\gamma \quad (2)$$

is called the spin tune.

If ν is close to ν_0 which satisfies a resonant condition

$$\nu_0 = n_x \nu_x + n_y \nu_y + n_s \nu_s + n \quad , \quad (3)$$

a small perturbing magnetic field will lead to a substantial deviation of the spin direction which then gives rise to depolarization. In Eq. (3), ν_x, ν_y and ν_s are the horizontal, vertical and synchrotron tunes of the orbital motion of the particles and n, n_x, n_y, n_s are integers.

The degree of this depolarization depends on the distance between ν from ν_0. The closer ν is to ν_0, the stronger is the depolarization. The width around ν_0 within which the polarization is reduced by >50% is defined to be the depolarization resonance width. It is designated by ϵ and it depends on the strength of the perturbing magnetic field.

The problem comes from the fact that the spin tune is proportional to the particle energy. As a result it varies as the beam is being accelerated and in doing so crosses resonances. We then need to calculate the amount of polarization lost due to each crossing. This is done by applying the Foissart-Stora equation[1]

$$P_{+\infty} = P_{-\infty} \left(2e^{-\pi\epsilon^2/2\alpha} - 1 \right) \quad (4)$$

where α is the crossing speed of ν relative to ν_0, $P_{-\infty}$ and $P_{+\infty}$ are the beam polarizations before and after crossing, respectively.

---

*Work supported by the Department of Energy, contract DE-AC03-76SF00515.

According to Eq. (4), there are two ways to assure a small polarization loss.[2-5] We either quickly cross a weak resonance ($\epsilon^2/\alpha \ll 1$) or slowly cross a strong resonance ($\epsilon^2/\alpha \gg 1$). In the later case, the polarization will be flipped after crossing. Anything in between the two extreme cases ($\epsilon^2/\alpha \simeq 1$) will be harmful for polarization.

The resonance crossing speed α is a strightforward kinematic quantity given by

$$\alpha = \frac{1}{2\pi}(a\Delta\gamma - \Delta\nu_0) \tag{5}$$

where $\Delta\gamma$ and $\Delta\nu_0$ are the changes of γ and ν_0 per revolution during acceleration. Fast crossing speed can be obtained by fast acceleration rate $\Delta\gamma$. For resonances that involve the orbital tunes, it can also be obtained by jumping the orbital tunes as the beam is being accelerated through the resonance by using pulsed quadrupole magnets.

The quantity in Eq. (4) that remains to be calculated is the resonance width ϵ. This needs to be done for all depolarization resonances crossed by the spin tune during the acceleration process.

3. Calculation of the Depolarization Resonance Widths

The basic equations of motion — the Thomas-BMT equation[6] — for the spin $\vec{S}$ in a magnetic field is

$$\dot{\vec{S}} = \vec{\Omega} \times \vec{S} \tag{6}$$

where

$$\vec{\Omega} = \frac{e}{m\gamma c}\left[(1+a\gamma)\vec{B}_\perp + (1+a)\vec{B}_\parallel\right]$$

where $\vec{B}_\parallel$ and $\vec{B}_\perp$ are the magnetic field components parallel and perpendicular to the instantaneous direction of motion, respectively. Note that the spin motion depends on the orbital motion since it is the orbital motion that determines the magnetic field seen by the particle. This is one reason why the spin motion is more difficult to analize than the orbital motion. The depolarization resonance width, for example, can be calculated only after the orbital motion has been analized.

More quantitatively, ϵ is defined as the Fourier component of $\vec{\Omega}$ at the resonance frequency,[1,7,8] i.e.,

$$\epsilon = \frac{1}{2\pi c}(\hat{x} - i\hat{z}) \cdot \int_{\pi 0}^{2} e^{i\nu_0\theta}\vec{\Omega}(\theta)\rho(\theta)d\theta \tag{7}$$

where $\hat{x}$ and $\hat{z}$ are the horizontal and longitudinal unit vectors, $\rho(\theta)$ is the bending radius and θ is the accumulated bending angle that increases by 2π every revolution.

It turns out that the strongest depolarization resonances belong to the two families[7-10]

$$\nu = n : \text{imperfection resonances excited by vertical closed orbit distortion;} \tag{8a}$$

$$\nu = nS \pm \nu_y : \text{intrinsic resonances excited by vertical betatron motion of particles (S is the periodicity of the accelerator.)} \tag{8b}$$

For these resonances, we need only to keep in $\vec{\Omega}$ those terms linear in the orbital coordinates.[8] When this is done, Eq. (7) becomes

$$\epsilon \approx \frac{1+a\gamma}{2\pi}\int_0^{2\pi} G(\theta)\, y(\theta)\, e^{i\nu_0\theta}\, \rho(\theta)\, d\theta \tag{9}$$

where G is the quadrupole strength. From Eq. (9), we see that most of the depolarization action results from having vertical excursion in quadrupoles. In deriving Eq. (9), we have assumed that the accelerator has a planar geometry by design and there is no solenoidal field.

The widths of the imperfection resonances are obtained by setting y=closed orbit distortion in Eq. (9). To do a calculation, it is necessary to input a Monte Carlo simulation of the closed orbit error with a certain given rms. The intrinsic resonance widths on the other hand are obtained by setting =betatron excursion in Eq. (9) and do not require any random number generations. Their strengths are calculated once the accelerator lattice is determined and the beam emittance is known.

Table 1 shows the number of imperfection and intrinsic resonances that the polarized proton beam has to cross for several accelerators. Obviously the high energy synchrotrons will have more resonances to cross than the lower energy ones.

Table 2 shows the intrinsic resonance widths[7] for three synchrotrons using the program DEPOL.[8] The results are for particles whose emittance is the average beam emittance. Resonance widths have also been calculated for ZGS,[9] KEK-PS[4] and SATURNE.[3]

4. Acceleration to High energies

To accelerate polarized proton beams to very high energies requires special effort. This is first of all due to the fact that the total number of resonances to be crossed is simply very large. (See Table 1.) Secondly, high energy synchrotrons typically have strong focussing lattices which tend to give stronger resonances than the weak focussing lattices. Thirdly, resonance widths tend to increase as energy goes higher.

Somewhat more quantitatively, let us take the resonance widths to be very roughly[11]

$$\epsilon_{\text{int}} \approx 0.02(E/25 \text{ GeV})^{1/2}$$
$$\epsilon_{\text{imp}} \approx 3 \times 10^{-4}(E/25 \text{ GeV}) \tag{10}$$

The imperfection resonance width is meant to be that after an orbit correction has been applied. If we then take the fast crossing approach, the total loss of polarization after crossing all resonances is

$$\frac{\Delta P}{P} \approx \sum_{i=\text{int}} \frac{\pi \epsilon_i^2}{\alpha_{\text{int}}} + \sum_{i=\text{imp}} \frac{\pi \epsilon_i^2}{\alpha_{\text{imp}}} \tag{11}$$

If we assume that the acceleration speed is determined by a given rf acceleration per unit distance while the ν_y-jump is determined by jumping ν_y by 0.2 in 1 μsec, then the resonance crossing speeds scale like

$$\alpha_{\text{int}} \approx 0.05(\hat{E}/25 \text{ GeV})$$
$$\alpha_{\text{imp}} \approx 5 \times 10^{-5}(\hat{E}/25 \text{ GeV}) \tag{12}$$

where $\hat{E}$ is the maximum energy of the synchrotron. Note that part of the increase in resonance widths with energy is compensated by the increase in crossing speed for higher energy synchrotrons.

Substituting Eqs. (10) and (12) into Eq. (11) and assuming a periodicity of $S=8$, we obtain

$$\frac{\Delta P}{P} \approx 6 \times 10^{-3} E(\text{GeV}) + 1.5 \times 10^{-4} E(\text{GeV})^2 \tag{13}$$

Note that for very high energy synchrotrons, the imperfection resonances dominate the depolarization.[12] If we demand $\Delta P/P < 50\%$, the maximum energy that a polarized proton beam can be accelerated to is then found to be about 40 GeV, which is not too much beyond the AGS energy.

Table 1

Accelerator	Energy (GeV)	Number of Intrinsic Resonances	Number of Imperfection Resonances
ZGS	12	10	22
SATURNE	3	2	6
FNAL (booster)	8	1	14
KEK-PS	12	10	21
AGS	30	9	56
CERN-PS	30	12	55
CERN-ISR	11.5-31.4	10	38
FNAL (main ring)	8-400	250	749
SPS	10-400	249	745
FNAL (Tevatron I)	150-1000	541	1626

Table 2

	CERN-PS			AGS			FERMILAB (main ring)*	
ν	γ	$|\epsilon|$	ν	γ	$|\epsilon|$	ν	γ	$|\epsilon|$
$10 - \nu_y$	2.06	.00005	$12 - \nu_y$	1.82	.0054	$0 + \nu_y$	10.84	.0256
$0 + \nu_y$	3.53	.00933	$0 + \nu_y$	4.89	.01535	$6 + \nu_y$	14.19	.0060
$20 - \nu_y$	7.64	.00045	$24 - \nu_y$	8.52	.00059	$12 + \nu_y$	17.54	.0035
$10 + \nu_y$	9.12	.00047	$12 + \nu_y$	11.59	.00539	$18 + \nu_y$	20.89	.0016
$30 - \nu_y$	13.23	.00087	$36 - \nu_y$	15.22	.01373	$24 + \nu_y$	24.25	.0025
$20 + \nu_y$	14.70	.00050	$24 + \nu_y$	18.29	.00101	$30 + \nu_y$	27.60	.0049
$40 - \nu_y$	18.82	.00077	$48 - \nu_y$	21.93	.00148	$36 + \nu_y$	30.95	.0062
$30 + \nu_y$	20.29	.00309	$36 + \nu_y$	25.00	.02663	$42 + \nu_y$	34.30	.0057
$50 - \nu_y$	24.40	.14192	$60 - \nu_y$	28.63	.15666	$48 + \nu_y$	37.65	.0028
$40 + \nu_y$	25.88	.00174	$48 + \nu_y$	31.70	.00233	$54 + \nu_y$	41.01	.0021
$60 - \nu_y$	29.99	.00195				$60 + \nu_y$	44.36	.0048
$50 + \nu_y$	31.64	.16773				$66 + \nu_y$	47.71	.0098
						$72 + \nu_y$	51.06	.0133
						$78 + \nu_y$	54.41	.0198
						$84 + \nu_y$	57.77	.0321
						$90 + \nu_y$	61.12	.0518
						$96 + \nu_y$	64.47	.1653
						$192 + \nu_y$	118.10	.0560
						$288 + \nu_y$	171.73	.2952
						$384 + \nu_y$	225.36	.0921
						$480 + \nu_y$	278.99	.2138
						$576 + \nu_y$	332.63	.0998
						$672 + \nu_y$	386.26	.2995
						$768 + \nu_y$	439.89	.0244

*Only dominant ones.

In view of this, a better way of acceleration to high energies is needed. One such possibility is called harmonic matching. Another is to install Siberian snake devices in the accelerator. These two topics are discussed in the following two sections.

5. Harmonic Matching

The idea of harmonic matching is to make $\epsilon = 0$ at the moment of crossing a resonance so that there will be no loss of polarization due to the crossing. For the imperfection resonances, the condition for achieving this is given by[13] (cf. Eq. (9))

$$\frac{1}{2\pi} \int_0^{2\pi} \exp(i\nu_0\theta)\, d\theta\, G(\theta) y_{co}(\theta) = 0 \tag{14}$$

where $y_{c.o.}$ is the vertical closed orbit distortion. Equation (14) is a Fourier harmonic of the vertical closed orbit, thus the name harmonic matching. Since (14) is a complex quantity, it imposes two conditions on $y_{c.o.}$ for each resonance to be crossed.

Before applying (14), there needs to be a good orbit correction in the rms manner. The orbit is then slightly changed to fulfill (14). The amount of change is rather small and is obtained not by the beam position monitor measurements but by empirically optimizing the polarization.

Harmonic matching for the imperfection resonances has been used successfully in ZGS and SATURNE. On-line controls have been applied so that a vertical orbit distortion with the right harmonics is generated shortly before crossing and switched off shortly after crossing the resonance. The control program makes orbit distortions in symphony with the acceleration process.

To accelerate polarized protons in high energy synchrtrons using the harmonic matching technique, however, one needs to do a more accurate matching than ZGS and SATURNE so that the imperfection resonance widths are much narrower than Eq. (10) gives.

To harmonic match the intrinsic resonances is more difficult. The quantity to be matched is[14]

$$\frac{1}{2\pi} \int_0^{2\pi} \exp\left[i(\nu_0\theta - \psi_y(\theta))\right] d\theta\, G(\theta) \sqrt{\beta_y(\theta)} = 0 \tag{15}$$

where ψ_y and β_y are the vertical betatron phase and beta-function, respectively.

At the moment of crossing, the quadrupole strengths are changed to satisfy (15). To minimize the effect on the betatron motions, this may have to be done keeping the tunes constant during the crossing process. Note that the quadrupole strengths are calculated by a lattice fitting routine rather than found empirically.

So far harmonic matching for intrinsic resonances has not been applied to existing synchrotrons. It is conceivable, however, that some intrinsic resonances (in the smaller synchrotrons for which Siberian snakes are not applicable) can be crossed this way.

6. Siberian Snakes

A Siberian snake[15] is a series of horizontal and vertical bending magnets that does two things:
1. Makes spin tune ν equal to 1/2, and
2. Does not affect the beam orbit outside the device.

A Siberian snake therefore makes the spin tune independent of the beam energy and thus eliminates the need to cross resonances.

There are two types of snakes:

Type I rotates the polarization by 180° around the longitudinal $\hat{z}$-axis, and

Type II rotates the polarization by 180° around the horizontal $\hat{x}$-axis.

One way to use the snake is to insert a type I snake at a symmetry point in the synchrotron. Another way is to have a type I snake at a symmetry point and another type II snake at the opposite symmetry point. The spin tune in both cases will be 1/2.

There are several possible designs of Siberian snakes. An explicit example of a type I snake is[16]

$$\begin{array}{ccccccccc} V & H & V & H & V & H & V & H & V \\ (45) & (45) & (-90) & (-90) & (90) & (90) & (-90) & (-45) & (45) \end{array} \tag{16}$$

where H and V mean horizontal and vertical bending magnets and the quantities in parentheses are the spin precession angles in degrees.

An explicit example of a type II snake is[16]

$$\begin{array}{ccccccc} V & H & V & H & V & H & V \\ (-90) & (90) & (90) & (-180) & (90) & (90) & (-90) \end{array} \tag{17}$$

To make spin to precess by fixed angles like the snake magnets do, the magnet strengths have to be independent of beam energy even during acceleration. One consequence of this is on the beam-stay-clear requirements. Although there is no net effect on the beam trajectory outside of the snake, the orbit distortion in the snake is not negligible, especially at the lower energies. (For this reason, Siberian snakes are not applicable to synchrotrons of low injection energy.) This beam-stay-clear requirement imposes strong constraints on the snake aperture and is a major concern facing the snake designers.

Error effects with snakes also need attention. Some analytical work has been performed,[17] which concludes that if a resonance existed before inserting the snake, then it may still have a remnant depolarization effect after the snake is inserted. In particular, if the accelerator has a resonance width ϵ in the absence of the snake, and if the two circular arcs of the accelerator contribute equally to the resonance width, then after the snake is inserted the resonance will shift the spin tune away from 1/2 according to[17,11]

$$\begin{aligned} \text{single snake} &: \nu = \frac{1}{2} + |\epsilon| \cos\phi \\ \text{double snake} &: \cos\pi\nu = \cos 2\phi \, \sin^2(\pi|\epsilon|/2) \end{aligned} \tag{18}$$

where ϕ is the phase factor of the complex resonance width. Clearly if the spin tune is shifted to a value that satisfies a resonant condition, then even the great snakes do not save the polarization. Numerical simulation of errors and tracking the spin motion crossing resonances in the presence of a snake will yield useful information here.

7. More studies

In this section, we mention a few additional effects that can be studied by numerical means.

(1) The imperfection and the intrinsic resonances are not the only ones encountered during acceleration. The other resonances, although weaker, can still contribute to loss of polarization. Among them are

$$\begin{aligned} \nu &= n \pm \nu_y &&: \text{excited by quadrupole field errors that destroy the periodicity} & (8c) \\ \nu &= n \pm \nu_x &&: \text{excited by horizontal excursion in skew quadrupoles} & (8d) \\ \nu &= nS \pm \nu_x \pm \nu_y &&: \text{excited by sextupoles} & (8e) \end{aligned}$$

Furthermore, multipole field errors will excite higher order resonances. Widths of these resonances need to be calculated. The programs that are used to calculate the imperfection and intrinsic resonances can be extended easily to calculate for resonances of type (8c) but not for types (8d) and (8e).

It should be pointed out that Eq. (9) is a result after linearization. To calculate the width of a nonlinear resonance, the more general result, Eq. (7), must be used in which nonlinear terms up to the proper order are included in $\vec{\Omega}$. One step toward this direction has been made in Ref. 18; the spin rotations are calculated to second order for quadrupoles and for sextupoles.

(2) The Froissart-Stora equation assumes a single isolated depolarization resonance. This assumption is not valid if the synchrotron oscillation plays a role during the resonance crossing.

According to the F-S equation, the polarization after crossing versus the resonance strength is a simple exponential function given by Eq. (4). What was observed in SATURNE for the imperfection resonance $\nu = 2$, however, looks like Fig. 1(a).[3] The discrepancy has been recently explained by the SATURNE group[19] using a tracking simulation taking into account of synchrotron oscillations. Part of their results is shown in Fig. 1(b). The agreement is rather convincing.

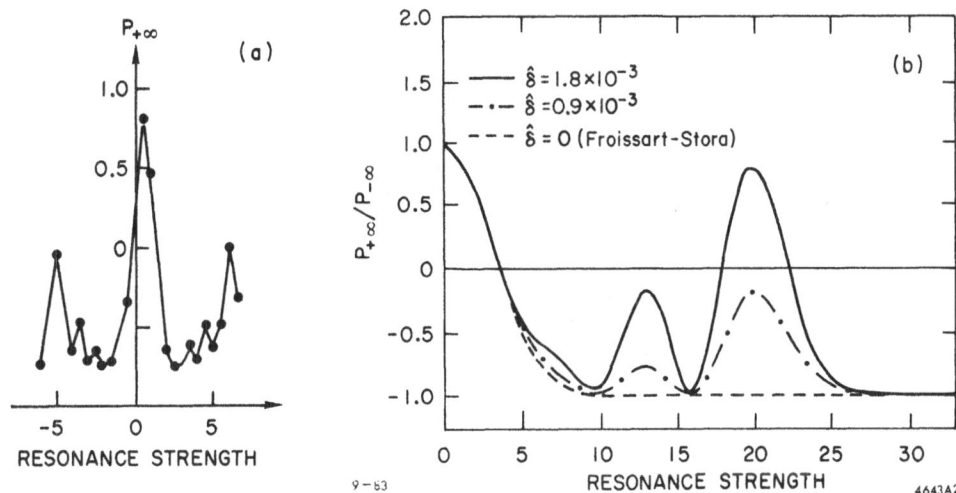

Fig.1. (a) measured polarization after crossing the $\nu = 2$ resonance in SATURNE versus the resonance strength which is controlled by exciting a vertical orbit distortion. (b) result of a simulation taking into account of synchrotron oscillations.

(3) In performing a slow crossing of an intrinsic resonance, particles with small emittance may not get flipped since their resonance widths are narrow. Simulation may help in estimating polarization loss due to this effect.

(4) Crossing is not necessarily done with a uniform speed from $\nu - \nu_0 = -\infty$ to $+\infty$. For instance, when crossing an intrinsic resonance, the crossing speed may be temporarily much enhanced by a ν_y-jump. This makes the Froissart-Stora equation not applicable.

Equivalents of the $F-S$ equation exist for a few special cases of crossing scheme.[8,20] However, one might still need to consider complicated crossing schemes in practice. One reason is that the pulsed quadrupoles generally have complicated time response. Another reason (perhaps minor) is that the crossing speed assumed in the $F-S$ equation is uniform in the variable θ (the accumulated bending angle), while in practice the energy accelerations are made at the rf cavities. Numerical tracking will of course be useful for general crossing schemes.

(5) When crossing an intrinsic resonance with a ν_y-jump, the pulsed quadrupoles alters the accelerator focussing lattice. Effects of this on the orbital motions – and thus on the depolarization resonance width — should be studied.

8. Polarization in an Electron Storage Ring

In a proton synchrotron, the main problem is to accelerate a beam crossing depolarization resonances. In an electron storage ring, we have a different problem. The beam energy is constant in time and depolarization comes from the noise associated with synchrotron radiation.

Another difference between protons and electrons is that a stored electron beam will slowly polarize itself through the Sokolov-Ternov mechanism.[21] If an unpolarized beam is injected into a storage ring, its polarization will build up exponentially along the vertical $\hat{y}$ direction according to

$$P(t) = \frac{P_0}{1+x}\left[1 - \exp\left(-\frac{(1+x)}{\tau_P} t\right)\right]$$

$$x = \frac{\tau_P}{\tau_D}$$

(19)

where $P_0 = 8/5\sqrt{3} = 92\%$ is the polarization level reachable in the absence of depolarization effects, τ_P is the polarization time constant given by

$$\tau_P = 99 \text{ sec } \frac{R(m)\,\rho(m)^2}{E(\text{GeV})^5}$$

(20)

with R the average ring radius and ρ the bending radius.

The depolarization effects are lumped into the parameter τ_D. In a planar ring without errors, there is no depolarization effects and $\tau_D = \infty$. Otherwise τ_D needs to be calculated. In the next two sections, we will describe a program SLIM that offers such a calculation.

9. SLIM Without Spin[22]

Before going on to discuss the polarization calculations, we will first describe the part of SLIM that calculates the orbital quantities regardless of spin since the technique used here is different from the conventional method and will be useful in describing the spin calculations later.

We begin with the vector[23]

$$X = \begin{bmatrix} x \\ x' \\ y \\ y' \\ z \\ \Delta E/E \end{bmatrix} = \begin{bmatrix} x_1 \\ x_2 \\ x_3 \\ x_4 \\ x_5 \\ x_6 \end{bmatrix}$$

(21)

that describes the orbital deviations of an electron. All beam-line elements (bends, quads, rf cavities, skew quads, solenoids and drifts) are then described by 6 × 6 matrices. Sextupoles are included by linearization around the closed-orbit.

The most distinct feature of SLIM is that it employs an eigen-analysis and all interested physical quantities are expressed in terms of the eigenvalues and eigenvectors resulting from the analysis. This is in contrast to the conventional technique which expresses the physical quantities in terms of the various machine functions (the β-functions and phases, the dispersion functions, etc). The adavantage is that by using a 6 × 6 formalism, all coupling effects among the three dimensions are included. As a comparison, the β-functions are undefined when there is x-y coupling. This advantage becomes critical when performing the polarization calculations because spin motion depends sensitively on the orbital motions and it is necessary to include the spin-orbit couplings between spin and all three orbital degrees of freedom.

Once the storage ring lattice — including the quadrupole misalignments and orbit correctors — is determined, a closed-orbit is calculated in the 6-dimensional phase space.

As shown in Fig.2, there are two places (indicated by dotted boxes) where SLIM makes detours to calculate the unperturbed and the perturbed machine functions. These are for display only and not used later.

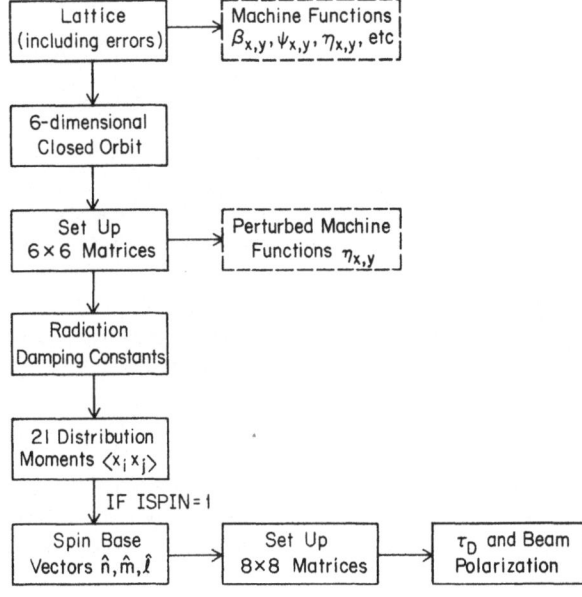

Fig. 2. Flow chart for SLIM.

Knowing the matrices of all the beam-line elements, we multiply them sequentially to obtain a total 6 × 6 matrix. The eigenvalues of this total matrix form three complex conjugate pairs. If the linear lattice is stable, all eigenvalues have absolute value of unity. Otherwise the motion is unstable and the program is stopped. The eigenvalues then give the three tunes according to

$$\lambda_k = \exp(\pm i 2\pi \nu_k) \qquad k = \text{I, II, III} \tag{22}$$

Radiation damping constants are calculated by slightly modifying the 6 × 6 matrices. For example, the rf cavity matrix will take into account of the reduction in x' and y' due to the acceleration by changing the 22- and the 44-elements from 1 to 1-eV/E. The bending magnets and the quadrupoles will also be similarly modified. After these modifications, we multiply matrices to obtain a total 6 × 6 matrix whose determinant is no longer unity. The eigenvalues of this matrix then give the radiation damping constants α_k:

$$\lambda_k = \exp(-\alpha_k \pm i 2\pi \nu_k) \qquad k = \text{I, II, III} \tag{23}$$

The beam is stable if all three damping constants are positive. Otherwise, the program is terminated.

The program also gives the beam distribution in the 6-dimensional phase space by calculating the second moments:

$$\langle x_i x_j \rangle(s) = 2 \sum_{k=\text{I,II,III}} \langle |A_k|^2 \rangle \, \text{Re} \, [E_{ki}(s) E_{kj}^*(s)] \tag{24}$$

with

$$\langle |A_k|^2 \rangle = 7.2 \times 10^{-28} \frac{\gamma^5}{\alpha_k} \oint ds \frac{|E_{k5}(s)|^2}{|\rho(s)|^3}$$

where $E_k(s)$ is the k-th eigenvector (complex) of the 6×6 matrix for one revolution around s and E_{ki} is the i-th component of E_k. We have assumed that the eigenvectors have been normalized and all lengths are expressed in meters. Eq.(24) of course contains information on the x- and y-emittances, bunch length, beam energy spread, tilting angles in $x-x'$, $y-y'$, $x-y$ planes, etc.

If the flag command for doing the spin calculations is on, we then proceed to the next section.

10. SLIM with Spin[24,25]

Knowing the electric and magnetic fields along the closed orbit, spin motion of a particle that follows the closed orbit can be determined by the Thomas-BMT equation. Each beam-line element is then associated a 3×3 rotation matrix that describe the spin precession in this element. Multiplying all these 3×3 matrices together gives a total rotation matrix T.

The beam polarization at equilibrium is going to lie along a direction $\hat{n}(s)$ at position s. It is given by the rotational axis of the rotation T, i.e.

$$T\hat{n} = \hat{n} \tag{25}$$

A fully polarized particle will have its spin along $\hat{n}$. Two axiliary unit vectors $\hat{m}$ and $\hat{\ell}$ are then defined so that $\hat{n}$, $\hat{m}$, and $\hat{\ell}$ form an orthogonal set and precess according to the Thomas-BMT precession along the closed orbit. In general, a slightly depolarized electron will have spin

$$\vec{S} = \hat{n} + \alpha \hat{m} + \beta \hat{\ell} \tag{26}$$

where $|\alpha, \beta| \ll 1$.

The spin part of SLIM is a generalization of the orbital part. The difference is that the spin degree of freedom is included in addition to the three orbital degrees of freedom. The vector instead of (21) is now

$$X = \begin{bmatrix} x \\ x' \\ y \\ y' \\ z \\ \Delta E/E \\ \alpha \\ \beta \end{bmatrix} \tag{27}$$

The next step is to form the corresponding 8×8 matrices for all beam-line elements that transform the vector (27). The upper-left corner of these matrices will be simply the 6×6 matrices used in the orbital calculations. The upper right 6×2 will be zero because spin motion does not affect orbital motions. The lower-right 2×2 gives the spin precession while the lower-left 2×6 gives the critical spin-orbit coupling coefficients. The matrix looks like

$$\begin{bmatrix} \text{TRANSPORT} & 0 \\ \text{spin-orbit coupling} & \text{spin precession} \end{bmatrix} \tag{28}$$

We then multiply all 8×8 matrices in sequence to obtain the total matrix and calculate its eigenvalues and eigenvectors. Three pairs of the eigenvalues and eigenvectors are the orbital ones obtained before. The fourth eigenvalue gives the spin precession tune, while the fourth eigenvector is used to calculate τ_D *

$$\tau_D^{-1}(\text{sec}^{-1}) = 8.64 \times 10^{-19} \frac{\gamma^5}{C} \oint \frac{ds}{|\rho(s)|^3} \left[\left(\text{Im} \sum_k E_{k5}^* E_{k7} \right)^2 + \left(\text{Im} \sum_k E_{k5}^* E_{k8} \right)^2 \right]_s \quad (29)$$

where C is the circumference of the ring. Eq. (29) is the final result. The procedure of spin calculation is illustrated in Fig. 2.

To simulate the depolarization effects, we start with a perturbed lattice with a known distribution of orbit distortion dipoles representing quadrupole misalignments. The resulting orbit distortion is then corrected by a set of orbit correctors to an rms value similar to that observed. Such a distorted lattice, including sextupoles, are fed into SLIM to calculate the equilibrium polarization, $P_0/(1 + x)$ of Eq. (19). The results are typically plotted while scanning the beam energy.

If the beam energy is such that the spin tune $\nu = a\gamma$ is close to one of the linear depolarization resonances

$$\nu \pm \nu_{x,y,s} = n \quad (30)$$

SLIM predicts a strong spin-orbit coupling and thus a loss of beam polarization. Due to the matrix technique used, SLIM does not give any information on the nonlinear resonances.

Figure 3 is the result of a simulation for the storage ring SPEAR. The rms orbit after correction is 1.2 mm. Figure 3 also shows the experimental results.[26] As can be seen, the agreement is satisfactory except that the nonlinear resonance $\nu - \nu_x + \nu_s = 3$ has been missed by the simulation.

It should be emphasized that the orbit distortion in the simulation is the one after correction. The Fourier contents of a corrected orbit and an uncorrected orbit are very different – yielding very different predictions on polarization – even if the two have the same rms value. See Fig. 4.

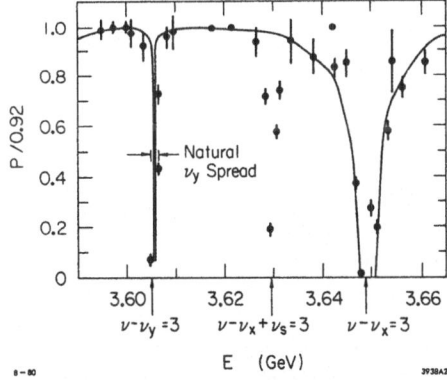

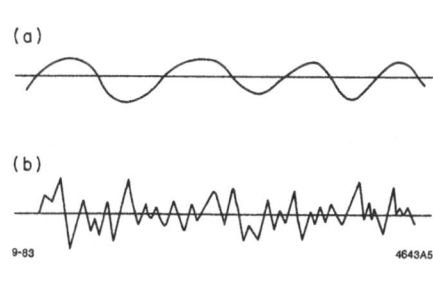

Fig. 3. SLIM simulation compared with the measured data for the storage ring SPEAR.

Fig. 4. (a) orbit before correction; (b) orbit after correction.

---

* More accurately what is calculated is the so-called spin-orbit coupling vector, or equivalently the spin chromaticity. Such details are omitted in this work.

Clearly calculations of ϵ for a proton synchrotron and τ_D for an electron storage ring are closely related. It has not been demonstrated explicitly how this can be done and how programs can be applied to both cases. If this is established, then for instance the eigen-analysis can be applied to proton synchrotrons as well and vise versa for the proton programs.

11. Nonlinear Effects

As mentioned before, one serious drawback of the SLIM technique is that it does not treat the nonlinear resonances. These include the sextupole resonances (8e) and the synchrotron sideband resonances $\nu \pm \nu_{x,y} \pm k\nu_s = n$ with $k \neq 0$. These resonances have been observed in SPEAR[26] and will only be stronger for the higher energy rings because

— the sextupoles tend to be stronger, and

— the energy spread is larger and synchrotron motion becomes important

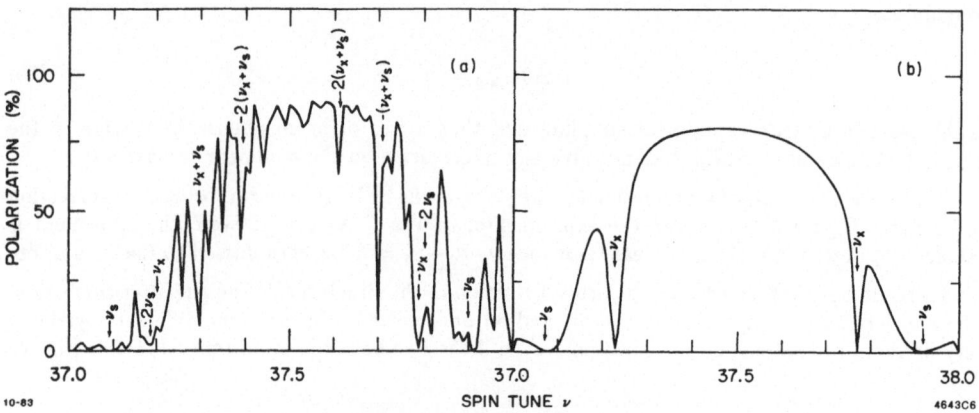

Fig. 5. (a) results of SITROS tracking for the storage ring PETRA;
(b) SLIM results for the same lattice.

One effective way of studying the nonlinear effects is by numerical simulations, as done by Kewisch using the program SITROS.[27] The result for the storage ring PETRA is shown in Fig.5(a). It clearly contains a structure of nonlinear depolarization resonances. Fig.5(b) gives the linear SLIM calculation for the same lattice. There is a general qualitative agreement between these results.

A simulation program can also be used to study the effect of beam instabilities on polarization. One might expect that if the beam intensity becomes high enough so there is a significant blow-up of the beam emittance, the effect on polarization will not be negligible.

One more possible application of a tracking program is to simulate the spin motion when two beams collide. The beam-beam interaction is a very nonlinear effect, potentially exciting depolarization resonances of high orders.[28] In this respect, there has been suggestions[29,30] of imposing a special set of spin matching conditions (see next section) exclusively for the beam-beam interaction. One such suggestion[30] is based on the argument that, although higher order resonances are excited, beam-beam depolarization is dominated by the linear $\nu \pm \nu_y = n$ resonance with the higher order resonances smaller by a factor of the order of τ_{rad}/τ_P (where τ_{rad} is the radiation damping time) which is $\ll 1$. If such a beam-beam spin matching can indeed be achieved, the beam-beam interaction will no longer cause depolarization. These expectations need to be confirmed by a large amount of tracking efforts.

12. Longitudinal Polarization and Spin Matching

Up to now, we have been talking about a vertical beam polarization. The experimenters, however, may request for a longitudinal polarization at the interaction points (IP) where beams collide. To do that, we install a "spin rotator" which consists of a few horizontal and vertical bending magnets on one side of the IP to rotate the polaeizarion from $\hat{y}$ to $\hat{z}$-direction and similarly on the other side of the IP to restore the polarization from $\hat{z}$- back to $\hat{y}$-direction. In the rest of the ring, polarization is along $\hat{y}$.

One design of spin rotator is the "mini-rotator" for HERA.[31] On one side of the IP, the rotator consists of

$$\text{arc} \leftarrow \begin{array}{cccccc} V & H & V & H & V & H \\ (-38.925) & (-62.3) & (77.85) & (90) & (-38.925) & (44.353) \end{array} \rightarrow \text{IP} \tag{31}$$

On the other side, the H magnets are symmetric with respect to the IP while the V magnets are anti-symmetric. The total length of the section (31) is 45.5 m. The H magnets are part of the normal bending in the ring geometry. There are no quadrupoles in the insertion.

Fig. 6 shows the lattice and beam envelopes for HERA with mini-rotator. A typical problem for the spin rotator designs is the orbit distortion inside of the rotator (just like for the Siberian snakes) and the associated stringent beam-stay-clear requirements. The mini-rotator has a 17 cm maximum vertical orbit displacement at 27.5 GeV and varies with the beam energy.

There is another more subtle problem: rotators are very strong depolarizing devices. To avoid their depolarization effects, it is necessary to fulfill a set of conditions on the quadrupole distribution of the ring called the spin matching conditions.[32] The exact number of these conditions vary with details of the rotator design. Typically there are 10 conditions with several of them automatically satisfied due to symmetry of the lattice.

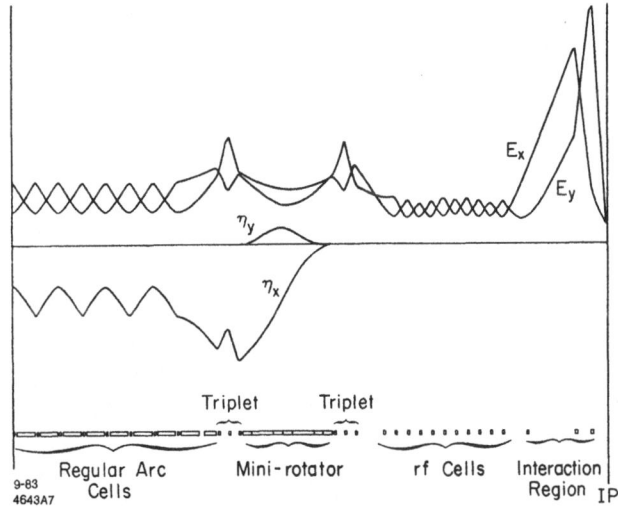

Fig. 6. HERA lattice with the mini-rotator.

For the mini-rotator, there are 4 nontrivial spin matching conditions to fulfill.[31] Two of them are imposed on the regular arc sections:

$$\int_{\text{arc}} G \sqrt{\beta_y} \cos \psi_{\text{spin}} \cos \psi_y \, ds = 0$$
$$\int_{\text{arc}} G \sqrt{\beta_y} \sin \psi_{\text{spin}} \sin \psi_y \, ds = 0 \tag{32}$$

where ψ_{spin} and ψ_y are the spin precession phase and the vertical betatron phase defined to be zero at the midpoint of the arc. The other two conditions are for the interaction region:

$$\int_{I.R.} G\sqrt{\beta_x}\,\cos\psi_x\,ds = 0$$
$$\int_{I.R.} G\sqrt{\beta_y}\,\cos\psi_y\,ds = 0 \quad (33)$$

where $\psi_{x,y}$ are set to zero at the IP.

A fitting program is needed to simultaneously perform the optical and spin matchings. For the mini-rotator, quadrupole strengths in the interaction region and in the arc are used as fitting variables. Such a fitting program eventually will become part of the on-line lattice control program.

13. Imperfections and Harmonic Matching

Depolarization resonances are excited by imperfections. It turns out that for each resonance, there exist two conditions that would eliminate its strength. (This is similar to the proton synchrotron case, remember that ϵ is a complex quantity and to make $\epsilon = 0$ requires two conditions.) In PETRA, the beam polarization has been shown to improve from 30% to 80% by varying a certain Fourier components of the vertical closed orbit.[13] Such a procedure is called harmonic matching. The change in orbit to reach the best polarization is hardly noticeable. These effects have been simulated on SLIM, yielding Fig.7.[33]

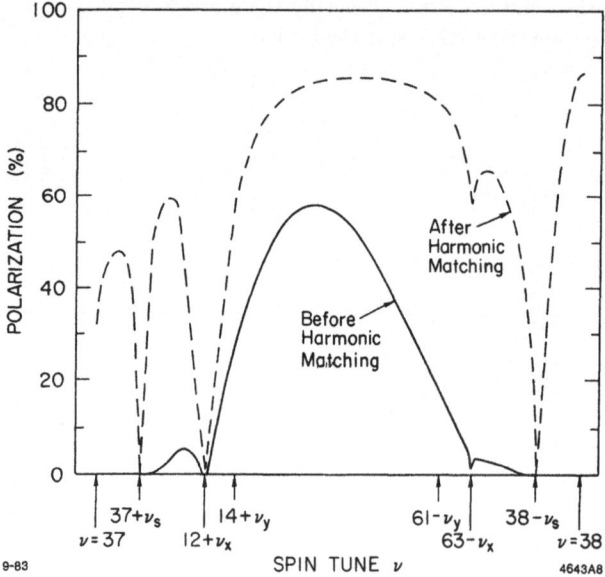

Fig. 7. Simulation of improving beam polarization by harmonic matching.

The spin matching and harmonic matching described above are linear effects. Strong nonlinearities may contaminate these matchings and cause depolarization. This is especially a question for the higher energy storage rings such as LEP, as explained before. A tracking program would provide useful information here.

Each depolarization resonance caused by imperfections requires two knobs for harmonic matching. The number of nearby resonances then determines the total number of knobs needed to optimize the polarization. Since the spin matching and the harmonic matching are both based on the principle of eliminating the resonance strengths, it is conceivable that some of the harmonic matching knobs can be conveniently provided by the quadupoles that are used to spin match the rotators. The right hand side of eqs.(32) and (33) are then replaced by values that an operator desires to tweak into the storage ring.

Acknowledgement

I would like to thank J. Laclare, E. Grorud, D. Barber, H. Mais, W. Bialowons, G. Ripken and J. Kewisch for providing me very useful information. Barber, Mais and Ripken also pointed out a mistake of mine in the oral report of this work.

References

1. M. Froissart and R. Stora, Nucl. Inst. Methods, 7, 297 (1960).
2. T. Cho et al., Conf. High Energy Physics with Polarized Beams and Polarized Targets, AIP Proc. No. 35, Argonne, 1976, page 396.
3. E. Grorud et al., Conf. High Energy Spin Physics, AIP Proc. No. 95, Brookhaven, 1982, page 407.
4. S. Hiramatsu and K. Muto, Conf. High Energy Physics with Polarized Beams and Polarized Targets, Lausanne, 1980, Experientia Supplementum, Vol. 38, page 475.
5. L. G. Ratner, Ref. 3, page 412.
6. V. Bargmann, L. Michel and V. L. Telegdi, Phys. Rev. Lett. 2, 435 (1959).
7. L. Teng, Conf. High Energy Physics with Polarized Beams and Polarized Targets, AIP Proc. No. 51, Argonne, 1978, page 248.
8. E. D. Courant and R. D. Ruth, Brookhaven Report BNL-51270 (1980).
9. T. Khoe et al., Part. Accel. 6, 213 (1975).
10. E. Grorud, J. L. Laclare and G. Leleux, IEEE Trans. Nucl. Sci., NS-26, 3209 (1979).
11. R. Ruth, 12th Int. Conf. High Energy Accel., Chicago, 1983.
12. E. D. Courant, Conf. High Energy Spin Physics, AIP Proc. No. 95, Brookhaven, 1982, page 388.
13. H. D. Bremer et al., Ref. 3, page 400.
14. A. W. Chao, Ref. 3, page 458.
15. Ya. S. Derbenev and A. M. Kondratenko, Proc. 10th Int. Conf. High Energy Accel.,Protvino, 1977, Vol.2, page 70.
16. K. Steffen, DESY Report 83-058 (1983).
17. E. D. Courant, Ref. 4, page 102.
18. D. Carey, Ref. 3, page 454.
19. J. Laclare, E. Grorud, private communications (1983).
20. A. Turrin, IEEE Trans. Nucl. Sci., NS-26, 3212 (1979). Also, A. Turrin, Ref. 3, page 461.
21. A. A. Sokolov and I. M. Ternov, Sov. Phys. Dokl. 8, 1203 (1964).
22. A. W. Chao, J. Appl. Phys., 50, 595 (1979).
23. A similar technique has been used in A. Garren and A. S. Kenny, SYNCH program.
24. A. W. Chao, Nucl. Instr. Methods 180, 29 (1981).
25. H. Mais and G. Ripken, DESY report 83-062 (1983).
26. Reports by J. Johnson and R. Schwitters, Workshop on Electron Storage Ring Polarization, DESY M-82/09 (1982).

27. J. Kewisch, DESY report 83-032 (1983).
28. A. M. Kondratenko, Sov. Phys. JETP 39, 592 (1974).
29. K. Steffen, Ref. 26.
30. J. Buon, Orsay report LAL/RT/83-04 (1983), also talk presented in the 12th Int. Conf. High Energy Accel., Fermi Lab., 1983.
31. K. Steffen, DESY notes HERA-83/09 and HERA-83/16 (1983), unpublished.
32. Reports by K. Steffen, R. Rossmanith, Y. Yokoya, S. Holmes, J. Buon and A. Chao, Ref. 26.
33. R. Schmidt, Ref. 26. Also, D. Barber and H. Mais, this conference.

Particle Tracking in Accelerators with
Higher Order Multipole Fields

A. Wrulich
Deutsches Elektronen-Synchrotron DESY
Notkestraße 85, 2000 Hamburg 52

1. Introduction

A good linear optics does not, in itself, guarantee a proper functioning of the accelerator. Deviations from the linearized equations may arise from geometrical effects of higher order, intrinsic field distortions due to the finite magnet size and higher order multipoles which may be artificially introduced as for instance the sextupoles to compensate the natural chromaticity.

In terms of general fields, the transverse motion of a particle in the nonlinear lattice can be expressed by

$$\frac{d^2x}{ds^2} - [k(s) - \frac{1}{\rho^2(s)}]x = +\frac{e}{p_0} B_z(x, z, s)$$

$$\frac{d^2z}{ds^2} + k(s) z = -\frac{e}{p_0} B_x(x, z, s)$$

(1)

where x and z are the horizontal and vertical coordinates respectively, k the focusing strength and ρ the radius of curvature.
B_x and B_z are the transverse components of the additional magnetic field that depends nonlinearly on particle coordinates.
It is practical to use a multipole expansion for the analytical description of this field

$$B_x + i B_z = B_0 \sum_{n=0}^{\infty} (b_n + i a_n)(x + i z)^n$$

(2)

with B_0 being the normal bending field and b_n and a_n the normal and skew multipole coefficients. If the multipole ansatz is compared with a Taylor expansion, the coefficients may be expressed as derivatives of the magnetic field components.

Since there exists no simple analytical solution for these equations anymore, a computer simulation must be performed to investigate the particle dynamic in the presence of arbitrary nonlinear fields.

2. Simulation Program (The program RACETRACK[1] is described as an example)

Equation (1) is solved piecewise for each element. For linear elements the solution may be expressed by a matrix transformation for the coordinate and its derivative. To make the nonlinear equation solvable, a thin magnet approximation is used

$$\frac{d^2x}{ds^2} = + \frac{e}{p} \delta(s) \cdot \int B_z(x, z, s) \, ds \quad (3)$$

That means the integrated effect is concentrated in a thin magnet element with zero length and the corresponding transformation merely changes the direction of the particle trajectory.
Thus the tracking is performed by transforming an ensemble of particles with different initial trajectory coordinates through the structure, corresponding to

$$\begin{pmatrix} x \\ x' \end{pmatrix}_1 = M(1|0) \begin{pmatrix} x \\ x' \end{pmatrix}_0 + \sum_{i=1}^{N} M(1|i) \begin{pmatrix} 0 \\ \delta_i(x_i, z_i) \end{pmatrix} \quad (4)$$

The index o indicates the initial coordinate vector and the index 1 the final vector after one revolution. δ is the deflection at the position of the nonlinear element; it depends on the particle coordinates at that position. M denotes the linear transformation matrices.

Initial coordinates

The initial coordinates are distributed on the phase plane ellipses, defined by the twiss parameters β and α at the starting point.

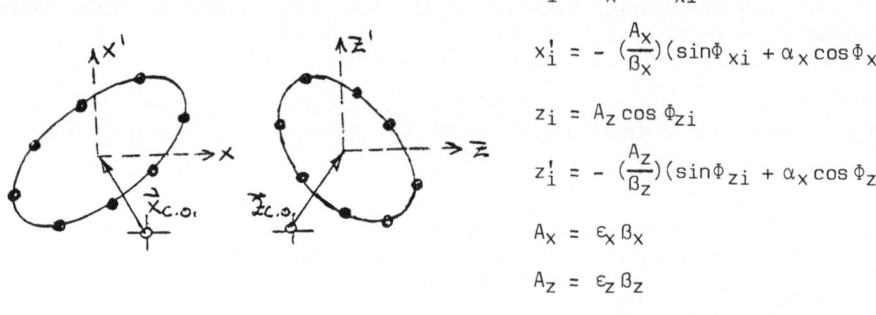

$$x_i = A_x \cos \Phi_{xi}$$
$$x'_i = -\left(\frac{A_x}{\beta_x}\right)(\sin \Phi_{xi} + \alpha_x \cos \Phi_{xi})$$
$$z_i = A_z \cos \Phi_{zi}$$
$$z'_i = -\left(\frac{A_z}{\beta_z}\right)(\sin \Phi_{zi} + \alpha_x \cos \Phi_{zi})$$
$$A_x = \varepsilon_x \beta_x$$
$$A_z = \varepsilon_z \beta_z$$
$$K = \varepsilon_z / \varepsilon_x$$

Regarding the initial position of horizontal and vertical beam motion (described by the emittance ratio K) an elliptical distribution should be taken for electrons and a rectangular distribution for protons. In case of closed orbit distortions the initial orbit vector will be added to the trajectory vectors.

An ensemble of particles is found to be stable if all particles during all revolutions remain within the physical aperture which is introduced in the program by aperture limiting insertions. They can be placed at any point in the structure and might be of rectangular or elliptical shape.

$$\text{rectangular:} \quad x_i < A_{px} \wedge z_i < A_{pz}$$
$$\text{elliptical:} \quad \left(\frac{x_i}{A_{px}}\right)^2 + \left(\frac{z_i}{A_{pz}}\right)^2 < 1$$

RACETRACK allows to vary optical parameters and introduce distortions in an easy way. Each task may be activated by introducing an additional data block or a switch in the input stream.

Closed orbit distortions:

Positions of dipole errors and of orbit displacement monitors may be defined in the lattice. A random set of dipole errors will be scaled to give desired rms orbit errors measured at the monitor positions.

Tune variation:

The tune may be adjusted to desired values by changing two quadrupole families.

Chromaticity adjustement:

The chromaticity can automatically be adjusted to zero or any other value by tuning two sextupole families.

Synchrotron oscillations:

Another possible feature is the variation of energy corresponding to the actual synchrotron oscillation. One or more cavities may be placed around the circumference.

3. Tracking results for the HERA proton ring

HERA[2]) is a large electron proton colliding beam machine where 820 GeV protons collide with 30 GeV electrons at four interaction points. The arcs are built of a FODO structure of 26 proton cells and 52 electron cells per octant. The proton cell has 3 bending magnets per half cell and sextupoles are introduced adjacent to the quadrupoles.

The acceptance of the proton ring will mainly be influenced by

- normal chromaticity correcting sextupoles
- field errors; since the field accuracy of superconducting magnets depends on mechanical tolerances of the coil, they tend to be larger than in conventional magnets
- at low energies by sextupoles induced by currents flowing in the superconductors; they are one or two orders stronger than in conventional magnets[3])

The maximum stable initial amplitude (or emittance) was found by tracking an ensemble of 16 particles through the structure. An amplitude was assumed to be unstable if at least one out of all 16 trajectories belonging to the same amplitude was unstable. Due to the fast particle loss at the stability limit, 100 revolutions were assumed to be sufficient for tracking. The half aperture was 30 mm in the regular part of the normal cell structure. The systematic part of the persistent sextupole was assumed to be compensated by a commonly powered coil in each magnet. Only a random fluctuation of 10 % of the normal strength was taken into account.

To get a feeling how the two nonlinearities given by normal cell sextupoles and multipole errors superimpose, the pure sextupole case was treated separately.

For the on energy case the influence of small tune changes on the dynamical aperture is shown in Fig. 1. The height of the mountain range is proportional to the maximum stable initial amplitude at the starting point, which is a horizontally focusing quadrupole in the normal cell. The decreasing amplitude at the right hand side comes from the intrinsic 3rd integer sextupole resonance $3 Q_H = 100$. By eliminating sextupoles in the straight sections, the driving term could be reduced by a factor of 4.

The optimal tunes were selected to be $Q_H = 33.15$ and $Q_V = 35.08$. For this best choice, the nonlinear chromaticity, i.e. the variation of tune with momentum (after correcting the linear part of the chromaticity) was calculated and plotted in Fig. 2. The variation is rather weak over the energy range of ± 1.5 ‰, which is the total energy spread of the proton beam at 40 GeV.

The number of revolutions for which tracking can be performed is limited by the available computer time. Note that the calculation of stable amplitude for one energy point with constant tune, random multipole set and orbit distortion takes about 10 minutes for HERA (in this case an ensemble of 16 particles is transformed through 100 revolutions and the amplitude resolution is determined by 8 amplitude steps).

In Fig. 3 is shown that an increase of revolution number does nearly not affect the results (if the tune is far away from a low order resonance). Once the particle starts beyond the stability limit, it gets lost very fast. Below the limit, an amplitude reduction of less than 1 mm increases the number of stable revolutions from 10 to more than 1500.

In the results discussed so far only sextupoles are considered. Multipole errors were introduced in the following way:
For an arbitrary set of random multipoles the same tune optimization as before was performed. Then for the optimal values 20 different sets of fluctuations were calculated as shown in Fig. 4. For each calculation the tunes were readjusted and the dipole errors were scaled to give an orbit rms value of 1 mm measured at the positions of normal cell sextupoles. One set out of the central part of the resulting distribution was used for subsequent calculations.

Fig. 5 shows the maximum stable initial amplitude as a function of relative energy deviation. The dotted line corresponds to the pure sextupole case, the solid line describes the case where multipoles and orbit distortions with 1 mm rms value are included. The single point at zero momentum deviation shows the result for multipoles only and chromaticity correcting sextupoles switched off. Thus the effects of sextupoles and multipoles on dynamical aperture are more or less comparable.

Fig. 6 shows the change of maximum stable amplitude with increasing orbit distortions. Horizontal and vertical orbit distortions were scaled simultaneously to the same rms value. There are two effects influencing the stable amplitude: the linear aperture reduction due to the orbit displacements and the modified multipole pattern. In the presence of orbit distortions each multipole excites all lower order components in addition.

Only parts of tracking calculations which have been performed for the HERA proton ring were shown here. In addition several specific problems had to be investigated.[4,5] For instance, the Sagitta effect of straight magnets, the tolerable systematic 14- and 18-pole, different cell lengths and different phase advances. Each modification step in the design period was and will be accompagnied by these tracking calculations.

References

1. A. Wrulich: RACETRACK, a Computer Code for Simulating Particle Motion in the Presence of Higher Order Multipole Fields, to be published

2. HERA, A Proposal for a Large Electron-Proton Colliding Beam Facility at DESY, DESY HERA 81/10, July 1981

3. A Report on the Design of the Fermi National Accelerator Laboratory, Superconducting Accelerator, May 1979

4. A. Wrulich: Aperture Limitations in the HERA Proton Ring due to Nonlinear Fields, DESY HERA 82-04, April 1982

5. A. Wrulich: Tracking Studies in HERA, DESY HERA 82-07, June 1982

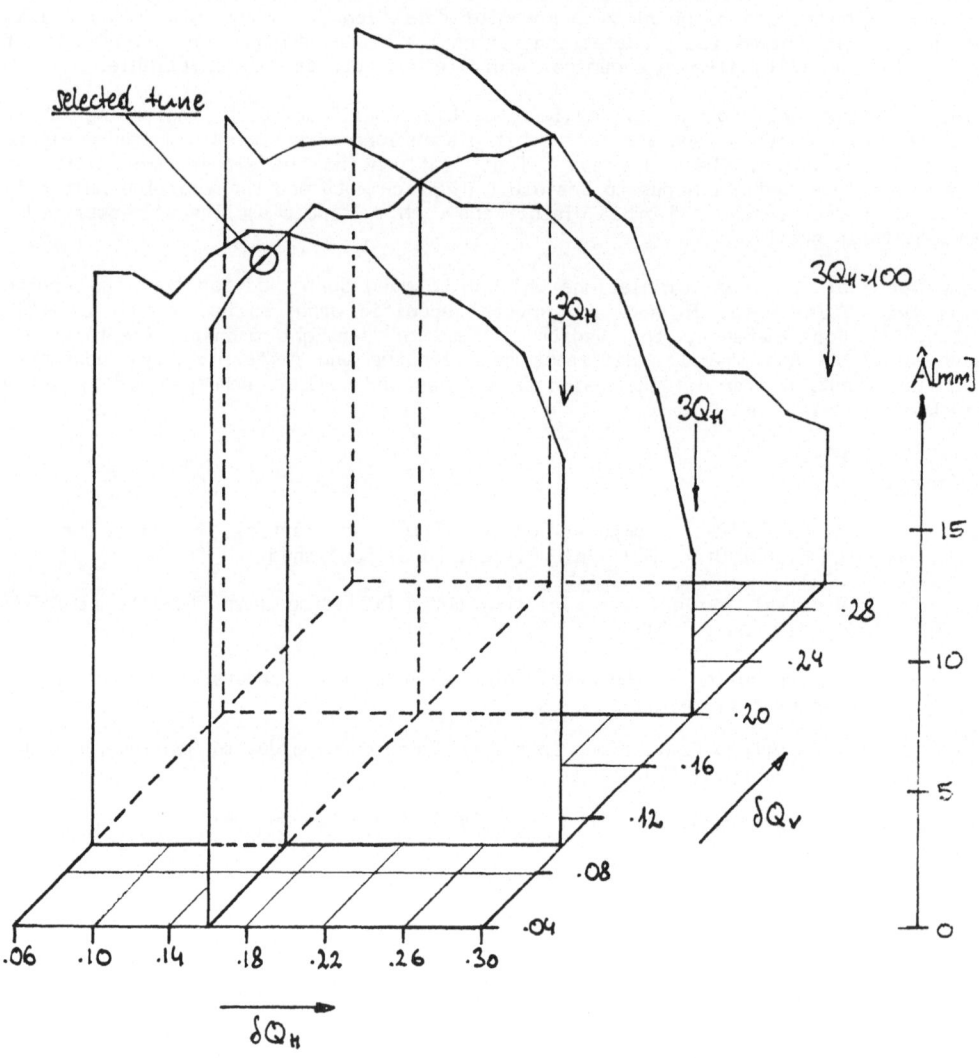

Fig. 1

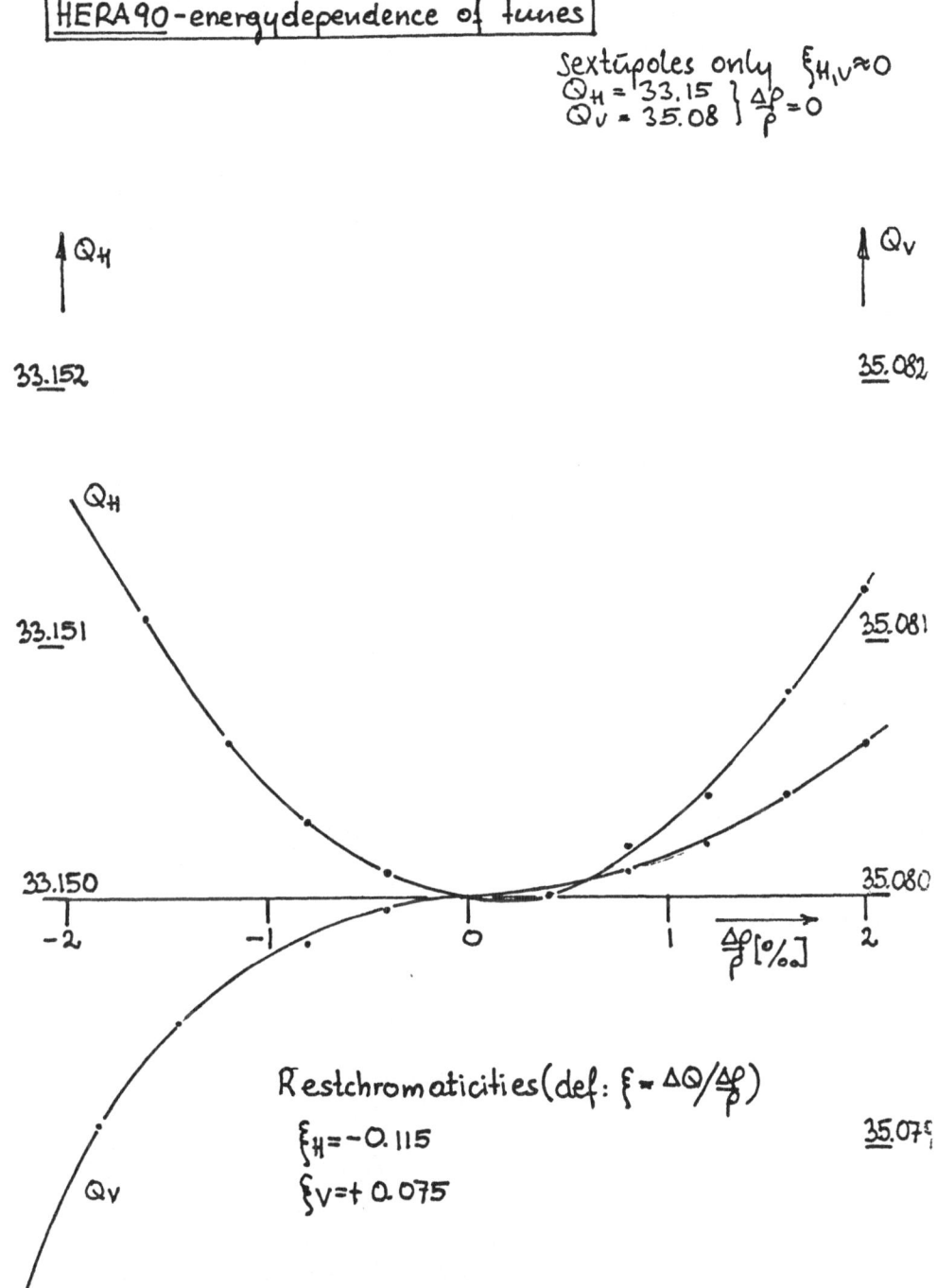

Fig.2

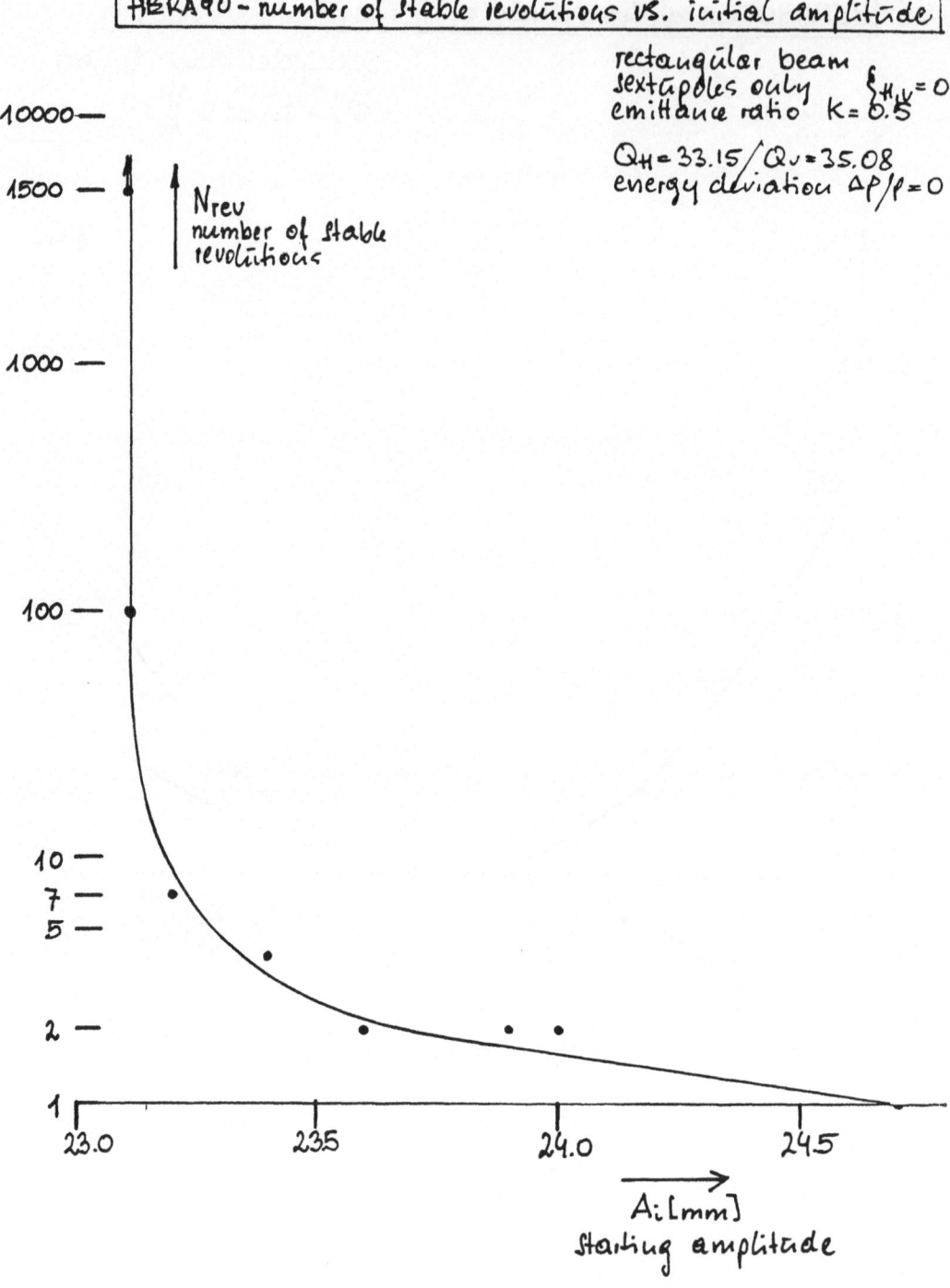

Fig.3

HERA90 – Variation of maximum stable amplitude with multipole statistics
rectangular beam
sextupoles and multipoles ξH,V=0
with closed orbit erros c.o.-rms = 1 mm
emittance ratio K = 0.5
energy deviation Δp/p = 0
NKV ≙ 100
A: QH=33.15/QV=35.18
10
15
20
(RN)o
Â[mm]
(RN)o = random multipole set selected for proceeding calculations
B: QH=33.15/QV=35.08
5
10
15
20

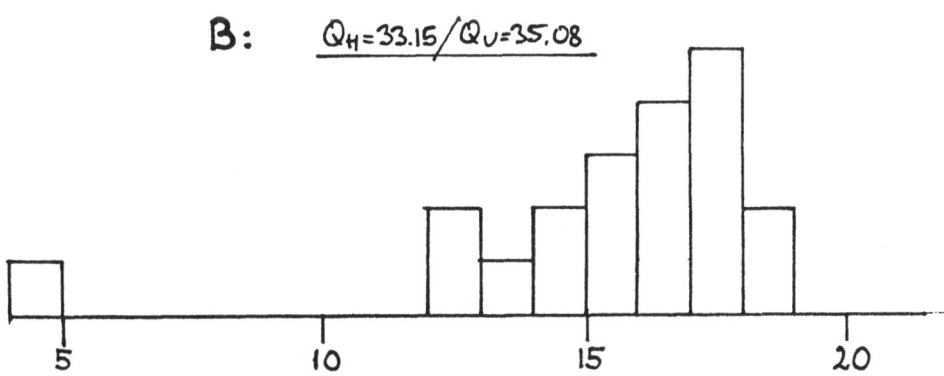

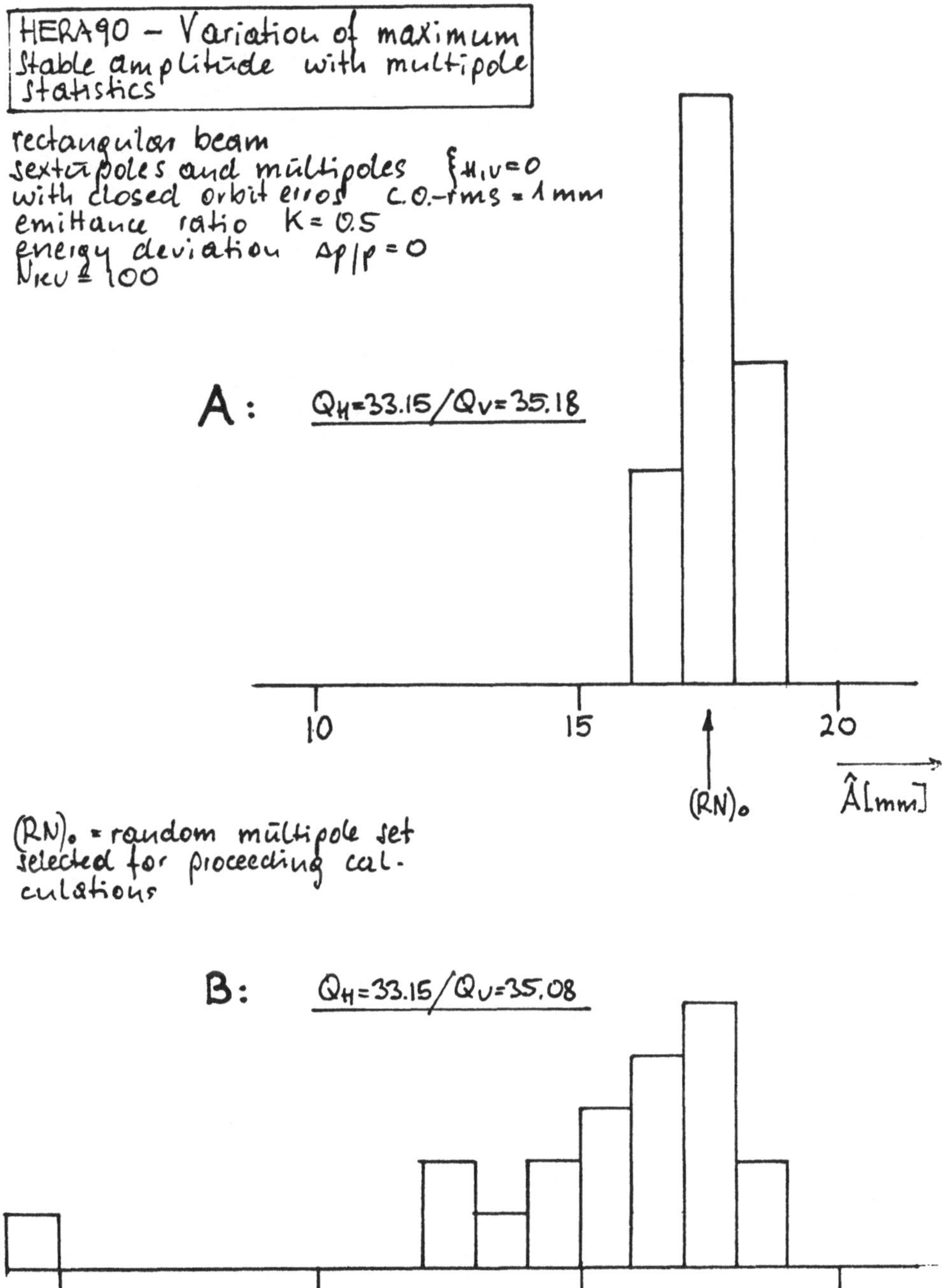

Fig.4

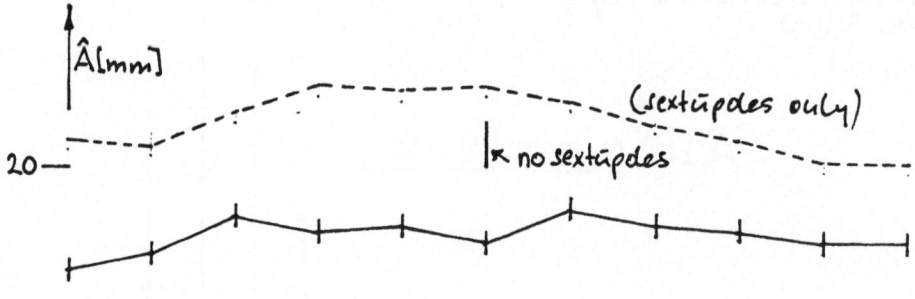
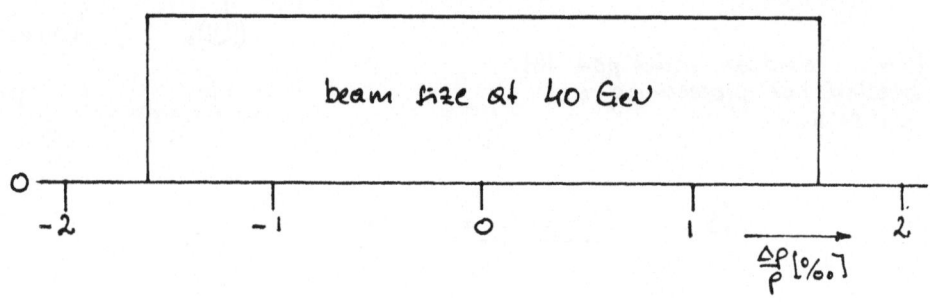

Fig. 5

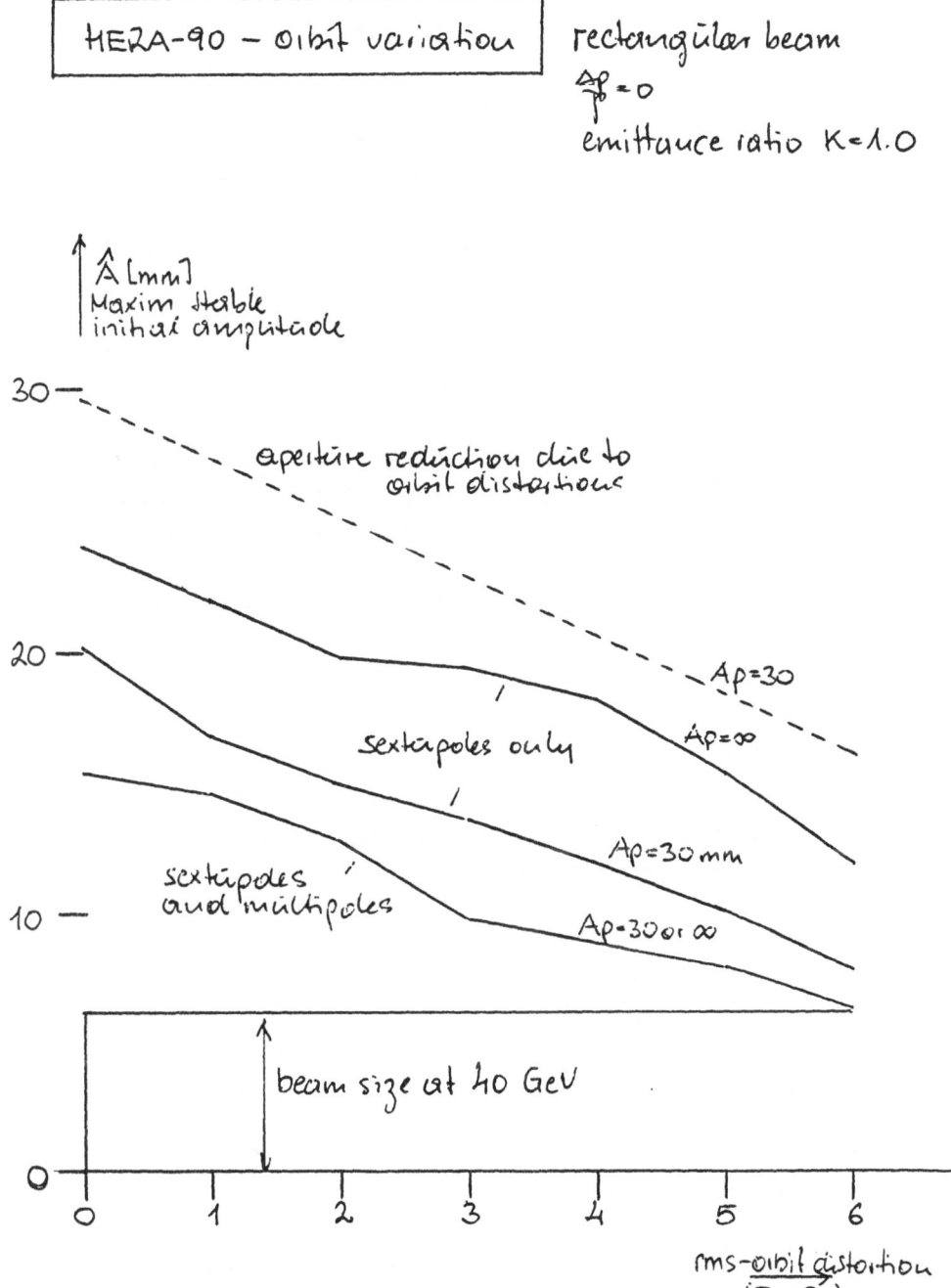

Fig. 6

PROGRAMS FOR DESIGNING THE ACCELERATING CAVITIES FOR LINEAR ACCELERATORS

S. Kuliński, L. Sawlewicz, J. Sekutowicz
Institute for Nuclear Science and Technology
05-400 Swierk, Poland

Recently two programs were developed for the calculation of the parameters of accelerating cavities for linear accelerators. The first - semianalytical - divides the cavity into few parts in such a manner that some parts can be treated analytically, the others numerically using the finite-difference overrelaxation method. Proper matching at the common boundaries yields the desired solution of the problem, although the number of equations can be considerably reduced.

The second approach uses the finite element method with curvilinear triangles to describe more precisely the boundaries of the cavity.

Both methods give good results in a reasonably short time. It seems also that proper connection of these two methods can still reduce the time of calculation.

THE PROBLEM

The problem of finding the electromagnetic field in the accelerating cavities of linear accelerators was treated by many authors [1-5]. Since usually the cavities have axial symmetry the problem can be conveniently formulated in terms of the potential $U = rH_\varphi$, where H_φ is the azimuthal component of the magnetic field. The potential U must fulfill the equation:

$$\frac{\partial^2 U}{\partial r^2} - \frac{1}{r}\frac{\partial U}{\partial r} + \frac{\partial^2 U}{\partial z^2} + k^2 U = 0 \qquad (1)$$

inside the cavity, together with boundary conditions stating that the derivative of U in the direction normal to the boundary is zero everywhere on the boundary except on the axis /cf. Fig. 1./ where U=0. / N.B. it can be shown that this normal derivative vanishes also on the axis /. Equation for the eigenvalue of the wave number k is usually obtained from the variational principle and can be written in the form:

$$k^2 = \min \frac{\int_D \frac{1}{r}\left[\left(\frac{\partial U}{\partial r}\right)^2 + \left(\frac{\partial U}{\partial z}\right)^2\right] dr\, dz}{\int_D \frac{1}{r} U^2 dr\, dz} \qquad (2)$$

D is the cross-section of an accelerating cell reduced in our case to 1/4 of the actual cell because of existing symmetries.

The most common method of numerical solution of the above stated problem is by transforming equations (1) and (2) together with boundary conditions into corresponding finite-difference equations which are then solved by some iterative procedure, e.g. succesive overrelaxation / SOR / method [1,2]. As a rule these methods require rather large memories of the computer and are time consuming. However, in many cases the compound analytical-numerical method can be used based on division of the cavity into few parts in such a way that some parts can be treated analytically, the others numerically [3,4]. Proper matching of partial solutions at the common boundaries gives the solution for the whole cavity. Depending upon the shape of the cavity this method can lead to substantial reduce of the computational time and required computer memory. A brief description of this method for the cavity of Fig. 1 . is given in the next section.

Recently, in connection with the development of finite element method, there is a trend to apply this method also to solve the problems connected with electromagnetic cavities [6,7] . As it is shown in the third section this method can be a competitive one to overrelaxation especially for the cases when many eigenvalues for the same cavity are desirable.

THE COMPOUND ANALYTICAL-NUMERICAL METHOD

This method is well suited for the cavity shapes with long rectilinear boundaries between which analytical solution is available. Such are for instance Alvarez type cavities with right angles in the outer regions. In that case it is enough to divide the cavity into two parts : one part in the vicinity of the axis with complicated boundary shapes where the numerical treatment is necessary and the other far from axis where analytic solution is possible. To assure quick convergence of matching between the analytical and numerical solutions it is convenient to make these parts overlapping and to use the procedure similar to the Schwartz Alternating Method for overlapping regions. The existing programs [3,4] written with the aid of this met-

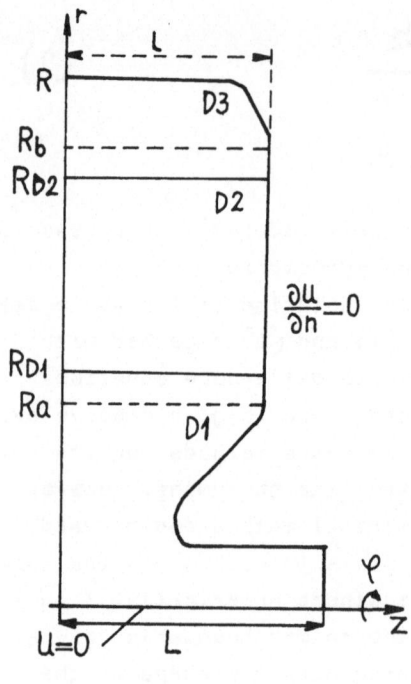

Fig. 1.

hod do really show some advantages such as small number of nodes in the grid /in fact the grid can be limited only to the lower part / and quick convergence of the iteration process caused both by smaller number of equations and overlapping of the regions.

The principles of this method can be also applied to more complicated cavities. Consider for instance the Ω shaped cavity presented on Fig. 1. To apply the method we divide the cavity into three domains: $D1: 0 \leq r \leq R_{D1}$, $D2: R_{01} \leq r < R_{D2}$, $D3: R_{D2} \leq r \leq R$

In the domains D1 and D3 the conventional finite-difference overrelaxation method of solution is used. For the domain D2 there exist analytical solutions of Eq. (1) of the form:

$$U(r,z) = r[c_0 J_1(kr) + d_0 Y_1(kr)] + \sum_{m=1} [c_m I_1(\gamma_m r) + d_m K_1(\gamma_m r)] \cos \beta_m z \quad (3)$$

where: $\beta_m = \dfrac{\pi m}{l}$, $\gamma_m^2 = \beta_m^2 - k^2$ (4)

J_1, Y_1, I_1, K_1 are Bessel and modified Bessel functions correspondingly. The solution to the problem is now obtained as follows:

1. The grids are imposed in the domains D1 and D3
2. Initial values of $U(r,z)$ and k are calculated by some approximate method
3. Some number of iterations / 10-12 / is done in the domains D1 and D3 keeping the values of U not changed along the lines $r=R_{D1}$ and $r=R_{D2}$ regarded as Dirichlet boundary lines
4. Fourier expansion cofficients are calculated for the domain D2 using the values of U along the lines $r=R_{D1}$ and $r=R_{D2}$, with the aid of equations:

$$U(R_a, z) = R_a \left(A_0 + \sum_{m=1} A_m \cos \beta_m z \right)$$
$$U(R_b, z) = R_b \left(B_0 + \sum_{m=1} B_m \cos \beta_m z \right) \quad (5)$$

Comparison of equations (3) and (5) allows for the calculation of finite number of coefficients c_m, d_m, $m=0,1,\ldots M$. Usually M is of the order of 10.

With the aid of Eq. (3) new values of U are calculated along the Dirichlet lines $r=R_{D1}$ and $r=R_{D2}$. Also the new value of k is obtained by use of Eq. (2) in which the integrals in the domains D1 and D3 are calculated numerically and in the domain D2 analytically. Analytical integration requires the knowledge of the integrals of the type

$$I(f_i, f_k) = \int_{R_1}^{R_2} r^2 f_i(dr) \cdot f_k(dr) \, dr$$

where f_i and f_k are the Bessel functions and modified Bessel functions of the order 0 and 1. The integrals of the above type are given as some algebraic expressions of $f_i \cdot f_k$ and are easy for calculations.

The operations of points (3) and (4) are repeated until the desired accuracy of k and U is obtained. Applying this method a modified version of the program used in [4] was written for CDC6600 computer. The test calculations showed that one can obtain decrease of necessary memory and computation time requirements. This decrease depends upon the relative participation of subspace D2 in the whole space D.

THE FINITE ELEMENT METHOD

The electromagnetic problem, as stated above, can be also solved by Finite Element Method /FEM /. The computer program JOLA1 using this method was specially written for this purpose. The potential function $U(r,z) = rH_\varphi$ is now treated as an element of the Sobolev space $H^1(D, 1/r)$, where D is 1/4 of the cavity cross-section and $1/r$ is a weight. Domain D is then divided into triangles. Some of these triangles have one curved side placed on the boundary ∂D of domain D. Typical triangulation of D is shown on Fig.2. All straight sides of triangles are represented exactly in the program. Curvilinear sides are approximated by third order Hermitian polynomials. The use of curvilinear triangles avoids the necessity of more dense

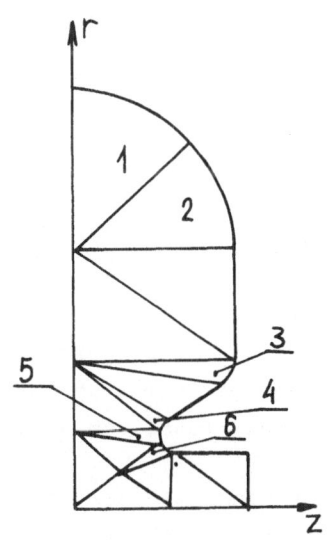

Fig. 2.

triangulation near the curved parts of D required for better boundary approximation if only straight lines are employed, allowing thus for diminishing the number of triangles.

Function $U(r,z)$ is approximated by the function $U^*(r,z) \in H^1(D,1/r) \cap C(\bar{D})$, where $C(\bar{D})$ is the space of continuous functions in D. Function U^* is constructed as follows [6]. All triangles are transformed onto standard triangle T_0 / cf. Fig. 3 /.

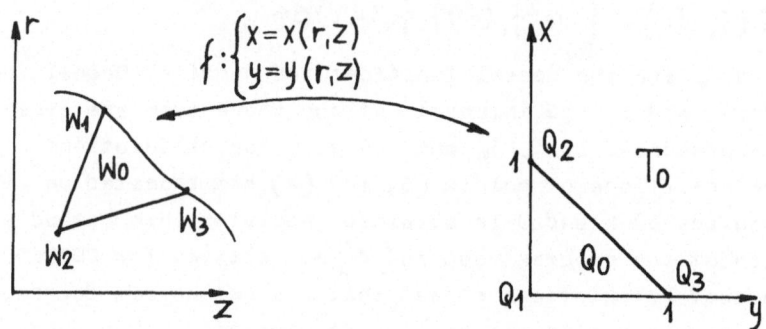

Fig. 3.

In triangle T_0 third order polynomial $w(x,y)$ is uniquely determined by ten parameters associated with the nodes w_0, w_1, w_2, w_3. The function $U^*(r,z)$ over each triangle is defined by the formula:

$$U^*(r,z) = w(x(r,z), y(r,z)) \qquad (6)$$

The function $U^*(r,z)$ is an element of finite-dimensional subspace of space $H^1(D,1/r)$, then the problem can be led to the solution of an eigenvalue problem:

$$Av - \lambda Bv = 0 \qquad (7)$$

The dimensions of matrices A and B depend on the number of triangles. The coordinates of eigenvector v are the values of functions U, $\frac{\partial U}{\partial r}$, $\frac{\partial U}{\partial z}$ in the interpolating nodes.

The above eigenvalue problem can be solved by different methods including iterative ones. The method chosen in the program transforms Eq.(7) into eigenvalue problem of tridiagonal matrix what allows for calculation of all eigenvalues without additional computations.

The program JOLA1 was tested on the CDC 6600 computer. In table I the results for a cylindrical cavity with the radius r=4cm /the theoretical value of the lowest frequency is f=2.86856283 GHz / are presented:

TABLE I

N matrix dimension	11	37	121	164
Error %	$4 \cdot 10^{-2}$	$2.6 \cdot 10^{-3}$	$1.7 \cdot 10^{-3}$	$3.6 \cdot 10^{-4}$
Exec. time t /sec/	2.2	10.7	82	166

Time t as given in the table is the time of computation of N resonant frequencies. Obviously the error decreases with dimensions of A and B matrices.

To test the influence of boundary approximation by straight lines and Hermitian polynomials the cavity treated by Hoyt [1] and Konrad [7] was analysed. Calculations were done for the triangulation given on Fig. 2 . The dimensions of the matrices were N=56. Results are given in table II. The first row contains the numbers of triangles for which the third order Hermitian polynomials were used. The second row contains the lowest resonant frequency which according to Konrad equals 808.799 ,and to Hoyt 804.8 MHz. Both Hoyt and Konrad used Lagrange interpolation for boundary approximation.

TABLE II

Triangle numbers	-	1,2,3,4	3,4,5,6	1-6
f [MHz]	834.812	801.948	832.999	801.162

The execution time was t=100 sec and the program JOLA1 computed 56 resonant frequencies and E and H field distribution for the lowest one.

As it is seen from table II application of Hermitian polynomials interpolation for curvilinear boundaries allows to decrease matrix dimension without loss of accuracy. To compare with the analytical numerical method the Alvarez cavity for 200 MHz was also computed. The results obtained with codes FITA and JOLA1 were 201.998 and 202.489 MHz and execution times 66 and 100 sec, correspondingly.

LITERATURE

1. C.H. Hoyt et al.   Rev. Sci. Instr. 37, 6 ,1966
2. A.Katz   SFFS TD 65/20
3. M.Martini, D.J.Warner   CERN Report 68-11, 1968
4. S.Kuliński   INR Report 1612/xxv/PL/A
5. K.Halbach et al.   Proc. 1976 Proton Lin. Acc. Conf., Chalk River
6. M.Zlamal   Lect. on the FEM, Techn. Hochschule KMS 1974
7. A.Konrad   Comp. Phys. Comm. 14, 1978

THE MAGMI PROGRAM FOR DOUBLE PASS ELECTRON LINEAR ACCELERATORS

T. Czosnyka, K. Deutschman, S. Kuliński, S. Zaremba

Institute for Nuclear Science and Technology
05-400 Swierk, Poland

The MAGMI / MAGnetic MIrror / program was written for designing the magnetic mirror for electron linear accelerators enabling electrons to be accelerated twice by the same accelerating structure.

Depending upon the incoming beam parameters such as: emittence of the beam, energy and admissible energy variation of electrons, the program minimizes the necessary number of magnets and optimizes their shape, distribution and magnetic field intensity to obtain a possibly perfect reflection of the beam by the magnetic mirror. After optimization the orbits of electrons are also calculated.

I. INTRODUCTION

The increasing interest in medical and industrial applications of linear electron accelerators has resulted in a number of new ideas and solutions. A very promising one is to employ magnet systems to deflect the outcoming beam back into accelerating structure to achieve further acceleration. Multipass system with magnetic mirrors has been proposed in 1967 [1] and its simplest variation, a double pass linear accelerator has been found to be especially attractive for medical purposes. Double acceleration enables to cover the energy range up to 30-40 MeV, i.e. the full range needed for medical treatment, using magnetron based rf power supplies. The whole device can be thus relatively compact and inexpensive, in addition providing flexible energy regulation in a wide range just by adjusting mirror-accelerating structure separation.

The magnetic mirror for a double pass accelerator should provide beam optics to reflect the beam acceptable for accelerating structure in a wide range of incident energies. Energy independence of mirror operation assures stability of the whole system despite the load increase with two beams being accelerated simultaneously. For the same reason different energy electrons time of flight should be kept approximately constant. The above requirements assure easy and simple accelerator operation in out-of-laboratory conditions.

The design of a magnet system providing the necessary beam transmission properties must be very precise. Widely used first and second order beam optics approximations are generally not applicable in our case, so the extensive computer simulation remains the only viable way to solve the problem. In this paper design methods for the magnetic mirror operating with the 10 MeV accelerator are presented.

II. PRELIMINARY DESIGN

The assumed energy-independent time of flight can be achieved by double focusing of different energy beams in median plane possible with a system of four magnets, as schematically shown at Fig. 1.

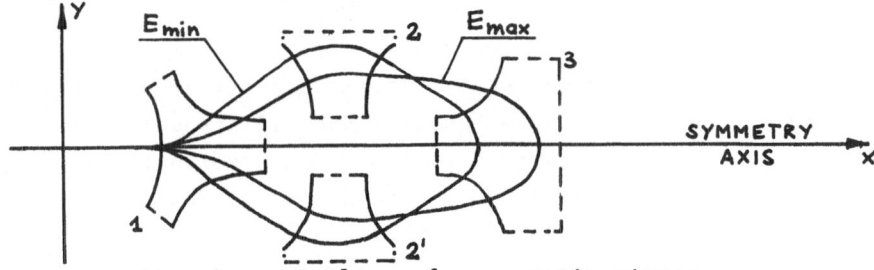

Fig. 1. Outline of a magnetic mirror.

Similar magnet system, a simplest one to provide isochronism, has been studied in Chalk River [2]. Initially, the problem has been treated as the transmission of a single electron beam spreaded energetically /8-11 MeV / around central trajectory of 9.5 MeV. This approach, although in most aspects inadequate, allows to determine the basic optical properties and to set the reasonable ranges of parameters for more detailed calculations. For central /9.5 MeV / trajectory the reflection condition can be expressed in terms of deflection angles of magnets 1, 2 and 3 - φ_1, φ_2 and φ_3, orbit radii of curvature inside magnets r_1, r_2, and r_3, and drift lengths l_{12} and l_{23} as:

$$r_3 \sin \frac{\varphi_3}{2} = l_{12} \tan \varphi_1 - l_{23} \cot \frac{\varphi_3}{2} + r_2 \left(\sin \frac{\varphi_3}{2} - \cos \varphi_1 \right) + r_1 \left(1 - \cos \varphi_1 \right)$$

$$2(\varphi_2 - \varphi_1) + \varphi_3 = \pi \qquad (1)$$

For the whole beam the reflection will take place if central trajectory intersects the symmetry axis at right angle and at this point the angular dispersion Θ vanishes. Using the first order approximation, one can express beam transmission through a magnet with matrix equation:

$$\begin{bmatrix} x_f \\ \Theta_f \\ \Delta E/E \end{bmatrix} = \begin{bmatrix} \cos\varphi & r\sin\varphi & r(1-\cos\varphi) \\ -\sin\varphi/r & \cos\varphi & \sin\varphi \\ 0 & 0 & 1 \end{bmatrix} \begin{bmatrix} x_o \\ \Theta_o \\ \Delta E/E \end{bmatrix} \quad (2)$$

Using realistic input parameters it was found that most promising configuration is obtained with $\varphi_1 \approx \pi/4$ and $\varphi_2 = \varphi_3 \approx \pi/2$. This configuration was further examined with CERN beam optics code TRANSPORT [3]. The calculations showed the sufficient focusing strength of the system even if the exit edges of magnets 1 and 2 above symmetry axis / Fig. 1 / were not curved. However, TRANSPORT results have been found inadequate for actual design due to too wide energy spread /second order corrections comparable with first order terms / and the well-known [4] parallel displacement of central trajectory if sharp cutoff of magnetic field is assumed instead of realistic distribution in the vicinity of magnet edges. TRANSPORT results were thus used only as input parameters for the computer design code MAGMI, written for this specific purpose.

III. PROGRAM MAGMI

The electron beam is represented in MAGMI by a finite number of electrons lying on the beam envelope and the one determining central trajectory. Relativistic equations of motion are numerically integrated using continuous, analytically given magnetic field distribution being a function of magnet shapes. With the magnet system shown on Fig. 2, it is defined by 14 parameters: 7 parameters of arc-shaped edges /only one variable for magnet 1/, 4 parameters of straight edges and 3 values of magnetic induction. Fringing field is calculated using empirical formula of Enge [4]:

$$B_z = B_z^{(o)} \cdot h(s) \quad (3)$$

where $B_z^{(o)}$ is maximum induction inside magnet and $h(s)$ is the magnet-dependent distribution function, s being the distance to the magnetic effective edge in units of magnet aperture. It was assumed that in the vicinity of magnet edges the lines of constant B_z are parallel to them. In this approximation one can find that $\mathrm{curl}\,\bar{B}=0$ yields

$$\begin{aligned} B_x &= B_z^{(o)} \cdot (dh(s)/ds) \cdot \cos\varphi \cdot (z/g) \\ B_y &= B_z^{(o)} \cdot (dh(s)/ds) \cdot \sin\varphi \cdot (z/g) \end{aligned} \quad (4)$$

where φ is azimuthal angle of a vector perpendicular to the magnet edge and g denotes the aperture. One can also extend (3) to the second order in z:

$$B_z = B_z^{(o)} \left(h(s) - \frac{z^2}{2g^2} \frac{d^2 h(s)}{ds^2} \right) \qquad (5)$$

With induction B defined by (4) and (5) the trajectories of the electrons are found by numerical integration of the system of relativistic differential equations of motion up to some user-defined in terms of x exit point. The ray-tracing module of the code is driven by minimization routines of well-known CERN program MINUITS [5] in order to minimize the penalty function F built of exit parameters of the electrons in a following way:

$$F = \sum_{i=1}^{N_{el}} \sqrt{y_i^2 + z_i^2} + \sum_{i=1}^{N_{el}} \sqrt{y_{ik}^2 + z_{ik}^2} \cdot \delta_i + \\ + w_z \sum_{i=1}^{N_{el}} \left(|z_{max}| - \frac{g}{2} \right)^2 \cdot \Delta_i + w_t \sum_{i=2}^{N_{el}} \left(t_i - t_1 \right)^2 \qquad (6)$$

where y_i and z_i are taken at given exit point, while y_{ik} and z_{ik} denote y and z coordinates of i-th electron after some time interval set by the user. δ_i assumes value of 0 if $z_{ik}^2 + y_{ik}^2 < R_k^2$ and value of 1 otherwise / R_k is connected with the aperture of accelerating structure and its focusing strength and is also specified by the user /. First term thus tends to minimize the spatial dimensions of the beam, while second term prevents beam divergence from becoming unacceptable. Third term of (6) enforces proper vertical focusing, maximum over z taken along the whole trajectory, with switch $\Delta_i = 0$ if $|z_{max}| < g/2$ and 1 otherwise. The last term of (6) is provided to assure isochronous properties of the mirror, electron 1 used conventionally as reference. User specified weights W_z and W_t allow to shift the relative importance of different effects at current stage of minimization.

Penalty function F is logically modified by the program if any of the electrons was not reflected back to the exit point and its value is defined to provide continuous minimization. At the initial stage of design simplex minimization [5] allows to treat not necessarily continuous F function, while close to the minimum variable metric gradient method [5] can be switched on. Dependent on the current stage of design there is also a possibility to choose sharp cutoff magnetic fields instead of tail distribution to accelerate initial optimization. The user can also switch off and on second order expression for B_z working in fringing fields mode. With all user-defined performance conditions the accuracy obtained can match current needs, thus essential amounts of computer time can be saved at initial optimization stages.

IV. RESULTS AND CONCLUSIONS

A number of trials with different starting parameters has been made to determine technologically most desirable solution. Several local minima have been found and examined. The design accepted for realisation fulfills the beam optics requirements with relatively low magnetic field / about 0.5, 0.4 and 0.85 Tesla in magnets 1, 2 and 3, respectively / and small system size /trajectory length of about 80 cm / . The magnetic mirror outline and different energy trajectories are shown on Fig. 2 . The system also displays satisfactory isochronism.

The analytic description of fringing fields is the only possible source of systematic error. The measurements of magnetic field of the magnet system manufactured according to the presented design showed however an excellent agreement of experimental and assumed values of this field. Therefore the methodics used has been proved to be adequate and well suited to the problem. The MAGMI code can thus be utilized to perform the magnetic mirror design calculations for the systems with various technological and performance requirements.

LITERATURE

1. A.A. Kolomensky   Proc. 6 Int. Conf. on High Energy Accelerators, Cambridge, 46, 1967
2. S.O. Schreiber, E.A. Heighway   IEEE Trans. Nucl. Sci. NS-22, No.3, 1975
3. K.L. Brown et al.   CERN Report 73-16, 1973
4. H.A. Enge   Rev. Sci. Instr. 35, No.3, 1964
5. F. James, M. Roos   CERN Computing Centre Report D-506

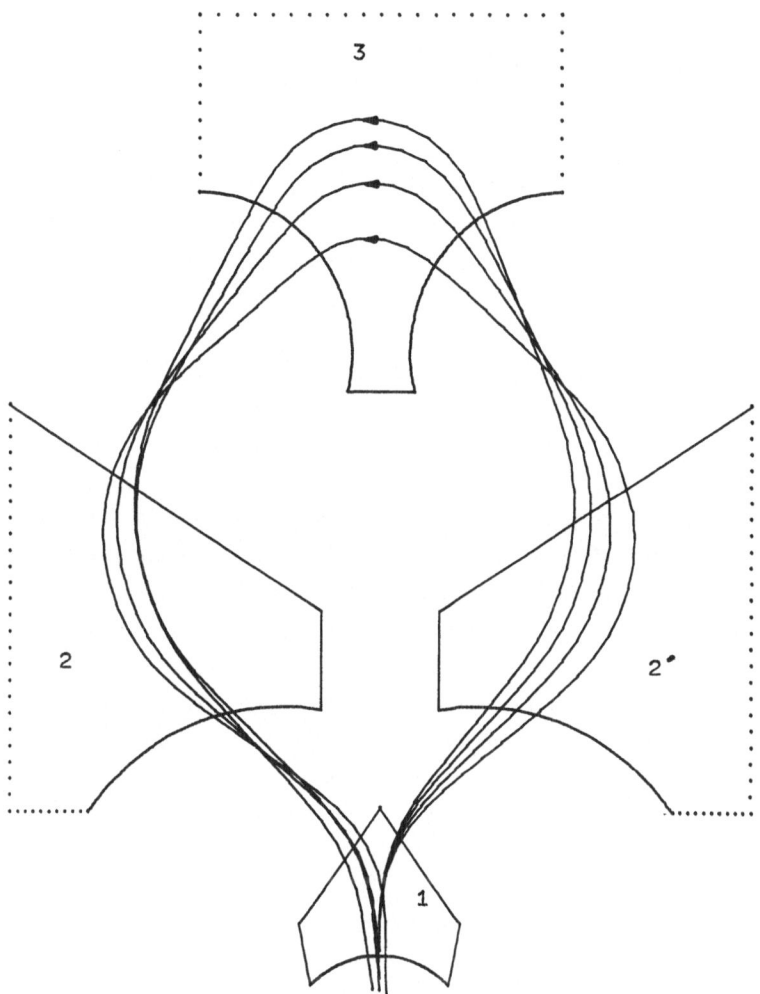

Fig. 2 .   Central trajectories of 8, 9, 10 and 11 MeV
electron beams in the magnetic mirror.
Irrelevant magnet edges are marked with
dotted line.

A Fortran Program (RELAX3D) to Solve the 3 Dimensional Poisson (Laplace) Equation

H. Houtman[*], C.J. Kost
TRIUMF, University of British Columbia
Vancouver, B.C., Canada

1) Introduction

RELAX3D is an efficient, user friendly, interactive FORTRAN program which solves the Poisson (Laplace) equation $\nabla^2 \phi = \rho$ for a general 3 dimensional geometry consisting of Dirichlet and Neumann boundaries approximated to lie on a regular 3 dimensional mesh. The finite difference equations at these nodes are solved using a successive point-iterative over-relaxation method [1,2]. A menu of commands, supplemented by an extensive HELP facility controls the dynamic loading of the subroutine describing the problem case, the iterations to converge to a solution, and the contour plotting of any desired slices etc.

Basic parts of the program were developed in 1973 [3] based on work by others[4,5] with the idea in mind that appropriate applications would be found for it at some future date when computers could run it in a practical manner. Fortunately, our view into the crystal ball turned out to be accurate and the program has found many problems to solve both at TRIUMF and elsewhere.

2) Method

Finite difference methods, dating back to Gauss [6] can be used to obtain numerical solutions, to any desired accuracy, for all static electric and magnetic fields in uniform media and time-varying fields where eddy currents can be neglected. We will restrict ourselves to problems given in TABLE 1. The partial differential equation for the field, being replaced by a set of finite difference equations at discrete points in a mesh occupying the volume of interest, results in a large number of equations connecting the potential ϕ at each point with those of adjacent points and is solved by an iterative procedure known as relaxation. The basis of the iteration procedure is that the potential ϕ at each grid point (that is allowed to vary) on each iteration is replaced using the finite - difference equation. The over-relaxation method accelerates the convergence by replacing the old ϕ by

---

[*]Current address Physics Department, University of British Columbia, Vancouver, B.C., Canada.

$$\emptyset_{old} + R * (\emptyset_0 - \emptyset_{old}) \text{ where usually } 1 \leq R \leq 2$$

The patterns of points used to form the finite difference equations are commonly called "templates", "molecules" or "stencils". Let the uniform grid spacing along the x, y, and z axes be h_x, h_y, and h_z respectively where in general these may all be different.

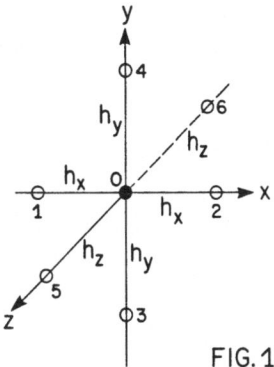

FIG. 1

The 3 dimensional molecule is shown in Fig. 1 for the Poisson equation $\nabla^2 \emptyset = \rho$ resulting in the finite difference equation

$$\emptyset_0 = W_x(\emptyset_1 + \emptyset_2) + W_y(\emptyset_3 + \emptyset_4) + W_z(\emptyset_5 + \emptyset_6) - \rho/C$$

where
$$C = \frac{2}{h_x^2} + \frac{2}{h_y^2} + \frac{2}{h_z^2}, \quad W_x = \frac{1}{Ch_x^2}, \quad W_y = \frac{1}{Ch_y^2}, \quad W_z = \frac{1}{Ch_z^2}$$

and $\rho = \rho(x_0, y_0, z_0)$

For cylindrical co-ordinates with rotational symmetry

$$\emptyset_0 = W_x(\emptyset_1 + \emptyset_2) + W_y(\emptyset_3 + \emptyset_4) + \frac{W_r}{R}(\emptyset_4 - \emptyset_3) - \rho/C$$

where
$$C = \frac{2}{h_x^2} + \frac{2}{h_y^2}, \quad W_x = \frac{1}{Ch_x^2}, \quad W_y = \frac{1}{Ch_y^2}$$

$$W_r = \frac{1}{2Ch_y^2} = \frac{W_y}{2}$$

and axis of symmetry is along x axis and radius R is along y axis, while for R = 0 (on axis)

$$\emptyset_0 = W_x(\emptyset_1 + \emptyset_2) + W_y \emptyset_4 - \rho/C$$

$$C = \frac{2}{h_x^2} + \frac{4}{h_y^2}, \quad W_x = \frac{1}{Ch_x^2}, \quad W_y = \frac{4}{Ch_y^2}$$

$$\rho = \rho(x_0, y_0) = \rho(z, R)$$

3. Boundaries

RELAX3D handles three types of boundaries. The first type is a DIRICHLET boundary, that is a boundary (usually metal) on which the potential ∅ is known (and hence fixed). These boundary points are always specified as negative in the user written subroutine BND (describing the user problem geometry) and consequently will not be allowed to change during the relaxation procedure.

The second type is a NEUMANN boundary. All or some of the points on the faces of the 3 dimensional box (cylinder) can be of this type (default). The charge density, permittivity (where applicable) and potential on adjacent points orthogonal to such a symmetry plane point (one inside the box and the other a "fictitious" one outside the box) are assumed to be equal and hence the normal gradients of charge density, permittivity and potential are zero for these symmetry boundary points. Thus equipotential lines always cross perpendicular to NEUMANN boundaries.

The third type is a dielectric boundary between two media of different permittivity (currently restricted to 2D problems). Although dielectric boundaries must be restricted to lie on the grid points their precise position is not very important to the fields a few mesh points away, only the existence of the boundary having the dominant effect.

Although restricting the boundaries to lie on the grid (mesh) points limits the resolution with which one can describe a particular geometry (given the maximum number of allowed grid points) this greatly simplifies the program structure and consequently its execution speed.

4) Rate of Convergence

The rate of convergence to where the sum over all mesh nodes of the absolute fractional value change is reduced below a user specified tolerance depends strongly on the over-relaxation factor and types of boundaries describing the geometry, Dirichlet boundaries being much better than Neumann type. A lot of time can be saved by making some initial guess, no matter how crude, of the potentials at each mesh point which is free to vary.

5) Variable Mesh Size

To speed up convergence the initial iteration can be done on a coarse mesh using the "REDU" command. Expansion to a finer mesh ("EXPA") by doubling the number of mesh points in each dimensions is simply accomplished by linear interpolation (this being of the same accuracy as the difference equations). This process is continued until a sufficiently fine mesh is attained (the initial specified mesh) over the region of particular interest. To improve convergence successive sweeps over the volume are performed in alternate directions.

The intrinsic error is proportional to the second power of the grid spacing. Where the solution of one problem can be used as the starting point for a similar problem much computational time can be saved.

A "typical" Laplace problem requires a few hundred sweeps over the volume to converge to <1% error (although in some cases involving largely Neumann boundary conditions or non-zero ρ several thousand sweeps may be required). For a 20,000 point grid a single sweep over all points takes about 1.0 sec on the VAX/780. The CPU time for other problem sizes will be in direct proportion.

6) <u>Friendly Features</u>

   Some friendly features built into RELAX3D are:
   a) Extensive HELP facility.
   b) Reprompts on bad user data making program virtually "bomb" proof.
   c) Ability to abort to main menu, e.g. contour plotting may be terminated at any stage.
   d) Display or write to a file the status of a node, a line of nodes, or a box of nodes. This feature is particularly useful in monitoring convergence in critical regions.
   e) Single command (ITER) to automatically converge to a user specified residue.
   f) Save or restore the problem so as to be able to continue at some later date
   g) User "hooks" to BND routine for more sophisticated control of boundaries (e.g. explore multiple geometries).

7) <u>Examples</u>

   The following is an example of the "textbook" Poisson equation $\nabla^2 \phi = xze^y$ in a 3 dim. region with Dirichlet boundary conditions that $\phi = xze^y$ at the edges of the region.

   The listing of the corresponding BND subroutine is shown below.

```
              SUBROUTINE BND2 (I,J,K,PHI,RHO,LL)
CCCCCCCCCCCCCCCCCCCCCCCCCCCCCCCCCCCCCCCCCCCCCCCCCCCCCCCCCCCCCCCCCCC
C       BND SUBROUTINE FOR RELAX3D                                 C
C                                                                  C
C       TO SOLVE THE EQUATION DEL**2(PHI)=X*Z*EXP(Y)               C
C                                                                  C
C       WITH PHI=X*Z*EXP(Y) ON THE BOUNDARIES.                     C
C                                                                  C
C       EXACT SOLUTION IS PHI=X*Z*EXP(Y)                           C
CCCCCCCCCCCCCCCCCCCCCCCCCCCCCCCCCCCCCCCCCCCCCCCCCCCCCCCCCCCCCCCCCCC

C==Initial values for interior of region:
        PHI=1.0
```

```
C==Dimensions of problem
          IMAX=21
          JMAX=21
          KMAX=21
C==Grid spacing:
          HX=.100000
          HY=.200000
          HZ=.400000
          X=(FLOAT(I)-1.0)*HX
          Y=(FLOAT(J)-1.0)*HY
          Z=(FLOAT(K)-1.0)*HZ
C
          RHO=X*Z*EXP(Y)
          IF(I.EQ.1.OR.J.EQ.1.OR.K.EQ.1.OR.
     #    I.EQ.IMAX. OR. J.EQ.JMAX .OR. K.EQ.KMAX) PHI=-(X*Z*EXP(Y))
C====Freeze zeros at boundary by setting small negative:
          IF(PHI.EQ.0.0)PHI=-1.0E-25
C==Exact solution:
      IF(LL.EQ.1)PHI=X*Z*EXP(Y))
      RETURN
      END
```

A simple parallel plates capacitor (infinitely repetitive) having voltage on the facing plates of +1.0 and -1.0 volts would have median plane symmetry so we need only examine the upper half. Actually we need only look at one quarter, but to test program symmetry we include entire upper half.

The user coded routine describing this geometry would be

```
          SUBROUTINE BND (I,J,K,PHI,RHO,LL)
          PHI=0.5
          IF (K .EQ. 1) PHI=-.000001   (freeze median plane)
          IF( (I.LT.12).OR.(I.GT.22).OR.(J.LT.12).OR.(J.GT.22) ) RETURN
    C   set voltage on a plate
          PHI=-1.
          RETURN
          END
```

The commands to the program to "solve" this problem and contour plot the J=16 slice were

```
          33 33 21         (specify problem size)
          OPT 1            (specify this is a Laplace problem)
          1 1 1            (specify hx=hy=hz=1)
          LOAD             (load the object module describing geometry)
          INIT             (initialize problem by calling BND)
          REDU             (reduce to coarse grid)
          ITER             (relax coarse grid)
          EXPA             (expand to fine grid)
          ITER             (relax fine solution)
          PLOT             (plot contours, desired list follows)
          16 J             (specify which slice)
          0                (terminate slice requests)
```

The equipotentials plot shown in Fig. 2 would then be produced.

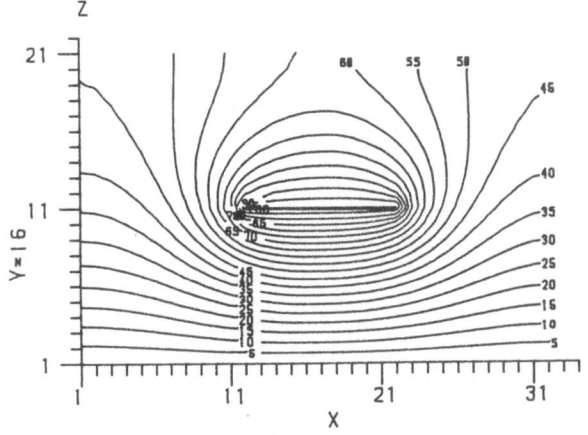

Fig. 2

	3D Cartesian	2D Cartesian	2D Cylindrical
LAPLACE	Yes	Yes	Yes
LAPLACE Dielectric	No	Yes	No
POISSON	Yes	Yes	Yes

TABLE 1   RELAX3D Capabilities

References

1. K.J. Binns, P.J. Lawrenson, "Analysis and Computation of Electric and Magnetic Field Problems", Chap.11,
2. Leon Lapidus, George F. Pinder, "Numerical Solution of Partial Differential Equation in Science and Engineering", Chap. 5, John Wiley & Sons (1982).
3. H. Houtman, TRIUMF Design Note (unpublished) TRI-DN-73-22 (1973).
4. D.L. Nelson, "Relaxation Code", University of Maryland, Technical Report 960 (1969).
5. R.J. Louis "A Guide to Relaxing at U.B.C.", TRIUMF Design Note (unpublished) TR-DN-69-13 (1970).
6. C.F. Gauss, Brief an Gerling, Werke 9, 278-281 (December 26, 1823).

Acknowledgment

The authors would like to thank Fred Jones for making significant upgrades to the program.

CALCULATION OF THREE DIMENSIONAL ELECTRIC FIELDS BY SUCCESSIVE
OVER-RELAXATION IN THE CENTRAL REGION OF A CYCLOTRON

S. Oh, R. Pogson and M. Yoon

University of Manitoba Cyclotron Laboratory,
Winnipeg, Manitoba, R3T 2N2, Canada

Summary

Successive Over-Relaxation has been used to calculate the detailed three dimensional electric field in the central region of the University of Manitoba Cyclotron. The potential near the dee tips was represented by a regular array of 256 X 128 X 13 in recently designed dee tips to accelerate $\overline{H}$ ions. A size of 0.25 millimeter mesh was used near the injection point to avoid time consuming interpolation near the boundary. Considerable reduction in computing time was obtained by assigning integer rather than floating point values of potential. The program is currently running on the VAX-11/750 computer.

Introduction

The simulation of the motion of the charged particles during the first several turns requires detailed knowledge of the electric and the magnetic fields in the central region of the cyclotron. The magnetic field can be obtained by measurement. The electric field, however, can not be calculated analytically, since it depends on the complicated geometry of the electrodes. Measuring electric fields, by mapping in an electrolytic tank, requires a model with a complicated electrode structure and a large number of manual measurements. Varying the design of the electrodes would be expensive and time-consuming. Moreover, random errors can occur during the measurements. The numerical solution of the Laplace equation by means of a digital computer avoids these difficulties. We developed a package of FORTRAN programmes to obtain the three dimensional electric fields in the central region of the University of Manitoba Cyclotron. With this method, specifying a three dimensional boundary numerically, there are difficulties such as, estimating accuracy, long computational times and obtaining the memory space needed to store the hundreds of thousands of potential values. The laboratory minicomputer permitted us to overcome most of these difficulties by providing a virtual memory computer architecture, 32 bit words and a speed of approximately one million instructions per second.

Relaxation method

Numerical solution of the Laplace equation can be obtained by changing the differential equation into the difference equation. By averaging each potential of the surrounding points, the potential at the point (i,j,k) is given by

$$V_{ijk} = 1/6 \, (V_{i-1} + V_{i+1} + V_{j-1} + V_{j+1} + V_{k-1} + V_{k+1})$$

where the higher order terms are neglected. The successive over-relaxation formula[1] can be described by the iterative sequence

$$V_{ijk}^{n+1} = V_{ijk}^{n} + C \, (1/6 \, (V_{i-1}^{n+1} + V_{i+1}^{n} + V_{j-1}^{n+1} + V_{j+1}^{n} + V_{k-1}^{n+1} + V_{k+1}^{n}) - V_{ijk}^{n})$$

where V_{ijk}^{n} is the nth calculated value of the potential at the point (i,j,k) and C is a constant which controls the convergence rate. To get the most rapid convergence, the value of C was taken 1.5 for the three dimensional case compared with the two dimensional case which is 1.9^2. With C=1.5 the coeffieicients become extremely well suited to the computation with a binary computer, division by 2 being simply a single bit position shift. C=1.5 gives

$$V_{ijk}^{n+1} = V_{ijk}^{n} + 1/2 \, (1/2 \, (V_{i-1}^{n+1} + V_{i+1}^{n} + V_{j-1}^{n+1} + V_{j+1}^{n} + V_{k-1}^{n+1} + V_{k+1}^{n}) - 3V_{ijk}^{n})$$

$$= 1/2 \, (1/2 \, (V_{i-1}^{n+1} + V_{i+1}^{n} + V_{j-1}^{n+1} + V_{j+1}^{n} + V_{k-1}^{n+1} + V_{k+1}^{n}) - V_{ijk}^{n})$$

Structure of the Program

The relaxation program consists of five main programs as described below. In addition, several specialized subprograms are incorporated; namely, an assembler language subroutine RLXFOR, a user supplied subroutine BOUND which describes the geometry of the boundary, and subroutines GET and PUT which read and write the data.

1. VINIT

This is the program to initialize all the potential array. Outer surfaces of the array are set up by Dirichlet and Neumann conditions. All non-boundary mesh points are arbitrarily assigned to a half of the maximum potential. If the mesh point lies in the boundary, as determined by calling the subroutine BOUND, then its potential value is marked by a set sign bit, and it is not changed during the relaxation process. The results are put into the subroutine PUT.

2. RLXCRNCH

This program first calls the subroutine GET to read the initial potential array. Then, subroutine RLXFOR is called repeatedly to do a three dimensional relaxation. This routine is very frequently executed and uses about 30 percent less time when coded in assembler language. The relaxation coefficient is taken to be 1.5 for convenience of coding with integer operations. A forward and a reverse pass are made at each call to RLXFOR so that the corrections near the boundary points propagate to the whole region rapidly. All the boundary points are indicated by a set sign bit, and are not altered during iterations.

3. EXPAND

This program makes a finer mesh over the potential in the region of interest close to the injection point. The original volume is reduced by a factor of eight, and the array dimensions are kept constant. Subroutine BOUND is called so that boundary points are re-initialized. Two-dimensional relaxation method is applied by calling subroutine RLX2D over the outermost surfaces of array except the median plane. Subroutine RLX2D is used on relaxations over planes not in contiguous memory locations by copying the desired plane to a temporary two dimensional array. This minimizes the paging in a virtual memory system.

4. LAYER

In the initial study of the beam orbit dynamics only the linear portion of the particle's vertical motion is of interest. Thus, keeping the data over the whole layers in the disk are not always necessary for computer simulation of the motion of the particle. To conserve disk storage, this program copies the potential values of selected layers from the whole array. The user is prompted to supply the number of layers copied. Usually the two layers consisting of the median plane and the one above it (k=2) are copied.

5. VLIST

The user should be able to confirm the results of the relaxation. This program generates a crude contour plot by expressing the potential at each point as a digit number or character corresponding to its potential value. The user has the choice whether only the potentials of the boundary points are displayed or those of both the boundary and the non-boundary points are displayed.

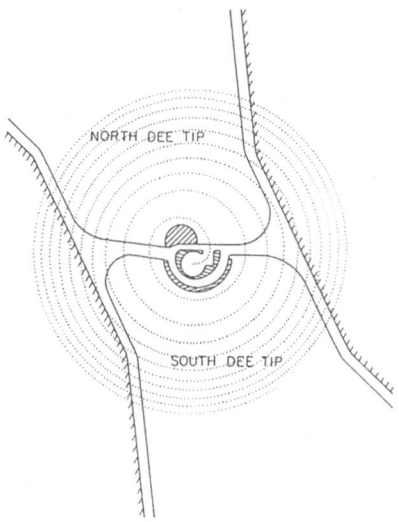

Fig. 1

Two dimensional geometry of the cyclotron central region on the median plane

Computational Details

To check the accuracy and the convergence, a test program which calculates the potential by concentric cylindrical shells was made. Since the values of the potential for this geometry can be obtained analytically, we can easily compare the relaxed value with the analytic one. Without changing the relaxation coefficient, the program calculates the average of the residuals and the average of the difference between the analytic and the relaxed value per each iteration. The accuracy was 10^{-1} % approximately when small mesh was used. Following table shows the average change per iteration.

	Two mm	One mm	Half mm	Quarter mm
Average Change	10^{-4}	10^{-5}	10^{-6}	10^{-6}

At the early stage of the study, the starting mesh size was 2 mm, which resulted in a dimension of K to be at least 20 to cover the whole boundary points in k direction. With this dimension of k and the starting mesh, the total time taken for relaxation and expansion up to the final mesh was about 12 hours CPU time. By taking 4 mm as the starting mesh size, and taking 13 as a dimension of K, the CPU time could be decreased down to around 6 hours. This number was taken so as to cover the whole vertical opening of the particle's first turn by the smallest meshes. Larger mesh size(such as 2 mm) is acceptable at a larger radius from the center of the cyclotron because the variation in the electric field becomes proportionately gradual with radius and also the effect of this electric field on the vertical focusing of the charged particles rapidly diminishes with radius. The two dimensional geometry on the median plane is given in Fig. 1. And the two dimensional vertical geometry for x= -29.5 mm is shown in Fig. 2. This plane includes the first accelerating gap. As can be seen in the figure, the quarter millimeter mesh points cover the whole vertical gap of the ground shield and the rf dee structure except the space between them and above z= 3 mm. The z component of electric field variation along the x= -29.5 mm and z= 1 mm line is shown in Fig. 3, which compares the two mm mesh with the quarter mm mesh.

Page faulting and computing time became excessive on the VAX-11/750, when the relaxation method was introduced. The enormous paging comes from the three dimensional iterations over the hundreds of thousands of mesh points which will not fit in one user's memory allotment. This makes the CPU time usage quite inefficient. Recently, the program was changed to test direct input and output access to minimize the paging though more time is used in read and write procedures in this case. Computing time was decreased by using integer rather than floating point arithmetic for the value of the potential at each mesh point. Integer arithmetic is four or five times faster than floating point arithmetic.

Since this is the relaxation with a finite mesh size, the boundary can not be fitted into the mesh points exactly. As mensioned above, the vertical dimension of the boundary is much smaller than the horizontal dimensions, thus, taking fine meshes in the region closer to the injection point rather than doing interpolation[3], can be regarded as the better means of the boundary correction. The systematic error involved in the method is proportional to the fourth power of the mesh size. Though the intrinsic error could be decreased by taking the weighted average[4] rather than the simple average, it is questionable that it would compensate for the additional computing time needed.

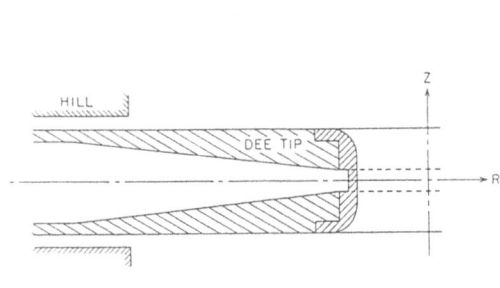
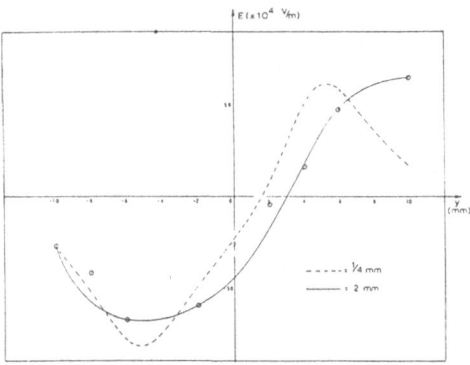

Fig. 2
Two dimensional geometry of the cyclotron median plane

Fig. 3
Electric field variation along the line perpendicular to the dee gap

The result of the relaxation is used for the computer simulation of the charged particle's motion in the cyclotron. Details of this program will be presented in future conference[5].

References

1. R. Louis, Ph.D. Thesis,TRIUMF,TRI-71-1,6-19, 1971
2. D. Nelson, H. Kim and M. Reiser, IEEE Trans. Nucl. Sci., NS-16, 766, 1969
3. Mao-bai Chen and D.A. Lind, IEEE Trans. Nucl. Sci., NS-28, 2636-2638, 1981
4. D. Nelson, Univ. of Maryland, Technical report No. 960, 1969
5. S. Oh et al., To be published, Tenth International conference on cyclotrons and their applications, 1984

THE DESIGN OF THE ACCELERATING CAVITY FOR SUSE WITH THE AID OF THE THREE-DIMENSIONAL CAVITY CALCULATION PROGRAM CAV3D

W. Wilhelm
Physik-Department der Technischen Universität München
D-8046 Garching, Germany

Abstract

SuSe is a project study at the Munich Accelerator Laboratory for a superconducting sector cyclotron to access the medium energy range up to 300 MeV/u for heavy ions. The RF-system for SuSe consists of two large cavities in opposite valleys. The cavities are driven in the TE 101-mode and have a radially increasing voltage distribution. The frequency can be tuned in a wide range between 59 MHz and 74 MHz by movable acceleration lips and two plunge tubes. It was a difficult task to find a cavity geometry that fits into the narrow place between the SuSe-magnets and has a high enough voltage at the injection-radius of o.4 m, a large frequency range and the best possible shunt impedance. A very valuable tool for optimizing the cavity geometry was the three-dimensional cavity calculation program CAV3D[2], together with model measurements. CAV3D calculates the frequency, voltage distribution, Q-factor and electromagnetic fields of low modes of arbitrarily formed three-dimensional cavities with an accuracy of several percent.

Introduction

Computer codes for RF cavities with constant cross section and, especially, cylindrical symmetry are used for a long time and established as a very fast and precise tool to calculate all characteristic data of such cavities. For the fully three dimensional case very few codes have been written up to now[1-3]. The main problems besides the enormous amount of memory space in the computer are edges in the cavity geometry, at which the field lines are sharply kinked. If such singularities are not properly handled, a drastic loss of accuracy is caused; therefore the code MAX3D[3] and, probably, also the code of Albani and Bernardi[1] can calculate only cavities with smooth boundaries without edges. In the code CAV3D[2], this problem has been overcome with the aid of a simple additional term in the difference equations for mesh points near edges. But the cavity geometry has to be approximated by a set of cubes, curved or oblique boundaries are stairlike approximated. Therefore the accuracy of CAV3D-results is only in the percent range, much worse than users of two-dimensional codes are accustomed. So, the designer of a fully three-dimensional cavity cannot do it completely without measurements on cavity models. But from the computer calculations he can get valuable hints what to do for a requested change in the cavity characteristics.

An improvement of the program CAV3D in comparison with the published version is the use of a penalty method. The eigenvalue problem is now

$$(\Delta + k^2 + s \cdot \text{grad div}) \underline{H} = 0,$$

where $k = \omega/c$ is the eigenvalue and s is a penalty parameter chosen between zero and 1.5. With this penalty term, solutions without physical meaning, i.e. div $\underline{H} \neq 0$, can be shifted away from the correct solution. This was very useful for the calculations of the SuSe-cavity, as a parasitic frequency was very close to the correct one without the penalty term, and the respective field solutions were mixed and not clean.

The RF-system for SuSe

SuSe is a project study at the Munich Accelerator Laboratory for a superconducting sector cyclotron as a booster for the 13 MV tandem to access the medium energy range up to 300 MeV/u for heavy ions[4]. The RF-system for SuSe consists of two large cavities in opposite valleys. The cavities have to provide a maximum accelerating voltage of 1 MV at the extraction radius of 2.40 m with a moderate power consumption, a frequency variability in the range between about 59 MHz and 74 MHz to accelerate all heavy ions to energies changeable within a factor of two at harmonic numbers between 5 and 16, and a radially increasing voltage distribution to get a phase compression. At the injection radius of 0.40 m, the voltage should be not lower than 20 percent of the maximum voltage to allow an injection system with an electrostatic deflector.

Fig. 1 shows a SuSe-cavity schematically. The shape is mainly determined by the space released by the SuSe-magnets. It is driven in the TE101-mode, so it has the wanted radially increasing voltage distribution. The frequency can be adjusted in a wide range by varying

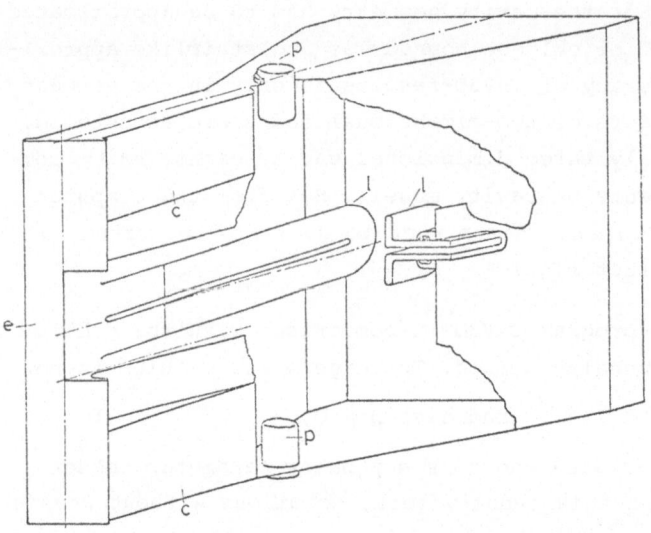

Fig. 1: SuSe-cavity drawn schematically. l-acceleration lips. p-plunge pistons for frequency fine tuning. e-ears and c-cheeks.

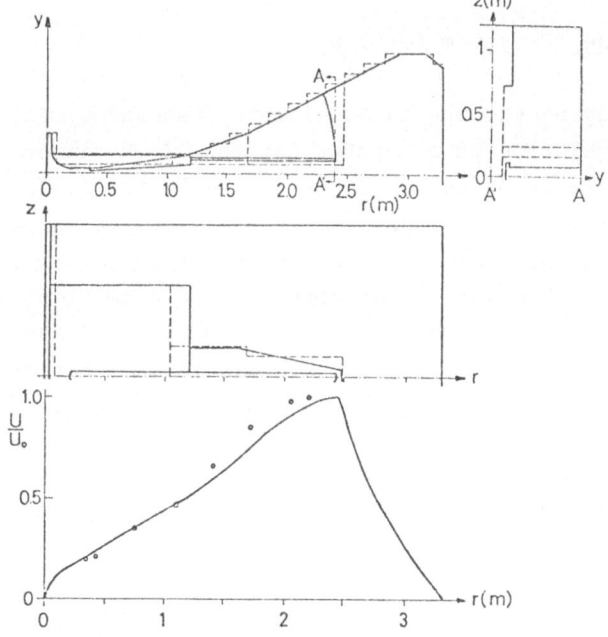

Fig.2a: Geometry of the SuSe-cavity with stair-like CAV3D-approximation (dashed). Calculated (drawn) and measured (circles) voltage distribution of the cavity with narrow acceleration gap.

the gap size between the capacitively loaded lips; the outer part of the acceleration lips is movable and can be turned around a vertical axis at r = 1.20 m (fig.1). The fine tuning is done with two plunge tubes at the top and the bottom of the cavity.

The frequency variation with the movable lips is very favourable for accelerating all sorts of heavy ions. With narrow acceleration gap the frequency is low and the transit time factor high, just as it is needed for the slower, heavier ions which are accelerated at high harmonic numbers. On the other hand, the cavity with wide acceleration gap is well adapted for the fast, light ions, which are accelerated at high frequencies and low harmonic numbers and which need the highest voltage.

As the acceleration gap is wide, nevertheless the electric field strength remains at moderate values.

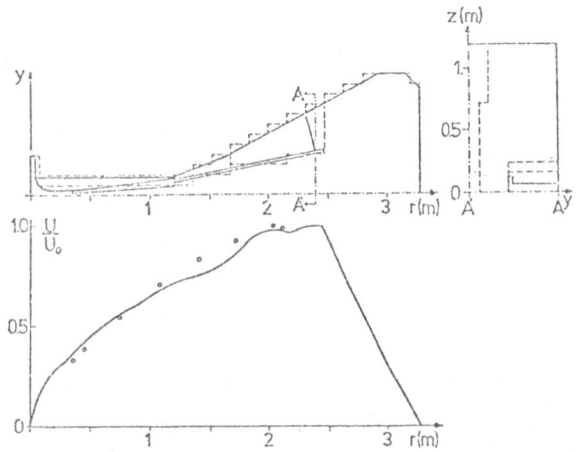

Fig. 2b: The same as fig. 2a for the cavity with wide acceleration gap.

The cavity design

The cavity geometry was optimized mainly with the aid of a small 1:8 scale model, which could be quickly changed, and CAV3D calculations. Fig. 2 shows the geometry and the stair-like CAV3D approximation of the final cavity version with narrow and wide acceleration gap, and the measured and calculated voltage distributions. A typical computer run for the SuSe-cavity with 3300 mesh points needs a computing time of 1100 CPU seconds on a Cyber 175. In table 1 the respective frequencies, Q-values and shunt impedances can be seen. The surfaces of this small cavity are not very good, so absolute measurements of the shunt impedance R_S are not possible. But the related quantity R_S/Q may be compared between

Table 1: Frequency ν, Q-factor Q, shunt impedance R_S and R_S/Q, as measured at the cavity model and calculated with CAV3D for narrow and wide acceleration gap. All values are scaled to the full size cavity. The measured Q and R_S are too low due to bad surfaces.

	CAV3D	Measurement
narrow gap	ν = 53.70 MHz Q = 21936 R_S = 3.01 MΩ R_S/Q = 137 Ω	ν = 60.2 MHz Q = 10160 R_S = 1.27 MΩ R_S/Q = 125 Ω
wide gap	ν = 65.0 MHz Q = 19832 R_S = 2.14 MΩ R_S/Q = 108 Ω	ν = 69.6 MHz Q = 7570 R_S = 0.68 MΩ R_S/Q = 90 Ω

calculation and measurement. The deviation between the measured and calculated numbers is in the order of ten percent, but the relative voltage distribution is well reproduced by the computer calculation.

The main problem was to get enough voltage at the injection, especially with narrow acceleration gap. Curve a in fig. 3 is the measured percentage U_{inj}/U_{max} versus the frequency of the first cavity version. This first cavity version had no "cheeks", smaller "ears" and capacitive plates at the acceleration lips as shown in fig. 1. The frequency range was large, but the injection voltage too low in the whole range. To enhance it, the cavity cross-section in the machine center was expanded into the neighbouring valleys as far as possible with larger "ears", and the cross-section in vertical direction was not kept constant but widened 0.7 m above and below the particle plane at the narrow region with "cheeks". Furthermore, the capacitive plates at the acceleration lips were made smaller with increasing radius (fig. 2a).

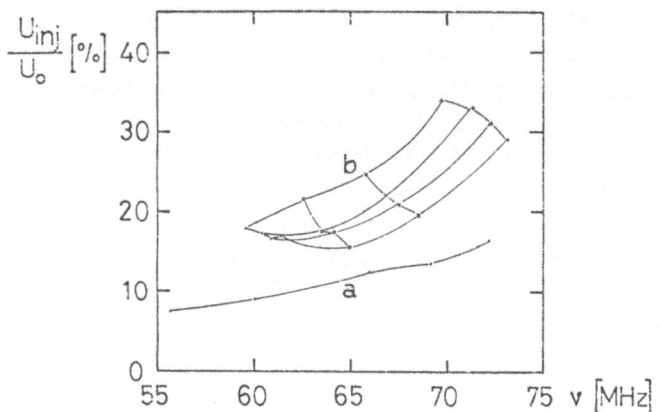

Fig. 3: Measured percentage U_{inj}/U_{max} versus the frequency.
a) First cavity version with small ears, large capacitive plates and without cheeks.
b) Final cavity version. The four curves respect to several positions of the plungers, between 5 and 50 cm wide into the full size cavity.

The curves b in fig. 3 show the resulting U_{inj}/U_{max} versus the frequency for several positions of the plunge tubes. Through the "cheeks" and "ears" more magnetic flux is fed into the machine center, so 18 percent of the maximum voltage are reached at the injection radius at the lowest frequency of 59 MHz. Inserting the plunge pistons lowers the injection voltage and raises the frequency, whereas the shunt impedance remains fairly constant over the whole frequency range.

From CAV3D calculations also quantities can be obtained, which are very difficult to measure with cavity models, especially the current - and power loss - distribution at the walls. In fig. 4 the calculated current distribution is indicated with arrows. Such maps are very valuable for the design of the welding seams and the water cooling.

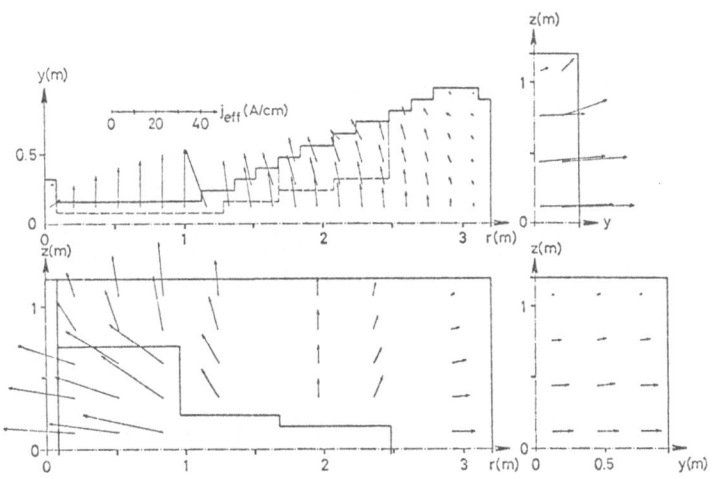

Fig. 4: Calculated current distribution on the cavity walls for a total power consumption of 100 kW.

Work supported by the Federal Government, BMFT.

References

1. M. Albani and P. Bernardi, IEEE Transactions on Microwave Theory and Techniques 22, 446 (1974).
2. W. Wilhelm, Particle Accelerators 12, 139 (1982).
3. M. Hara, T. Wada, T. Fukasawa and F. Kikuchi, IPCR Wako Saitama 351 Japan, program MAX3D (1983).
4. W. Schott, E. Zech, contribution A2/15 to this conference.

THE FURTHER DEVELOPMENT
OF THE CALCULATION OF THE THREE DIMENSIONAL ELECTRIC FIELD
IN THE CENTRAL REGION OF THE INR CYCLOTRON

Mao-bai Chen and Wen-bin Sen

Institute of Nuclear Research, Shanghai, China

Abstract

A number of special problems with general sense are given arising from the calculation of the three dimensional electric field in the central region of the INR cyclotron.

INTRODUCTION   In the previous paper[1] a simpler geometric central region of the Colorado cyclotron was calculated. Recently, a more complicated configulation for the INR cyclotron was computed with 20° rotation of the ion source-puller position with respect to the dee-dummy dee gap line. A number of special problems arising from such configuration were to be solved in the calculation of the three dimensional electric field. This paper will give a brief discussion about what wasn't mentioned in the previous paper and has been developed since then.

SELECTION OF COORDINATES   The origin of the mesh coordinate is usually set at the center of the machine to coincide with the one of the magnetic field. With the rotation of the ion source having rectangular periphery, a large number of boundary smoothing corrections would be involved in such coordinate. Then a so called " Ion Source Coordinate (X,Y) " is adopted to keep a large part of the ion source-puller boundary away from the heavy correction, the origin of this coordinate is located at the center of the exit slit of the ion source as shown in Fig 1. Two additional " Chamber Coordinate (XV,YV) " and " Dee Coordinate (XX,YY) " are also introduced for the convenience of positioning the relaxed mesh points relative to the dee structure, their origins are both located at the center of the machine. Through the transformation among these coordinates, the complex calculation is thus simplified; the program is shortten and much calculation time is saved.

ENLARGEMENT OF THE CALCULABLE REGION   Results from our calculated map[2]

indicated that for our geometry at a distance inside the dee or dummy dee
of about 1.2d and 1.5d, the potential was about 99 percent and 99.9 percent
of a constant value respectively. Accordingly, the dee and dummy dee should both be truncated at 2d, from beyond it the exact uniform field region
is reached. However, after adopting the " Ion Source Coordinate ", the
practical calculable region, along abscissa Y (Fig 1), is restricted by
the above requirement. The limited region with 1 mm mesh size is too small
to proceed to the centeral region research. Therefore, an enlarged field
region using 4 mm mesh size is first calculated. Subsequently, a smaller
region around ion source-puller is re-calculated using 1 mm mesh size and
taking the map value of the calculated large field region at appropriate
points, where it is assumed that the potential is not sensitive to the
trivial geometric change of the ion source-puller, as the specified boundary value of small field region being calculated. In so doing, not only
the calculable region is enlarged, also around the ion source-puller region, where the field is important for the central region study, a more
accurate field map is guaranteed by using a fine mesh size.

DEFINITION OF THE BOUNDARY VALUE   Potentials are specified for all boundaries. As mentioned earlier[1], one can specify an artificial appropriate
boundary either with a given potential as is the case described above,
or with an uniform field assumption. For the boundaries at both ends of
abscissa Y (Fig 1), it is assumed that the equi-potential lines are parallel to the dee-dummy dee gap line and the potential at U'(1) is equal
to U(1) instead of U"(1). (Fig 2)

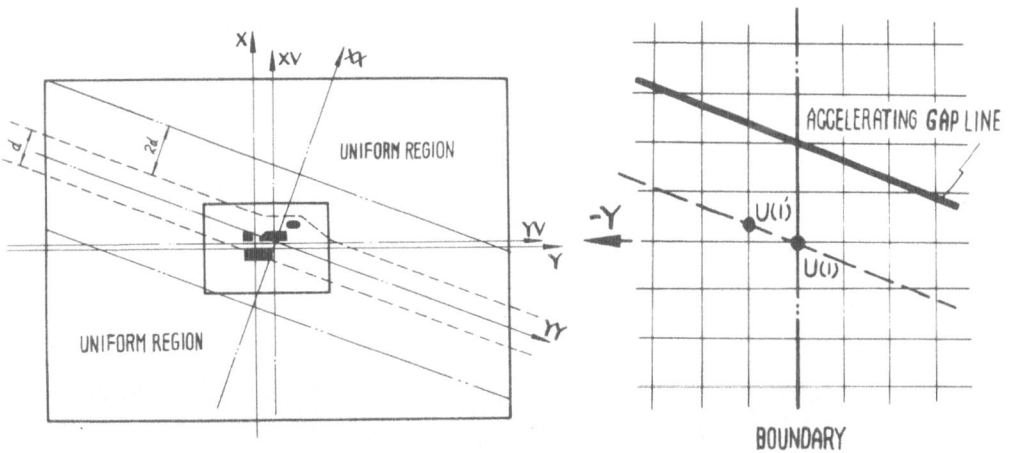

Fig. 1   Fig. 2

GENERAL BOUNDARY SMOOTHING CORRECTION COEFFICIENTS The most trouble problems involved in the relaxation method are the " boundary smoothing correction " and the " higher order correction " required for the precise result. Here the general boundary smoothing correction coefficients suitable to any mesh size are formulated corresponding to the three different methods[1]. The relation

$$\frac{\partial^2 V}{\partial x^2}+\frac{\partial^2 V}{\partial y^2}+\frac{\partial^2 V}{\partial z^2} = A_x V_a + A_y V_b + A_z V_c + B_x V_d + B_y V_e + B_z V_f - (C_x + C_y + C_z) V_o \qquad (1)$$

is the finite difference three dimensional equation for Laplaces equation with boundary smoothing correction coefficients:

a). Southwell's method sets:

$$A_x = \frac{hH}{Si} \ ; \qquad A_y = \frac{hH}{Sj} \ ; \qquad A_z = \frac{hH}{Sk} \ ;$$

$$B_x = 1.0H; \qquad B_y = 1.0H; \qquad B_z = 1.0H;$$

$$C_x = H + A_x \ ; \qquad C_y = H + A_y \ ; \qquad C_z = H + A_z \ ;$$

of which h is the mesh size and Si, Sj, Sk are the fraction of the mesh size (Fig 3). ($H = 1/h^2$).

b). Fox's method sets:

$$A_x = \frac{2}{Si(Si+h)} \ ; \qquad A_y = \frac{2}{Sj(Sj+h)} \ ; \qquad A_z = \frac{2}{Sk(Sk+h)} \ ;$$

$$B_x = \frac{2}{h(Si+h)} \ ; \qquad B_y = \frac{2}{h(Sj+h)} \ ; \qquad B_z = \frac{2}{h(Sk+h)} \ ;$$

$$C_x = \frac{2}{Sih} \ ; \qquad C_y = \frac{2}{Sjh} \ ; \qquad C_z = \frac{2}{Skh} \ ;$$

c). In our modified method[1] for which we set:

$$A_x = \frac{1}{Si^2} \ ; \qquad A_y = \frac{1}{Sj^2} \ ; \qquad A_z = \frac{1}{Sk^2} \ ;$$

$$B_x = \frac{1}{hSi} \ ; \qquad B_y = \frac{1}{hSj} \ ; \qquad B_z = \frac{1}{hSk} \ ;$$

$$C_x = A_x + B_x \ ; \qquad C_y = A_y + B_y \ ; \qquad C_z = A_z + B_z \ ;$$

In the program whichever method is available. However, the third method is preferred, since this method yielded the best results[1] yet needn't have higher order correction, which makes it possible for a small computer such as PDP-11 to be applied in the calculation of the three dimensional field, because more memories for data storage and much time for reading-writting data storaged are substantially saved.

GENERAL FORMULATION FOR DIFFERENT MESH SIZE CORRECTION   By the method c), the general finite-difference equation for Laplaces equation is derived with different mesh size correction at points encompassed by different mesh size: (only one dimension is shown)   (Fig 3(d))

$$\frac{\partial^2 V}{\partial z^2} = \frac{Va}{h^2} + \frac{Vb}{hSk} - (\frac{1}{h^2} + \frac{1}{hSk})Vo \qquad (2)$$

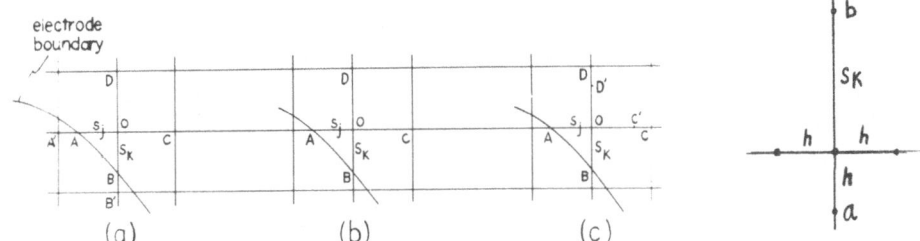

Fig 3   BOUNDARY MESH POINT INTERPOLATION SCHEMES   Fig 3(d)

DETERMINATION OF THE RELAXATION FACTOR   In the relaxation method, the value of Vo at the relaxed point is corrected by the value of fR/6. R is the residual value and f is a relaxation factor which determines the convergent rate. However, our calculation with above correction indicated that f value varied from 0.04 to 1.4 depending on the special situation. It is impossible to make the trial to get an appropriate f in the tedious three dimensional calculation. We, therefore, use fR/C instead of fR/6, where C= Cx+Cy+Cz is the coefficient of the Vo in expression (1), and f can then be fixed at value of 1.4 or 1.5 without overflow occurred in running code.

JUDGEMENT OF THE RELAXATION RESULT   The best way for the judgement of the relaxation result is the magnitude of the residual value R, which is defined as the value of the right side in equation (1). According to the physical meaning of the residual R,[1] the residual R must be zero at all non-boundary mesh points through relaxation to meet Laplaces equation. The smaller the residual R is, the more accurate the results will be. Nevertheless, such concept was sometime confused by the fact that for a three dimensional central region the maximum residual R value among all mesh points after completing the calculation swung from 10^{+1} to 10^{+3} other than approaching to zero value. It is attributed to those mesh points, which

are closely adjacent to the boundary with distance Si or Sj or Sk much
smaller than 1, thus the corresponding coefficients of equation (1) become
much bigger than 1. Consequently, the unnormal R value is resulted even
for the resonably small potential data error. Therefore, a code called
CHANGE was written to serve as a judgement mean as discribed earlier[1].

IMPROVEMENT OF THE PROGRAMS   The codes for the calculation of the three
dimensional electric field have been extended and improved recently. It
consists of thirteen sub-programs with versatile function. The flow chart
of the whole calculation process is shown in Fig 4. One iteration calculation should take about 5 minutes on the computer PDP-11/70 and one complete calculation should proceed about 300-400 iterations.

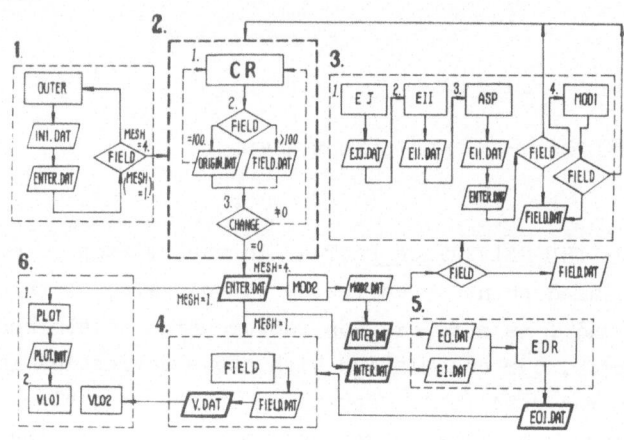

Fig 4

RESULTS   The distribution of the calculated equi-potential lines around
the ion source-puller region at both vertical plane and horizental plane
are shown in Fig 5 and Fig 6 respectively. The obtained three dimensional
field map has been used in the orbit programming for the remodelling of
the INR cyclotron[3,4].

Acknowledgement
We are indebted to Dr. Sen-lin Xu for his helpful discussion.

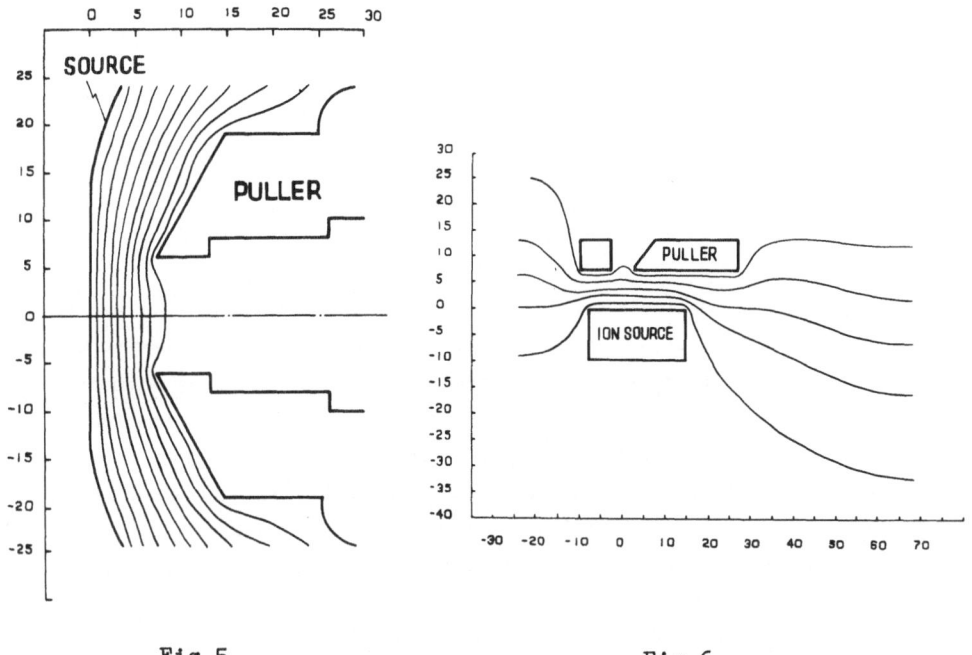

Fig 5   Fig 6

Reference

1. Mao-bai Chen and D. Lind   I.E.E.E. Trans NS-28 (1981), 2636
2. Mao-bai Chen   CU Internal Report (1981)
3. Mao-bai Chen and Sen-lin Xu   The Proc. of 2nd China-Japan Symposium on Accelerators (1983)
4. Mao-bai Chen, Sen-lin Xu and Wen-bin Sen   This Conference

PARTICLE TRACKING USING LIE ALGEBRAIC METHODS*

Alex J. Dragt
University of Maryland, College Park, Maryland 20742

David R. Douglas
University of California, Berkeley, California 94720

I. Introduction

A study of the nonlinear stability of orbits in an accelerator or storage ring lattice usually includes particle tracking simulations. Programs such as PATRICIA[1] typically trace rays through linear and nonlinear lattice elements by numerically evaluating linear matrix or impulsive nonlinear transformations. For large lattices with many nonlinear elements, this can be time consuming because of the necessity of evaluating a separate transformation for each group of linear elements and every individual nonlinear element. Moreover, the validity of the impulsive approximation for nonlinear elements can be questioned. Finally, "linear" elements are really not linear.

An alternate approach is to trace rays using the first and second order information about trajectories through beamline elements as provided, for example, by the second order matrix code TRANSPORT[2]. DIMAT[3] works this way. Such an approach has the advantage of being exact through second order. Moreover, it offers the possibility of trying to represent the effect of several elements by a single transfer map thereby increasing the speed of tracking computations. However, because the second order matrix transformations for single elements and collections of elements are in fact generally only correct through second order, they are generally not symplectic (canonical) beyond second order. The errors due to third and higher order violations of the symplectic condition, which condition must hold for Hamiltonian systems, may in some cases build up over long tracking runs, and may, if not properly understood, lead to improper conclusions.

It should be emphasized that the failure to be symplectic beyond second order is not due to any error in second order codes. Rather, it is simply a consequence of the fact that the symplectic condition generally requires the retention of terms of arbitrarily high order if it is to be satisfied.[4]

II. The Program MARYLIE

The purpose of this paper is to describe a third approach and its realization in MARYLIE, a Fortran-language beam transport and tracking code developed at the

---

*Work supported in part by the U.S. Department of Energy under Contract Nos. DE-AC03-76SF00098 and DE-AS05-80ER10666.A001.

University of Maryland.[5] MARYLIE employs algorithms based on a Lie algebraic formalism, and is designed to compute transfer maps for and trace rays through single or multiple beam-line elements. This is done <u>without</u> the use of numerical integration or traditional matrix methods; currently all nonlinearities (including chromatic effects) through third (octupole) order are included. Thus, MARYLIE includes effects one order higher than those usually handled by existing matrix-based programs. Moreover, if desired, MARYLIE can be run in a mode which maintains the symplectic condition to all orders.

Presently, the following idealized beam-line elements are described by MARYLIE:

drifts	magnetic sextupoles
normal-entry and parallel-faced dipole bends	magnetic octupoles
hard-edged fringe fields for dipoles	electrostatic octupoles
magnetic quadrupoles	axial rotations
hard-edged fringe fields for quadrupoles	radio frequency bunchers

In addition, there are provisions for user specified transfer maps (through nonlinear terms of degree 3) that make it possible to handle real fringe fields and non-ideal magnets.

A beam transport or tracking code must compute and manipulate some representation of the transfer map describing particle trajectories through beam lines. Matrix codes, for example, do this by computing coefficients in the Taylor series expansion of the transfer map for a beam line. MARYLIE deals directly with the nonlinear transfer map itself; it computes the following approximate representation for the transfer map,

$$M = \exp(:f_2:) \exp(:f_3:) \exp(:f_4:) \ . \tag{1}$$

The lowest order part of any transfer map is described by the factor $\exp(:f_2:)$. In MARYLIE (as in matrix codes) this portion is represented by a real 6x6 matrix. The nonlinear behavior of a transfer map is specified (through third order) by the polynomials f_3 and f_4.

When operated in a transport mode, MARYLIE will provide the user with the linear matrix and polynomials f_3 and f_4 for any beamline element or collection of beamline elements. Alternatively, if desired, the code will provide second and third order transfer matrices.[6] Also, upon request, MARYLIE will provide tunes, first and second order chromaticities, first, second, and third order eta functions, etc. At present, MARYLIE has no provisions for fitting.

When operated in a tracking mode, MARYLIE will perform ray traces which are correct through first, second, or third order. If desired, with a slight increase in computation time, MARYLIE will perform ray traces which are correct through third order and symplectic through all orders. Tracking may be carried out either element by element, or through various collections of elements with each collection treated as a lumped element. This flexibility in types of ray traces and in collections of elements to be lumped together facilitates the study of tracking errors, the importance of "cross coupling" between nonlinearities, and the importance of maintaining the symplectic condition.[7]

In summary, MARYLIE contains remedies for deficiencies which can be present in other tracking procedures. First, if desired, it is completely symplectic. No spurious damping or growth of emittance (phase space) will then occur. Secondly, it incorporates the nonlinear aspects of the so-called linear elements (drifts, dipoles, and quadrupoles). Thirdly, it uses finite length rather than impulsive nonlinear elements. Finally, if desired, single transformations may be employed for groups of linear and nonlinear elements. When sufficiently accurate, this procedure can increase the speed of computation significantly.

III. Examples

A. Los Alamos LAMPF II

One lattice under consideration for a proposed fast cycling synchrotron at Los Alamos would accelerate protons from the LAMPF linac to 32 GeV. This lattice has 60 cells (each consisting of 4 combined-function bends and 4 sextupoles). One finds that trajectories in this lattice can be tracked without <u>obvious</u> difficulty using second order matrix methods provided the tracking is carried out element by element. If, however, single element maps are combined to compute the second order transfer map for an entire cell, and if this cell transfer map is used in second order tracking, the result exhibits a spurious growth in the single-particle phase space. Figure 1 illustrates this growth in the vertical phase space during 200 turns of cell-to-cell tracking.

This striking effect is purely an artifact of the tracking procedure. As explained earlier, it is due to the fact that second order transformations for single elements and collections of elements are, in general, only symplectic through second order. Indeed, one expects a close examination of second order tracking even when done element by element would also reveal a spurious (but much smaller scale) phase space behavior.

Figure 2 shows the result of tracking with MARYLIE for 48,000 turns. The initial conditions are the same as in Figure 1. The results of Figure 2 agree closely with those obtained from element by element tracking in second order. However, the MARYLIE tracking was carried out using a transfer map which lumped 120 cells (2 complete turns) together! Because MARYLIE is completely symplectic, no spurious growth of the vertical phase space is seen. Because it is third order, it is possible to track 2 complete turns per transformation (with an attendant increase in speed) without loss of relevant cross-couplings amongst elements.

We therefore conclude that there are cases for which it is very important to maintain the symplectic condition to higher than second order.

B. Prototype SSC Lattice

Preliminary studies have been made of orbits in a proposed prototype SSC lattice. The lattice contains 6 interaction points.[8] Each interaction point is surrounded by an insertion composed of dispersion suppressor cells, dipoles that produce crossing of the two beams, special quadrupoles to reduce the betatron functions at the interaction point, drifts, and assorted sextupoles. Each beam is transported between interaction regions (insertions) by a regular lattice consisting of 72 cells. Each such cell contains 2 quadrupoles (focussing and defocussing), 2 sextupoles, 12 dipoles, and various drifts between magnets.

For simplicity, we have devoted our initial studies to the behavior of particles in a storage ring composed entirely of those cells which make up the regular portion of the lattice. (A later study will include as well all the special elements surrounding the interaction point.) Figure 3 shows the results in horizontal phase space arising from a MARYLIE cell by cell tracking through 24,000 cells for a particular initial condition. Figure 4 shows the result of using MARYLIE to combine the effect of 139 cells into a __single__ map and then tracking through this "lumped element" 173 ($\simeq$ 24,000/139) times. Evidently, the results of figures 3 and 4 are in excellent agreement. This example illustrates that it may well be possible with MARYLIE to represent the effect of all the elements between any two interaction points in the SSC by a single map! (When insertions are included, the number of elements between interaction points is about the same as the number in 139 cells.) Thus, there appear to be circumstances where MARYLIE is considerably faster as well as more accurate than impulsive codes such as PATRICIA.

As a final example, figures 5 and 6 show a dynamic aperture study made for the regular part of the SSC lattice using MARYLIE. It was performed by combining 67 cells into a single map, and then tracking through this lumped element 24,000 times for a variety of initial conditions. The vertical scale is in radians, and the horizontal scale is in meters. The outer horiztonal phase space tracing has the same initial condition as that of figures 3 and 4, and illustrates the complicated behavior orbits exhibit when the betatron amplitude is not sufficiently small.

IV. References

1. Wiedemann, H., User's Guide for PATRICIA, Stanford Linear Accelerator Center, PTM-230 (February 1981).

2. Brown, K.L., SLAC-75, revision 3 (1975).

3. Servranckx. R., DIMAT, future SLAC PEP note.

4. Dragt, A.J. and D. Douglas, in Proceedings of the Workshop on Accelerator Orbit and Particle Tracking Programs, Brookhaven National Laboratory, BNL-31761 (1982).

5. Dragt, A.J., Lectures on Nonlinear Orbit Dynamics, A.I.P Conference Proceedings, 87 (1981); Douglas, D.R., Ph.D. Dissertation, University of Maryland, 1982 (unpublished); Dragt, A.J. and E. Forest, Computation of Nonlinear Behavior of Hamiltonian Systems Using Lie Algebraic Methods, to appear in J. Math. Phys., December 1983. Dragt, A.J., R. Ryne and D. Douglas, MARYLIE, A Program for Charged Particle Beam Transport Based on Lie Algebraic Methods (in preparation). Douglas, D.R. and A.J. Dragt, IEEE Trans. Nuc. Sci., NS-30, p. 2442 (1983). Douglas, D.R. and A.J. Dragt, Lie Algebraic Methods for Particle Tracking Calculations, to appear in proceedings of 12th International Conference on High Energy Accelerators (1983).

6. In all cases encountered to date, it is found that the second order transfer matrices as given by TRANSPORT and MARYLIE are in complete agreement. This is reassuring for the correctness of both programs, because their contents were derived independently by two completely different procedures.

7. It is widely known from experience with tracking simulations, and with the well-known programs TRANSPORT and TURTLE, that nonlinear lattice elements of a given order "cross-couple" to generate effects of even higher order. See, for example, the paper of K. Adams, IEEE Trans. Nuc. Sci., NS-30, p. 2436 (1983). MARYLIE presently truncates products of Lie transformations to the form (1). Hence, cross-couplings of fourth and higher order are eliminated. The number of elements which can be described by a single transformation is therefore ultimately limited. Criteria to quantitatively specify this limit have not yet been developed. Experience with tracking simulations for different types of storage rings indicates, however, that low emmitance, very strongly focussing rings (with strong nonlinearities) require more care in this respect than do large emittance rings with weaker focussing and weaker nonlinearities. See reference 4 above.

8. Garren, A., 20 TeV Collider Lattices with Low Beta Insertions, to appear in proceedings of 12th International Conference on High Energy Accelerations (1983).

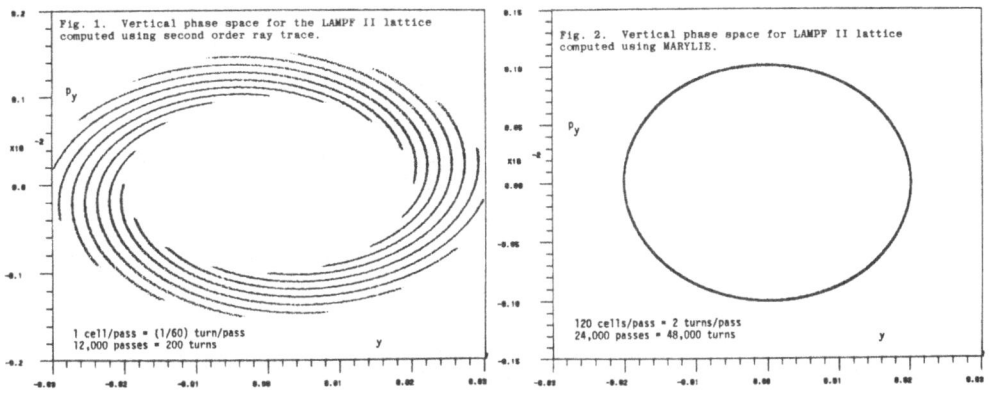

Fig. 1. Vertical phase space for the LAMPF II lattice computed using second order ray trace.

Fig. 2. Vertical phase space for LAMPF II lattice computed using MARYLIE.

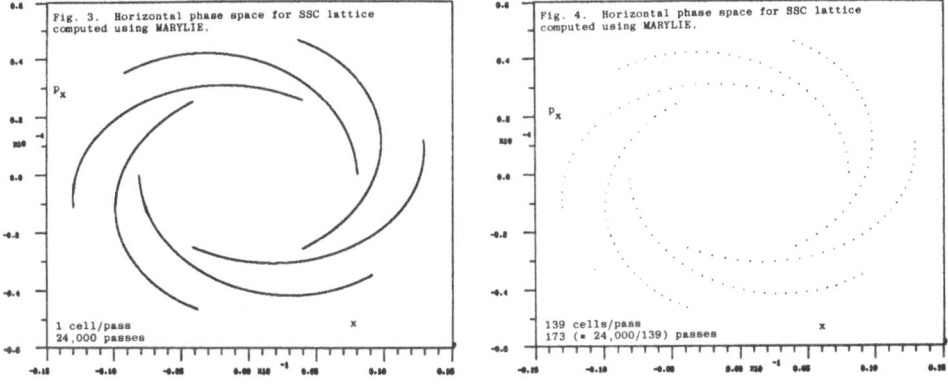

Fig. 3. Horizontal phase space for SSC lattice computed using MARYLIE.

Fig. 4. Horizontal phase space for SSC lattice computed using MARYLIE.

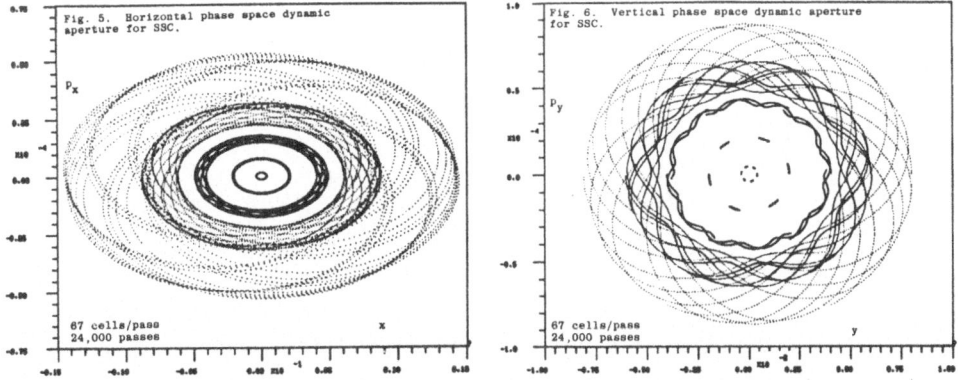

Fig. 5. Horizontal phase space dynamic aperture for SSC.

Fig. 6. Vertical phase space dynamic aperture for SSC.

NUMERICAL INVESTIGATION OF BUNCH-MERGING IN A HEAVY-ION-SYNCHROTRON

I. Bozsik, I. Hofmann, A. Jahnke
Max-Planck-Institut für Plasmaphysik, D-8046 Garching, Germany

R.W. Müller
GSI, P.O.B. 110541, D-6100 Darmstadt, Germany

Summary

We study longitudinal merging of two bunches without phase space dilution by slowly switching to half the RF-frequency. By repeating this process, all particles accelerated in a heavy-ion-synchrotron are merged in a single bunch.
The problem is to determine the rate of change of RF-amplitudes, such that the merging of buckets is sufficiently adiabatic and the phase space dilution is minimized.
The use of a simulation program permits to take into account the effect of intra-bunch and bunch-bunch space charge forces.

Introduction

Reducing the number of bunches in a heavy-ion-synchrotron by merging the bunches without longitudinal phase space dilution leads to an increased peak current. For conservation of the longitudinal emittance a slow variation of the RF-amplitudes is necessary.
A possible way of switching the RF-voltage to half of its initial frequency is a linear change of amplitudes by keeping the total voltage constant (Fig.1).

$$U(t) = U_2 + U_1 \sim (1 - \frac{t}{T}) \sin(2\varphi) + 2\frac{t}{T}\sin\varphi \quad \text{for} \quad 0 < t < T$$

$$U(t) = U(T) \quad \text{for} \quad T < t$$

This gives a simultaneous decrease of U_2 and increase of U_1.
The resulting change of the shape of the RF-potential is shown in figure 1.

The shape of the longitudinal phase space ellipse is determined by the ratio $\Delta p/\Delta \varphi$. After the bunch merging is completed the value of this ratio equals its initial value, if the RF-amplitudes satisfy the following relation

$$U_1(t > T) = 2\, U_2(t = 0) \;.$$

For ideally merged bunches this means that the phase width $\Delta \varphi$ or the bunch length has increased by a factor of $\sqrt{2}$ and so has the momentum spread Δp.

$$\left. \begin{array}{l} \Delta p_T = \sqrt{2} \cdot \Delta p_o \\ \Delta \varphi_T = \sqrt{2} \cdot \Delta \varphi_o \end{array} \right\} \qquad \mathcal{E}_{L,T} = 2 \cdot \mathcal{E}_{L,o}$$

Between $t = 0$ and $t = T$ the RF-amplitudes may change other than linearly, as long as the variation is slow compared to the phase oscillation period.

The time needed for an adiabatic compression of a single bunch has been estimated [1]. For the merging of two bunches however, more time is required, since the synchrotron frequency at the center of the two bunches approaches zero when $2\, U_1 = U_2$.

Results and Discussion

Assuming a linear change of the RF-amplitudes with time, computation of the particle motion has been performed with the particle-in-cell simulation program SCOP-RZ [2]. A fast bunch merging that takes place during one phase oscillation with respect to the initial holding voltage $U_2(t = 0)$ simply moves the two phase ellipses on top of each other. The fast acceleration and deceleration of the bunches leads to noticeable trapping of empty phase space between the bunches, hence emittance dilution (Fig.2).

A plot of the relative longitudinal emittance versus the merging time shows that five synchrotron oscillations are sufficient to obtain adiabatic merging with negligible phase space dilution (Fig.4). The case without space charge even shows a slight reduction of the r.m.s. value of the longitudinal emittance. This is not surprising, since the r.m.s. emittance is only an approximate measure.

Another criterion for the degree of adiabaticity is the number of
particles that are contained by a phase space ellipse enclosing
an area that equals the sum of the initially separated ellipse
areas (Fig.3).

When space charge is taken into account, it is useful to define
a space charge parameter Σ. $(1 + \Sigma)$ is the factor by which
the holding voltage has to be increased in order to compensate
space charge repulsion.

$$e\, E_0 \propto \frac{z}{z_m} = \frac{A\, m_0\, c^2\, \beta^2}{q \cdot \gamma}\, \frac{\varepsilon^2}{z_m^3}\, (1 + \Sigma) \qquad \qquad 2)$$

The results in Fig.3 and Fig.4 are calculated without space charge
($\Sigma = 0$) and with $\Sigma = 1$ for the initially separated bunches.
The time scale is in units of the space charge depressed
synchrotron oscillation period for the separated bunches.

As for the case $\Sigma = 1$ a higher voltage is applied, the particle
velocity in z-direction rises above the value for $\Sigma = 0$ during
the bunch merging. This reduces the average synchrotron period by
up to 17 % for $\Sigma = 1$ and by about 50 % for $\Sigma = 5$.
Therefore a common synchrotron period for the merged bunches does
not exist.

Concerning the time dependence of the bunch merging general trends
have to be pointed out. The effect of a too fast varying RF-voltage
has already been described (Fig.2). The longer the merging process
becomes the more the final phase space figure is twisted and the
smaller is the distance in phase space between the two bunches
(Fig.5 a, b, c). The total phase space dilution caused by a bunch
merging taking more time than five synchrotron periods is
practically independent of the merging time.

The structure of the final phase space figure is smeared out by
space charge forces which also produce a halo (Fig.6). We have
also calculated a case with stronger space charge corresponding
to $\Sigma = 5$, and we found that during five synchrotron periods still
85 % of the total intensity lies within the initially available
phase space area. This justifies to assume that five synchrotron
oscillation periods are sufficient for adiabatic bunch merging,
independent from the intensity.

Conclusion

Bunch merging by slowly varying the amplitudes of two harmonics RF-voltages was investigated. Calculations with a particle-in-cell simulation code show an adiabatic merging for RF-switching times longer than five synchrotron oscillations in the initially separated bunches.
A faster bunch merging leads to a considerable phase space dilution. This result holds for space charge parameters from $\Sigma = 0$ to $\Sigma = 5$.

References

[1] G. Guignard, Selection of Formulae Concerning Proton Storage Rings, CERN 77 - 10, p. 16

[2] I. Hofmann and I. Bozsik, Multi-Dimensional Computer Simulation of Intense Beams, Proc. of the Symposium on Accelerator Aspects of Heavy Ion Fusion, GSI - 82 - 8, Darmstadt, 1982, p. 181

Legend to Figures

Fig.1   Potentials generated by the time-dependent mixing of two harmonics RF-holding-voltages

Fig.2   Two bunches shown in the longitudinal phase space before and after merging during one synchrotron oscillation

Fig.3   Intensity in a phase space area representing a merged bunch without dilution as a function of merging time

Fig.4   The r.m.s. longitudinal emittance of the merged bunch as a function of merging time in units of initial synchrotron oscillation periods

Fig.5   Merged bunches in the longitudinal phase space with a merging time of 4.5 6.5 and 12.5 synchrotron oscillations and space charge parameter $\Sigma = 0$

Fig.6   same as Fig.5 (first case) with $\Sigma = 1$

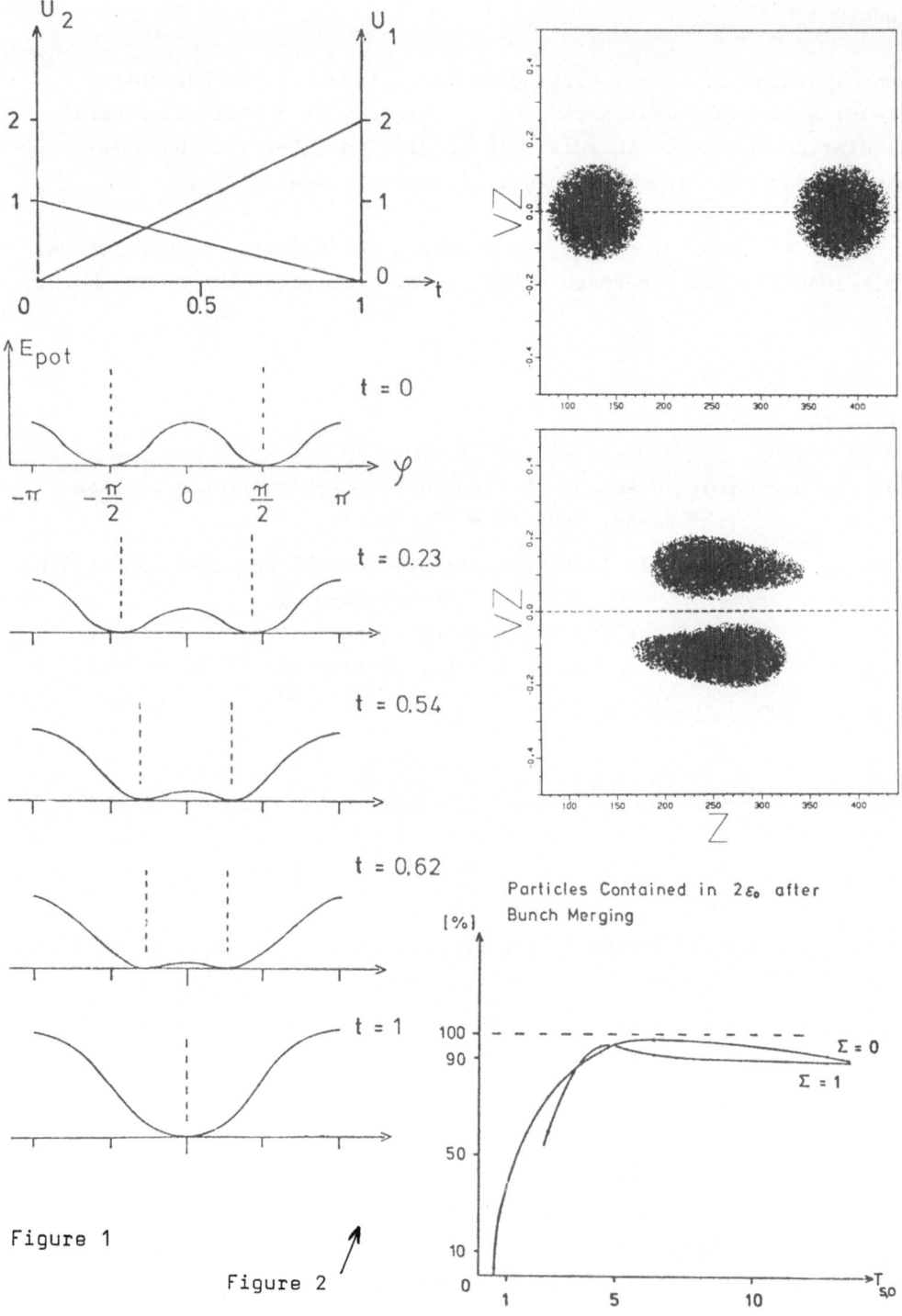

Figure 1

Figure 2

Figure 3

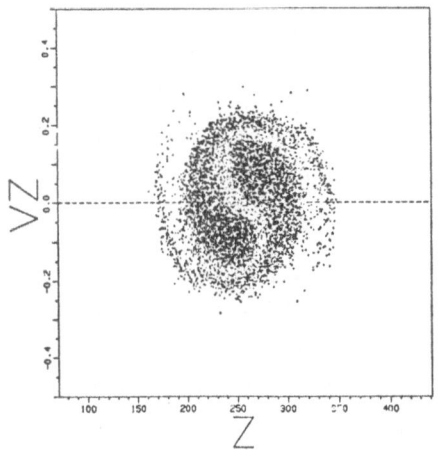

Figure 6

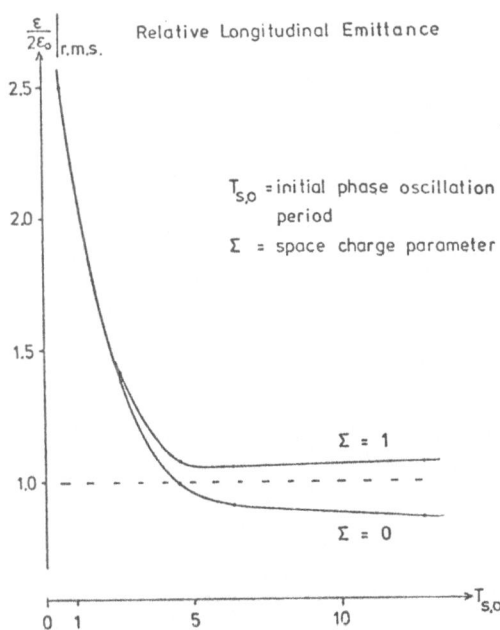

Figure 4

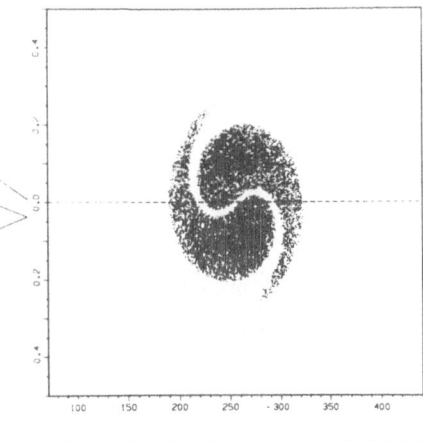
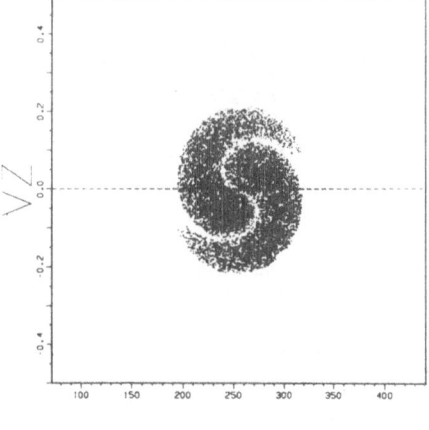
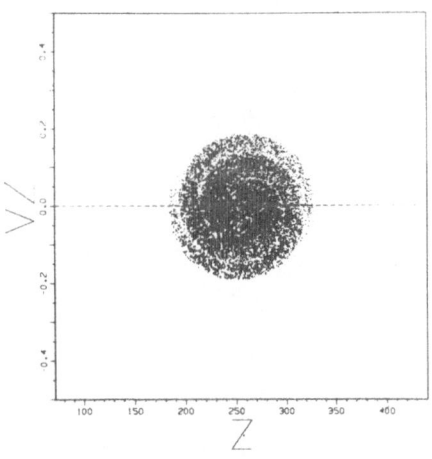

Figure 5 a
         b
         c

NONLINEAR ASPECTS OF LANDAU DAMPING IN COMPUTER SIMULATION OF THE MICROWAVE INSTABILITY

I. Hofmann

Gesellschaft für Schwerionenforschung mbH - GSI
Postfach 110541, D - 61 Darmstadt 11, Germany

Computer simulation with the multi-dimensional particle-in-cell code SCOP-RZ is applied to the study of nonlinear saturation phenomena of the resistive microwave instability of a coasting high-current ion beam. Due to the option of a fast Poisson-solver employing r-z geometry the code is particularly suitable for long beams with strong space charge effects due to non-relativistic energy. It is shown here that for currents far above the Keil-Schnell limit the instability due to a broad-band resonator near teh cut-off frequency saturates at a relatively low level without severe degradation of the momentum distribution, provided that the resistive part of the impedance is small compared with the reactive (space charge dominated) part: The final distribution shows Landau damping due to a thin stabilizing tail towards smaller momenta.

I. INTRODUCTION

The linearized theory of microwave instability (see for instance Ref. 1) describes exponential growth of a given harmonic starting from infinitesimally small initial amplitudes. The dispersion equation for a mode with mode number n and frequency ω can be written in the frequently used form[2] using a normalized dispersion integral and the scaled impedances U', V'. The stability boundary in U', V' resulting from a Gaussian momentum distribution is shown in Fig. 1. De-scaling to the impedance Z(n) is performed according to:

$$(1) \quad V' + i U' = \frac{2 I q}{\pi M_o c^2 \beta^2 \gamma |n| (\frac{\Delta p}{p})^2} (ReZ/n + i\, ImZ/n)$$

For vanishing momentum spread the dispersion integral yields

$$(2) \quad (\omega - n\Omega_o)^2 = \frac{n^2 \Omega_o^2 n\, Iq}{2\pi M_o c^2 \beta^2 \gamma} (Im\, Z/n - i\, Re\, Z/n)$$

Assuming that the impedance is only due to a resonator at frequency ω with quality Q

$$(3) \quad Z = \frac{R_s}{1 + i Q (\frac{\omega}{\omega_r} - \frac{\omega_r}{\omega})}$$

we find for the growth rate $\gamma \sim (Re\, Z)^{1/2}$ at resonance, where ImZ is negligible. Such a case has been investigated by simulation in connection with the ISR in Ref. 3. The growth rate predicted from linear theory agreed well with the simulation. In addition, it was found that an initially unstable beam levels off at a final momentum width - after saturation of the instability - which is worse than the threshold predicted from linear theory (Keil-Schnell limit). This "overshoot" phenomena which provides more Landau damping than actually necessary is a nonlinear effect that was also confirmed experimentally in the ISR.

An entirely different situation can arise for a non-relativistic beam, for which space charge gives a large capacitative contribution to the impedance of the order of $k\Omega$'s. In the long wave-length limit it can be written for a perfectly conducting smooth vacuum chamber of radius h as :

(4) $$\text{Im } Z/n = \frac{377 \; \Omega}{2\beta\gamma^2} \cdot g$$

with the geometry factor $g \equiv 1 + 2 \ln h/b$. For $\beta\gamma^2 < 1$ and the assumption $\text{Re } Z \ll \text{Im } Z$ we obtain from Equ. (2) a large real frequency shift of the microwave mode $\sim \text{Im } Z/n$ and a growth rate which is small due to a reduction factor :

(5) $$\gamma \sim (\text{Re } Z)^{1/2} \cdot (\text{Re } Z / \text{Im } Z)^{1/2}$$

It is the purpose of this work to show by means of computer simulation that with this assumption new features appear in the nonlinear behaviour of the microwave instability. The problem has been stimulated by performance studies of high-current heavy ion storage rings for inertial fusion drivers when the question was raised[4] whether several e-foldings due to a broad band impedance would lead to deterioration of an initial momentum distribution whose width is typically a factor 5 below the Keil-Schnell limit.

II. SIMULATION PROGRAM

The multi-dimensional simulation program SCOP-RZ is capable of tracking particles in 6D phase space along with their self-consistent field calculated in r-z geometry. It is an extension of the previously developed transverse simulation program SCOP-2 [5] and is based on the particle-in-cell method using a fast Poisson solver. This method is advantageous for the simulation of a large number of super-particles ($10^4 \ldots 10^5$) in order to reduce statistical noise effects. A mesh in r and z is defined within a conducting pipe of circular cross section. For each time-step the charge of a particle is distributed to the nearest mesh points and Poisson's equation is solved to determine the direct self-field. In addition to this the beam density can be Fourier analyzed and the effect of resonators is then incorporated by using the resonator impedance function $Z(n)/n$. Arbitrary bunching voltages and (rotationally symmetric) transverse focusing forces of an earlier version of SCOP-RZ without the resonator option are described in Ref.'s 6, 7.

The input beam distribution in 6D phase space can be produced with a random number generator and a large number of options for the distributions in transverse and longitudinal phase space. For problems where no longitudinal-transverse coupling is expected it is also possible to turn off the transverse motion. Output options allow for phase space projections into arbitrary 2D-planes, line densities, momentum distributions and Fourier spectrum of the line density. The code employs periodic boundary conditions in z and is thus suitable for bunched or coasting beams in circular accelerators. The actual computation is performed in the rest frame of the beam. Obviously the periodicity length depends on the number of harmonics that are to be treated simultaneously. An increase in periodicity length is at the expense of computing time, since it requires a larger mesh and - for a coasting beam - more simulation particles. At our best we have made runs with an r-z mesh of 16 x 1024 cells and 10^5 super-particles. This requires over an hour of computer time on the Cray 1 at the Max-Planck-Institut für Plasmaphysik, Garching, if the final state of microwave saturation for a non-relativistic beam with small growth rate was to be observed.

III. RESULTS

Our main concern is to study the effect of a broad band ($Q \simeq 1$) resonator centered at about the lowest magnetic cut-off frequency $\omega_c = \gamma \frac{R}{h} \Omega_0$ in order to describe the effect of many cross section variations (bellows, tanks etc.). The corresponding impedance has a (resistive) real part of $\frac{Zn}{n}$ which can be typically of the order of 10 Ω. We consider a coasting beam of nonrelativistic ions (β typically $\simeq$ 0.3), which has a large reactive space charge impedance of a few kΩ and choose a geometry where the period length exceeds sufficiently the beam pipe circumference (= cut-off wave length). In the examples shown below the cut-off frequency is the 17th harmonic of the fundamental mode, which is equivalent to a pipe diameter : length ratio of 1 : 53. The broad band (Q = 1) resonator has been centered around the 12th harmonic. Obviously the spacing of permissible harmonics is not as dense here as in a real machine, where the cut-off frequency is at harmonic 100 ... 1000 of the revolution frequency. The choice of a mesh with 512 cells in z direction allows the resolution of as many as 256 Fourier harmonics. In the problems shown all harmonics above about the 60th are not distinguishable from statistical noise due to the relatively limited number of simulation particles. To start the simulation independent of the noise level we found it practical to impose on the beam a sinusoidal density perturbation of 10% on the 12th harmonic, where the resistive impedance has its maximum. The values of U' and V' for the cases considered in this paper are shown in Fig. 1.

Case A : weak space charge effect (Fig. 2a, 3a, 4a)

This case is as usually considered for relativistic energies. We have assumed a ratio 1 : 2 for the resistive : reactive impedance at harmonic n = 12. To give an example: for a broad band resonator impedance of Re Z_n/n = 10 Ω this corresponds to a space charge (reactive) impedance of 20 Ω or, according to Equ. (4) with g = 2, a β = 0.97 (γ = 4.4 assumed to be below transition). The initial momentum width was a factor 2.2 below the Keil-Schnell threshold. The microwave instability sets in rapidly with a wave velocity outside of the distribution of particles velocities. Hence there is no Landau damping initially, but the wave starts decelerating particles down to its own shifted velocity, where they get trapped and build up a new, but broadened distribution with enough Landau damping. The final momentum width (fwhm) is about doubled. There is no evidence for noticeable overshoot in contrast with Ref. 3; we attribute this to the presence of the reactive impedance, which is responsible also for the shifted wave velocity (coherent frequency shift).

Case B : strong space charge effect (Fig. 2b, 3b, 4b)

Moving more in direction of non-relativistic energies we now assume a ratio of 1 : 16 for resistive : reactive impedance (at n = 12). For the previous example, this would correspond to β = 0.81 (γ = 1.7). We choose an initial momentum width, which is even a factor 4 below the Keil-Schnell threshold and find a very different result now: again there are particles trapped at the shifted coherent wave velocity, but this is a relatively small fraction, whereas the main part of the distribution changes very little only. The tail apparently provides enough Landau damping to stop further microwave growth. Note that for the starting distribution the linear theory e-folding period is 330 μsec, hence the total run shows good stability over as many as 13 such periods. We have also simulated a case with twice the resistivity as in B and found again a tail developing, yet with about twice as many particles.

IV. INTERPRETATION OF RESULTS

The early saturation of the broad band resistive microwave instability for a non-relativistic beam found by simulation can be explained as follows: as is seen in Fig. 4b the coupling of the initial amplitude to higher harmonics is faster than resistive growth. Since these harmonics are above cut-off, their impedance is much less than according to Equ. (4) if n >> n_c. Hence, their corresponding point in Fig. 1 is shifted downwards towards the Landau-damping regime and thus the initial perturbation energy dies out. The resulting tail is also linearly stabilized by Landau-damping as was shown in Ref. 7. An experimental verification of this stabilizing effect would be very desirable, in order to use it with confidence for the design of heavy ion fusion storage rings.

ACKNOWLEDGEMENT : The author is grateful to I. Boszik for performing the computational work.

REFERENCES
1. V.K. Neil and A.M. Sessler, Rev. Sci. Instr. 36, 429 (1965)
2. K. Hübner and V.G. Vaccaro, CERN-Rep. ISR-TH/70-44 (1970)
3. E. Keil and E. Messerschmid, Nucl. Instr. and Meth. 128, 203 (1975)
4. D. Möhl, Proc. Heavy Ion Fusion Workshop, Berkeley, LBL-Rep. 10301, 315 (1979) and J.L. Laclare, Proc. of the Symposium on Accel. Aspects of HIF, Darmstadt, GSI-Rep. 82-8, p.278 (1982)
5. I. Boszik and I. Hofmann, Nucl. Instr. and Meth. 187, 305 (1981)
6. I. Hofmann and I. Boszik, Proc. of the Symposium on Accel. Aspects of HIF, GSI Darmstadt, p. 181 (1982)
7. I. Hofmann, I. Boszik and A. Jahnke, IEEE Trans. Nucl. Sci. NS-30, 2546 (1983)

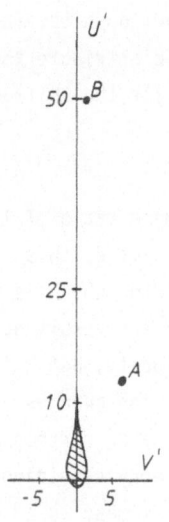

Fig. 1 (left): Stability boundary (dashed area stable) in scaled impedance plane for a Gaussian momentum distribution, showing cases A, B at resonator center.

Fig. 2 (bottom, left): Momentum distribution for case A with linear e-folding time 40 μsec. Bump at 150 μsec indicates shifted coherent wave velocity.

(bottom, right): case B with e-folding time 320 μsec.

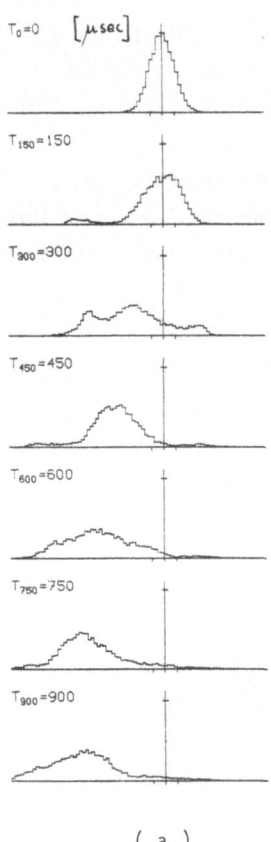

(a)

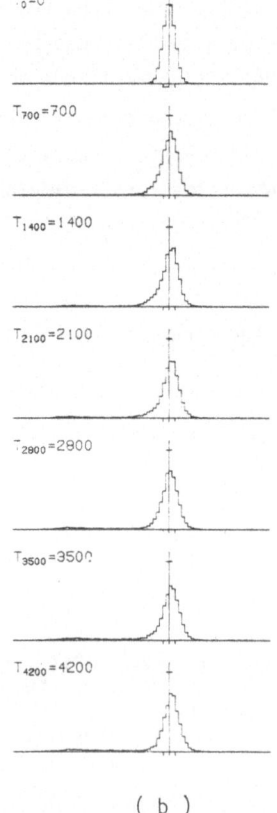

(b)

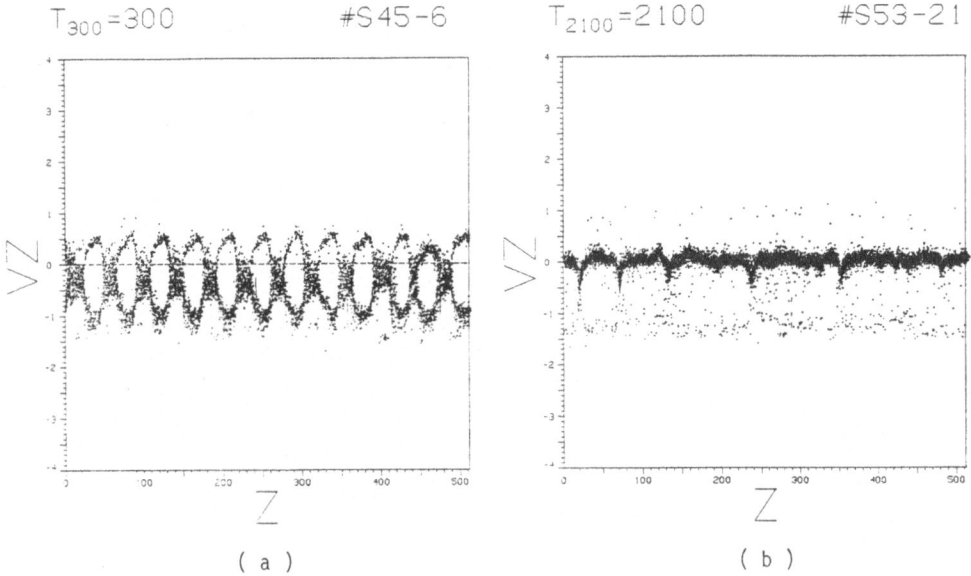

Fig. 3: Momentum space projections after about 7 linear theory e-foldings for case A (left) and B (right)

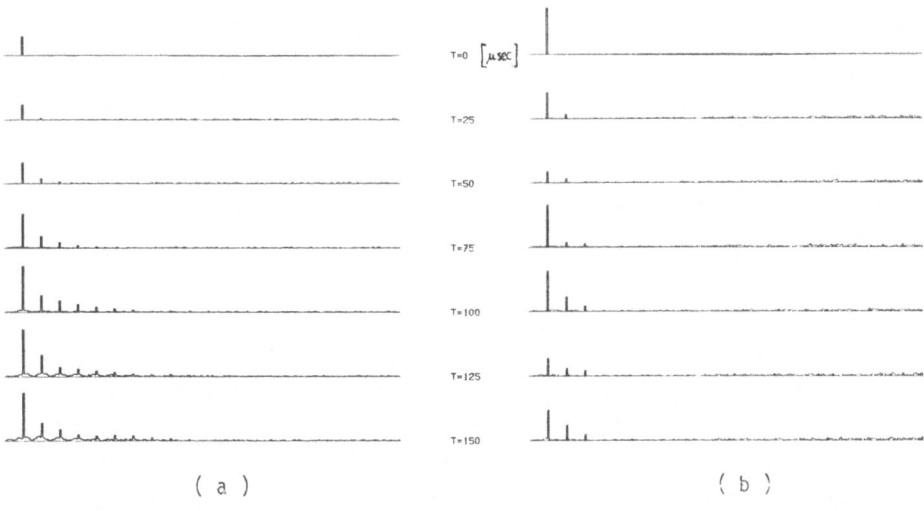

Fig. 4: Fourier spectrum during first 150 µsec showing exponential growth for case A (left) and dominance of coupling for case B (right).
Note initial 12th harmonic modulation in both cases 10% (scales differ)

THE TRANSPORT THEORY OF PARTICLE BEAM-CONGREGATION IN SIX-DIMENSIONAL PHASE SPACE

Cao Qing Xi   Guan Xia Ling

Tandem accelerator laboratory, Institute of Atomic energy, Academia Sinica. Beijing

Abstract

By way of the analysis of dynamics of particle beam-congregation for a tandem Van de Graff accelerator, the equations of motion of particle in a six-dimensional phase space are derived. The first order transfer of the pulse beam-optics and the -matrix transmission are also discussed. The matrix formulae describing the transfer of beam congregation passing through various optical components are given.

1. Introduction

Since 1977, it has been in progress for us to study systematically the transport theory of particle beam congregation and some papers have been published.[1],[2],[3],[4] Although the similar discussion has been made, for example by I. Benzvi,[5] we consider it is convenient to use "t" and "$t' = dt/dz$" as longitudinal coordinates in 6-dimensional phase space to describe the beam behaviour in longitudinal as well as transverse motions. With a different coordinate system and a different method, a study in detail about motion equation in this phase space and its solution for some devices being used in tandem are given.

2. Basic equations

For a weak beam, the phase point motion can be described in a 6-dimansional phase space. When the Cartesian coordinate system is used as the real space of particle motion, a axial direction variable Z is taken as the independent variable, and x,x',y,y',t,t' as the coordinates of this phase space. The mation equations of phase point are :

$$\begin{cases} x_1' = \frac{dx}{dz} = x' \; ; \quad x_3' = \frac{dy}{dz} = y' \; ; \quad x_5' = \frac{dt}{dz} = t' \; ; \\ x_2' = \frac{dx'}{dz} = [F_x - x'F_z] t'^2/m \; ; \\ x_4' = \frac{dy'}{dz} = [F_y - y'F_z] t'^2/m \; ; \\ x_6' = \frac{dt'}{dz} = \left(-F_z \cdot (1+\beta^2(x'^2+y'^2)/c^2 t'^2) + (F_x x' + F_y y') \cdot \gamma^2/c^2 t'^2 \right) t'^3/mv^2 \end{cases} \quad (1)$$

where $m = \gamma m_0$; $\gamma = (1-\beta^2)^{\frac{1}{2}}$; $\beta = (1+x'^2+y'^2)^{\frac{1}{2}}/ct'$

If the plane curve coordinate system [2] is used for such devices characterized by bending the trajectory of particle and with the middle symmetry plane of field distribution, it is convenient to take "l" as the independent variable. Under this case the motion equation are:

$$\begin{cases} X_2' = \dfrac{dx'}{dl} = [F_x - x'F_\ell/(1+hx)]t'^2/m + h(1+hx) + x'\left[\dfrac{2x'h+xh'}{1+hx} - \gamma^2 hx(h'x+x'h)/c^2t'^2\right] \\[4pt] X_4' = \dfrac{dy'}{dl} = [F_y - y'F_\ell/(1+hx)]t'^2/m + y'\left[\dfrac{2x'h+xh'}{1+hx} - \gamma^2 hx(h'x+x'h)/c^2t'^2\right] \\[4pt] X_6' = \dfrac{dt'}{dl} = -\big[F_\ell t'^2/m - (2hx'+xh') - \gamma^2(1+hx)(h'x+x'h)/c^2t'^2\big]\cdot\big[1+\gamma^2(x'^2+y'^2)/ (c^2t'^2)\big]\cdot t'/[\gamma^2(1+hx)] + x'\big[F_x t'^2/m + (1+hx)[h - x'\gamma^2(hx+x'h)/(c^2t'^2)]\big]/ \\[4pt] (c^2t') + y'[F_y t'^2/m - y'\gamma^2(1+hx)(h'x+x'h)/c^2t'^2]/c^2t' \end{cases} \quad (2)$$

where $\beta = \sqrt{x'^2+y'^2+(1+hx)^2}/ct'$

It is seldom possible to obtain exact analytic solutions of equations (1) and (2) except for a few simplest conditions, so the numerical solutions and series solutions are generally applied. The linear series solutions can be obtained by means of expanding the right hand side of motion equation and taking the first order approximation only. In case of the phase space vectors of an arbitrary particle with respect to the central particle expressed by X,

$$X^T = (x-x_c, x'-x'_c, y-y_c, y'-y'_c, t-t_c, t'-t'_c)$$

we have the matrix equation of motion as follows:
$$X' = FX \qquad (3)$$

where F is a first order coefficient matrix with its elements expressed as $F_{ij} = (\partial x_i'/\partial x_j)_c$. By introducing of transfer matrix R of optical device with its elements $R_{ij} = (\partial x_i/\partial x_{j_0})_c$, the motion equation (3) is simplified as

$$R' - FR = 0 \quad ; \quad R(0) = 1 \qquad (4)$$

The motion equation and the transport equation of first order mean square envelope for arbitary particle are:

$$\begin{cases} \dfrac{d}{d\xi}\langle X\cdot X^T\rangle = F\langle X\cdot X^T\rangle + \langle X\cdot X^T\rangle F^T \\[4pt] \langle X\cdot X^T\rangle = R\langle X(0)\cdot X^T(0)\rangle R^T \end{cases} \qquad (5)$$

where ξ is the independent variable.

So long as the boundary of the phase space is finite and the phase diagram of the particle congregation can be described by -matrix phase ellipsoid, we have

$$\begin{cases} \sigma = R\,\sigma(o)\,R^T \\ d\sigma/dz = F\sigma + \sigma F^T \end{cases} \tag{6}$$

Since the matrix " F " may simply be obtained with the use of equation (3), the transfer matrix R can be obtained by means of the solution of equation(4) in case the external field is known.

3. Some Transfer matrixes

To specify the application of above dynamics theory in 6-dimensional phase spase, the transfer matrices of some devices in tandem, shch as buncher, inclined field, stripper, bending magnet and quadrupole lens, will be given as follows.

1). Buncher with sinusoidal wave

With constant phase during particle through the gap, the transfer matrix of buncher with single gap is expressed as:

$$\begin{cases} R_{11} = R_{33} = [(1+I_1)chK_o S + \tfrac{1}{2}P_S((1-I_1)shK_oS/K_o + I_2 chK_oS)](1-\tfrac{1}{4}\delta) \\ R_{12} = R_{34} = [(1-I_1)shK_oS/K_o + I_2 chK_oS](1-\tfrac{1}{4}\delta) \\ R_{21} = R_{43} = [\tfrac{1}{2}P_S((1+I_1)chK_oS + I_3 shK_oS) + K_o((1+I_1)shK_oS + I_3 chK_oS) + \\ \qquad (\tfrac{P_S}{2})^2(I_2 chK_oS + (1-I_1)/K_o \cdot chK_oS) + \tfrac{P_S}{2}K_o[I_2 shK_oS + (1-I_1)chK_oS/K_o]] \\ \qquad (1-\tfrac{1}{4}\delta) \\ R_{22} = R_{44} = [\tfrac{P_S}{2K_o}(I_2 K_o chK_oS + (1-I_1)shK_oS) + (K_o I_2 chK_oS + (1-I_1)shK_oS)](1-\tfrac{\delta}{4}) \\ R_{55} = R_{66} = 1 \\ R_{65} = -\dfrac{2\pi\,q\,V_m}{\beta\lambda} \cdot \dfrac{\sin\varphi_o}{\varphi_o} \end{cases} \tag{7}$$

where $K_o^2 = (\pi V_m/\beta\lambda S)\cdot(\cos\varphi_o/2w_o) - 3(V_m \sin\varphi_o/4w_o S)^2$; w_o is the initial energy of central particle; " φ_o " the initial phase; S: the gap length; $\delta = \sin\varphi_o \cdot (q \cdot V_m/w_o)$; V_m: the amplitude of modulating voltage; q : the charge of particle;

$$P_S = -\delta/2s\ ;\quad I_1 = \int_0^S K_o shK_o z \cdot chK_o z \cdot \delta(z)dz\ ;\quad I_2 = \int_0^S sh^2 K_o z \cdot \delta'(z)dz\ ;$$

$$I_3 = -K_o \int_0^S ch^2 K_o z \cdot \delta'(z)dz\ ;\quad \delta'(z) = K_o' z \cdot \delta/K_o\ ;\quad K_o' = \tfrac{3}{4}\cdot\dfrac{\pi V_m}{\beta\lambda s w_o}\cos\varphi_o - \tfrac{3}{8}(\tfrac{\delta}{S})^2\ ;$$

The remanent elements of R are all zero.

2). Inclined field region.

We only consider the two-dimensional inclined field with an uniform field E_o making an angle α with respect to the axis "Z" of system. Under the assumption of central particle entering into inclined field along the axis Z and taking the first order approximation the elements of transfer matrix of inclined field are:

$$\begin{cases} R_{11} = R_{33} = R_{55} = 1 \\ R_{12} = 2L\cos^2\alpha [\sqrt{Q}(1+x'_{co}tg\alpha) - tg\alpha(x'_{co}-tg\alpha)]/[\sqrt{Q}(\sqrt{Q}-1)(1+x'_{co}tg\alpha)^2] \\ R_{16} = -2L\cos^2\alpha(\sqrt{Q}-1)x'_{ic}/[\sqrt{Q}(\sqrt{Q}+1)t'_{co}] \\ R_{22} = [Q(1+x'_{co}tg\alpha)+tg\alpha(tg\alpha-x'_{co})]/[\cos^2\alpha\cdot\sqrt{Q}(\sqrt{Q}-x'_{ic}tg\alpha)^2(1+x'_{co}tg\alpha)^3] \\ R_{26} = (1-Q)x'_{ic}/[\cos^2\alpha\cdot\sqrt{Q}\cdot(\sqrt{Q}-x'_{ic}tg\alpha)^2 t'_{co}] \\ R_{34} = 2L/[(\sqrt{Q}+1)(1+x'_{co}tg\alpha)] \\ R_{44} = 1/[\cos^2\alpha\cdot(\sqrt{Q}-x'_{ic}tg\alpha)(1+x'_{co}tg\alpha)] \\ R_{52} = -2L\cos\alpha\, tg\alpha\cdot t'_{ic}/[\sqrt{Q}(\sqrt{Q}+1)(1+x'_{co}tg\alpha)] \\ R_{56} = 2L\cos\alpha\, t'_{i\circ}/[\sqrt{Q}(\sqrt{Q}+1)t'_{co}] \\ R_{62} = tg\alpha[(1+x'_{co}tg\alpha)(\sqrt{Q}+tg\alpha\, x'_{ic}-1)-\sqrt{Q}/\cos^2\alpha]\, t'_{ic}/[\cos\alpha\cdot(1+x'_{co}tg\alpha)^2\sqrt{Q}\cdot(\sqrt{Q}-x'_{ic}tg\alpha)^2] \\ R_{66} = (1-\sqrt{Q}\, tg\alpha\, x'_{ic})/[\sqrt{Q}\cdot(\sqrt{Q}-x'_{ic}tg\alpha)(1+x'_{co}tg\alpha)\cos^2\alpha] \end{cases} \quad (8)$$

where $Q = 1 + 2L\cos\alpha\cdot q\cdot E_o\, t'_i/m$; $\quad t'_i = t'_o/[\cos\alpha(1+x'_o tg\alpha)]$;

$x'_i = (x_o - tg\alpha)/(1+x'_o tg\alpha)$; $\quad y'_i = y'_o/[\cos\alpha\cdot(1+x'_o tg\alpha)]$.

L is the length of field; q the charge of particle.

The remanent elements of matrix are all zero. The central particle parameters X_{ci} can be obtained by means of the solution of non-linear equation as follows:

$$\begin{cases} x_c = x_o + L\cos^2\alpha[2x'_i/(\sqrt{Q}+1) + tg\alpha] \\ x'_c = (x'_i + \sqrt{Q}\, tg\alpha)/(\sqrt{Q}-x'_i tg\alpha) \\ y_c = y_o + 2L\cos\alpha\, y'_i/(\sqrt{Q}+1) \\ y'_c = y'_i/[\cos\alpha(\sqrt{Q}-x'_i tg\alpha)] \\ t_c = t_o + 2L\cos\alpha\, t'_i/(\sqrt{Q}+1) \\ t'_c = t'_i/[\cos\alpha(\sqrt{Q}-x'_i tg\alpha)] \end{cases} \quad (9)$$

3). Stripper

The transverse matrix of stripper have been treated by some paper [7],[8]. By virture of the postulated most probable distribution theorem of incoherently scattered particle, taking $R(L/2)$ represent a drift matrix for half the length, $\sigma_-(0)$: the intial beam matrix and $\sigma_-(s)$: the scattering beam matrix, then we have:

$$\sigma = R(\tfrac{L}{2})\, \sigma(0)\, R^T(\tfrac{L}{2}) + \sigma(s) \qquad (10)$$

$$\sigma(s) = \begin{pmatrix} \frac{\theta_s^2 L^3}{24}, & 0, & 0, & 0 & 0 & 0 \\ 0, & \frac{\theta_s L}{2}, & 0, & 0 & 0 & 0 \\ 0, & 0, & \frac{\theta_s^2 L^3}{24} & 0 & 0 & 0 \\ 0, & 0, & 0, & \frac{\theta_s L}{2} & 0 & 0 \\ 0, & 0, & 0 & 0 & (\frac{5\pi\Delta E\, L}{6\beta\lambda w_i})^2 & -\tfrac{1}{2}(\frac{\pi\Delta E}{\beta\lambda w_i})^2 \\ 0, & 0, & 0 & 0 & -\tfrac{1}{2}(\frac{\pi\Delta E}{\beta\lambda w_i})^2 & (\frac{\pi\Delta E}{\beta\lambda w_i})^2 \end{pmatrix}$$

where $\theta_s = 0.25\,(Z_s(Z_s+1)/A_s)(Z_i/w_i)^2\,\rho$ with the quantities defined in same way as referece (8); λ is the wave length of buncher drive voltage; ΔE: energy spread coming from stripper.

4). Magnet analyzer

The solution of the equation (4) for magnet field gives the transfer matrix of bending magnet.

$$R = \begin{pmatrix} \cos K_x \ell, & \tfrac{1}{K_x}\sin K_x \ell, & 0, & 0, & 0, & -\frac{h\gamma^2}{K_x^2 t_o'}(1-\cos K_x \ell), \\ -K_x \sin K_x \ell, & \cos K_x \ell, & 0, & 0, & 0, & -\frac{h\gamma^2}{K_x t_o'}\sin K_x \ell, \\ 0 & 0 & \cos K_y \ell, & \tfrac{1}{K_y}\sin K_y \ell, & 0, & 0, \\ 0 & 0 & -K_y \sin K_y \ell, & \cos K_y \ell, & 0, & 0, \\ \frac{h t_o'}{K_x}\sin K_x \ell, & \frac{h t_o'}{K_x^2}(1-\cos K_x \ell), & 0, & 0, & 1, & \ell - \frac{h^2\gamma^2}{K_x^2}(\ell - \frac{\sin K_x \ell}{K_x}), \\ 0, & 0, & 0, & 0, & 0, & 1 \end{pmatrix}$$

where $K_x^2 = h^2(1-N);\ K_y^2 = h^2 N;$ N: the magnet field index.

5). Quadrupole lens

The transfer matrix of a quadrupole field in 6-dimensional phase space is

$$R = \begin{bmatrix} \cos K_0 z & \frac{1}{K_0}\sin K_0 z & 0 & 0 & 0 & 0 \\ -K_0 \sin K_0 z & \cos K_0 z & 0 & 0 & 0 & 0 \\ 0 & 0 & ch K_0 z & \frac{1}{K_0} sh K_0 z & 0 & 0 \\ 0 & 0 & K_0 sh K_0 z & ch K_0 z & 0 & 0 \\ 0 & 0 & 0 & 0 & 1 & z \\ 0 & 0 & 0 & 0 & 0 & 1 \end{bmatrix}$$

where $K_0 = qB/m_0 c^2 \gamma \beta$

4. Conclusion

In above discussion, the strict and general equation of motion in 6-dimensional phase space are given without any approximation. By means of the particle beam congregation transfer theorem, the linear matrices for some major optical diveces of tandem have been given. Using these results we can integrate the longitutinal and transverse theory in the dynamics of accelerator. Since the longitudinal coordinates t and t' are used, it is possible to ensure the rigor for the motion equation.

The authors gratefully acknowledge Prof. Chen Jarer, Beijing university, for his great help and thank prof. Wang Chuan-Ying and Xie-Xi comments on this mamuscript.

Reference:

(1) Cao Qingxi and Guan Xialing,
    Chinese J. " particle accelerator "   (1977)   415
(2) Cao Qingxi and Guan Xialing,
    Chinese. J. Nucl. phys. 2.1   (1980)   67
(3) Cao Qingxi Nucl. Instr. and Meth. 184   (1981)   269
(4) Cao Qingxi and Guan Xialing,
    Chinese J. Nucl. phys. 4.4   (1982)   363; 5.1 (1983) 83
(5) I. BEN-Zvi and Z. segalov particle Accelerator
    10 (1979) 31
(6) A.L.Dymnikov at al. Nucl. Iustr. and Meth. 148 (1978) 567
(7) J.Joy Nucl. Instr. and Meth.   106   (1973)   237
(8) J.D.Larson. Instr. and Meth.   122   (1974)   53
(9) A.B.Banford " the transport of charged particle beam " 5.2
(10) K.L.Brown SLAC-75   (1972).

THE MAD PROGRAM

F. Christoph Iselin
CERN, Geneva, Switzerland

SUMMARY

The program "MAD" (Methodical Accelerator Design) has been written at CERN to provide a standard user interface for solving problems arising in accelerator design. Its aim is to solve these problems using a common input data structure. The program performs basic computations like computing Twiss parameters, chromaticities, or machine geometry; and it is able to feed other programs with the basic results. It also provides flexible ways of matching various parameters of the machine.

MAD uses a format-free input language which is easy to learn and yet flexible enough to describe even very large accelerators in a few lines. Once an accelerator structure is defined, solving a new problem just requires a few lines of additional information.

BASIC COMPUTATIONS AVAILABLE IN MAD

<u>Twiss Parameters</u>. MAD tracks the Twiss parameters for any arbitrary beam line. The initial conditions are normally given by the periodic solution. They may also be specified by the user. If a periodic beam line is symmetric, only half of it need to be entered, and the computing time is halved. The user is free to select any position(s) in the beam line where the Twiss parameters will be printed. MAD also computes chromaticities, transition energy, as well as the largest values of β and dispersion for each plane. Optionally MAD also computes the chromatic functions[1,2].

The transfer matrices are computed by the methods of the program TRANSPORT[3]. Momentum-dependent effects are treated by using the second-order terms delivered by the TRANSPORT formalism. Should this prove inadequate, it is easy to plug in a different method.

<u>Particle Tracking</u>. At the time of writing, investigations are being made on the most appropriate methods for particle tracking to be added in MAD. It is planned to include tracking in the presence of synchrotron oscillations, and of field and positioning errors. Simulation of synchrotron radiation may also be provided.

<u>Survey (Machine Geometry)</u>. For large accelerators it is a tedious task to compute the position in cartesian space for all magnets. MAD computes these positions as well as the direction angles.

DEFINITION COMMANDS

When running the MAD program, the user enters a set of commands which are obeyed in the order read. Some commands serve to define beam elements:

    DRIFT,D1,L=5
    QUAD,QF,L=2,K1=-0.01345
    SBEND,B,L=4,ANGLE=0.0113

The above commands define a drift space D1, a quadrupole QF, and a sector bending magnet B respectively. Blanks may be used freely in commands, and commands can continue on as many input lines as required. An ampersand (&) denotes continuation on the next line. Having defined a set of beam elements, the user may combine them with the LINE command to form sequences:

    LINE,PER=(D1,B,D2,QF,D3,SF,D1,B,D2,QD,D3,SD)

This command defines a simple FODO cell. For more complex sequences, beam lines can be nested to any level and/or repeated:

    LINE,INS1=(...)
    LINE,INS2=(...)
    LINE,SUP=(INS1,8*PER,INS2)

This defines a superperiod SUP consisting of 8 cells with the name PER, sandwiched between two insertions called INS1 and INS2. The definitions of INS1 and INS2 must of course be filled in.

Frequently the same beam line is used several times with small variations. Assume that in the above example the odd-numbered cells contain the sextupoles SF1 and SD1, while the even-numbered cells contain the sextupoles SF2 and SD2. To avoid defining two different cells, the sextupoles in the line PER are made formal arguments:

    LINE,PER(SF,SD)=(D1,B,D2,QF,D3,SF,D1,B,D2,QD,D3,SD)

When this beam line is used, the arguments are replaced by the corresponding actual arguments:

    LINE,SUP=(INS1,4*(PER(SF1,SD1),PER(SF2,SD2)),INS2)

This will be expanded to eight cells, as defined above.

Using the above facilities and suitable element definitions, the complete element sequence for the LEP machine can be written in 14 beam line definitions:

    LINE,LEP=(4*SUP)
    LINE,SUP=(OCT,-OCT)
    LINE,OCT=(LOBS,RFS,DISS,ARC,DISL,RFL,LOBL)
    LINE,LOBS=(L1,QS1,L2,QS2,L3,QS3,L4,QS4)
    LINE,RFS=(L5,QS5,L5,QS6,L5,2*(QS7,L5,QS8,L5))
    LINE,DISS=(QS11,L25,BW,L22,QS12,L25,B4,L22,QS13,L25,B4,L22,&
            QS14,L25,B4,L31,QS15,L25,B4,L32,SF,L23,QS16)
    LINE,ARC=(L21,B6,L22,SD,L23,QD,7*(CELL(SF1,SD1),CELL(SF,SD)),&
            CELL(SF1,SD1),L24,B6,L41,QF,L21,B6,L22,SD4,L23,QD,&

```
              7*(CELL(SF4,SD3),CELL(SF3,SD4)),CELL(SF4,SD3),&
              L24,B6,L22,SF3,L23)
    LINE,DISL=(QL16,L34,B4,L22,QL15,L33,B4,L22,QL14,L25,B4,L22,&
              QL13,L25,B4,L22,QL12,L25,BW,L22,QL11)
    LINE,RFL=(2*(L5,QL8,L5,QL7),L5,QL6,L5,QL5,L5)
    LINE,LOBL=(QL4,L14,QL3,L13,QL2,L12,QL1,L11)
    LINE,BW=(2*W,L26,2*W)
    LINE,B4=(2*B,L26,2*B)
    LINE,B6=(2*B,L26,2*B,L26,2*B)
    LINE,CELL(SF,SD)=(L24,B6,L22,SF,L23,QF,L21,B6,L22,SD,L23,QD)
```

The beam lines BW, B4 and B6 represent strings of bending magnets. For preliminary calculations one will rather define them as single magnets. Having found a suitable design it is easy to replace them by the actual sequences.

ACTION COMMANDS

Another set of commands requests specific actions on the machine structure. In order to do any computations, the user first specifies which beam line is to be studied. This is done by a command like

```
    USE,OCT,SUPER=4,SYMM
```

All subsequent computations refer to the octant OCT, made symmetric by appending its mirror image, and repeated 4 times.

Twiss Parameters and Chromatic Functions. The actual computations are requested by a TWISS command. This command computes the Twiss parameters for several user-specified values of $\delta p/p$. By default the program prints results at both ends of the beam line. The user may however request additional positions for printing like

```
    PRINT,QD[1/15]
    PRINT,#27/30
```

These are requests for print-out after the 1st to 15th QD, and after the 27th to 30th physical element. As an option, MAD also prints the chromatic functions for the same beam line and the same values of $\delta p/p$. The print positions are selected by the same mechanism as for the Twiss parameters.

Survey. After a USE command the user may also enter a SURVEY command. This command causes MAD to print a table of positions and angles in cartesian space for all magnets in the accelerator. The print positions are selected by the same mechanism as for the TWISS command. The initial position and direction can be chosen freely.

Matching a Cell. Having defined a machine period PERIOD, one may match this beam line as a period:

```
CELL,PERIOD
VARY,QD[K1]
VARY,QF[K1]
CONSTR,#E,MUX=0.25,MUZ=0.166666667
MIGRAD,TOLER=1.0E-10
ENDMATCH
```

The CELL command initiates matching of a periodic structure. The strengths of the quadrupoles QF and QD are varied such as to obtain phase advances of one quarter and one sixth of 2π in the horizontal and vertical plane respectively at the end (position #E) of the period. The matching methods available are in essence the same as the ones used in the MINUIT program[9]. The user makes his choice by entering e.g. MIGRAD or SIMPLEX. Matching mode is left by entering the ENDMATCH command.

<u>Matching an Insertion</u>. A simple insertion INSERT may be matched as follows:

```
MATCH,INSERT,LINE=CELL
CONSTR,#E,LINE=CELL1
VARY,Q1[K1]
! OTHER VARY COMMANDS REQUIRED
SIMPLEX
MIGRAD
ENDMATCH
```

The MATCH command initiates insertion matching mode and defines the beam line and initial conditions to be used. The above CONSTR command imposes conditions at the end (position #E). Thus the line INSERT is adjusted such that it fits between the two periodic lines CELL and CELL1. There is no need to enter numeric values for Twiss parameters at either end. The program computes these values and fills in the constraints. It is however permissible to enter numeric values instead of phrases like "LINE=CELL". Here the user wants to run first the SIMPLEX method, followed by the MIGRAD method. The ENDMATCH command terminates matching.

Sometimes it is desired to make part of an insertion look like a period, i.e. the Twiss parameters should be equal in two positions. This is requested by the command

```
COUPLE,M1,M2
```

where M1 and M2 identify the two positions. They may both be any names occurring in the beam line. The phase advances from M1 to M2 can also be fixed with this command.

ADDITIONAL FACILITIES

<u>Auxiliary Input</u>. MAD has a mechanism to imbed predefined input files in its input stream. This allows to keep the accelerator structure in a data base, and to enter only the actions to be performed.

Ancillary Programs. A link is available to run the HARMON program[56] as a subroutine. Any modified sextupole strengths will be replaced in the MAD data tables. All tables generated by MAD may be written out on coded disk files. A set of other programs is available to read these files and to perform additional calculations on the lattice. This idea had already been realized with the AGS program[7]. This program however writes the Twiss parameters in binary format. To allow easier transfer of results between different computers, MAD uses coded files.

AN EXAMPLE FOR MATCHING

A complete example for a matching run is shown below. Element definitions have been left out in order not to overload the example.

```
! DEFINE BEAM LINES
LINE,CELL=(QDH,L24,B6,L22,SF,L23,QF,L21,B6,L22,SD,L23,QDH)
MARKER,MARK
LINE,DISL=(QL16,L34,B4,L24,QL15,L33,B4,L24,QL14,L25,B4,L24,&
          QL13,L25,B4,L24,QL12,L25,BW,L24,QL11)
LINE,RFL=(2*(L5,QL8,L5,QL7H,MARK,QL7H),L5,QL6,L5,QL5,L5)
LINE,LOBL=(QL4,L14,QL3,L13,QL2,L12,QL1,L11)
LINE,INSL=(QDH,L24,B6,L22,SF3,L23,DISL,RFL,LOBL)
! ENTER MATCHING MODE AND DEFINE INITIAL VALUES
MATCH,INSL,LINE=CELL
! VARIABLE PARAMETERS
VARY,QL1[K1],STEP=0.001
...
VARY,QL16[K1],STEP=0.001
! REPLACE DEFAULT MATCHING WEIGHTS
WEIGHT,BETX=10,BETZ=50,ALFX=10,ALFZ=10,MUX=10,MUZ=10,DX=10,DX'=100
! CONSTRAINTS FOR INTERACTION POINT
CONSTR,#E,BETX=3.2,BETZ=0.2,ALFX=0,ALFZ=0,MUX=1.771875,MUZ=2.0125,DX=0,DX'=0
! CONSTRAINTS AT TWO INTERMEDIATE POINTS (MARK OCCURS TWICE IN RFL)
CONSTR,MARK,ALFX=0,ALFZ=0,DX=0,DX'=0
! LIMITS FOR DISPERSION IN BEAM LINE DISL
CONSTR,DISL,DX>0,DX<1.25
! PERFORM MATCH AND LEAVE MATCHING MODE
MIGRAD,TOLER=1E-9
ENDMATCH
```

HOW TO GET ACCESS TO MAD

The MAD program is written in standard FORTRAN 77. The source code is available from the author. The documentation consists of a reference manual[8] and a primer[9], both of which are available in print or in machine-readable form from the author. MAD has been tested successfully on the CERN CDC 7600 computer as well as on the CERN IBM 3081 / Siemens 7880 system. The only restriction is that HARMON cannot be run on the CDC machine due to space limitations. It is hoped that this restriction can be overcome with future systems.

REFERENCES

1. B. Autin and A. Verdier, Focusing Perturbations in Alternating Gradient Structures, CERN ISR-LTD/76-14.

2. B. Montague, Linear Optics for Improved Chromaticity Correction, CERN, LEP Note 165, 30.7.1979.

3. K. Brown, D.C. Carey, Ch. Iselin and F. Rothacker, TRANSPORT, a Program to Compute Charged Particle Beam Transport Systems, published simultaneously as CERN 73-16, FNAL 91, SLAC 91, revised as CERN 80-4.

4. F. James and M. Roos, MINUIT, A Package of Programs to Minimize a Function of n Variables, Compute the Covariance Matrix, and Find the True Errors, CERN Program Library code D 506.

5. M.H.R. Donald, P.L. Morton and H. Wiedemann, Proc. Particle Accelerator Conference, Chicago, 1977, (IEEE Trans. Nucl. Sci. NS-24, No. 3 (1977) 1200).

6. M. Donald and D. Schofield, A User's Guide to the HARMON Program, CERN, LEP Note 420, 9.12.1982.

7. E. Keil, Y. Marti, B.W. Montague and A. Sudboe, AGS, The ISR Computer Program for Synchrotron Design, Orbit Analysis, and Insertion Matching, CERN 75-13, 1975.

8. Ch. Iselin, MAD, Methodical Accelerator Design, Reference Manual. Available from the author.

9. Ch. Iselin, MAD, Methodical Accelerator Design, A Primer. To be published.

ANALOGUE COMPUTER DISPLAY OF ACCELERATOR BEAM OPTICS

K. Brand
Ruhr-Universität Bochum
Dynamitron-Tandem-Laboratorium
Universitätsstraße 150

4630 Bochum 1, Germany

Introduction

Analogue computers have been used years ago by several authors (1 - 4) for the design of magnetic beam handling systems. At Bochum a small analogue/hybrid computer was combined with a particular analogue expansion and logic control unit for beam transport work (5). This apparatus was very successful in the design and setup of the beam handling system of the tandem accelerator. The center of the stripper canal was the object point for the calculations, instead of the high energy acceleration tube a drift length was inserted into the program neglecting the weak focusing action of the tube.
In the course of the installation of a second injector for heavy ions it became necessary to do better calculations. A simple method was found to represent accelerating sections on the computer and a particular way to simulate thin lenses was adopted.

Mathematics

The motion of a charged particle through the different ion optical elements is described in the first order theory by a set of linear differential equations listed in table 1.
An actual beam transport system consists of a sequence of elements - drift spaces, lenses, acceleration sections as well as quadrupole and dipole magnets. The calculation of a particle track through such a system requires a sequence of the said differential equations to be solved corresponding to the arrangement of the elements. The values of the displacement (x,y) and the divergence (x',y') of the particle after passage through one element are the initial conditions for the particle track in the next element and so on.

Analogue circuit

All the equations of table 1 are second order differential equations of the form

$$z'' + az' \pm bz \pm c = 0$$

z stands for x or y.

To obtain the analogue circuit the equation is solved for the highest derivative

$$z'' = -az' \pm bz \pm c$$

This derivative is integrated twice to obtain the lower order terms -z' and z (refer to figures 1 and 2). The lower order terms are fed into the appropriate components as called for by the equations. Their output is connected to the input of the first inte-

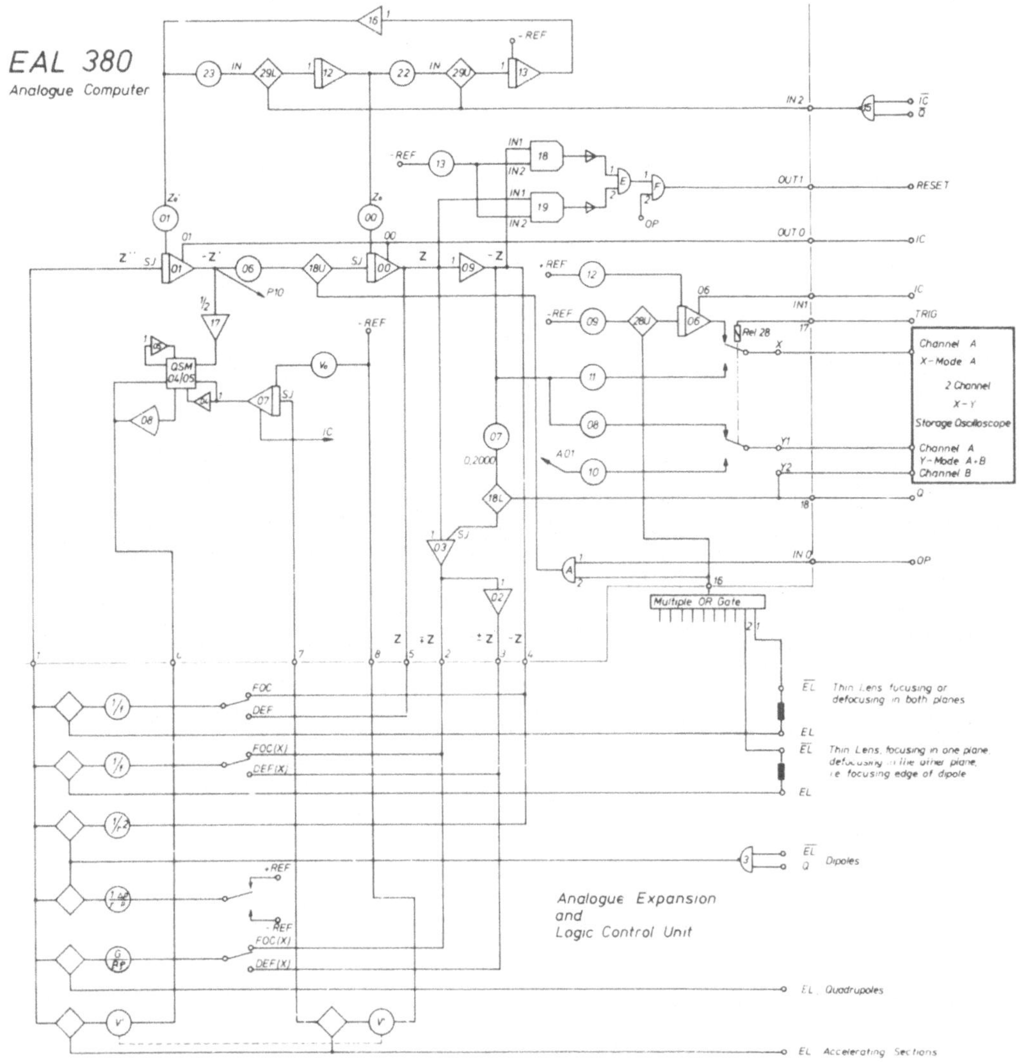

Figure 1. Analogue circuitry

grator via the individual element switch. In case no element switch is closed the two integrators 00, 01 solve the equation of the drift space.

The control signals to execute the program are shown in fig. 3.

At the begin of each cycle the integrators accept the initial conditions under the control of the signal IC.

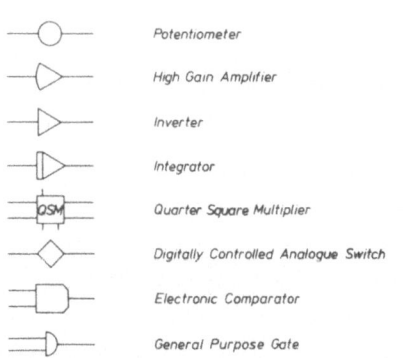

Figure 2. Symbols

Element	x - plane	y - plane	Comments		
Drift space	$\frac{d^2x}{ds^2} = 0$	$\frac{d^2y}{ds^2} = 0$	x,y	Displacement	(cm)
			s	Distance along direction of beam	(cm)
Quadrupole, focusing in x	$\frac{d^2x}{ds^2} + \frac{G}{B\rho} x = 0$	$\frac{d^2y}{ds^2} - \frac{G}{B\rho} y = 0$			
			$B\rho$	Stiffness	(kG·cm)
			G	Gradient	(kG/cm)
Quadrupole, defocusing in x	$\frac{d^2x}{ds^2} - \frac{G}{B\rho} x = 0$	$\frac{d^2y}{ds^2} + \frac{G}{B\rho} y = 0$			
Dipole	$\frac{d^2x}{ds^2} + \frac{1}{r^2} x = \pm \frac{\Delta p}{p}$	$\frac{d^2y}{ds^2} = 0$	r	Radius of curvature	(cm)
			$\frac{\Delta p}{p}$	Momentum Deviation	
Field edge of dipole	Thin lens with $f = -\frac{r}{tg(\alpha,\beta)}$	$f = \frac{r}{tg(\alpha,\beta)}$	f	Focal length	(cm)
			α,β	Entrance and exit angles	
Accelerating section	$\frac{d^2x}{ds^2} + \frac{V'}{2V} \frac{dx}{ds} = 0$	$\frac{d^2y}{ds^2} + \frac{V'}{2V} \frac{dy}{ds} = 0$	Equation of motion for linear potential rise $V(s) = V_0 + V' \cdot s$ V' = constant field on axis (kV/cm) Entrance and exit fringe fields are approximated by thin lenses (6)		
Thin lens	$x_1 = x_0$ $x'_1 = -\frac{x_0}{f} + x'_0$	$y_1 = y_0$ $y'_1 = -\frac{y_0}{f} + y'_0$	Artificial element of length O. No equation of motion. $x_0(x'_0), y_0(y'_0), x_1(x'_1), y_1(y'_1)$ displacement and divergence before (o) and after lens (1). f = focal length		

Table 1. Equations of motion in the first order theory

The signal OP controls the switch between the two main integrators defining the end of the computing time.

Up to 22 elements signals EL can be placed within the limits set by the signals IC and OP. The signals OP and EL are adjustable in length in order to correspond to the lengths of the individual beam transport elements and the total length of the system under cosideration.

The plane signal Q which is alternating from cycle to cycle reverses the polarity of the analogue signal feeding into the element potentiometers of quadrupoles and lenses which are focusing in one plane and defocusing in the other (output of inverters O2, O3).

Thus the computer calculates the particle

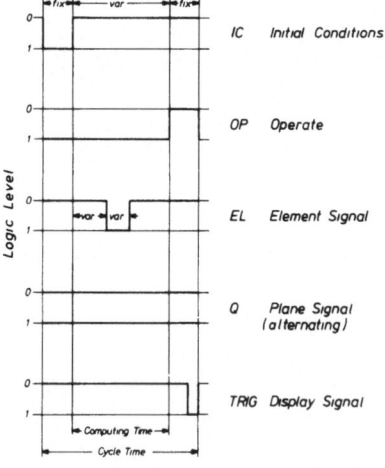

Fig. 3. Logic control signals

track with the given initial conditions alternately in the two planes. In case of dipoles the signal Q suppresses the element signal via gate 3 during the y - plane run.

The display circuit consists of integrator O6 producing the sweep proportional to s and the potentiometers O6 through 11. These potentiometers determine the scales of z, z' and s together with the sensitivities of the oscilloscope inputs. The traces of the two planes are separated from each other on the oscilloscope screen by the signal Q which is added to the computer output. The signal TRIG switches from ray tracing mode to phase space mode at the end of each cycle.

Whereas quadrupoles and dipoles are represented by potentiometers (i.e. constant coefficients in the equation of motion), accelerating sections are described by the term az' with the variable coefficient

$$a = \frac{V'}{2V} = \frac{V'}{2(V_0 + V's)}$$

The axial potential V is generated by integrator O7. $\frac{z'}{2}$ at the output of amplifier 17 is divided by V using the multiplier (QSM) and then multiplied by the constant axial field V' to obtain $\frac{V'}{2V} z'$.

Since V' is required twice a tandem potentiometer is provided.

Thin lenses are not described by an equation of motion. The coordinates z_0, z_0' of a particle change after passage through the lens to

$$z_1 = z_0 \qquad z_1' = -\frac{1}{f} z_0 + z_0'$$

This is represented on the computer by opening the switch 18U between the two integrators at the time of arrival of the particle. Thus the value of z calculated for this position is stored. It is then fed into the element potentiometer to generate $\pm \frac{z}{f}$. While the element switch is closed the value of z' stored in integrator O1 is changed accordingly. The time required for this change, i.e. the fictitious length k of the lens must be subtracted from the total length of the system. The s - sweep for the display oscilloscope is stopped for the same amount of time.

In order to study the behaviour of a full beam the initial conditions are automatically altered along the phase space ellipse at the origin. This is done using the sine/cosine generator composed of integrators 12,13. The circuit it controlled by the signals IC and OP in such a way that tracks with the same initial condions are calculated in both planes before the circuit moves to the next point on the ellipse. The number of points on the ellipse is determined by the potentiometers 22,23.

Overloading of the operational amplifiers is avoided by the circuit composed of the comparators 18,19. In case the variable z reaches the preset value (pot 13) in either polarity the control unit is reset and a new calculation is started with different initial conditions.

Technical realization

As can be seen from figure 1 the commercial analogue/hybrid computer contains all the components of the basic analogue circuitry. For the readout of potentiometer settings and amplifier outputs it is equipped with a digital voltmeter.

The analogue expansion and logic control unit provides the analogue components for the 22 element loops as well as the logic control signals (fig. 3). These signals are taken from a four decade counter counting the 100 kHz clock pulses which are generated by a quartz oscillator in the main computer. Begin and end of the element signals as well as the end of the computing time are set by four decade thumbwheel switches. The fixed length signals are proportional to the maximum computing time which can be selected in steps from 0.1 sec up to 40 sec for recorder output.

The display oscilloscope is a 2-channel x,y-storage oscilloscope with variable persistence. This characteristic makes it the ideal instrument since the persistence can be adjusted to a few seconds which is about the time required for a full run around the phase space ellipse.

Scaling

Scaling, i.e. relating the computer variables voltage and time to the problem variables displacement, divergence, element length, gradient etc. is required to make the computer circuit consistent with the original problem equations.

The result of the scaling procedure is the specification of the gains of the different amplifiers and integrators and a list of scale factors expressed in terms of the maximum values of the problem variables per unit computer reference.

Potentiometer settings and amplifier outputs are always present as the ratio of the actual value of a problem variable to its maximum value.

Example

Figures 4-7 show the computer output for the optics of a small single stage accelerator.

Scales:

Particle tracks: 5 mm/div   vertical   Phase space: 5 mrad/div vertical
                    75 cm/div   horizontal   2   mm/div horizontal

The top diagrams on each figure are related to the x - plane.

The system consists of the ion source followed by an einzel lens, double focusing $90°$ - analyzing magnet, acceleration tube and electrostatic quadrupole triplett. The particles are extracted from the ion source at an energy of 30 keV and accelerated up to a final energy of 350 keV.

24 individual tracks were calculated in either plane in the fastest mode of the computer requiring about 3.5 seconds for a complete run.

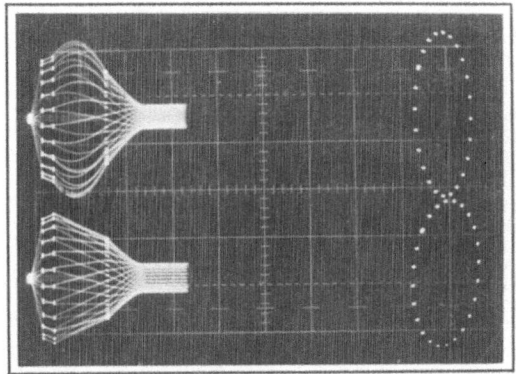

Fig.4 Source - Analyzing Slits

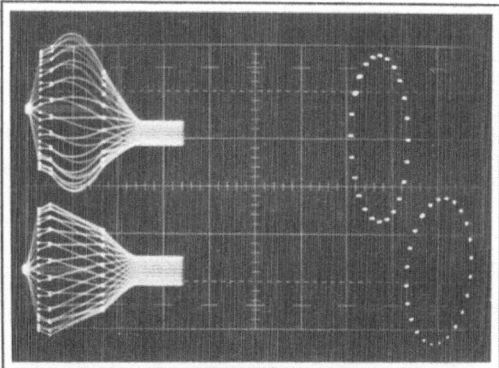

Fig.6 Source - Target
       Out of focus

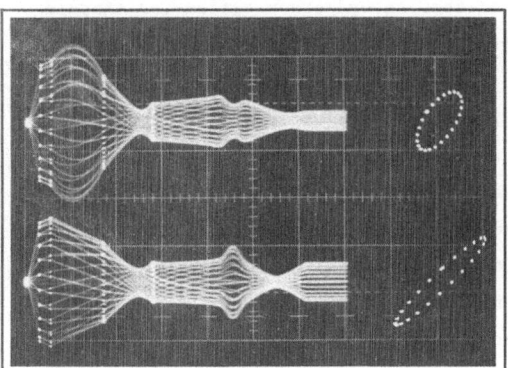

Fig.5 Source - Analyzing Slits
       $\frac{M}{\Delta M} = 250$

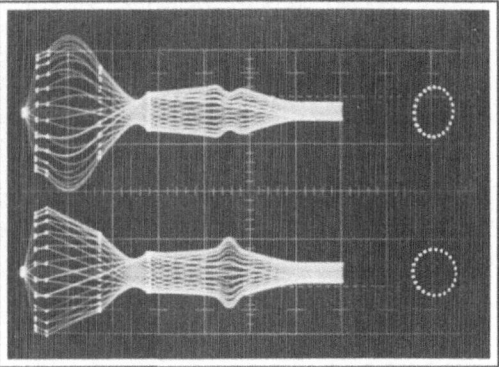

Fig.7 Source - Target
       Focused

Conclusion

The analogue computer system proved its usefulness in the design and in studies of the characteristics of different accelerator installations over many years. The results of the calculations are in very good agreement with real accelerator data.

The apparatus is the ideal tool to demonstrate beam optics to students and accelerator operators since the effect of a change of any of the parameters is immediately visible on the oscilloscope.

References

1. R.H. Good, O. Piccioni, Rev. Sci.Instr. 31(1960) 1035
2. K.G. Steffen, High energy beam optics (Interscience Publ.,New York, 1965)
3. R.N. Hansford, R.I. Aspley, Proc.2nd Intern. Conf. Magnet Technology (Oxford 1967)
4. AEG/TELEFUNKEN ADB 009 1068(1968)
5. K. Brand, Nucl. Instr. Meth. 84, 21(1970)
6. C.J. Davisson, C.J. Calbick, Phys.Rev.,42,580(1932)

A Monte Carlo Beam Transport Program, REVMOC

C. Kost
TRIUMF, University of British Columbia,
Vancouver, B.C. Canada.

P.A.Reeve
TRIUMF, University of Victoria,
Victoria, B.C. Canada

1) Introduction

REVMOC is a Monte Carlo program which does second order optics for charged particle beam lines and includes the effects of multiple scattering, decay, nuclear scattering and energy loss. The program is similar to TURTLE[1] but includes more sophisticated algorithms for atomic and nuclear scattering. Also user defined or internally generated 5 dimensional phase space distributions, such as uniform, Binomial[2], Sangford Wang[3], etc., can be specified. Program output includes beam loss bookkeeping and plots or multi-dimensional histograms of beam distributions, for both the original and/or particles which have undergone interactions or decay. Histogram correlations up to a maximum of 3 dimensions can be output at or between any labelled elements in the beam line.

2) Beam Optics

The beam optics is done using second order matrices, including cross coupled produced second order terms. Available beam elements include drift, quadrupoles, dipoles, fringe fields, pole edge rotations and curvature, sextupoles, solenoids, separators, filters, alignment and arbitrary transfer matrix. A dipole magnet can have second order pole surface shaping. Fringing fields of bending magnets include linear drop, clamped Rogowski and square edged. For sextupoles the actual momentum of the particles is used, therefore the chromatic abberations are accurately calculated. Separators can be crossed field devices, with D.C. or R/F operation. Transfer matrices are first order only, although the actual momentum of the particle is used. The phase of the fundamental and harmonic voltages can be set to correspond to the time of arrival of the beam at this element by knowing the central path length from the target to this element and the central ray velocity. The filter element is used to remove unwanted particles from the calculations, mainly to save CPU time. The rotate element, rotates the coordinates at a given point in the beam line

allowing the user to orient a magnet differently from its normal configuration
or to simulate a rotational misalignment of an element about the beam axis. The
centroid shift element shifts the beam coordinates of each particle by adding
the shifts to the current values. Thus a single beam line element may be
misaligned by putting a beam shift before and after the element. Drift lenghts
can include degraders or collimators, and both atomic and nuclear effects are
calculated. A collimator has two apertures, an inner aperture which contains
vacuum and an outer aperture which is opaque, with a region of matter between
them. A collimator can have cross-sectional shapes which are rectangular, or
elliptical, and along the beam direction be straight or cone shaped. The user
can control the maximum step size in tracking through collimator material.

3) Atomic Effects

For multiple scattering calculations, both scattering angles and
displacements are calculated in both planes and are chosen randomly
from distributions deduced by Moliere[4]. The program assumes that the
particle energy is constant and equal to its value half way through the
scattering region. Up to three atomic species may be specified in each region,
and the calculations take account of the Hungerford[5] relativistic
correction and are done up to and including the second order correction terms in
the Moliere distribution.

For energy loss calculations the program uses either a Gaussian, or a
Landau[6] distribution, depending upon the effective thickness of the
degrading material. As the particles are traced through the system,
the ionization energy loss for each region is selected randomly from an
internally generated table, the accuracy of the table is optimized for 600 MeV
protons and 200 MeV pions, for a wide range of materials.

For electrons the radiative energy loss is calculated using a formula due to
Rossi[7]. Any photons or secondary electrons which might be emitted in the
radiative energy loss process are neglected. The program assumes that the
central momentum of the beam does not change during energy loss process, and it
does not calculate energy loss for decay products. Spin polarisations for
muons can be set and tracked through beam systems.

4) Nuclear Effects

The decay length of particles is calculated from the lifetime and central
momentum of the beam. A two body final state is assumed to result from
particle decay. The decay angle is chosen randomly from a uniform distribution
in the center of mass of the decaying particle. This is correct only if the
final state has orbital angular momentum of zero. The energies of the decay

products are then determined from their masses, which are specified as input data. Only the first decay product is tracked through the rest of the system. It is assumed that this decay product undergoes no absorption, nuclear scattering, energy loss or decay in traversing the rest of the system. Multiple scattering, however, is allowed to occur. Up to two different two body decay modes may be specified. The branching ratios and masses for the decay products are given as input data.

For each region in the system, a nuclear absorption length is calculated. For each atomic species, the absorption cross section per nucleus is calculated from an empirical formula by Williams[8]. From the density of material in the region and the proportions by weight of each atomic species, which are specified as input data, the absorption length is calculated. The nuclear scattering length Λ_e is obtained from the nuclear absorption Λ_a by means of the relation.

$$\Lambda_e = (\sigma_a/\sigma_e)\Lambda_a$$

where σ_e is the nuclear scattering cross section per nucleon in mb and σ_a is the absorption cross section per nucleus in mb. Thus beam elements of different materials will have different elastic scattering cross-sections. The angular distribution of the elastically scattered particles is assumed to be a forward diffracting peak of the form:

$$d\sigma/d\omega = \alpha e^{-(\theta_{lab}/\theta_o)^2}$$

where
$$\theta_o^2 = 0.333\ A^{0.666}\ (P_o/M_\pi c)^2$$

M_π = mass of pion

A = atomic weight of nucleus

This distribution is in reasonable good agreement with experimental data in the GeV energy range. When a particle is being traced through the system and a nuclear scattering occurs, the atomic species on which the scattering takes place is determined with probabilities given by the proportion by weight and total elastic cross section for each species. The scattered particle is then traced through the rest of the system, in which further interactions such as decay, absorption, scattering or energy loss may occur.

For each region of the system a total interaction length is defined by the relation

$$\Lambda^{-1} = \Lambda_d^{-1} + \Lambda_a^{-1} + \Lambda_e^{-1}$$

As a particle is traced through the system the distance L to the next interaction is determined at the beginning of each region by the usual Monte Carlo technique, from the distribution $\exp(-L/\Lambda)$. If L is greater then

the length of the region, then no interaction is assumed to occur, and the
particle is traced through to the beginning of the next region. If L is less
than the length of the region, then the particle is traced up to L, where an
interaction is asssumed to occur. The type of interaction is decided with
probabilities determined by the relative magnitudes of the interaction
lengths.

5) Input/Output Data

For the input beam parameters, the default distribution is a uniform
distribution in momentum and position and angle, but other distributions
such as the BINOMIAL can be specified. The binomial distribution can be used
to produce hollow, uniform, quadratic and Gaussian shapes. Sangford Wang
distributions for pion and kaon production with protons can also be used. Both
the central position and the widths of beam distributions can be set. The
initial coordinates for each particle are chosen randomly within the specified
input distributions. The seed for the first random number may be chosen if
required. The number of particles to be tracked is also chosen on input. For
checking optics parameters or to continue a previous case the starting ray
coordinates may be read from an input file while the phase space coordinates
of the rays at any element(s) may be written to an output data file. Also a
fitting routine may be invoked to determine the 1st, 2nd and the important 3rd
order aberration coefficients (transfer matrix coefficients), from the Monte
Carlo data.

Data can be output in several different ways, ranging from the phase space
coordinates of individual rays to plots and histograms. Data can also
be output in up to three dimensions, which may be all at the same position
in the system; or one or more dimensions may be at different positions.
For example the spatial coordinates at an experimental target, can be outputed
against an angular coordinate at a production target at the beginning
of the beam line. Also the type of particle can be selected, it can be all
particles, the initial particles only, or the decay and scattered particles
only. Also for the decay particles the user can chose between only decay
particles which get to the end of the system, or the distribution of all decay
particles and their parent particles, whether they reach the end of the system
or not, etc.

6) Program Flow Chart

The order in which the subprograms are called are as follows. The MAIN
program calls READ which reads in all the data from unit 5 (and performs most of
the error checking done by the program). Next the MAIN program calls SETUP.

REVMOC: A Monte Carlo Beam Transport Program

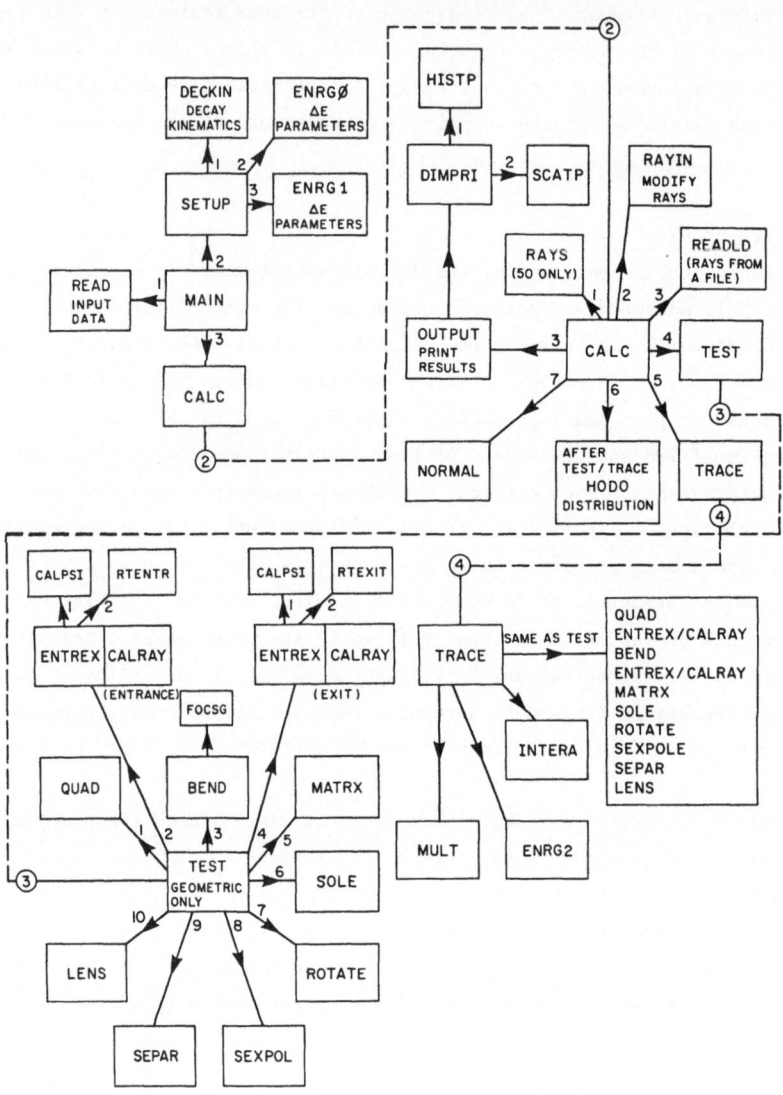

REVMOC FLOW DIAGRAM

It prints out the beam line setup if in the long ouput mode, and it calculates the energy loss distributions, multiple scattering distributions, decay, and other interaction parameters for every beam element. It calls the three routines DECKIN,ENRG0,ENRG1 to handle decay kinematics and energy losses. SETUP returns back to MAIN which then calls CALC where individual particle tracing begins by the selection distribution, a read from unit 3, or specified by a call to a user supplied routine RAYIN which may respecify some or all of the previously specified coordinates of the ray.)

First, there is an initial run with no decay, multiple scattering, absorption, or nuclear scattering. The task of tracing one particle down the entire beam line is done by TEST (called from CALC). Then there is the final run where decay and all interactions can occur (collimators and absorbers material have an influence). The task of tracing one particle down the entire beam line being done by TRACE again called from CALC. If binning is requested (for output distribution printing) subroutine HODO is called immediately after calls to TEST and TRACE from CALC. The optics is handled by RTENTR, BEND, RTEXET, QUAD, MATRX, SOLE, ROTATE, SEPAR, and SEXPOL called by both TEST and TRACE. TRACE also calls MULT for multiple scattering, INTERA to handle decay, and nuclear scattering and absorption and calls ENRG2 for energy losses. After CALC is finished with tracing it calls OUTPUT which prints out all the outputs except histograms and 2D-bins demanded by NSPACE cards, these being done by DIMPRI called by OUTPUT.

The program then returns to MAIN which again calls READD to handle possible replacement cards to the current case. If none exist the program terminates. If they do exist then READD updates the current status of the case to be run and runs this new case as before. More details on the program are available in ref.9).

7) Acknowledgments

The authors would like to thank their colleages at TRIUMF, in particular, B. Henin, W. Hsieh, P. Kitching, D. Lobb, and G. Stinson for making significant contributions to the development of this program, and to NSERC for funding.

References

1) K.L. Brown, Ch. Iselin, 'DECAY TURTLE' CERN 74-2 (1974).
2) W. Joho, Representation of Beam Ellipses for Transport Calculations, SIN-REPORT TM-11-14 (1980) 18-30
3) J.R. Sanford, C.L. Wang, BNL-11479 (1967)
4) G. Moliere, Z. Naturforsch. 2a (1974) 133,3a (1948) 18, 10a (1955) 177;Nuovo Cimento 7 (1958) 720, Z Physik 156 (1959) 318.
5) E.V. Hungerford et al. Nuclear Physics A197 (1972) 515-528
6) L. Landau, J. Phys. (USSR) 8 (1944) 201
7) B. Rossi, High Energy Particles, Ch. 5 (Prentice-Hall, New York (1952)
8) R.W. Williams, Reb. Mod. Phys. 36 (1964) 815
9) C. Kost, P. Reeve 'REVMOC A Monte Carlo Beam Transport Program' TRI-DN-82-28

MULTIPARTICLE CODES DEVELOPED AT GANIL

Mrs. J. SAURET, Mr. A. CHABERT, M. PROME

GANIL. BP 5027. 14021 CAEN-CEDEX. FRANCE

ABSTRACT : A general multiparticle code was developed for studying particle motion in cyclotrons. Related to the structure of the GANIL separated sector cyclotrons, it could be adapted to other configurations. A general description of this code is given and its usefulness illustrated by some results concerning beam dynamics into the separated sector cyclotrons of GANIL.

I - INTRODUCTION

The three dimensional multiparticle code NAJO was developed in the case of a 4 separated sector cyclotron, in fact it can accommodate any magnetic field configuration and its main limitation comes from the shape of the accelerating gaps which are presently restricted to radial ones. Accelerating field effects are expressed as kicks applied at the gap centers allowing for a complete decoupling in the treatment of the magnetic and electric fields.

Simplified versions have been derived, restricted either to the median plane (JOAN) or to a single particle in this plane (ANJO). In its most general version the code takes into account space charge effects. A more precise description of these codes including listings and examples is given in an internal report[1].

II - THE 3 DIMENSIONAL MULTIPARTICLE CODE NAJO

II.1 - The magnetic field treatment

Equations of the motion of particles in the magnetic field are written in cylindrical coordinates, the azimuth θ being the independant variable :

$$
1 \quad \left|
\begin{array}{l}
dt/d\theta = r/(v \cdot p_\theta) \\
dr/d\theta = r \cdot p_r/p_\theta \\
dz/d\theta = r \cdot p_z/p_\theta \\
dp_r/d\theta = p_\theta - q/(mv) \cdot \{B_z - P_z/p_\theta \cdot B_\theta\} \cdot r \\
dp_z/d\theta = - q/(mv) \cdot \{P_r/p_\theta \cdot B_\theta - B_r\} \cdot r
\end{array}
\right.
$$

with $p_r = v_r/v$; $p_\theta = v_\theta/v$; $p_z = v_z/v$; $p_r^2 + p_\theta^2 + p_z^2 = 1$

Field components are restricted to first order in z :

$\{B_z\}_{z=0}$; $B_r = z \cdot \{\partial B_z/\partial r\}_{z=0}$; $B_\theta = z \cdot \{\partial B_z/\partial \theta\}_{z=0}$

so that we just need the median plane field $B_z(r,\theta)$.

Equations are integrated using the 4th order Runge-Kutta method with a constant integration step equal to 1° ; note that this method requires field values at the middle of the integration step, that means every 0.5°.

The magnetic field is introduced point by point in the median plane using an ad-hoc subroutine "LECB" which gives B (I, J) every 0.5° over 360° (J = 1,721) on a set of equally spaced radii (presently a maximum of 90 radii : I = 1,90).
At each azimuth, field components have to be calculated for each particle of the

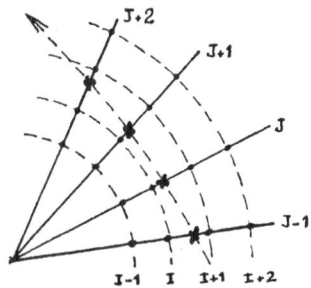

Figure 1

bunch according to its radial position. This is done using a third order interpolation in r for B_z and B_r (between I-1 and I+2) and two successive linear interpolations for B_θ (in θ and r, respectively between J-1 and J+1 and between I and I+1). It proves to be very important to use at a given azimuth the same polynomia for all the particles, so forth their coefficients are calculated for the central one; moreover these coefficients are computed once by integration step using for the 3 azimuths the value of I determined at the beginning of the step. Note that the field introduced by means of "LECB" takes into account all the magnetic elements present in the cyclotron (injection, ejection) as well as all the field perturbations ; field values come either from measurements (on models or on the real elements) or from an analytical description of these elements or of the field perturbations.

II.2 - The gap crossing treatment

Kicks at gap crossing are calculated according to the thin lens approximation (constant velocity and position). With a linear variation of V along the gap given by :
$V = V_0 + V_p (r-r_6)$, we obtain with $\phi = \omega t + \phi_0$:

$$2 \quad \begin{vmatrix} \delta W = T \cdot ch(\eta z) \cdot V(r) \cdot \cos\phi \\ \delta(mv_r) = -T \cdot ch(\eta z) \cdot V_p/\omega \cdot \sin\phi \\ \delta(mv_z) = -T \cdot \eta \, sh(\eta z) \cdot V(r)/\omega \cdot \sin\phi \end{vmatrix}$$

In order to preserve the coherence of this formalism T and η have to be the same for all the particles. They are calculated for the central one :

$$3 \quad \begin{vmatrix} \eta = \frac{\omega}{v_\theta} \sqrt{1-\beta^2} \quad ; \quad T = \frac{Q}{A} \cdot \frac{\sin(\omega\theta_g r/v_\theta)}{\omega\theta_g r/v_\theta \, ch(\eta a_g)} \end{vmatrix}$$

$2\theta_g$ and $2a_g$ being the azimuthal and vertical apertures of the gap. Potential V(r) is introduced point by point at 12 equally spaced radii for 9 equally spaced frequencies so that we have first to interpolate between frequencies to find V(r) at the frequency we use and then a linear interpolation gives V_p and V(r) for each particle.
All these operations are done in the subroutine "CAVHF" which is called each time the end of an integration step fits with the azimuth of a gap center (gap center azimuths are then restricted to an exact number of degrees).

II.3 - The injection of particles

Beam dimensions are always defined as two times the RMS value of the related distribution, that means :

$$\Delta x = 2\sigma \text{ with } \sigma = \int_{-\infty}^{+\infty} u^2 p(u) \, du \, / \, \int_{-\infty}^{+\infty} p(u) \, du \quad (4)$$

On the same way, two dimensional emittances are defined by :

$$\varepsilon_x / \pi = 4 \{ (\overline{\Delta x^2} - \overline{\Delta x}^2)(\overline{\Delta x'^2} - \overline{\Delta x'}^2) - (\overline{\Delta x . \Delta x'} - \overline{\Delta x} . \overline{\Delta x'})^2 \}^{1/2} \quad (5)$$

Particles are injected at a given azimuth (usually the center of a valley) where the 6 coordinates of the central one (W_0, ϕ_0, r_0, r'_0, z_0, z'_0) and the characteristics of the uncorrelated bunch ($\Delta W/W$, $\Delta \phi$, ε_r, Δr, ε_z, Δz) are given.

The 6 dimensional phase space is filled with particles whose coordinates are randomly distributed inside ellipsoids of axis equal to 1 (usualy 2 distributions, one in the ΔW, $\Delta \phi$, Δr, $\Delta r'$ phase space and the other in the Δz, $\Delta z'$ space). By means of linear transformations, the coordinates are centered and the distributions corrected to obtain a unitary covariant matrix. Then we introduce the dimensions of the bunch to obtain r.m.s. sizes and emittances equal to the input values. With this method a number of 40 particles is enough to obtain bunch sizes and emittances with a good precision, that means than variations of these values between different cases are clearly seen. We then introduce correlations in the following order (subscript "dc" being related to an uncorrelated value :

$$\begin{aligned}
\Delta W/W &= (\Delta W/W)dc + C_{W.\phi} (\Delta \phi)dc \\
\Delta r &= (\Delta r)dc + C_{r.r'} (\Delta p_r)dc + 0.5 \cdot C_{W.r} (\Delta W/W) \\
\Delta \phi &= (\Delta \phi)dc - C_{W.r} (\Delta p_r) \, dc \\
Z &= (Z)dc + C_{z.z'} (\Delta p_z) \, dc
\end{aligned} \quad (6)$$

All these operations are done inside the subroutine "REMPLI".

II.4 - The outputs of the code

The characteristics of the accelerator and of the injected beam being now defined, the code simulates the acceleration of a bunch of particles ; the integration step being 1°, the beam characteristics can be obtained every degree. Usually ouputs are asked each time the bunch crosses a particular point (center of a valley or of a magnetic sector). Anyway at the end of integration step we can obtain, calling for the subroutine "STATIS", the various characteristics of the beam at this azimuth(size , emittances) and the coordinates of the central particle. Graphical displays of the bunch projection on various planes are available on request by calling for the subroutine "DGRAF" and "GRAFPR". At the end of the acceleration process, a graphical display of various parameters against the number of turns at the azimuth 0° can be asked for (subroutines "DGRAF" and "GRAFEV").

II.5 - Space charge effects simulation/[2]/

Space charge effects were treated by two methods : the particle to particle interaction method and the equivalent continuous distribution method, these two methods taking into account neighbouring bunches as well as image effects. Space charge forces were computed 4 times per turn. Results given by the particle to particle interaction method (used with 100 particles) are rather noisy, the other method gives smoother results and consequently was used four our investigations.

III. - SOME RESULTS ON BEAM DYNAMICS INTO THE SEPARATED SECTOR CYCLOTRONS OF GANIL

All numerical simulations of the acceleration of ions in the cyclotrons of GANIL were made using these codes. Exact parameters and tolerances of the various elements were checked, characteristics of the expected beams determined. Moreover we were able to confirm various important theoretical effects such as the influence of the correlations at injection and even to exhibit small effects, such as axial coupling effects, prior to their theoretical description. GANIL accelerates particles since June 1982 and beam characteristics as given by our beam probes are always very close to numerical results.

Various paper describe in details the results we obtained. Let us shortly report some of them.

III.1 - Betatron oscillations as simulated on computer and as measured on the beam

Figure 2 shows betatron oscillations of an Ar^{+16} beam accelerated in SSC2 (80 first turns) as measured by radial probes on the axis of the magnetic sectors and as computed using ANJO (the field introduced in the code is derived from field mappings on the cyclotron). As can be seen there is a very good agreement (Fitting is obtained by varying the initial conditions of the injected beam : r and r'). The difference between the two ($\delta r_2 - \delta r_4$) curves comes from a slight difference in the field level of the corresponding sectors. A small residual field perturbation due to injection elements accounts for the sharp variation around turn number 40.

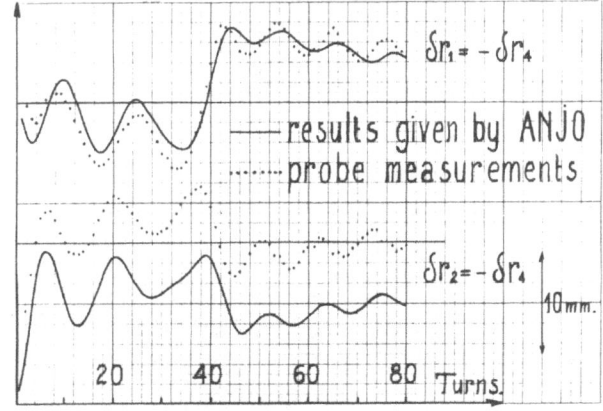

Figure 2

III.2 - Emittance growth associated with phase compression /³/

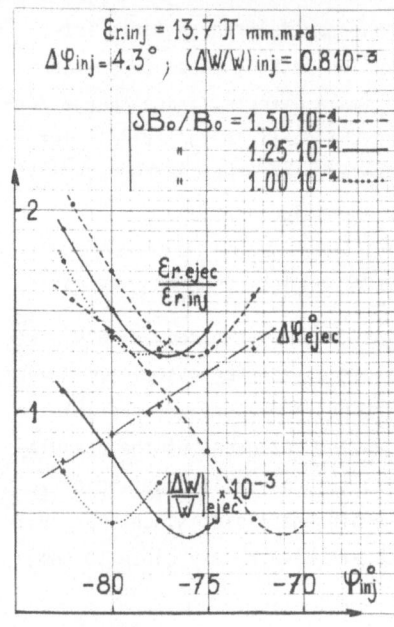

Figure 3

At high energy (ν_r depends on W) and with phase compression a special type of coupling appears between radial and azimuthal motions due to the fact that we need a precessional injection.

The figure 3 (obtained from results given by JOAN) shows the influence of the injection phase on the phase and energy spreads and on the radial emittance at the end of acceleration : an ad-hoc main field correction allows to obtain a good coincidence of the two ΔW and ε_r minima giving thus a good tuning. Tolerances on the field level B_0 and on the phase of injection ϕ_{inj} are clearly seen.

III.3 - Axial emittance effects on median plane phase space /³/

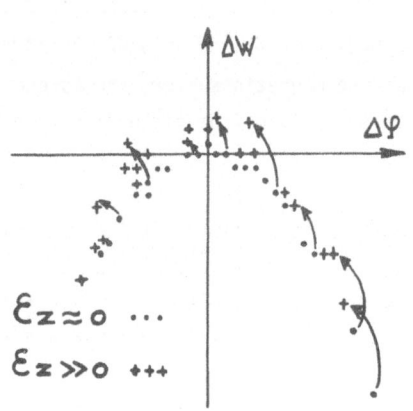

Figure 4

Due to the Z energy gain dependence there is a coupling between axial and median plane phase spaces. Using NAJO we are able to see this effect. On figure 4 we show the the $\Delta\phi$-ΔW distributions at the end of acceleration for particles having the same initial median plane distribution and two different axial emittances ($Z,Z' \gg 0$ and $Z,Z'=0$) : arrows indicate the shift for each particle in the $\Delta\phi$-ΔW plane.
With usual values of axial emittance this effect is very small and at least in the case of GANIL there is an almost total decoupling between axial and median plane so that most of our investigations were restricted to median plane using JOAN.

III.4 - Space charge effects /⁴/

Using NAJO with space charge forces included, we were able to simulate longitudinal and radial space charge effects. Typical results are shown on figure 5 : for the 2 cases emittances at injection are : $\varepsilon L = 7.5° * 5.10^{-3} \pi$ ($\Delta\phi.\Delta W/W$), $\varepsilon r = 41.8$ πmm.mrd $\varepsilon z = 30.2 \pi$ mm.mrd.

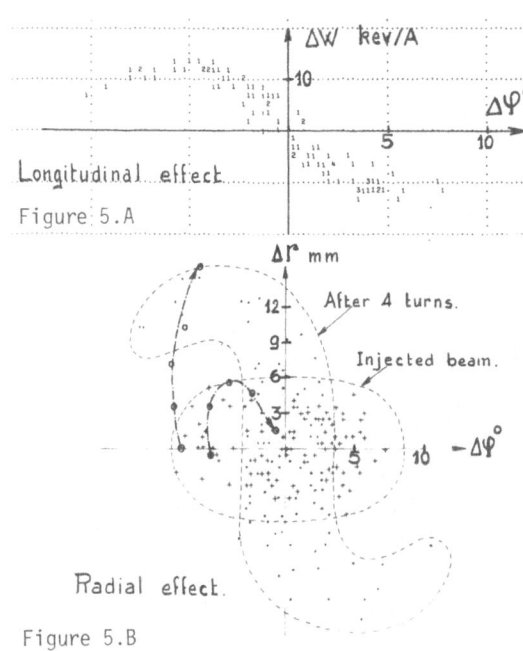

Figure 5.A
200 µA beam of C^{+2} accelerated from 0.33 to 5.3 MeV/A and simulated by 100 particles.

Isochronism is preserved but due to space charge forces front and tail particles gain or lose energy as seen from the tilt of the ΔW-$\Delta \phi$ distribution.

Figure 5.B
200 µA beam of U^{+6} accelerated from 16 to 33 kW/A and simulated by 100 particles.

Under strong radial space charge forces, outer particles of the bunch move toward the tail and inner ones toward the head producing during acceleration alternate phase and radial expansion or compression.

Figure 5

IV.- CONCLUSION

These codes were very usefull during the design period of GANIL. Now, due to the very good agreement between simulated and measured beam characteristics. The same codes are currently used to obtain, from beam measurements, more precise machine properties and then deduce corrections to apply in order to increase beam quality.

Acknowledgments

Mr GENDREAU and Mr LAPOSTOLLE took a major part in the theoretical studies and the exploitation of the code results.

References

/1/ Groupe Théorie-Paramètres : "Les programmes ANJO, JOAN, NAJO".
    GANIL 80 R/132/TP/06. Octobre 1980.

/2/ A. Chabert, T.T. Luong, M. Promé : "Separate sector cyclotron beam dynamics with space charge". I.E.E.E. Vol. NS -22 n°3 June 1975, p.1930.

/3/ P. Lapostolle : "Recent developments on beam dynamics in cyclotrons" 9th Int. Conf. on Cyclotrons and their Applications. 1981. CAEN, France.

/4/ A. Chabert, T.T. Luong, M. Promé : "Beam dynamics in separate sector cyclotrons" 7th Int. Conf. on Cyclotrons and their Applications. 1975. ZURICH, Switzerland.

MIRKO - AN INTERACTIVE PROGRAM FOR BEAM LINES AND SYNCHROTRONS

B. Franczak

GSI, Gesellschaft für Schwerionenforschung
Postfach 110541, D-6100 Darmstadt 11, Fed. Rep. of Germany

1. Introduction and Summary

The ion-optical design of beam lines and synchrotrons is usually not done by a single run of one program. It takes many iterations of calculation, examination of results, and modification of input data. In most cases the first order design has to be followed by the investigation of higher order effects, i.e. chromatic and geometrical aberrations or resonance phenomena.

The interactive computer program MIRKO is operated from a terminal and has a command structure, which enables the user to edit data, perform calculations, and to obtain alpha or graphics output on the terminal in any desired sequence. With graphics one can recognize the properties of an optical system much faster than with numbers only. Thus modifications of input data depending on the results of calculations can be made easily without stopping and restarting the program.

Higher order effects can sometimes influence the first order design. Therefore, particle tracking capability was included in MIRKO as well as the calculation of stop band widths for synchrotrons. Consequently a large variety of phenomena can be studied with one program in one session based upon exactly the same data for the optical system and the possibility of fast switching between the different features.

2. Linear Optics

Optical elements as well as the beam are represented by 6x6 matrices. The theory has been described elsewhere[1]. The following elements can be treated: drift space, quadrupole, dipole, pole face rotation, beam rotation, misalignment, and arbitary matrices. For nondispersive systems the matrix dimensions can be set to 4, thus reducing the required CPU-time considerably.

As an alternative to the ellipse notation single particles can also be transformed through the system. This allows the investigation of the behavior of particles with different momenta in dispersive systems.

It is easy to calculate the properties of a given optical structure, but difficult to find parameters (i.e. settings or positions of quadrupoles) to achieve given properties. Therefore, up to 6 parameters of the optical elements can be defined as variables to meet some fitting constraints. These can be elements of the beam matrix at the end of the system, elements of the transfer matrix, or their sum or difference, and several others. Of course any desired coupling of variables is possible, e.g. symmetric triplet or variable position of a triplet at constant total length.

Beam envelopes are calculated along the system and can be displayed on a graphics terminal. Fig.1 shows the horizontal envelope through a focussing channel with triplets, one of them beeing misaligned.

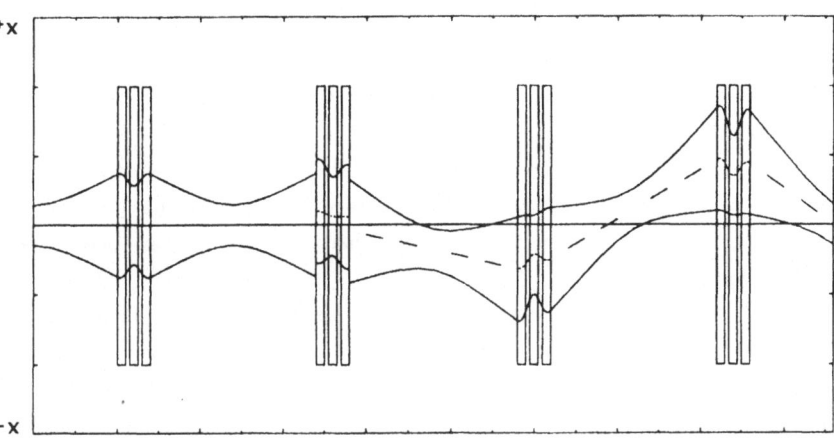

Fig.1: Horizontal envelope in a triplet channel, the second triplet is misaligned, and the dashed line shows the center of the beam.

The vertical envelope can also be displayed, alone or combined with the h.e., as well as the coordinates of a single particle, or ellipses in transverse phase space anywhere in the system just by entering the appropriate commands.

Apertures of the optical elements define the acceptance of the system. It is calculated if desired and can be shown as a polygon on the screen. To modify the acceptance, slits can be included in the sequence of optical elements.

Cyclic Accelerators consist basically of beam lines with a periodic structure. Therefore, an optical system calculated with MIRKO can be considered as a superperiod of a synchrotron. Now the frequency of betatron oscillations (Q-value), the transition energy, and the stationary ellipsoid (alpha and beta functions) can be evaluated and set as the input beam. It is possible to tune the machine to given Q-values using the strengths of the quadrupoles, or anything else, as variables.

At high currents the electrostatic forces between the particles cannot be neglected. In the case of constant current density over the cross section, the defocussing forces become linear, and under this assumption MIRKO can calculate the effects of space charge. Thin lenses with strengths depending on the beam dimensions and current are inserted in short intervals and included in the system. An application is the calculation of the incoherent tune shift in a synchrotron; numerical results are in good agreement with the well known Laslett formula[2].

3. Aberrations and Chromaticity

The chromaticity is defined as the derivative of the tune (Q-value) with respect to the momentum deviation. It is determined by aberrations of second order in the system, while higher order contributions cause the chromaticity itself to be dependent on momentum.

To deal with all these effects in a correct way, one can define a reference particle in MIRKO, which is tracked through the system and takes into account all aberrations according to its actual coordinates in six dimensional phase space. The transfer matrix of the system then refers to this particle rather than the one on the geometrical axis and is in general slightly different.

In a quadrupole the only aberration of second order is the chromatic one, and it is treated as follows: momentum deviation of the reference particle changes the magnetic rigidity used for calculation of the transfer matrix. All higher order geometric aberrations are not considered.

In a dipole with pole face rotation three effects of at least second order take place: the lengths of drift space adjacent to the dipoles changes according to transverse coordinates of the reference particle. The effective rotation angle contains the slope of that particle at the magnet entrance or exit. The radius of curvature depends on momentum, as well as the effective bending angle. All these effects are treated with simple geometrical methods and lead to an exact transformation of the reference particle. The matrices are calculated using these modified lengths, angles, and radii.

Drift spaces have third order aberrations, which can also be calculated. They do not contribute to chromaticity, but a little to its derivative.

For chromaticity evaluation a particle with small momentum deviation is put on its closed orbit. The matrix calculated on this orbit as described above gives the correct tune for this particle. With the tune for a particle of correct momentum the chromaticity can be found easily.

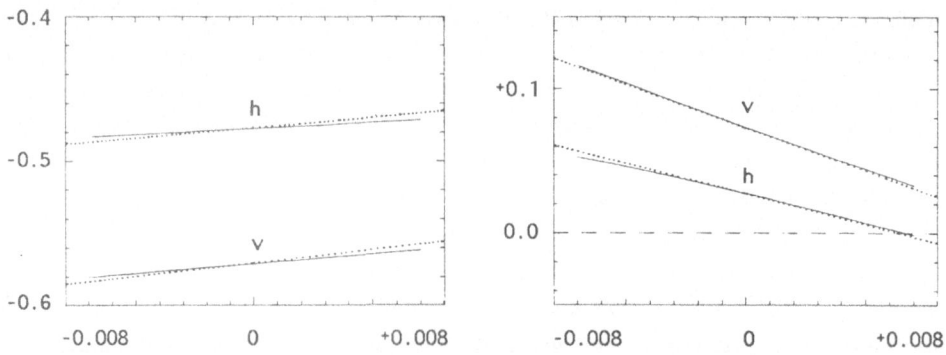

Fig.2: Horizontal (h) and vertical (v) chromaticities for Los Alamos proton storage ring calculated with MIRKO for momentum spread $\Delta p/p = \pm 8 \times 10^{-3}$. Left: Natural chromaticity, right: corrected with two families of sextupoles. The dotted line indicates chromaticity and its derivative as found in ref.[3].

A. Dragt describes a method for exact numerical calculation of chromaticity[3]. He also gives detailed lattice data for the proposed proton storage ring at Los Alamos. Using these data chromaticity calculations with MIRKO differ less than 10^{-4} from his results. Also with sextupoles for chromaticity correction the difference is about 10^{-3}. How sextupoles are treated in MIRKO will be described in the next Chapter. Some of the results are shown on Fig.2. It should be noted that the slope of chromaticity is calculated from the 7th or 8th significant digit of the tune and since no aberrations of quadrupoles are considered, the agreement with [3] seems to be quite satisfactory.

4. Tracking through Nonlinear Elements

If an optical element is sufficiently described by a first order matrix, then a series of such elements will be represented by the product of matrices that is also of first order. If second order terms have to be included in the description of an element, then the multiplication of such matrices generates higher order terms, which may be of considerable importance, but have to be neglected if one is resticted to second order in matrices.

Therefore, in MIRKO no attempt was made to adopt a second order matrix formalism. Instead, normal elements are represented by first order matrices only and all multipoles are introduced by thin nonlinear lenses. This approximation seems to be tolerable, since multipoles are usually weak compared to the normal elements. Such a thin lens leaves the transverse spatial coordinates of a particle unchanged, while a contribution depending on these is added to the slopes (kick). Thus phase space is conserved when passing a thin multipole.

The magnetic potential of a multipole of order n is $\sim i(x + iy)^n$, where the real part corresponds to skew and the imaginary part to normal multipoles. The forces are obtained by differentiating with respect to the transverse coordinates. So the equations to calculate the kick can be obtained. For a normal sextupole (n=3) one gets: $\Delta x' = w \times (x^2 - y^2)$ and $\Delta y' = w \times (-2xy)$. w is a constant proportional to the sextupole strength B". Multipoles from dipole (n=1) to dodecapole (n=6) can be treated in MIRKO using the coordinates of the reference particle.

If one is interested in the behaviour of particles in the vicinity of the reference particle, the matrix of the multipole has to be evaluated. It is done by partial differentiation of the kicks with respect to the transverse coordinates. For n>2 the result is a matrix, that contains the coordinates and is, therefore, valid only for these particles since different particles produce different matrices. Usually the matrix contains coupling terms between x and y, but it is always symplectic. It is used for the calculation of chromaticities in presence of sextupoles, or eigenvectors of the motion close to an unstable fixed point.

Since the optical properties of all linear elements do not change for different particles, all elements between two multipoles are replaced by a single matrix, if tracking of many particles has to be performed. Thus a lot of CPU-time can be saved, but in this case no aberrations of dipoles can be taken into account.

In a synchroton there are invariants of transverse motion, which are ellipses in the linear case, and more complicated closed or open curves if the machine contains multipoles. If one starts with an arbitrary set of coordinates on one of these curves and performs one or more transformations around the machine, then the coordinates are different, but still on the same curve. Now the transformation can be repeated once more with these coordinates as input and so on. So the whole curve can be scanned, if a sufficient number of transformations is made.

MIRKO provides interactive graphics for investigation of problems concerning fixed points, separatrices, slow extraction, nonlinear motion etc. The starting coordinates in tranverse phase space can be set with the cursor and a key stroke represents a command i.e. to transform and display several times, to look for fixed points, generate pattern of input points, etc. Fig.3 shows as an example the particle behavior close to a third integer resonance used for extraction and inside a half integer stopband stabilized with octupoles.

If details of the particle motion obtained with tracking are not important, it is also possible to calculate stopband widths due to multipoles based on the theory of Guignard[4] using unperturbed beta functions and multipole strengths.

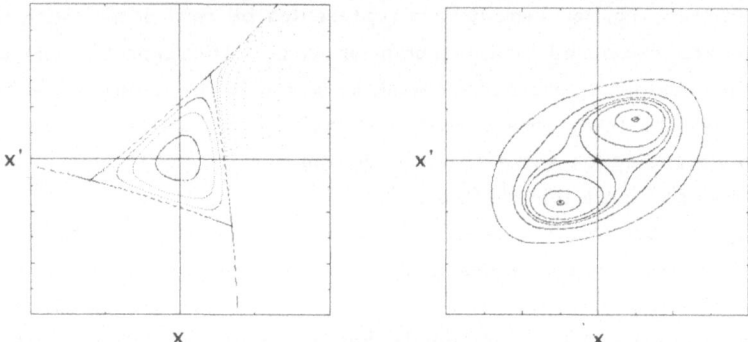

Fig.3: Particle tracking in horizontal phase space. Left: close to a third integer resonance, three unstable fixed points with separatrices and several closed and open lines for invariants of motion, dots represent particle coordinates after 3 revolutions. Right: inside a half integer resonance exited by a perturbing quadrupole, stabilized for higher amplitudes by octupoles.

5. Commands and Operation

All actions of the program, such as editing data, starting calculations, displaying data, etc. are performed by entering the appropriate commands and parameters if necessary. There is no menu or selection table. This has the advantage, that all functions can be accessed in arbitrary sequence and the user is not restricted to follow any fixed procedure. On the other hand this freedom means, that one has to know, what the program can do and how to explain to the program, what he wants to do.

To make the beginning easier, an information system is included in MIRKO, which is activated any time the user types a "?" at the terminal. If it is typed immediately behind a command, this will be explained instead of beeing executed. Typing it as response to prompting for parameters will cause these to be explained. A "?" instead of a command gives information about the command executed before or, if repeated, a general introduction to MIRKO.

The input format of the commands is flexible; abbrevations are permitted, prompting for parameters can be skipped by entering them in advance. One may type as many commands at a time, as fit into one input line. If sequences of commands have to be repeated several times, the definition of command procedures is possible. At present about 110 commands are available and the following list gives an impression of what can be done with MIRKO:

- Editing of optical systems - input of new elements, copy, move, delete groups of elements, change parameters, define variables, list elements on the terminal
- Optimization and fitting - define constraints, list on terminal, perform fitting calculations, restore previous status if iteration failed
- Graphics - plot of ellipses, envelopes on a graphics display, fit envelopes using the cursor, define mode of envelopes
- Alpha output on terminal - envelopes, ellipses at the end, kind of particle, beam current, behavior of reference particle, momentum resolution in phase space, transfer matrix, elapsed CPU-time
- Synchrotron design - define normal period, number of periods, evaluate tune, chromaticity, transition energy, closed orbit, tune the machine to given values, for multipoles: calculate stopband widths, perform particle tracking, find fixed points and separatrices
- Documentation - list all data describing the system on the line printer, plot envelopes and ellipses on the plotter, prepare data for auxliary programs: plot of tune and chromaticity vs. momentum, transition energy vs. tune
- Data storage - save system and beam on direct access dataset, save quadrupole settings only, save procedures, read system, rename, delete in dataset, skip to another system, list all stored systems.

References

1 - K.L.Brown, SLAC-Report 75 (1969)
2 - L.J.Laslett, Proc. 1963 Brookhaven Summer Study on Storage Rings, pp. 324-367
3 - A.J.Dragt, Exact Numerical Calculation of Chromaticity in Small Rings, Particle Accelerators Vol.12 (1982), pp.205-218
4 - G.Guignard, A General Treatment of Resonances in Accelerators, CERN 78-11

APERTURE STUDIES OF THE BNL COLLIDING BEAM ACCELERATOR
WITH REDUCED SUPERPERIODICITY*

G.F. Dell

Brookhaven National Laboratory
Upton, New York 11973

Summary

Chromatic properties of the Brookhaven CBA (Colliding Beam Accelerator) with one low β insertion in each of the three superperiods have been studied using the PATRICIA particle tracking program. Systematic multipoles of order $5 < n \leq 10$ as well as random multipoles of order $1 \leq n \leq 10$ are, along with random closed orbit errors and sagitta effects, included when determining the aperture of the lattice.

1.0 Introduction

Aperture studies on the CBA have been made with the PATRICIA[1] particle tracking program. Results for the machine operated in the standard mode have been reported previously.[2] The present report documents changes in aperture that result for operation of the CBA in a mode having three low β insertions designed to produce luminosities in the 10^{33} cm^{-2} sec^{-1} range.

The CBA lattice has 366 dipoles, 174 quadrupoles, and 540 multipole elements that are used to represent the higher order fields of the dipoles and quadrupoles. In addition, all dipoles in the 42 regular cells have either a focusing or defocusing chromaticity correcting sextupole at each end.

Separate sets of multipole coefficients are input for focusing quadrupoles, the defocusing quadrupoles, and the dipoles of the regular cells as well as for the dipoles and quadrupoles of the insertions. Each multipole set contains mean values as well as rms deviations for both normal and skew coefficients. The values of the multipole coefficients were obtained from the CBA Parameter List[3]. The average values of all multipoles of order $n \leq 5$ have been set to zero using the consideration that correction coils will be used to cancel the contributions of these multipoles.

For each magnet a set of normal and skew coefficients was generated according to a gaussian distribution that extends to three standard deviations on each side of the mean. The normal and skew coefficients were then combined to yield complex multipole coefficients used by PATRICIA.

Random closed orbit errors were generated according to a gaussian distribution whose modulus varied as $k\beta^{1/2}$. The rms values of the resulting errors were

_____
*Work performed under the auspices of the U.S. Dept. of Energy.

determined at the focusing and defocusing quadrupoles of the regular cells, and
the factor k was adjusted to give an rms closed orbit error of 0.5 mm in the x
and y directions at the focusing and defocusing quadrupoles, respectively. In
the particle tracking the treatment of the closed orbit errors was limited to the
extra kick associated with them.

2.0 CBA Aperture Scans

2.1 Standard Configuration

Aperture scans were made for the standard CBA having no low beta insertions
at a working point of ν_x = 22.631 and ν_y = 22.620 with a chromaticity of 2.0 in
both the x and y planes at $\Delta P/P$ = 0.0 and with the requirement of a 2 percent momentum aperture at the injection energy of 30 GeV. The working line for the linear machine was determined for momentum deviations up to ± 1 percent. This
working line was used as a reference for all subsequent aperture scans. An
aperture scan was made for the CBA without multipoles. A second scan was made
for which the influences of magnetic multipoles and closed orbit errors were
included. For this scan the strengths of the quadrupoles in the regular cells
were adjusted to bring the tunes at each $\Delta P/P$ back to the tunes of the reference
working line.

At each $\Delta P/P$ particles having increasingly large x and y amplitudes and zero
x' and y' were tracked. An emittance ratio of $\varepsilon_y/\varepsilon_x$ = 1.0 was maintained as the
amplitudes in the x and y directions were increased until a point was reached
where the particle survived the requested number of turns at one amplitude but was
lost within that number of turns at the next amplitude. Particles were tracked
for 300 turns. A particle was considered lost when its amplitude $(x^2 + y^2)^{1/2}$
exceeded the machine aperture of 40 mm in any of the multipole elements.

The results of these scans are shown in Fig. 1. The bars on each point are
used to signify that the particle survived the requested number of turns at the
lower amplitude but was lost within that number of turns at the larger amplitude.
The solid line denotes the expected physical aperture defined by beam size in the
insertion quadrupoles for $-0.6 \leq \Delta P/P \leq 0.6$ percent and by dispersion in the regular cells for $|\Delta P/P| > 0.6$ percent. The aperture obtained by tracking without
magnet multipoles and closed orbit errors agrees with the physical aperture
derived from the machine functions. When the effects of magnet multipoles and
closed orbit errors are included, the aperture is reduced by $\sim$ 4.5 mm. The
dashed line has been included to identify the data for the scan with multipoles.

2.2 Low β Configuration

To provide luminosities of $\sim 10^{33}$ cm^{-2} sec^{-1}, insertions have been designed
with sets of dipoles between the crossing point and the first quadrupoles. These
dipoles can be energized to reduce the crossing angle between the two beams, and

the insertion quadrupoles can be adjusted so that either the standard or low beta options are possible. One insertion capable of operating in either the standard or low beta mode is included in each of the three superperiods.

As before, aperture scans were made with and without multipoles. At each momentum the quadrupoles in the regular cells were adjusted to make the tune equal to that of the reference working line. The resulting aperture scans are shown in Fig. 2. As before, the solid line delineates the physical aperture obtained from the machine functions at each value of $\Delta P/P$. The reduced aperture results from the large beta functions in the quadrupoles nearest the crossing points as well as from the strong momentum dependence of these beta functions- see Fig. 3.

The aperture at $\Delta P/P = 0.0$ is reduced by a factor of ~ 1.7 for operation in the low beta mode. Since the aperture varies as $\varepsilon^{-1/2}$, and ε varies inversely as the energy, the operation of the CBA in the low beta mode at ~ 95 GeV would give the same aperture as that of the standard CBA at 30 GeV. Hence injection into the normal CBA at 30 GeV and dynamic switching to the low beta mode at ~ 100 GeV is anticipated.

To assess the limitation to the aperture by the quadrupoles adjacent to the crossing points, scans were also made when the radial apertures of the first two quadrupoles on each side of the low beta crossing point were increased from 40 to

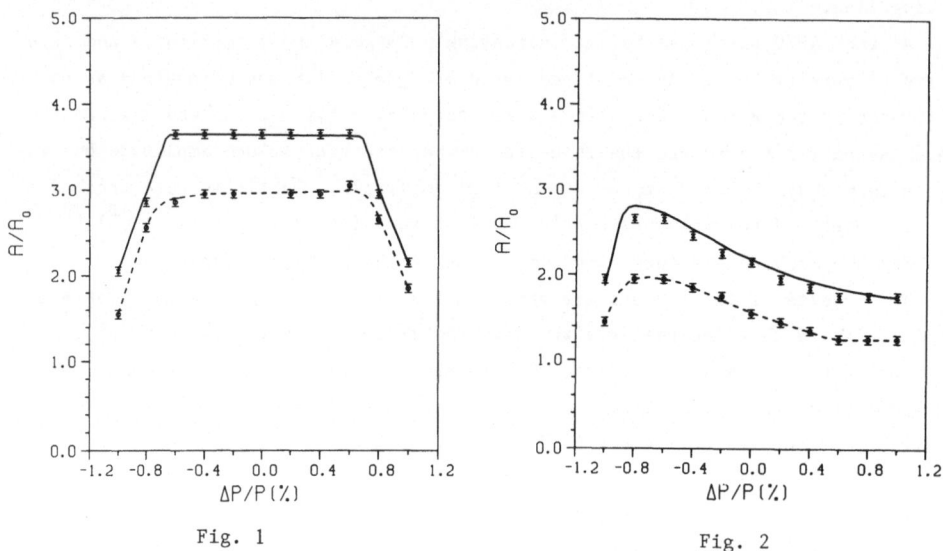

Fig. 1. Fig. 2.

Fig. 1. Aperture scan for the CBA operated in the standard mode. A/A_0 is the relative amplitude with $A_0 = \varepsilon_0 (\beta_x + \beta_y)^{1/2}$ and with $\varepsilon_0 = 0.5\pi \cdot 10^{-6}$m radians, the nominal beam emittance. The upper curve denotes the aperture scan without multipoles; the lower curve denotes the aperture scan when multipoles and closed orbit errors are included.

Fig. 2. Aperture scan without (upper curve) and with (lower curve) multipoles for the CBA in the low beta mode of operation. The radial aperture of all machine elements is 40 mm.

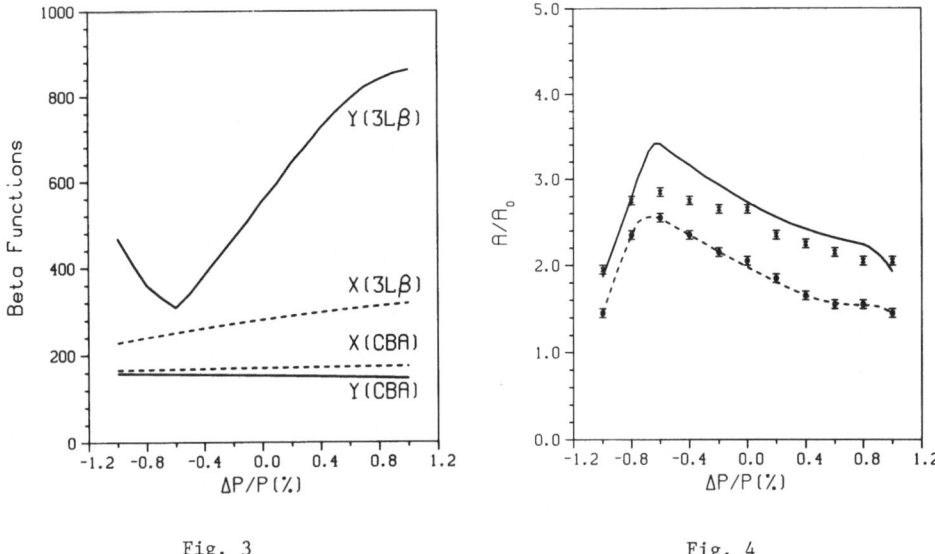

Fig. 3.

Fig. 4.

Fig. 3. Maximum values of β_x and β_y in the insertions for normal (CBA) and low beta (3Lβ) modes of operation.

Fig. 4. Aperture scan without (upper curve) and with (lower curve) multipoles for the CBA in the low beta mode of operation when the radial aperture of the first two quadrupoles on each side of the crossing points is 50 mm.

50 mm. These results are shown in Fig. 4. In this case the low beta option has the same aperture at ~ 60 GeV that the standard CBA has at 30 GeV.

3.0 Aperture Reduction From Sagitta

The magnetic length of the CBA dipoles is 4.36 m, and the radius of curvature in these dipoles is 252.5 m. The sagitta in the dipoles is 9.8 mm; an additional 3.4 mm reduction results from the projection of the bore tube beyond the magnetic length of the dipole. The dipoles are positioned to divide this aperture reduction equally relative to the trajectory of an on momentum particle; the reduction is 6.6 mm.

At each multipole element a test is made to determine whether or not the particle has exceeded the radial aperture of the associated magnet. These multipole elements are located at the center of quadrupoles and at the downstream end of dipoles. The position of a particle is measured relative to an on momentum particle having zero amplitude rather than with respect to the centerline of the magnet. Since an on momentum particle makes an excursion of 6.6 mm on each side of the center line of a dipole, the effect of sagitta on reducing the physical aperture of a dipole has been treated by increasing $|x|$ by 6.6 mm when the test for aperture is made. Hence the condition for staying within the dipole aperture of 40 mm is:

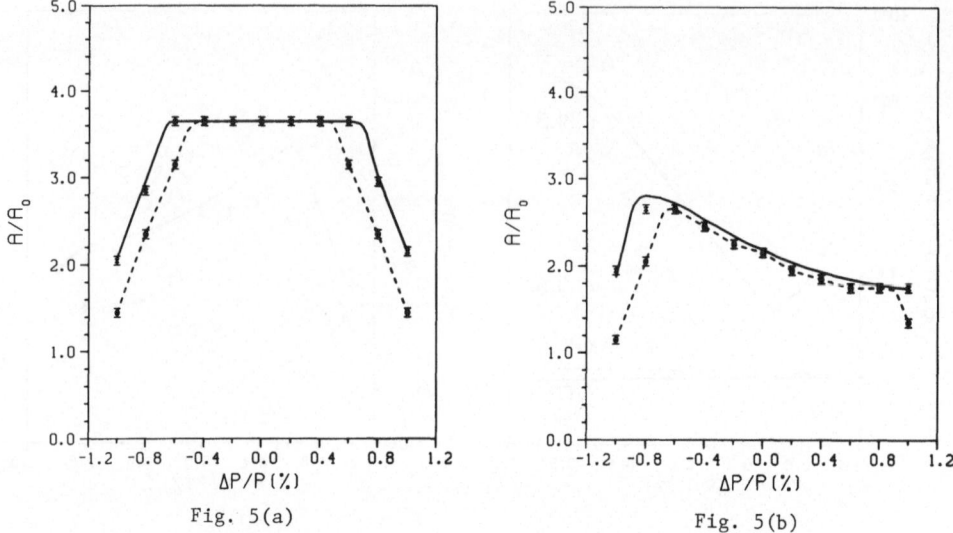

Fig. 5(a). Aperture scan for the CBA without multipoles. Solid curve denotes aperture without sagitta; dashed curve denotes aperture when sagitta is included.

Fig. 5(b). Aperture scan for the CBA in the low beta mode of operation. Solid curve denotes the aperture scan without sagitta; dashed curve denotes the aperture scan when sagitta is included.

$$(|x| + 6.6)^2 + y^2 \leq 1600 \tag{1}$$

A partial aperture scan for the standard CBA was made for $|\Delta P/P| \geq 0.4$ percent, the region where the aperture is limited by dispersion in the regular cells. A significant reduction was observed and is shown in Fig. 5(a). In the CBA the aperture reduction due to bore tube projection in the dipoles nearest the horizontally focusing quadrupoles is reduced nearly a factor of two by rotating these dipoles ~ 1.5 mrad about their vertical axis. This correction has not been included in the present studies; consequently the reduction of aperture from sagitta and bore tube extension is approximately half that indicated in Figs. 5(a) and 5(b).

4.0 Conclusion

The present studies document the expected aperture when the CBA is operated in the low beta mode. To maintain the full aperture indicated in Fig. 1, injection at 30 GeV in the standard mode and switching at ~ 100 GeV to the low beta mode is anticipated. Increasing the radial aperture of the first two quadrupoles on each side of the crossing point in the low beta insertion from 40 to 50 mm increases the aperture so that the low beta option could be used at an energy as low as 60 GeV. In Fig. 4 the expected aperture (solid curve) disagrees with the aperture obtained by tracking in the region $-0.8 < \Delta P/P < 0.0$ percent. This may be an indication that the aperture is limited by the dynamic aperture rather than the physical aperture in this region; this point has not yet been investigated.

It should be noted that the quantity A/A_o in all the aperture scans denotes the amplitude A relative to the amplitude A_o that corresponds to the nominal beam emittance of $0.5\pi \cdot 10^{-6}$ m radians and that the CBA aperture is limited to $2.0\pi \cdot 10^{-6}$ m radians ($A/A_o = 2.0$) by an injection septum. Hence, the aperture in the low beta mode without multipoles (Fig. 2) is almost adequate for operation at 30 GeV. Possible increases in aperture by reducing the strong momentum dependence of β_y (Fig. 3) by including extra families of sextupoles is being studied. Finally, aperture reduction from sagitta effects may become important at a $2\pi \cdot 10^{-6}$ m radian emittance for $|\Delta p/p| > 0.8$ percent.

References

1. H. Wiedemann, PEP-220, Stanford Linear Accelerator Center, September, 1976.
2. G.F. Dell, IEEE Transactions on Nuclear Science, Vol. NS-30, No. 4, August 1983, 2469-2471.
3. H. Hahn, CBA Parameter List, Brookhaven National Laboratory, Upton, NY. July 1, 1982.

THE STUDY OF MISALIGNMENTAL CHARACTERISTICS OF BEAM OPTICAL COMPONENTS OF HI-13 TANDEM

Guan Xia-ling   Cao Qing-xi

Tandem accelerator laboratory, Institute of Atomic Energy. Academia Sinica. Beijing, China.

Abstract

In this paper, the principle and method of alignment error design of tandem accelerator are discussed. The transfer matrices of alignment error for the optical components are given. The alignment error criteria are determined by means of statistical method. Finaly the numerical results of alignment error of HI-13 tandem are calculated with the aid of the optic- program modified to meet the needs of the alignment investigation.

1. Introduction

A tandem accelerator HI-13 [1] with terminal voltage 13 Mv is going to be installed in Institute of Atomic Energy, Beijing, by High Voltage Engineering Corporation. In order to guarantee that the installation is done in such a way that the machine has a batter operation performance, the alignment error design of accelerator will play a important role. Some people, for example D. E. lobb [2], have given certain contributions about this problems. In this paper aimed at the HI-13 tandem, we shall discuss the principle and method of alignment error design. The numerical results for this machine are given in detail. Here we consider the first-order alignment error of component only.

2. Content and method

We shall consider all possible alignment error of various optical devices of which tandem is composed. But the inherent tolerance of devices itself and the electritic error will not be included in this paper. The following five kinds of misalignment are included in our discussion namely: the horizontal parallel displacement ΔX, the horizontal inclination increase $\Delta X'$, the vertical displacement ΔY, the vertical inclination increase $\Delta Y'$, and the rotation α around the optical axis.

For convenience of consideration, at first let us set up both element coordinate system, X,Y,Z, with its Z axis coinciding with the axis of element, and space coordinate system, x,y,z, with its z axis coinciding with one of the ideal beam axis of accelerator system. Under perfect alignmental circumstance, these two coordinate systems coincide exactly each other; otherwise they are separate, as shown in fig. 1.

Let us take V_i as a coordinate vector of phase point with respect to the space coordinate system at the input element,

$$V_i^T = (X_i, X_i', Y_i, Y_i', \Delta P/P, 1),$$

and T_i as a coordinate vector of that phase point with respect to the element coordinate system at the same point.

$$T_i^T = (x_i, x_i', Y_i, Y_i', \Delta P/P, 1).$$

Under first order approximation, vector T_i is a linear function of vector V_i, i,e,
$T_i = \Theta_i V_i$, in which the conversion matrix Θ_i depends upon the misalignment of element only. By the subscript "i" we mean the input end of an element.

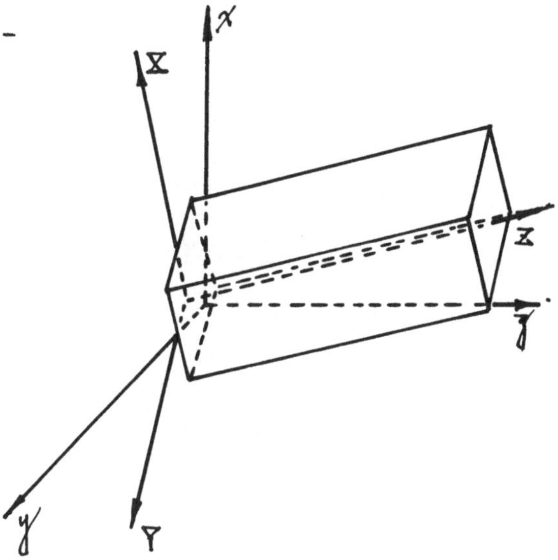

fig. 1.

We refer to "T_o" as the coordinate vector of corresponding phase point with respect to the element system at the output element, and we have

$$T_o = R T_i$$

where R is the transfer matrix of this element. By the subscript " o ", we mean the output end of an element. If we take V_o as the coordinate vector with respect to the space system at the output element, in the same way as input end, we have the linear relation $V_o = \Theta_o T_o$, with matrix Θ_o, in general, being different from matrix Θ_i.

By coordinate transformation, matrices Θ_i and Θ_o can be expressed as follows [2];

$$\Theta_o = \begin{pmatrix} \cos\alpha_i & 0 & \sin\alpha_i & 0 & 0 & \Delta X_i \\ 0 & \cos\alpha_i & 0 & \sin\alpha_i & 0 & \Delta X_i' \\ -\sin\alpha_i & 0 & \cos\alpha_i & 0 & 0 & \Delta Y_i \\ 0 & -\sin\alpha_i & 0 & \cos\alpha_i & 0 & \Delta Y_i' \\ 0 & 0 & 0 & 0 & 1 & 0 \\ 0 & 0 & 0 & 0 & 0 & 1 \end{pmatrix}$$

$$\Theta_0 = \begin{pmatrix} \cos\alpha_i & 0 & -\sin\alpha_i & 0 & 0 & -\Delta X_i - L\Delta X_i' \\ 0 & \cos\alpha_i & 0 & -\sin\alpha_i & 0 & -\Delta X_i' \\ \sin\alpha_i & 0 & \cos\alpha_i & 0 & 0 & -\Delta Y_i - L\Delta Y_i' \\ 0 & \sin\alpha_i & 0 & \cos\alpha_i & 0 & -\Delta Y_i' \\ 0 & 0 & 0 & 0 & 1 & 0 \\ 0 & 0 & 0 & 0 & 0 & 1 \end{pmatrix}$$

for non-bending element;

$$\Theta_0 = \begin{pmatrix} \cos\psi & 0 & -\sin\psi & 0 & 0 & -\Delta X_i \cos\psi - \rho\Delta X_i'\sin\psi \\ 0 & \cos\psi & 0 & -\sin\psi & 0 & -\Delta X_i' \\ \sin\psi & 0 & \cos\psi & 0 & 0 & -\Delta Y_i - \rho\Delta Y_i'\sin\psi + \rho(1-\cos\psi)\sin\alpha_i \\ 0 & \sin\psi & 0 & \cos\psi & 0 & -\Delta Y_i' + \sin\psi \cdot \alpha_i \\ 0 & 0 & 0 & 0 & 1 & 0 \\ 0 & 0 & 0 & 0 & 0 & 1 \end{pmatrix}$$

for bending element.

In above matrices, $\sin\psi = \sin\alpha_i \cdot \cos\varphi$; φ is the total bending angle of central ray; ρ: the radius of curvature of central ray and L_i the effective length of element.

If we take V_{oo} as the coordinate vector with respect to the space system at the output element under the cass of ideal alignment, the deviation increase resulting from misalignment of element is expressed as:

$$\Delta V_0 = V_0 - V_{oo} = [\Theta_i \cdot R \cdot \Theta_i - R] \cdot V_i$$

3. Criteria of alignment error

In the case of HI-13 tandem, there is a beam line as long as 90 meters from ion source to target, as shown in fig. 2. The beam line can be divided into four section according to their beam optical characteristics, namely:
--- injector section: from ion source to object aperture before tandem tube;
--- low energy accelerating section: from the object point before tank to the terminal stripper;
--- high energy accelerating section: from the stripper to object slit of analyzing magnet;
--- post transport section: to target.

The deviation increases, ΔX and ΔY, of beam central particle at the end of each section in horizontal and vertical plane are the functions of the total effect of various possible misalignment of all components. They are

$$\begin{cases} \Delta X = F_1(\Sigma_{1x}, \Sigma_{2x}, \cdots \Sigma_{Nx}) \\ \Delta Y = F_2(\Sigma_{1y}, \Sigma_{2y}, \cdots \Sigma_{Ny}) \end{cases}$$

where Σ_{ix} and Σ_{iy} represent the total effect of various misalignment error for i-th component in this section. The factors of i-th component itself in ΔX and ΔY are expressed as ΔX_i and ΔY_i, they are

$$\begin{cases} \Delta X_i = F_1(0, \cdots \Sigma_{ix}, \cdots 0) \\ \Delta Y_i = F_2(0, \cdots \Sigma_{iy}, \cdots 0) \end{cases}$$

For various optical components of accelerator, since the alignment error is a random error, we have

$$\begin{cases} \Delta X = (\Delta X_1^2 + \Delta X_2^2 + \cdots + \Delta X_N^2)^{\frac{1}{2}} \\ \Delta Y = (\Delta Y_1^2 + \Delta Y_2^2 + \cdots + \Delta Y_N^2)^{\frac{1}{2}} \end{cases}$$

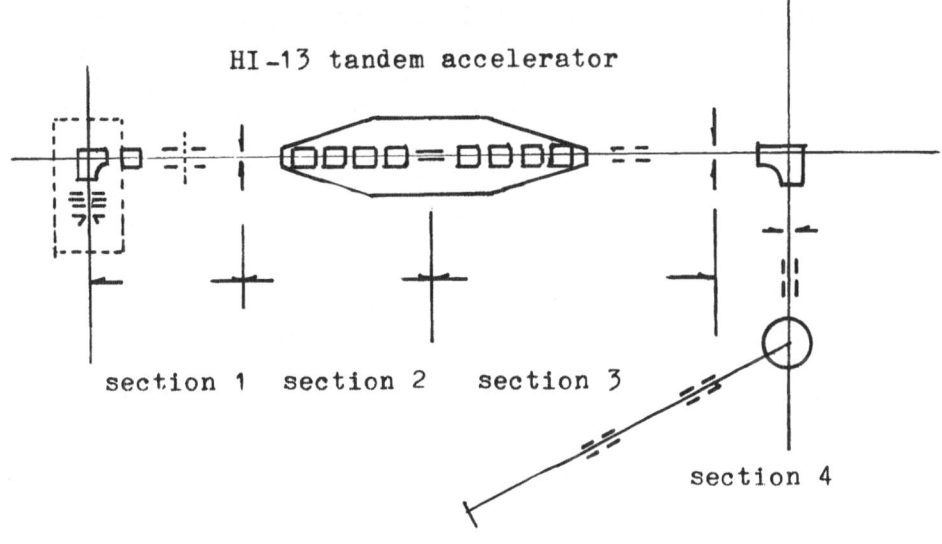

fig. 2.

In order to derive the criteria of alignment error in installation of tandem, we assume that each component in a section have the identical probability of the deviation of central particle resulting from misalignment mounting. That is

$$\Delta X_1 = \Delta X_2 = \Delta X_3 = \cdots = \Delta X_0 = \Delta X / \sqrt{N}$$

We usually define the standard deviation as one third of limiting deviation. It is made as a rule that the limiting deviation of central particle coming from the accumulation of alignment error for all components taken together should not go beyond the radius of the aperture located at the end of the section and that the allowable maximum standard deviation for each component, corresponding with what has been said above, is taken as the alignment error criterion, i. e. the allowable deviations of central particle at the end of section for each component are

$$\begin{cases} \Delta X_i = \frac{R_x}{3\sqrt{N}} \\ \Delta Y_i = \frac{R_y}{3\sqrt{N}} \end{cases}$$

4. Calculation results

According to the principle and criteria of alignment error design, we have calculated the alignment error of HI-13 tandem with the aid of optic- program modified to meet the needs of misalignment investigation.

1). Injector section

There are six components in this section, namely: ion source, trim lens, bending magnet, preacceleration tube, gridded lens and object aperture with a diameter of 0.374" . The allowble alignment errors for each component in the section are listed in table I.

Table I

component	x (in)	x'(mrad)	y (in)	y'(mrad)	α(mrad)
ion source	0.010	0.5	0.010	0.5	---
trim lens	0.010	3.0	0.010	3.0	---
bending magnet	0.005	0.8	0.005	0.1	0.1
preacele. tube	0.015	1.3	0.015	1.3	---
gridded lens	0.010	5	0.010	5	---
object apert.	0.020	5	0.020	5	---

2). Low nenrgy accelerating section

This section is composed of four tubes and a stripper with diameter of 0.34 inch. The results are shown in table II.

Table II

component	x (in)	x'(mrad)	y (in)	y'(mrad)	α(mrad)
1 tube	0.005	0.15	0.009	0.25	5
2 tube	0.013	0.35	0.015	0.42	5
3,4 tubes	0.015	0.42	0.015	0.42	5
stripper	0.012	0.5	0.012	0.5	---

3). High energy accelerating section

In this section, there are seven components namely: a quadrupole triplet lenses in terminal, four tubes, a quadrupole doublet lenses and a object slit of magnet. The allowable alignment error can be found in table III.

Table III

component	x(in)	x'(mrad)	y(in)	y'(mrad)	α(mrad)
Q triplet	0.01	1.0	0.01	1.0	5
5 tube	0.013	0.38	0.015	0.42	5
6,7,8 tubes	0.015	0.42	0.015	0.42	5
Q. doublet	0.002	0.04	0.005	0.04	5
slit	0.007	---	0.01	---	5

4). Post transport section:
   This section consists of 90 analyzing magnet, quadrupole doublet lenses, switch magnent and quadrupole doublet lenses. The results of calculation are listed in table IV.

Table IV

component	x(in)	x'(mrad)	y(in)	y'(mrad)	α(mrad)
magnent	0.002	0.3	0.004	0.1	0.1
Q. doublet	0.005	0.1	0.005	0.1	5
switch magnet	0.004	0.1	---	---	0.1
Q. doublet	0.005	0.1	0.005	0.1	5

5. Discussion

   As a matter of fact, each component in beam line has alignment error with its random distribution. The influence on beam comes from the accumulation of error for various components taken together. Regarding the values calculated in the preceding paragraph as the standard deviations of the normal distribution random error of every component for all of the misalignment, the displacements of central particle at the end of section have calculated. The numerical results show that, as was expected, the displacements are around one thrid of the radius of aperture.

   Under the first order approximation, all of the alignmental error of element, except rotation around the optical axis, affect the displacement of central particle only but not the beam envelope. The rotation error of element can affect both the displacement of central particle and the beam envelope.

Referance

(1) Yu Jne-xian Nucl. Instr. and Meth.   184 (1981) 157

(2) D.E. Lobb. Nucl. Instr. and Meth.   87   (1970) 59

CALCULATIONS FOR THE DESIGN AND MODIFICATION OF

THE 2 CYCLOTRONS OF S.A.R.A.

P.S. Albrand, J.L. Belmont, ⁕F. Ripouteau

Institut des Sciences Nucléaires, 53, avenue des Martyrs
38026 GRENOBLE CEDEX, France

⁕ Present address :
Grand Accélérateur National d'Ions Lourds, B.P. 5027
14021 CAEN CEDEX, France

INTRODUCTION

S.A.R.A. (1,2) is a heavy ion accelerator constituted by 2 cyclotrons. The second cyclotron (post-accelerator) was entirely calculated at the I.S.N. The pole tips of the first cyclotron which is much older, have recently been modified.

An almost identical procedure was used for the calculation of each element of the post-accelerator of S.A.R.A. and also for the modifications to the first cyclotron.

1) - The basic design is predetermined by analytic means.

2) - The fundamental choice of parameters is made after calculation of beam dynamics with the programme (TRANSPORT). The matrices which result from this programme can also be used in certain analytic calculations. With a few precautions we obtain within a few per cent the definitive values, for example the angles of the magnets, and of their faces, ν_r, ν_z, admittances etc.

3) - The measurement of field maps permits the simulation of the movements of ions subjected to these fields with a precision better than one millimetre.
The superposition of electric fields or of additional magnetic fields is performed analytically by the programmes (TRAJ 30, ANJO etc...). The quality of the acceleration depends on the quality of the field measurements.

CALCULATION OF THE MAGNET SHAPE

The shape of the elements with an axis of revolution is predicted with the aid of the programmes MAGNET and POISSON. We made evaluations of the complementary field flux which cannot be calculated by the above programmes and thus completed the magnetic circuits.

In the case of the principal magnets these elementary calculations had to be verified and completed by measurements on a model (1/7). Althoutgh the use of a model is fruitful. The fine adjustment of the shape of the pole is done by successive shimmings.

THE FIELD MAPS

After acquisition and pretreatment (programmes CARP and NEWETAL) we obtain a field map B(azimuth, radius) in Tesla for regular steps of the azimuth and radius. Qualitative analysis and hence judgement of the analytical studies is however difficult with an ensemble of around 7000 pts (for 1/4 of the accelerator). We often worked with a hard-edge equivalent of the magnets, obtained from the calculation of mean fields.

The use of mean fields <B> also permitted the calculation of the magnet shims, and the correcting coils. The biggest difficulty lies in the definition of the mean field. The best would be to average along an isochronous orbit, but these are known only after the calculation, thus we used trajectoiries of simplified geometry ("hard-edge") or simply along the arcs of circles.

For four different levels of the magnetic field we nedd
- The field maps of the magnet without any correcting coils
- The field maps of the effect of each correcting coil (11 and 15 coils).

The accuracy of the field measurements must be such that no smoothing is necessary, for each treatment of the data reduces the credibility of the results.

The aim of the calculations of the magnets of the second cyclotron is to obtain a synchronous field for a particular ion, without correcting coils. We chose the ion O^{8+} at 32 MeV/A.

Starting with the measured field map one calculates the "isochronous" maps (by homothetic variation of the field $B(\theta)$ using the programmes TRAJ22 or ORBISO. By comparison of these 2 maps, or rather, the mean fields obtained under the same conditions, we calculate the shims and the currents required in the correcting coils.

TAILORING OF THE SHIMS OF THE PRINCIPAL MAGNETS OF THE POST-ACCELERATOR

From the difference between the <B> obtained along the same path in an uncorrected field and that obtained in an "isochronous" field we obtain the average variations <ΔB> (R) which should then be redistributed along the trajectory to obtain the desired synchronous field. However this variation <ΔB> (R) is aplied only locally at the position of the shim. The trajectory obtained with this local modification is different from that previously calculated. Moreover the local modification of a gap where the trajectory passes influences the neighbouring trajectory a little. The continuity of trajectory and fields (the absence of an abrupt perturbation) leads to a convergence of an iterative process of modification of the shims and calculation.
A second problem encountered is that the local effect of the shim is not very precisely known. We precalculated this effect using MAGNET then mesured it with a model. One must take into account the apparent length of the shim under the considered trajectory.
An error of several per cent on <ΔB> is unavoidable in the calculation of a perturbation of <ΔB> /<B> of one per cent.
The process converges quickly and after 5 iterations the precision of the calculation was better than that possible for the machining of the steel.

CALCULATION OF INTENSITIES OF CORRECTING COILS

A similar procedure to that described above is followed, to calculate the corrections which must be applied for a particular ion.
From the field measurements we obtain the effect $dB(I_j,r)$ of a current I_j in the jth correcting coil at a particular point r.
The sum of the effect at radius R, of all the correcting coils is given by :

$$\Delta B(R) = \sum_{j=1}^{n} dB(I_j, R)$$

Where n = 12 for the first cyclotron
Where n = 16 for the second cyclotron
The nth coil represents the adjustement of the principal field.

In matrix form for radius R.

$$\Delta B(R) = \sum_{j=1}^{n} T(R, I_j) \times I_j$$

The effect was measured for 58 and 54 values of R for the two cyclotrons respectively.
Let
- B_{1i} = the mean field calculated along a path which follows the coil shape, in the measured field.
- B_{2i} = the mean field calculated under the same conditions but in the isochronous field.
- I_j = the current in the jth coil.
- T_{ij} = the matrix of the effects of unit current in the jth coil at the point R_i.

We seek to minimise

$$S = \frac{1}{N} \sum_{i=1}^{N} \varepsilon_i^2$$

where

$$\varepsilon_i = B_{2i} - \left(B_{1i} + \sum_{j=1}^{n} T_{ij} \times I_j \right)$$

The solutions found for I_j (by the programme REPEND) can in general be directly applied to the correcting coils.

CALCULATION OF THE INJECTION AND EXTRACTION

As already indicated, the calculations using TRANSPORT were completed by the exact calculations of trajectories in magnetic and electric fields.
In order to determine the limiting conditions of injected or extracted beams we seek equilibrium orbits with the programmes ANJO or TRAJ30. These orbits are calculated either in the synchronous field (results of ORBISO or TRAJ22) or in the reconstituted corrected field (result of REPEND).
The shape of the septa and the diaphragms was determined by the observation of the shadow of these elements in phase space (programme OMBRE).

TABULATION OF OPERATING CONDITIONS

- The tabulation of correcting coil currents is the result of a systematic use of the programmes described above, starting from the 4 measured incorrected fields and the measured effects of the correcting.
- The geometry of the trajectories transferred between the machines is constant, so currents in optical elements obey a linear relation.

PROGRAMMES USED

- TRAJ22, TRAJ30
  J. Fermé - Ganil, Caen.

- ORBISO
  J. Sauret, A. Chabert - Ganil, Caen.

- ANJO
  Groupe Théorie - Ganil, Caen.

- POISSON, MAGNET, TRANSPORT
  C.E.R.N.

- REPEND, OMBRE, CARP, NEWETAL
  I.S.N., Grenoble.

REFERENCES

- 1. M. Lieuvin - "SARA a low cost heavy ion accelerator for 10 to 40 MeV/A"
  IEEE - Vol NS - 30, N° 4, p. 2072 - 1983.

- 2. M. Lieuvin - "SARA Grenoble status report"
  9th int. Conf. of Cyclotrons - Caen 1981.

- 3. J.L. Belmont, G. Bizouard, F. Ripouteau - "Design of injection, extraction and magnetic fields in SARA".
  9th int. Conf. of Cyclotrons - Caen 1981

MAGNETIC FIELD OPTIMIZATION AND BEAM
DYNAMICS CALCULATIONS FOR SUSE[+]

W. Schott, E. Zech, N. Rösch
Physik-Department der Technischen Universität München,
Garching, W.-Germany

Abstract

The superconducting sector magnets of SuSe have been designed using the newly developed program GFUNFIT. By means of this program the deviation of the desired magnetic field from the actual value, which is produced by the magnetic system, can be minimized to within 0.5 Gauss by variation of several parameters, such as conductor current, geometry of conductors and of iron. Varying these parameters, fields with an absolute accuracy of $\Delta B/B \lesssim 1$ % have been obtained using the finite-element programs GFUN3D and GETM400. The code GOC has been used to study the beam dynamics in these fields. Two examples for fully stripped heavy ions (q/A = 0.5) with maximum and half maximum final energies (T_2/A = 300 MeV/n and 150 MeV/n) are given.

Introduction

SuSe is a superconducting heavy ion cyclotron which is planned as an additional facility for the accelerator laboratory of the two Munich universities. The cyclotron will be the booster of the 13 MV-tandem. The combined accelerator system will deliver heavy ion beams in the whole mass range of rather high energy and intensity (e.g. about 4 pµA C^{6+}). The energy T_2/A at the mean extraction radius r_2 = 2.4 m will be T_2/A = 450 MeV/n for very light particles, e.g. ^{3}He. For the very heavy particles, e.g. $^{238}U^{38+}$, T_2/A = 24 MeV/n is expected. More details on SuSe are described in refs. 1 - 7. In fig. 1 the essential components of SuSe, i.e. four superconducting sector magnets and two rf-cavities are sketched. One sector magnet consists of a superconducting main coil and a system of 23 superconducting correcting coils on each side of the median plane. Furthermore, there are two cold iron poles within the main coils and a warm iron yoke.

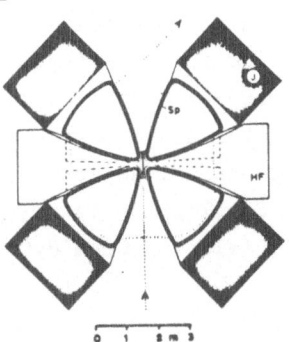

Fig.1: Plan view of SuSe

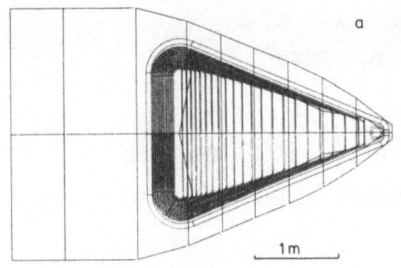

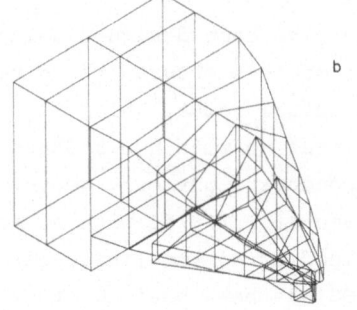

Fig.2:
a. Plan view of one SuSe-magnet
b. Perspective view of half a magnet showing yoke and pole. The iron is subdivided into 60 elements by means of the program GUNFIT.

In fig. 2 the plan view of a magnet with coils and a perspective view of the magnet upper half with the coil system removed are shown. The coils have a triangular shape with round edges. The minimum distance from the median plane is 6.5 cm. The main coil data is: maximum radial extension, measured in the vertical magnet symmetry plane, 3.08 m, sector angle $45°$, axial height a = 47.5 cm, radial width b = 6 cm. The main coil is subdivided into four winding sections. The one close to the median plane is axially 13 cm wide, the three others are 8.5 cm wide, leaving gaps between the sections of 3.5 cm and twice 2.75 cm for mechanical support. Within each main coil there are 22 large correcting coils with a = 9 cm, b = 1.2 cm and one small correcting coil close to the machine center with a = 13 cm, b = 2.5 cm.

By means of the superconducting coils large fields can be produced which cause an almost complete saturation of the poles and the yoke. Thus, the sector coils behave similar to air coils yielding high fields within the magnets and large reversed fields outside. For particles with the specific charge $q/A = 0.5$ and the maximum final energy T_2/A = 300 MeV/n, e.g., the magnetic guiding field B reaches the value 4.9 T in the vertical magnet symmetry plane at r_2, whereas between the magnets -0.6 T are obtained. Corresponding to this large field variation a high flutter factor is found (for the present example: $F(r_2) = 1.4$). Therefore, the acceleration of light heavy ions to the above mentioned energies by means of radial sector fields with zero spiral angle becomes feasible. In section 2 the computer program development is reviewed. In section 3 some results of our magnetic field and beam dynamics calculations are presented.

Computer Program Development

Some software work had to be done in order to use the program GFUN3D[8] as a serious tool for designing the SuSe magnet system.

The ANSI-version of GFUN3D of September 1976 including release 3 (July 1979) was implemented at a Cyber 175. The graphics is used offline. The magnetization is calculated by means of GETM400. When the number of iron elements exceeds 40, a linear equation solver with partitioned submatrices of the size 60 x 60 is used. By means of GETM400 and GFUN3D detailed field maps in the median plane are calculated. For the SuSe magnet system (c.f. fig. 1 and 2) with a subdivision of the magnet iron into 60 elements GETM400 takes 40 min. for 50 iterations, GFUN3D takes about 200 min. for 1200 points in a 45°-sector. Comparing these results with a calculation based on 245 iron elements the absolute error of the field $\Delta B/B$ is estimated to about 1 %.

The program GFUNFIT has been developed in order to optimize coil and iron parameters in such a way that the deviations from a given field are minimized. Coil currents or linear dimensions of current and iron elements may be taken as fitting parameters. As fitting objects either magnetic field components at fixed points or derived functions, e.g. field integrals, fourier components or a combination of these quantities can be used. GFUNFIT is based on a general minimization routine with under-relaxation[9] which uses a strongly reduced version of GFUN3D for computing the magnetic fields. By means of this version the unique part of the iron, which cannot be represented by mirror reflections (i.e. one fourth of one SuSe magnet), is subdivided into 90 iron elements at maximum. Up to 1000 fixed points and 30 parameters can be specified. The calculation of one SuSe field map takes several CPU-hours at the Cyber 175. Therefore, for an optimization with variation of many parameters time consuming intermediate results must be put on mass storage. The coefficients, which describe the geometry dependent effect of the current and iron elements at the fixed points, are calculated and stored first. For a given coil and magnet design and for a given choice of fixed points these coefficients are constants. Different field shapes can be fitted using these constants, only adjusting currents. At the Cyber 175 the calculation of the coefficients for the above mentioned magnet (c.f. fig. 2) and 54 fixed points takes 105 min., the fitting of a field with four iterations 40 min.

A typical result is given in fig. 3. The azimuthally averaged field $\bar{B}$ has been fitted to the desired mean field $\bar{B}_s$ in a radial width

$\Delta r = 0.38$ m...2.70 m. Δr is slightly larger than the range of particle orbits. At small radius the deviation $\Delta \bar{B} = \bar{B} - \bar{B}_s$ reaches 100 Gauss and changes sign between the radial positions of two adjacent correcting coils. This is tolerable because the turn separation is large. For $r \gtrsim 1.6$ m $\Delta \bar{B}$ is compensated to $\Delta \bar{B} \leq 0.5$ Gauss which is extremely small permitting an operation of the cyclotron even at very high harmonic numbers ($h \gtrsim 16$).

Because of the relatively small core available at the Cyber 175 (128 K words at maximum) running GFUNFIT requires an intense communication with background storage devices. Thus, for the above mentioned fit of 2400 s CPU time 10,000 SRU are necessary. To overcome these I/O problems GFUNFIT has now been implemented at a Cray-1 computer. To exploit the array processing capabilities of the Cray-1 subroutines which are often used, such as DBEND and STRBAR, had to be rewritten into a completely "vectorized" form by avoiding inner loops. The above mentioned coefficient computation is now done in 12.5 min. Most of the computing time during a fit run is used for the iterative determination of the magnetization. This is done by solving a general system of linear equations many times. By means of a fully "vectorized" solver[10] the time for one fit iteration is reduced to 35 sec. This improvement is especially important for computing isochronous mean fields. Thereby, corrections to a given $\bar{B}$ are calculated by the beam dynamics program GOC[11] in a superimposed iteration cycle. In the subsequent run of

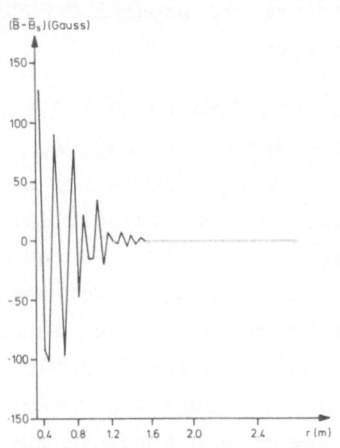

Fig.3:
Deviation of the produced mean field $\bar{B}$ (azimuthal average with r = const) from the desired field $\bar{B}_S$; $\bar{B}_S = B_O \cdot \gamma$ for r = 0.38...2.7 m with $B_O = B$ (r = 0, q/A = 0.5, T_2/A = 300 MeV/n) versus r.

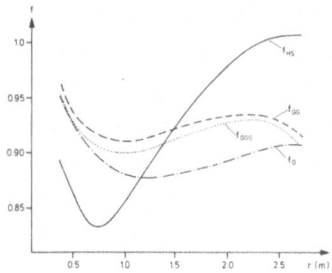

Fig.4:
Field correction factors versus r. $f_{HS} = \bar{B}_{HS}/\bar{B}_S$, $\bar{B}_{HS}$ results from Hagedoorn, Schulte[12]. $f_G = \bar{B}_G/\bar{B}_S$, $f_{GG} = \bar{B}_{GG}/\bar{B}_S$ etc. B_G is the corrected field after applying GOC once, $\bar{B}_{GG}$ after two iterations etc.

GFUNFIT corrected coil currents are calculated yielding $\bar{B}$ which again is corrected for by GOC and so forth. In this procedure only fit iterations are done with GFUNFIT which are by a factor of 16 faster on the Cray-1 than on the Cyber 175. In fig. 4 the correcting factors f to $\bar{B}$, which were computed by GOC, are shown for four iteration cycles. The iteration converges nicely. The difference in f after applying GOC twice and three times is less than one percent indicating that f_{GGG} is already close to the final correcting value.

Magnetic Field and Beam Dynamics Calculations

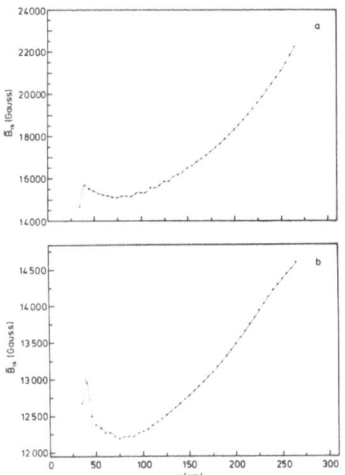

Fig.5:

Isochronous mean fields $\bar{B}_{is}$ versus r for heavy ions with q/A = 0.5.

a. T_2/A = 300 MeV/n
b. T_2/A = 150 MeV/n

In fig. 5 the resulting isochronous mean fields $\bar{B}_{is}$ are drawn for fully stripped heavy ions with maximum and half maximum T_2/A, respectively. The sharp rise of $\bar{B}_{is}$ at the injection radius r_1 = 0.4 m demands a mean current density in the main coil of $\bar{j} \simeq 12000$ A/cm^2. For all other coils $\bar{j} \leq 6000$ A/cm^2 results. The corresponding ν_r- and ν_z-curves are shown in fig. 6 and fig. 7. The phase shifts which are caused by the wiggles in $\bar{B}_{is}$ at small r, are tolerable even at very high harmonic number h. The π-stop band ν_r = 2 (c.f. fig. 6b) is reached at r < 2.4 m and T_2/A > 300 MeV/n. T_2/A (r_2 = 2.4 m) = 300 MeV/n will be achieved by a slight overall reduction of the field. Thereby the cyclotron frequency in a GOC run is varied suitably and the resulting isochronization factors are taken into account in a further GFUNFIT calculation.

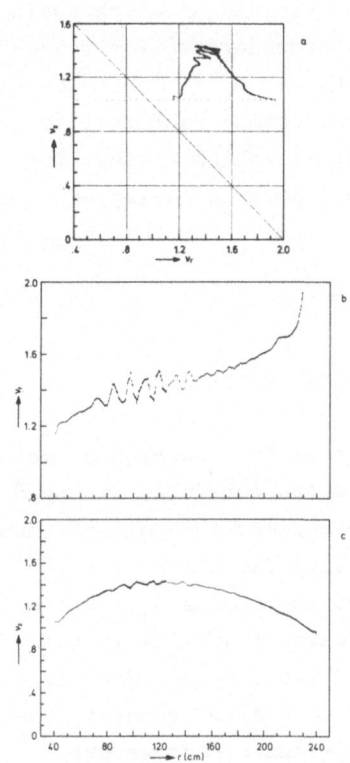 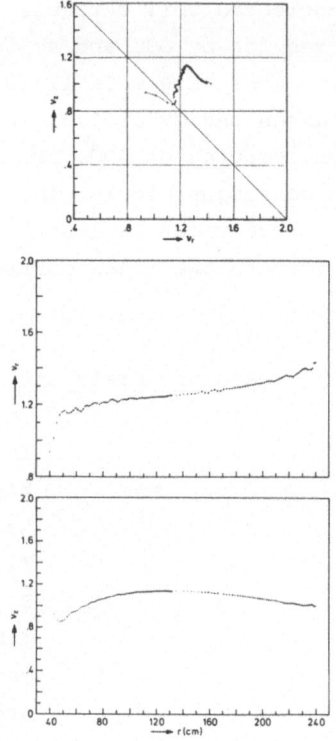

Fig.6: ν_r and ν_z for $q/A = 0.5$, $T_2/A = 300$ MeV/n
a. $\nu_z(\nu_r)$, b. $\nu_r(r)$, c. $\nu_z(r)$

Fig.7: Analog fig. 6 for $T_2/A = 150$ MeV/n

+Supported by the German Ministry of Research and Technology

References

1. H. Daniel et al., IEEE Trans. on Nucl. Science, NS-28, 2107 (1981)
2. W. Schott, Proc. Ninth Int. Conf. on Cyclotrons and their Applications, Caen, 177 (1981)
3. W. Schott et al., IEEE Trans. on Nucl. Science, NS-30, 2624 (1983)
4. G. Hinderer et al., op. cit., p. 2812
5. U. Trinks et al., op. cit., p. 2108
6. W. Wilhelm et al., op. cit., p. 3450
7. U. Trinks et al., Proc. Eighth Int. Conf. on Magnet Technology, Grenoble (1983), to be published in Journal de Physique
8. A.G.A.M. Armstrong et al., Report RL-76-029/A, Rutherford Laboratory (1976), unpublished
9. C. Reinsch, private communication
10. S. Piper, private communication
11. L.B. Maddox, G.S. McNeilly, Rep. ORNL/CSD/TM-53, Oak Ridge National Laboratory (1979)
12. W.M. Schulte, H.L. Hagedoorn, Nucl. Instr. and Meth., 137 (1976) 583

"DFLKTR" THE CODE FOR DESIGNING THE ELECTROSTATIC EXTRACTION SYSTEM FOR CYCLOTRONS

R. C. Sethi* and A. S. Divatia

VEC Centre, Bhabha Atomic Research Centre
Calcutta - 700064, India

* present address: Hahn-Meitner-Institut, Berlin

1. Introduction

The main consideration in the design of any extraction system is to obtain as high extraction efficiency as possible for multienergies and multiparticle machines. The major deciding factors which effect the extraction efficiency are, the quality, the turn separation of the beam and the configuration of the channel. The acceleration history of the beam inside the cyclotron decides the quality and the turn separation of the beam whereas configuration of the channel is mainly decided by the effects prevailing at the extraction. By properly evaluating the above mentioned factors, an optimum configuration of the channel for getting reasonable extraction efficiency, can be obtained. For the given set of conditions the program DFLKTR evaluates the above mentioned factors and gives a suitable configuration of the channel.

2. Description of the Procedure

The basic requirements for the design procedure are to determin the proper radial and azimuthal position, the length, the profile, max. electric field required and its max. height to gap ratio, to ascertain the proper field in the vertical plane. The demands for the better efficiency requires positioning the channel in a region of high turn separation which has to meet the conflicting requirements of beam stability and quality. Whereas the other factors are evaluated mainly by the beam emittance, max. rigidity of the beam and the dispersion effects prevailing in this region.

(a) Position of the channel:

The efficiency of the system η defined by

$$\eta \simeq 1 - \sigma/\Delta_r \qquad (1)$$

requires the turn separation Δ_r to be as large as possible to achiev better efficiency. Under the precessional extraction mechanism, Δ_r(total) is given by (1,2)

$$\Delta_r = \Delta_r(Acc.) + \Delta_r(Prec.)$$

$$\Delta_r(Prec.) = 2\pi A / 1 - \nu_r /$$

$$\Delta_r(Acc.) = R \cdot (\Delta T / T)(\gamma / \gamma + 1)(1/K+1) \qquad (2)$$

$$\gamma = 1 + T/E_0$$

$$\text{and} \quad K = (\partial \langle B \rangle / \partial r)(r / \langle B \rangle) \simeq \nu_r^2 - 1$$

For this, the equation (2) demands a region where ν_r is less than unity, which corresponds to a region where the field falls off with the radius. But this region can give rise to radial and vertical instabilities and beam may blow up under certain situations, specially when it encounters $\nu_r=1$ and $\nu_r=2\nu_z$ resonances. The other problem is the R. F. phase slipping in this non-isochronous region. If the field fall off is not controlled properly, the de-acceleration of the beam may take place. First, the field profile is decided by the amount of tolerable phase slip and then by using this field profile, the beam properties; the stability, the max. amplitude, turn separation etc. are evaluated right from the inner radii ($< R\nu_r=1$) to the outermost radii. On the basis of this evaluation, DFLKTR determins the suitable position or region for placing the channel.

(b) Profile of the channel:

After getting the position of the channel, a working point (R, θ) and its angle are selected just outside the cyclotron in such a way that the beam after passing through the extraction system, should come approximately at that position. Since the region of extraction system being a region of field fall off, the dispersion will play a dominant role. The beam will go on getting wider and wider as it passes through the extraction system and the region beyond that. To accommodate such a situation, the gap between the electrodes is made to follow a series function of the form,

$$g\theta = g_0 + g_1(\theta-\theta_0) + g_2(\theta-\theta_0)^2 + g_3(\theta-\theta_0)^3 + \ldots \qquad (3)$$

here g_0 is the gap at the start of the channel and g_1, g_2, g_3,... etc. are the flaring parameters. g_0 is decided by the amount of radial amplitude of the beam. The approximate length l , of the channel in the beginning is decided by the approximate required amount of deflection δ .

$$\delta \simeq (ze/2m)[E-B\beta](cl^2\beta^2) \qquad (4)$$

where B is the magnetic field and β is the velocity of the particle.The widening or flare of the channel is made to obey the sparking limit criteria (V.E.law),to avoid impractical solutions of voltages for the channel. Under these initial conditions and constraints the program calculates the first optimum parameters, l, Vd, σ, g_0, g_1, g_2, g_3,..., and gives a suitable profile, which is further optimised for as high transmission as possible. This profile is then used as an initial configuration for different energies and different particles and by optimising further and further a final configuration giving reasonable extraction efficiency is obtained. Once the profile has been finalysed, it calculates the field distribution inside the channel by solving the Laplacian Equation by the bindary value approach. This information is used to get the required minimum height to gap ratio of the channel to accomodate the beam in the vertical plane by confining the field deviations in this plane to such a value so that the beam is not effected adversly.

3. Brief Description of the Program:

The program DFLKTR consists of various routines , the function of some of them is explained below. The basic input parameters to the program are the particle, energy, charge state, Dee voltage and the field profile.

(A) ORBIT :It computes equilibrium orbit parameters as well as the information under the accelerated orbit approach.

(B) MATRIX :Computes the submatrices of the system at various stages, wherever this information is required.these matrices are particularly helpful in determining the dispersion effects.

(C) EORBIT :It calculates orbits under crossed electric and magnetic fields.

(D) SRCH :This optimises the various parameters under the given conditions and constraints.

(E) CHANEL :This prints the profile of the channel.

(F) EFCNCY :The efficiency calculations are done through this routine.

(G) LAPLAC :This computes the electric field distribution and the electric fields for a given geometry in the vertical plane.

4. Discussion and Results:

For checking the functioning of the program, the studies were done by using the field profile of the α-beam. A working point at 150^0 with radius equal to 1·84/m and an angle of 0.238 rad with respect to cyclotron axis, was selected as the point, outside the cyclotron, for the reference orbit. The behaviour of the internal orbit, till extraction radius is shown in fig. 1(A, B, C).

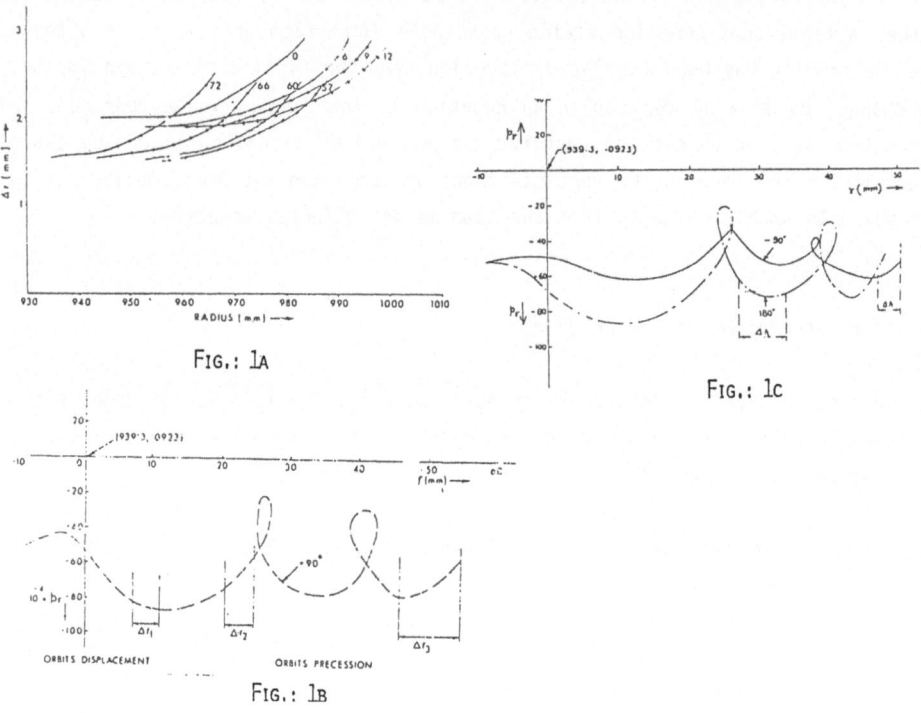

FIG.: 1A

FIG.: 1C

FIG.: 1B

Fig. 1: A shows the turn separation vs. Radius for different azimuthal angles. B and C show the orbits in the (r,pr) space under accelerated approach, including the effects of first harmonic at different azimuths.

Fig. 1(B,C) show that the orbits are quite stable till the extraction radius. These orbits include the effect of first harmonic which is located at radial position of $\nu_r = 1$. The enhancement in turn seperation due to precession is clearly evident. Fig. 2A gives the required min. angular length and the electric field for the channel. Fig. 2B shows the typical profile of the channel. The typical required angular length for 130 MeV α beam is $\simeq 100°$ for an electric field of 125 KV/cm. The whole channel can be made to have radial movement and also the whole length can be divided into two or three parts. This will help in adjusting the channel with the orbits of different energy and particles.

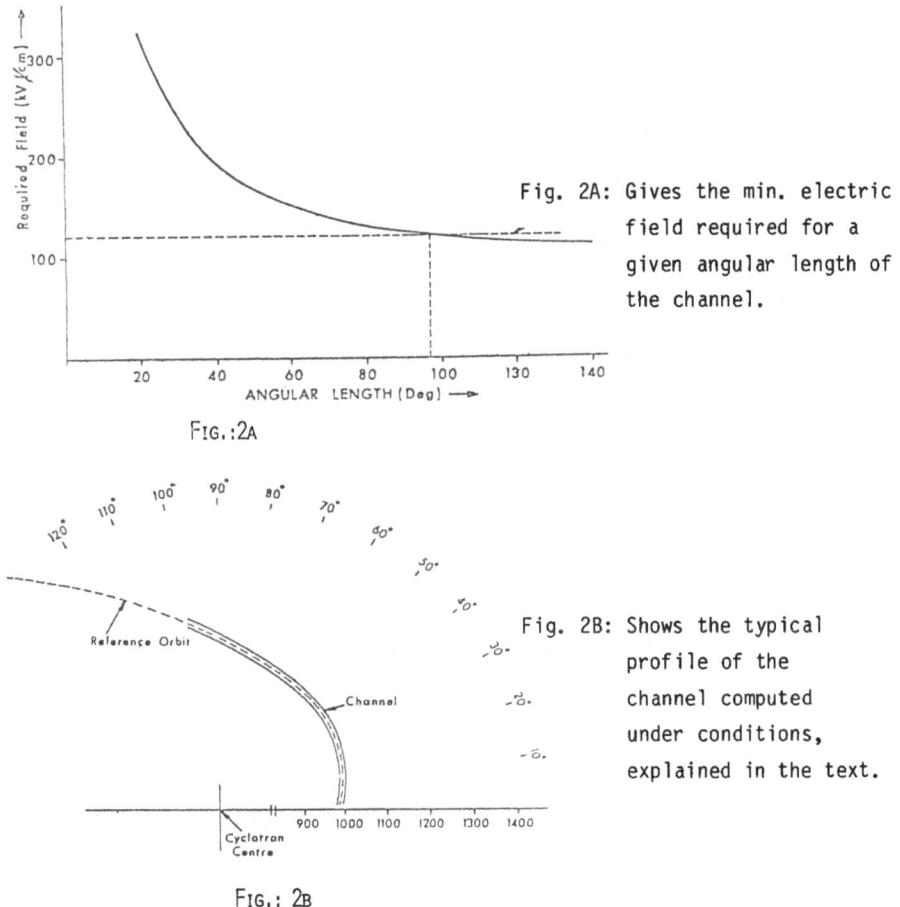

Fig. 2A: Gives the min. electric field required for a given angular length of the channel.

Fig. 2B: Shows the typical profile of the channel computed under conditions, explained in the text.

Fig. 3A and 3B shows the transmission as a function of emittance and as well as the energy spread.

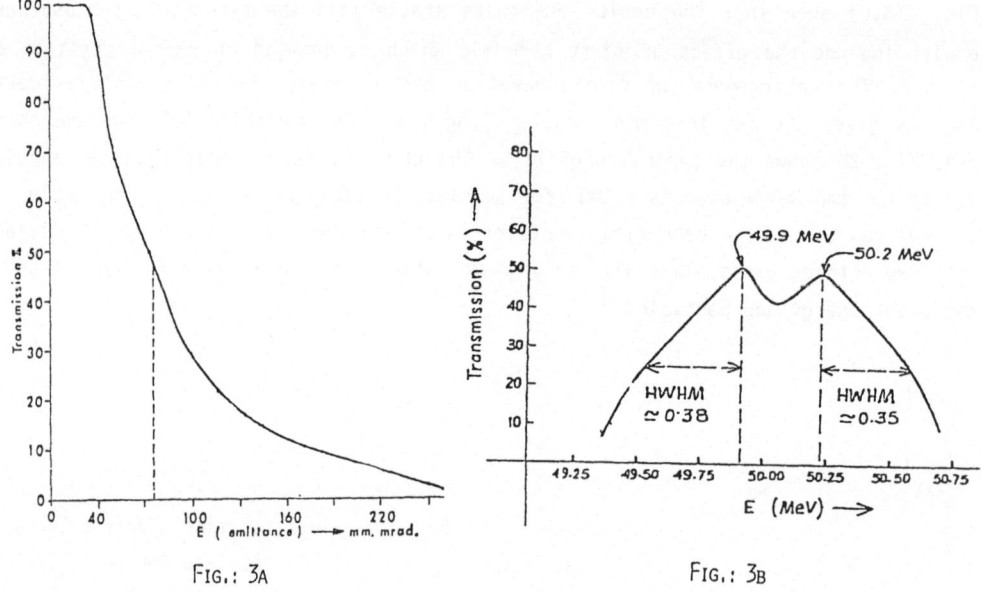

Fig.: 3A   Fig.: 3B

Fig. 3: A shows the transmission of the channel as a function of emittance.
B gives the energy width acceptance of the channel.

For the channel shown in fig. 2B, the typical transmission for 50 MeV α and for an emittance of about 70mm • mrad is about 50 %. The energy acceptance of the channel exhibits a double hump behaviour, showing better transmissions for 49 • 9 & 50 • 2MeV particles, each having almost the same HWHM. The electric field distribution inside and around the channel in the vertical plane is shown in the fig. 4a. Fig. 4b gives the percentage deviation of the electric field with respect to the central field, as a function of height of the channel. For a gap of 13 mm and a deflector voltage of 50 kV, the min height required for the channel is about 47 mm. This allows about 0 • 5 % deviations in the electric field with respect to the central field inside the channel. This is tolerable, because the angular deviations introduced in the beam are still small and acceptable.

The program was used for setting the parameters (3) and for extracting the beam from the cyclotron.

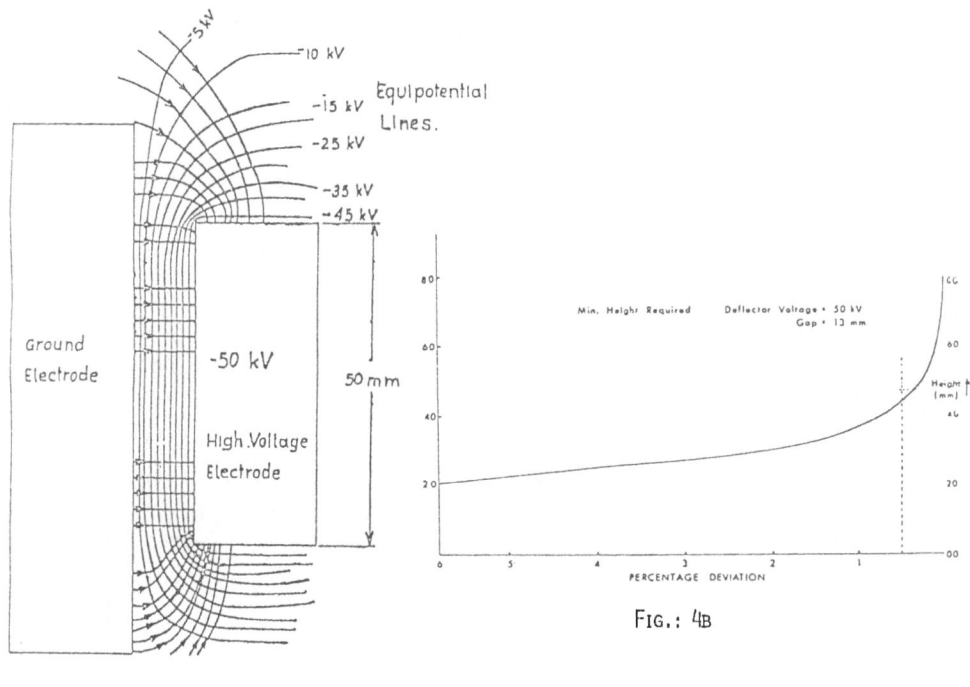

Fig. 4: A gives the electric field distribution in the vertical plane
B gives required min. height of the channel as a function of percentage deviation of the field w.r.t. the central field.

5. References:

1. F.J.M. Farley, Progress in nuclear techniques and instrumentation, vol 1, north holland publishing co., Amsterdam, 1965.
2. G. E. Tripard and W. Joho, Nucl.Instr. & Meth. 79 (1970) 293
3. A. Jain, R.C. Sethi and A.S. Divatia, 9th int. conf. on cyclotrons and their applications, held at Caen, sept. 1981.

RFQ DESIGN CONSIDERATIONS *

P. Junior, H. Deitinghoff, K.D. Halfmann, A. Schempp, N. Zoubek

Institut für Angewandte Physik, Universität Frankfurt am Main,
Robert-Mayer-Str. 2 - 4, 6000 Frankfurt am Main

For the determination of the appropriate electrode geometry within the buncher and accelerator part of a Radio Frequency Quadrupole (RFQ) structure the following steps are performed by corresponding computer programs:
1. Iterative computation of that state of undepressed phase advances, where longitudinal and transverse beam current limits turn out identical, giving the maximum beam current transportable.
2. Generation of corresponding accelerating and focusing parameters.
3. Determination of the electrode geometry required by above parameters.
The availability of our design principle is demonstrated by the construction of a compact proton linac (10 - 300 keV, 108 MHz) and its successfull beam performance.

Introductory Remarks

RFQ linacs are well suited for the acceleration of slow particles ($\beta \simeq 10^{-3}$) and offer a significant progress in the handling of high beam currents. This capability has mainly two reasons, the feasibility of adiabatic bunching and the possibility of simultaneous strong focusing and effective acceleration. In this paper a design principle with regard to maximum beam current and corresponding computer programs are presented.

IMAX is concerned with the correlation between maximum beam current and undepressed transverse phase advance σ_o and is based on the postulate that all three phase advances per $\beta\lambda$ period and maximum beam bunch dimensions are kept invariant in all buncher sections. With given data e/m - specific charge of ion to be accelerated, V - electrode voltage, $\omega = 2\pi f$ - radio frequency, ϕ_o - synchronous phase at buncher input, ϕ - at last buncher section (in the following quoted as reference section), v_o - synchronous velocity at buncher input and the invariances mentioned the beam current represents an invariant as well and is calculated in the reference section, where the final synchronous phase and the final electrode aperture are arrived at.

GENERATION considers two parts of the accelerator, i. e. the buncher, where three degrees of freedom in constructing a section namely the in-

* Work supported by BMFT

ner aperture R, the modulation m and the synchronous phase match the invariances given. In the linac part the degree of freedom left is utilized such that the modulation matches the still unchanged transverse phase advance. Thus a diagram like fig. 1 calculated with the IMAX code informs on beam current limits at given σ_o, where the optimum σ_o can be taken as the basis of GENERATION. Input data must meet limitations, such as peak surface fields [1] or tune depressions [2]. For small phase advances analytical estimations of I_{max} and σ_o can be assigned [3,4]. The exact design, however, requires the computer. This paper has to restrict on foundations, a more detailed discussion will be reported elsewhere.

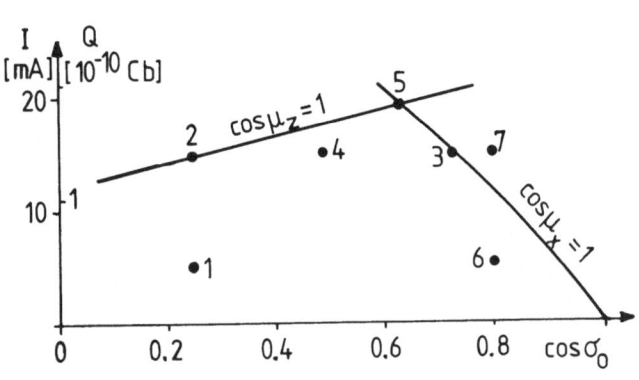

Fig. 1 Current limits vs. undepressed transverse phase advance leading to the proton RFQ

1. Formulas used

The following linear equations form a complete set for our considerations:

a) transverse motion $\quad \frac{d^2 {x \atop y}}{dt^2} + f_{x \atop y}(t){x \atop y} = 0 \quad$ (1)

with $f_{x \atop y}(t) = -\frac{e}{m}[\frac{VA\omega^2\sin\phi}{8v(t)^2} - \frac{VX}{R^2}\sin(\omega t - \phi) + \frac{3QF_{x \atop y}(a_x,a_y,a_z)}{4\pi\varepsilon_o a_x a_y a_z}]\quad$ (1a)

b) longitudinal motion $\quad \frac{d^2 z}{dt^2} + g(t)z = 0 \quad$ (2)

with $g(t) = \frac{e}{m}[\frac{VA\omega^2\sin\phi(t)}{4v(t)^2} - \frac{3QF_z(a_x,a_y,a_z)}{4\pi\varepsilon_o a_x a_y a_z}] \quad$ [5] (2a)

Symbols stand for: ε_o - dielectric constant, Q - charge of bunch ellipsoid with uniform charge density, semiaxes $a_x a_y a_z$ and corresponding formfactors $F_x F_y F_z$ being given by elliptic integrals [6]. Space charge is expressed by the KV setting [7] extended to three dimensions with the usual reservations in selfconsistence, giving the mean beam current

$$I = Q\omega/2\pi \quad (3)$$

With input velocity v_{in} output and mean velocities v_{out} resp. $\bar{v}$ are in sufficient approximation

$$v_{out} = v_{in} + \frac{eVA\pi\cos\phi}{2mv_{in}} \quad (4a) \qquad \bar{v} = v_{in} + \frac{eVA\pi\cos\phi}{4mv_{in}} \quad (4b)$$

Acceleration and focusing parameters A and X for the ideal profile depend on modulation [5]

$$A = \frac{m^2 - 1}{m^2 I_o(kR) + I_o(mkR)} \quad (5a) \qquad X = 1 - AI_o(kR) \quad (5b)$$

Maximum envelope radii are taken as

$$a_{x,max \atop y,max} = R \quad (6) \qquad a_{z,max} = \frac{2\bar{v}}{\omega}\sqrt{1 - \phi\operatorname{ctg}\phi} \quad (7)$$

where equ. (7) represents the elliptic approximation of the separatrix. The invariance in longitudinal phase advance means for any two sections i and k with equs. (2), (2a) and (4b)

$$\frac{A_i \sin\phi_i}{(v_i + \frac{eVA_i\pi\cos\phi_i}{4mv_i})^2} = \frac{A_k \sin\phi_k}{(v_k + \frac{eVA_k\pi\cos\phi_k}{4mv_k})^2} \quad (8)$$

as the bunch length equ. (7) is claimed constant, equ. (4b) gives

$$(v_i + \frac{eVA_i\pi\cos\phi_i}{4mv_i})\sqrt{1 - \phi_i\operatorname{ctg}\phi_i} = (v_k + \frac{eVA_k\pi\cos\phi_k}{4mv_k})\sqrt{1 - \phi_k\operatorname{ctg}\phi_k} \quad (9)$$

2. IMAX

With input parameters R, V, σ, ω, v_o, φ, ϕ_o (the index $_o$ refering to the first buncher - the omitted index to the reference section) and regarding fig. 2 the program starts at statement ①, where A and Q correspond to a supposed starting situation of fig. 1. Then a statement (8) x (9) follows, where the two equs. (8), (9) are solved with respect to two unknown quantities behind OUT. BEAM determines the undepressed transverse phase advance of the reference section, where for the first time both the RFQ inner aperture and the maximum transverse beam envelope are identical. This is not the case for the first buncher section, the conversion of velocities using (8) x (9) is inevitable therefore. The subroutines BEAM, ACCEPT and ITER will be topics of the next chapter. Then all depressed acceptances and phase advances together with the matched beam envelopes are calculated by ACCEPT at statement ②. Supposing a situation 1 of fig. 1 with logical FLAG1 "false" ITER73 iterates the bunch charge Q towards that longitudinal stability limit, where $\cos\mu_z = 1$ (situation 2 in fig. 1). Symbols μ stand for depressed phase advances. As this situation generally does not correspond to the transverse limit already, FLAG1 is set true and ITERJ4 iterates the transverse phase ad-

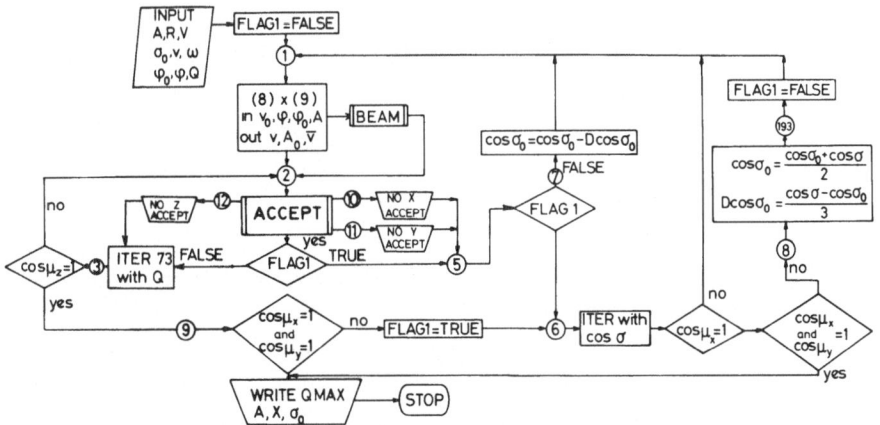

Fig. 2   Flow chart of code IMAX

vance such that point 3 in fig. 1 is reached. Now via statement (8) (fig. 2) the distance between points 2 and 3 (fig. 1) is divided into halves leading to point 4. In a normal run the peak 5 of fig.1 corresponding to the maximum beam current transportable is thus obtained stepwisely. However, when we start at situation 6 of fig. 1 with FLAG1 = "false" statement (9) with $\cos\mu_z = 1$ cannot be reached, because ACCEPT cannot find a transverse acceptance and situation 7 of fig. 1 arises. Now via statements (10) or (11), (15) and (7) σ_0 is successively altered, until 4 (fig. 1) is obtained and the run normally proceeds towards 5. As fig. 2 indicates logical FLAG1 decides, if Q or σ_0 is varied. A favourable starting situation 1 or 6 can easily be found, when smooth approximation formulas [5] are used.

3. BEAM, ACCEPT, ITER

BEAM determines the transverse phase advance per period at Q = 0 with given parameters eV/m, R, A, v, ϕ, ω for Mathieu's equ. (1) using equs. (4b), (5a), (5b). Here the time of flight through a section corresponds to $\beta\lambda$ and is subdivided into many time elements (90 in our calculations). The time dependent coefficient in equ. (1a) is approximated by a stepfunction. The matrix product of all 90 submatrices then leads to the phase advance.

ACCEPT at first considers anyone of the three components of motion and determines the corresponding envelope function along this section at given envelope slopes of the other two components. The routine starts with zero space charge and calculates the elements of the transfer matrix in the same way as in BEAM. From these the β-function along the section for the matched beam follows, then an acceptance and an envelope function

$a_0(t)$ result [8]. After space charge is switched on this $a_0(t)$ together with the other two unchanged envelope slopes give a new acceptance and a new envelope function both depressed. Formfactors now appearing in equ. (1a) and (2a) are evaluated with a Gaussian procedure [9], where the envelope functions are also approximated by stepfunctions. The process of improving the envelope function is stopped, when successively following acceptances agree within 1 %. As this still involves approximations of the other two components of envelope motion, where for a zero order start all envelopes are set constant according to equs. (6) and (7), the iteration of all time dependent envelope functions alternates successively between the transverse and the longitudinal motions, ending up, when simultaneously all three depressed acceptances resp. agree with those of the foregoing step. When space charge causes instability, an extraordinary return to either ⑩, ⑪ or ⑫ of fig. 2 goes ahead.

ITER is based on the regula falsi, where the zero root of a numerical function is successively determined to given accuracy. So when a decision - say $\cos\sigma_0 = 1$ YES (fig. 3) - appears, iteration stops.

4. GENERATION

This program determines the geometrical parameter set successively from one $\beta\lambda$ section to the next. The flow is illustrated in fig. 3. Starting with data as in IMAX, the reference section is iterated via ①. Then statement ② for K = 0 is reached with the correct A_0 of the first bun-

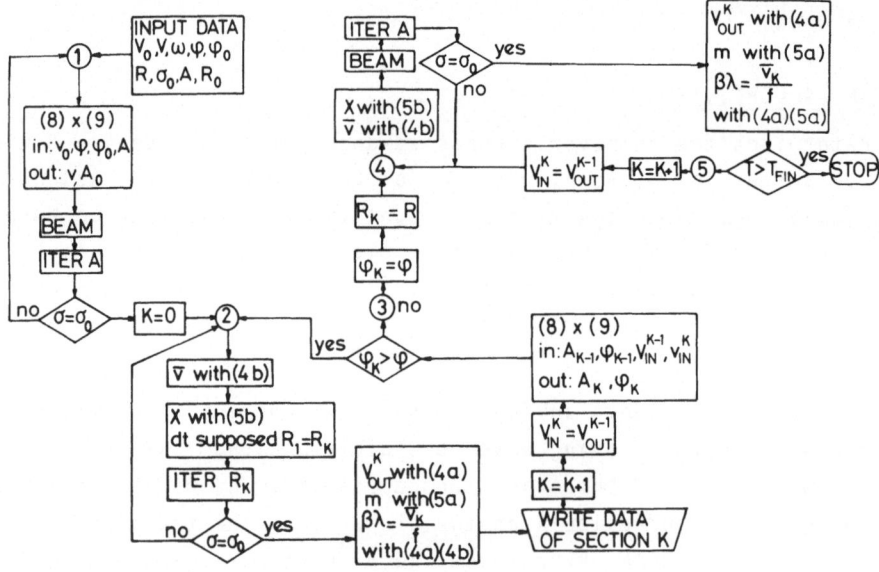

Fig. 3   Flow chart of code GENERATION

cher section. Now R is iterated at given A_o and σ_o. Thereafter data of this section are written and the next section K+1 is investigated. Simultaneously keeping both the longitudinal phase advance and - aperture constant at statement (8) x (9) the new A_K and ϕ_K are calculated and via ② the iteration towards given σ_o delivers data of this new section. The process of successive determination of phase and geometry in the buncher section ends, when the final synchronous phase ϕ is obtained. Now via ③, ④, R and ϕ are kept constant, the only degree of freedom left being m. After iteration via ④ data of sections are again successively printed until the designed final energy T_{FIN} is reached.

5. Example of a Linac

Fig. 4 View of Frankfurt RFQ

The photo of fig. 4 shows a proton linac (10 - 300 keV) built in our institute, where the profile was chosen in trapezoidal form as a suitable approximation for the ideal profile. Thus the electrodes have easily been manufactured on a lathe. Details on successfull operation and beam tests are given in [10]. Present computational work concerns investigations of higher field harmonics caused by boundary conditions deviating from those with ideal profile [1].

Computations were carried out at Hochschulrechenzentrum.

References

[1] P. Junior et al., IEEE Trans. Nucl. Sci. NS-30 (1983) p. 2639
[2] I. Hoffmann, IEEE Trans. Nucl. Sci. NS-28 (1981) p. 2399
[3] T.P. Wangler, LA 8388 (1980)
[4] P. Junior, Part. Acc. Vol. 13 (1983) p. 229, in print
[5] K.R. Crandall et al., BNL 51134 (1979) p. 205
[6] O.D. Kellogg, Foundations of Potential Theory. Springer: Berlin, Heidelberg, New York 1967, p. 192
[7] I.M. Kapchinskij, V.V. Vladimirskij, Proc. Intern. Conf. High Energy Acc. Geneve, CERN 1979, p. 274
[8] E.D. Courant, H.S. Snyder, Ann. Phys. 3 (1958)
[9] P. Bru, MPS/LIN/70-10, CERN Geneve, Nov. 1970
[10] A. Schempp et al., IEEE Trans. Nucl. Sci. NS-30 (1983) p. 3536

EFFECTS OF HIGHER ORDER MULTIPOLE FIELDS ON HIGH CURRENT RFQ ACCELERATOR DESIGN

G.E. McMichael and B.G. Chidley

Atomic Energy of Canada Limited
Chalk River Nuclear Laboratories
Chalk River, Ontario, Canada K0J 1J0

Abstract

Focusing and accelerating fields in a radiofrequency quadrupole (RFQ) accelerator are usually obtained from a potential function limited to two terms. The higher order multipole terms are not included because they are difficult to calculate and their effects on beam dynamics are usually negligible. However, if an accelerator is designed with a large aperture for high current, or if the vane tips are shaped to reduce sparking, the two terms may not be sufficient to accurately describe the potential. This paper describes the design of RFQ1, a 75 mA, 100% duty factor (cw), 600 keV proton accelerator, for which the two-term approximation was not sufficient. The effect on the beam dynamics of including the higher order terms, and the selection of a vane tip profile to minimize critical terms, is discussed.

Introduction

The ZEBRA[1] (Zero Energy BReeder Accelerator) program at Chalk River is the first stage of a development program leading to an accelerator breeder which could produce fissile fuel for nuclear power reactors. Because the breeder must be competitive with the other sources of fissile fuel, high efficiency and reliability are essential requirements for the accelerator. Thus for ZEBRA, and for the development work preceding ZEBRA, the goal is to accelerate high currents, in highly beam loaded structures, that are operated with reasonable safety margins to minimize surface sparking from excessive surface fields.

The reference design[2] for ZEBRA, includes (a) a 75 keV, 375 mA dc proton injector, (b) a 2 MeV output energy, 305 mA cw RFQ accelerator, and (c) a 10 MeV output energy, 300 mA cw Alvarez accelerator. To obtain pertinent design information and operating experience prior to ZEBRA RFQ construction, RFQ1[3] is being built to accelerate 75 mA of protons from 50 to 600 keV. RFQ1, operating at 2.5 times the ZEBRA frequency and at a lower surface electric field, will have relatively similar space charge and beam loading to ZEBRA. RFQ1 has been designed for conservative electric fields and modest currents so that operation near the space charge limit can be conveniently studied.

This paper investigates the design of high current RFQ accelerators from a beam dynamics viewpoint. The design of RFQ1 is used as an illustrative example.

RFQ Dynamics Design

Designing a high current RFQ accelerator is a multi-step process. Using basically a linear analysis of the RFQ, Crandall et al[4] showed how to define principal RFQ parameters as functions of distance "z" along the accelerator axis. Their programs CURLI and RFQUIK provide a simple means to design an RFQ for beam currents with small space-charge forces.

For RFQs with appreciable space charge forces in the beam, CURLI and RFQUIK do not always produce good designs. Instead, over the region of parameter space where the linear analysis predicts possible designs, CURLI/RFQUIK are used to generate many "best linear" designs, and the multi-particle simulation code PARMTEQ is used to select which is actually the best. The principal parameters of the RFQ (intervane voltage "V", bore "a(z)", vane modulation "m(z)" and synchronous phase "$\phi_s(z)$", calculated by CURLI/RFQUIK, are input to PARMTEQ. An idealized vane shape is assumed such that the electric potential is a simple algebraic function of these parameters. Vanes with this idealized shape are unacceptable[5] for an accelerator, whereas actual vanes have a more complicated potential that requires considerable computer calculations. Near the beam axis (i.e., for "r" small relative to the cell length), the simple function is a good approximation for the real potential. However, for large bore, high current RFQ's such as RFQ1, it may be necessary to use a more accurate approximation for the potential, making the RFQ design more difficult.

The Electric Potential in the Beam Region of an RFQ

In the quasi-static approximation, the space dependent part of the electric potential in the vicinity of the beam axis of an RFQ can be written as follows[5]:

$$\Phi(r,\theta,z) = \sum_{s=0}^{\infty} C_{0s} (kr)^n \cos(n\theta) + \sum_{m=1}^{\infty} \cos(kmz) \sum_{s=0}^{\infty} C_{ms} I_n(kmr) \cos(n\theta)$$

where: $k = \pi/L$ (L = cell length = $\beta\lambda/2$)

$n = 4s$ if m is odd, $n = (4s + 2)$ if m is even

I_n = modified Bessel function of order n.

For the idealized vanes assumed in PARMTEQ, CURLI and RFQUIK, only the C_{00} and C_{10} terms are non-zero, ("two-term potential vanes") and:

$$\Phi(r,\theta,z) = C_{00}(kr)^2 \cos(2\theta) + C_{10} I_0(kr) \cos(kz).$$

PARMTEQ dynamics calculations have been modified to use the 8 lowest order terms (C_{00}, C_{10}, C_{01}, C_{11}, C_{20}, C_{21}, C_{30}, C_{31}) when they are included as part of the input data. This modification makes it possible to study the effects of complicated potentials on beam dynamics and to optimize the vane shape for both minimum sparking and "best" dynamics.

RFQ1 Initial Design

The initial design of RFQ1 used the procedure outlined above. Approximately 150 possible designs suggested by CURLI/RFQUIK were checked with PARMTEQ. From beam dynamics considerations, the choice was reduced to four designs with best transmission (but with different input energies or surface electric fields). Manufacturing and operating considerations influenced the selection of the lowest field, lowest input energy design as the reference design for RFQ1. Its basic parameters are as follows:

Input Energy	50 keV
Output Energy	600 keV
Input Current	90 mA
Vane Length	2.32 m
Intervane Voltage (peak) V	73 kV
Average Bore Radius r_0	0.464 cm
Maximum Electric Field	20.72 MV/m (1.25*Kilpatrick[6] limit)
Input Emittance (normalized)	3 π (mm-mrad)

Vane Tip Profile Selection

Initially, it was assumed that differences between the real potential and the two-term approximation would not influence the beam dynamics, so a vane cross-section was sought that would reduce the peak surface electric field and minimize the probability of sparking[5]. Ease of machining was an additional consideration. A profile that appeared acceptable was one with a fixed tip radius of curvature which was 0.75 of the mean bore radius r_0. The enhancement factor "κ" (the ratio of peak surface field to V/r_0) for such vanes is lower than that for the vanes used on the first operating RFQ, POP[7] at LANL (which more closely approximated two-term potential vanes). Higher vane voltages are possible with a lower "κ", which should yield better transmission. However, the expected improvement in transmission was not found when the actual potential function for such vanes (calculated using the program CHARG3D[8]) was used in PARMTEQ calculations. Further investigation showed that the

$$C_{11}\ I_4(kr)\ \cos(kz)\ \cos(4k\theta)$$

term caused the degradation in transmission. The computer program POTRFQ[5] was used to plot vane profiles which would give a potential with C_{00} and C_{10} terms identical to the two-term potential, but with $C_{11} = 0$. An approximation to this profile, that is simpler to visualize and machine, has a tip with a centre of curvature a fixed distance ($1.75*r_0$) from the beam axis and a longitudinal profile identical to POP style vanes.

Calculated output currents as functions of input current for three different RFQ1 vane profiles are shown in Figure 1. The vane voltage was adjusted for each profile to give a maximum surface field of 1.25 *Kilpatrick limit. Included for comparison are the two-term potential predictions with V chosen to give 1.25*Kilpatrick by assuming $\kappa = 1.355$ (the value used in the initial design stage by CURLI). For each case, the modulation was adjusted to keep the first order ratio of longitudinal to transverse focusing (C_{10}/C_{00}) the same as the ratio for the two-term potential. Current plans are to use the constant center-of-curvature style vanes for RFQ1.

Discussion

The necessity of including the multipole harmonic terms of the potential in beam dynamics calculations for an accelerator like RFQ1 is evident from Figure 1. However further studies are required to determine why their effect is greater for some designs than for others and to be able to predict when the extra terms must be included. Averaged beam properties for the different profiles and potentials show that although beam emittance along the RFQ is rather similar, beam loss is noticeably different. PARMTEQ has been modified to plot trajectories of two particles each starting with identical momentum and position, but with the two-term potential used for one particle and the eight-term potential for the other. Figure 2 shows an example of the loss at cell 240 of the particle seeing the eight-term potential. After about 150 cells, the cumulative effects of the different forces have caused large differences in the particle trajectories. An investigation of many such pairs shows that the probability of being lost is lower for those particles seeing only the two-term potential, but does not explain why.

For both the constant radius and POP style tips, beam loss decreased significantly when the C_{11} term was set to zero for the RFQ1 calculations. However, for the constant centre-of-curvature tip vane calculations, whose other terms are intermediate to the two cases above, setting $C_{11} = 0$ resulted in an increase in beam loss. The higher order terms may be beneficial or detrimental depending on other details of the design. The FMIT RFQ, which is also a large bore radius accelerator, was found to be adequately described by dynamics calculations using the two-term potential[9]. It would appear that there is no simple way to predict the significance of the additional terms without performing detailed dynamics calculations.

Conclusions

RFQ1 is designed to accelerate as much current as possible at a low surface electric field. It was found that a beam dynamics program which approximated the potential by only two terms did not give sufficient accuracy. It was also found for RFQ1 that a vane tip shape in which the tip has a centre-of-curvature a constant distance from the beam axis, combines good beam dynamics with low sparking probability.

Acknowledgements

The assistance provided by the AT-1 group at Los Alamos National Laboratories, and in particular that of K.R. Crandall and T.P. Wangler, is gratefully acknowledged.

References

1. S.O. Schriber, "The ZEBRA (Zero Energy BReeder Accelerator) Program at CRNL - 300 mA, 10 MeV Proton Linac", Proc. 1981 Linear Accel. Conf., Los Alamos National Lab., Report LA-9234-C, 363 (1981).
2. B.G. Chidley, et al., "Design and Constraints for the ZEBRA Injector, RFQ and DTL", ibid., 49.
3. M.R. Shubaly, et al., "RFQ1: A 600 keV, 75 mA CW Proton Accelerator", IEEE Trans. Nucl. Sci., NS-30, No. 2, 1428 (1983).
4. K.R. Crandall, R.H. Stokes and T.P. Wangler, "RF Quadrupole Beam Dynamics Design Studies", Proc. 1979 Linear Accel. Conf., Brookhaven National Lab., Report BNL-51134, 205 (1979).
5. B.G. Chidley, G.E. McMichael and G.E. Lee-Whiting, "Design of RFQ Vane Tip to Minimize Sparking Problems", IEEE Trans. Nucl. Sci., NS-30, No. 4, 3560 (1983).
6. W.D. Kilpatrick, "Criterion for Vacuum Sparking Designed to Include Both rf and dc", Rev. Sci. Instr. 28, No. 10, 824 (1957).
7. J.E. Stovall, K.R. Crandall and R.W. Hamm, "Performance Characteristics of a 425 MHz RFQ Linac", IEEE Trans. Nucl. Sci., NS-28, No. 2, 1508 (1981).
8. K.R. Crandall, "Effects of Vane-Tip Geometry on the Electric Fields in Radio-Frequency Quadrupole Linacs", Los Alamos National Lab., Report LA-9695-MS (1983).
9. K.R. Crandall, R.S. Mills and T.P. Wangler, "Radio-Frequency Quadrupole Vane-Tip Geometries", IEEE Trans. Nucl. Sci., NS-30, No. 4, 3554 (1983).

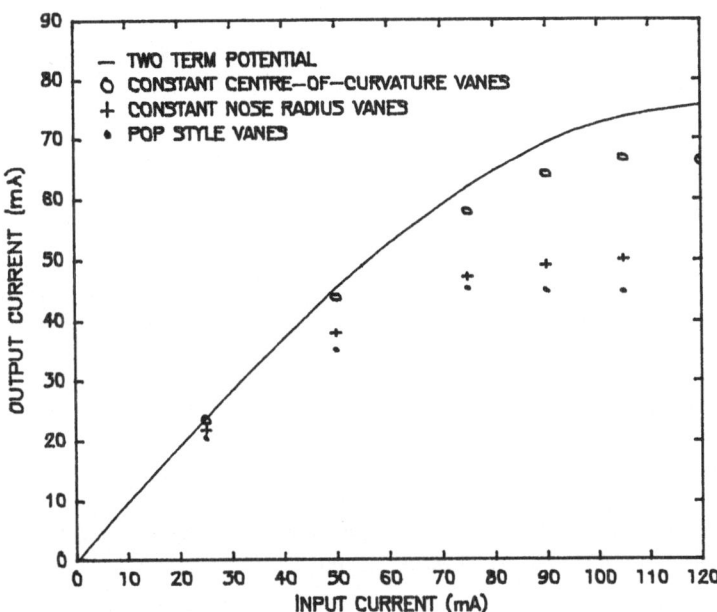

Fig. 1 Transmission of RFQ1 for different vane tip profiles. Calculated with PARMTEQ using an 8-term potential with the vane voltage adjusted to make the peak field = 1.25*Kilpatrick limit.

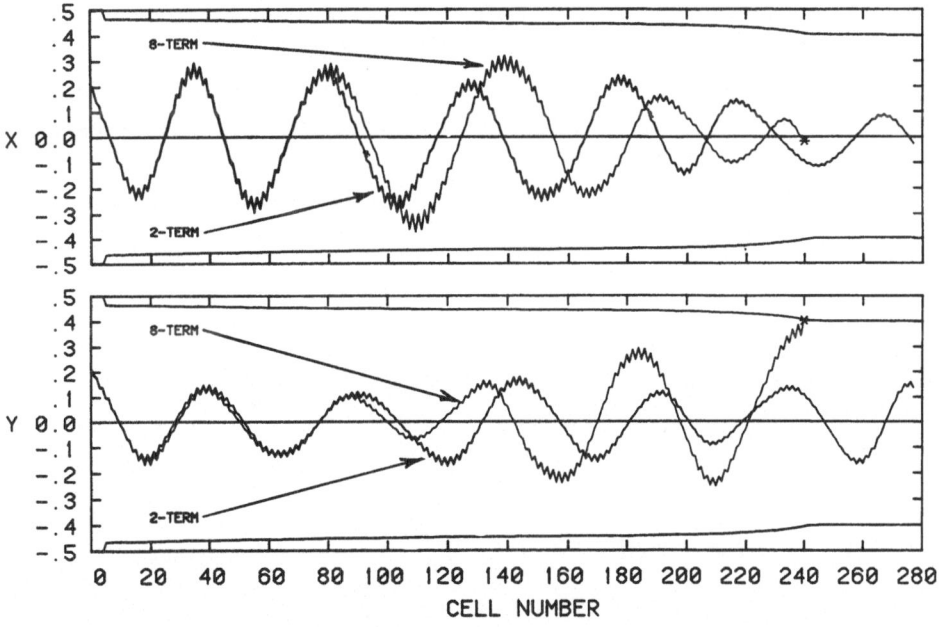

Fig. 2 X and Y trajectories for idential input particles experiencing either 2 or 8-term potentials. Space-charge effects from the rest of the beam are included.

VERSATILE CODES AND EFFECTIVE METHOD FOR ORBIT PROGRAMMING
WITH ACTUALLY EXISTING FIRST HARMONICS IN CYCLOTRON

Mao-bai Chen, Sen-lin Xu and Wen-bin Sen
Institute of Nuclear Research, Shanghai, China

A brief description of the program peculiar features is given. the effective method and systematic procedure for orbit programming with actually existing first harmonics by using above codes are summarized.

Description and features of the codes

A series of versatile codes suitable to the small computer PDP-11 have recently been developed with the characteristics of MSU's, but with a lot of improvements. The motivation behind developing these codes is that only small computers PDP-11 are available in our laboratory, consequently the present advanced routines such as MSU's can not straightforwardly be applied on such computers. Several new considerations in orbit programming also make it necessary to develope our own codes:

(a). It was found that in an actual three dimensional electric field, the assumption of the linearity in axial motion is not so appropriate as the case in a two dimensional electric field (Fig 1). Therefore, the three dimensional electric field calculated by using the relaxation method[1] is carried in ZMOTION routine.

(b). It was measured that the natural first harmonics in the cyclotron

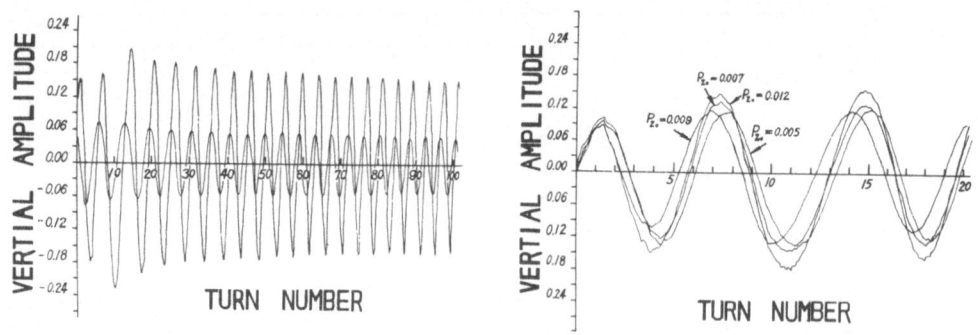

Fig1. Non-linearity of Z motion in 3-dimensional field: a) $Z_2(0)=2Z_1(0)$, but $Z_2(r)=3Z_1(r)$; b) almost the same $Z(r)$ for the different initial $P_z(0)$.

could be cancelled down to less than 0.1 gauss at all radii by using valley coils. The residual first harmonics, though small, have an lasting effect on the particle motion. Hence, all first harmonics existing in the cyclotron including both static one and acceleration one are incorporated into the main magnetic field, of which the static first harmonics are caused by magnet, sectors, coils and magnetic extraction devices, and the acceleration one are produced by the electric gap crossing resonance and the asymmetric dee voltage during the successive gap crossings.

(c). Generally speaking, the initial conditions of the particles emerged from the ion source are determined by five parameters $X_o, Y_o, T_o,$ and B_o, Q_s (excluding space charge) (Fig 2), X_o, Y_o--the location of the source exit slit; Q_s--the rotating angle of the source exit surface with respect to the dee-dummy dee gap; T_o--initial r.f time; B_o--magnetic field along the emerged particle path. However, we have to mainly rely on the inner valley coils instead of varing the two parameters X_o and Q_s, to center the beam by adjusting the current magnitude I_a and argument I_q of the inner valley coils, since the ion source in our case can only be moved along Y axis. A program VATUNE was then developed in which I_a and I_q were taken as the amplitude and argument of an equivalent first harmonic.

(d). Without the existing static first harmonics considered and without the gap crossing resonance involved, the approach of the beam centering can be achieved by using the so called "matrix method",[2] which in principle is the same method as in finding out an equilibrium orbit[3]. With the matrix method, a program RSP is used to reset the ion source position. However, with the gap crossing involved, and the actually static first harmonics considered, the above method is not appropriate since the real first harmonics make a lasting and extending effect on the particles. Especially the beam will be centered mainly by the valley coils as the case in our cyclotron, the usual way in finding the so called "quasi-fixed point"[4] or

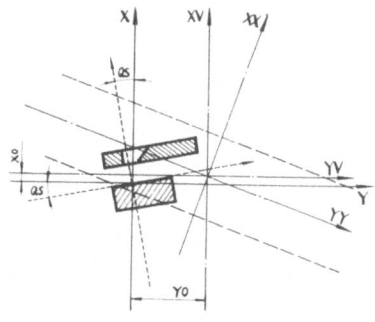

Fig 2. The scheme of the central region.

"AEO"[5] is also not proper due to the undetermined valley coil current and argument. Then the trial and error method has to be applied even though much time has to be taken.

(e). In view of the difficulty in adjusting the source position, the option of a constant orbit geometry is preferred. According to the similarity theorems[6], if $L_1^2 B_1^2 (e/A)_1/U_1 = L_2^2 B_2^2 (e/A)_2/U_2$ the ion trajectories in both cases are similar. Apparently it is just appropriate to the uniform magnetic field. Since the isochronous field with radius are different for both similar orbits required, so $B_1(r)/B_2(r) \neq$ constant and it is not proper to get U_2 from U_1 through the above relation. For exemple, the difference in R between the "reference orbit" and the "similar orbit" would be 7 mm at R=55cm, if the central field of both orbits were substituted for B_1 and B_2 in above relation. Therefore a program DEEVP has been written, in which a central ray of the "reference orbit" is simultaneously accelerated both in its own isochronous field and in the one of the "similar orbit" to be calculated. The average values $\overline{B}_1$ and $\overline{B}_2$ of fields experienced in each integrated step by this particle are substituted for B_1 and B_2. Then the obtained U_2 will trace the central ray of the "similar orbit" in its own field with excellent similarities[7] on the computers.

(f). To make beam energy focusing and to make the "similar orbit" have almost the same phase history at specified points as the one of the "reference orbit", the "energy focusing" relation[4] putforward by M.M.Gordon was introduced in the program, where $SINF = \int_0^E \sin Q/\cos^3 Q \, dE$ and Q is the phase as particles are crossing the accelerating gaps.

(g). Since the plotter is not available in our laboratory, so a series of plotting routines have to be written to meet the needs of the orbit programming.

Apart from these peculiar features, the device of the codes is similar to the one of MSU's both in mathematic mode and in physics mode[2]. This group of programs consist of seven major programs. Of which programs LINEAR and RMOTION are the two main routines. The program LINEAR is used to investigate the static behaviour of the particle and to estimate the features of the magnetic field. The program RMOTION is used to study the particle dynamics in the median plane. This program consists of three parts of the cyclotron, and can start and end at any part of the three parts, also can run backward. A simplification of the PART 1 of RMOTION, called SP is used to fix the relative position between the ion source and puller, and to set the preliminary source-puller position with respect to the center of the cyclotron. A complete description of the programs can be found in an internal report[2].

Method and procedure of orbit programming

With the above codes, the work of the orbit programming has just been completed for the remodelling of the INR cyclotron. Thus an effective method and systematic programming procedure have been explored and are summarized in following discussion.

(a). Static correction of the radio frequency

Running LINEAR to examine a measured magnetic field, it would show that the deviation of the isochronism is a few of 10^{-4} rather than a few of 10^{-5} as in the calculated field trimmed. Changing the central magnetic field Bo would make the isochronism of the measured field better than $1*10^{-4}$ corresponding to an adjacent isochronous field, thus the radio frequency fo ($f_o=QB_o/2\pi mc$) is preliminarily corrected.

(b). Running program SP

To select the favourable initial phase To according to the maximum energy gain ε during first gap crossing by the particle emerged from the center of the source slit. The To should be set at the flat part of the curve of Fig3 to get minimum energy spread.

To fix the relative position (Fig2) between the source and puller by passing through particles emerged from two sides of the source slit with maximum emerged angle of $30°$;

To set the preliminary source-puller position (in Y direction) with respect to the center of the machine at 3/4 R1, where R1 is the average radius of the first quarter turn (Fig 4).

(c). Determination of the bump field

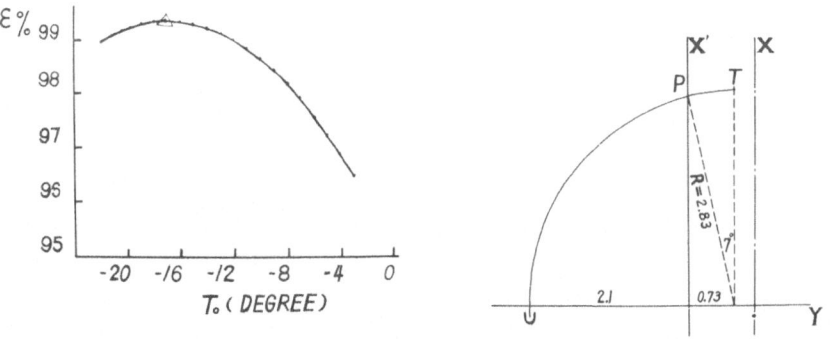

Fig 3. Initial phase vs energy gain. Fig 4. The first quarter turn.

The bump field trimmed can analytically be expressed with approximate form of $B_{bump}(r) = B_0 A e^{-f(r/1.5)^2}$, where the magnitude A is set at such a value that makes the particle have positive average phase during first turns to provide enough electric focusing, and the f--the shape factor of the bump is selected by making the phase of the particles approaching to $0°$ phase when entering the isochronous region to get maximum energy gain. It can be obtained by running PART 1 and PART 2 of program RMOTION.

It should be pointed out that the achievement of positive average phase not only depends on the bump magnitude, also the rotation angle of the ion source. In our case the source has been fixedly rotated for $20°$, and the bump magnitude A is restricted by both the amount of the phase shift required within bump region and practical capacity. So sometime it is necessary to change the initial phase.

(d). Positioning the ion source-puller

Beam centering is obtained either by positioning ion source or by adjusting inner valley coils. However, even though the latter is adopted, this procedure is still necessary for:

a. furnishing the initial conditions for the calculation of the outer valley coils to accomplish the precessional extraction.

b. determining the source-puller position in Y direction.

c. changing the initial phase to compensate the unchangable Qs which ought to be changed.

d. making acceleration correction of the radio frequency.

In so doing, all static first harmonics are temporarily set at zero in program, on the other words, are presumably cancelled by valley coils. Then the "matrix method" can be approximately applied to find out the accelerated equilibrium orbit due to the small effect of the gap crossing effect, the center error is determined by compareing the r and pr of the accelerated orbit with the one of the correspoding static equilibrium orbit at three azimuth of $120°$ interval.

After determining Xo, Yo and Qs, the beam will well be centered at large radius, from where the extraction calculation starts.

(e). Acceleration correction of the radio frequency

Since an accelerated equilibrium orbit by no means coincides with the static equilibrium orbit at all radii, consequently, the radio frequency should be modified according to the "energy focusing" to make SINF as small as possible.

(f). Adjusting the outer valley coils

Owing to the extensive effect of the outer valley coils on the inner centered beam, the outer valley coils should first be calculated according to the demand of the precessional extraction by using program VATUNE with all first harmonics involved. In such calculation, the trial and error method has to be adopted, and either the results of the procedure (d) or the data of a static equilibrium orbit with appropriate energy are taken as the initial conditions.

(g). Adjusting the inner valley coils

After the outer valley coils are fixed, with Xo,Qs set at mechanically prescribed values (in our case Xo=-0.3cm and Qs=0°) and with To changed to appropriate value, the inner valley coils can then be calculated to find out an accelerated equilibrium orbit by using VATUNE code.

(h). Optimizing the radio frequency

Due to the coupling effect between the radial motion and longitudinal motion, the radio frequency should be once again modified to obtain the "energy focusing". Thus far, an accelerated equilibrium orbit with good behaviour has been achieved.

(i). Exact similarities of the orbits

To obtain the constant orbit with exact similarities, the program DEEVP is used to get dee voltage required. Furthermore, the fine adjustment of the frequency would make the "similar orbit" have the same phase history at specified points as the "reference orbit". Since the first harmonics are not included in the similarity theorems, and are different for both orbits, it is clear that both inner and outer valley coils should be readjusted for the "similar orbit" by repeating the procedure (f) and (g).

(j). After getting an accelerated equilibrium orbit (AEO), all aspects of the beam behaviour can then be studied, and the orbit programming can be proceeded. Their results have been discussed in paper[7].

We would like to give credit to Prof.M.M.Gordon of Michigan State University. A lot of his talent ideas have been incorporated in this work.

Reference

1. Mao-bai Chen and Lind. D I.E.E.E. Trans NS-28 (1981), 2636
   Mao-bai Chen and Wen-bin Sen This Conference
2. Sen-lin Xu and Mao-bai Chen INR Internal Report (1983)
3. Gordon.M.M and Welton.T.A. ORNL-2765 (1959)
4. Gordon.M.M I.E.E.E. Trans NS-13 (1966), 48
5. Gordon.M.M N.I.M. 169 (1980), 332
6. Reiser.M The CFS Conference (1962) 370
7. Mao-bai Chen and Sen-lin Xu The Proceeding of 2nd China-Japan Symposium on Accelerators (1983)

CALCULATIONS OF THE HEAVY ION SACLAY TANDEM POST ACCELERATOR BEAMS

S. Valero, B. Cauvin, J.P. Fouan
CEN Saclay, France

P.M. Lapostolle
GANIL Caen, France

1. INTRODUCTION

The funding for the construction of a superconducting linear postaccelerator for the Saclay 9 MV FN Tandem started mid 1981. The output energies from ^{12}C (14 MeV/A) to ^{72}Ge (7 MeV/A) and the beam qualities will be equivalent to those of a 25 MV Tandem with two strippers. Excellent energy or time resolution are also available from such independent cavities linac. The original cavity design and development was done by and with the Karlsruhe group. It consists in a solid niobium tapered helix structure (with f_o = 135 MHz, β_o = 0.08 and dW = 560 keV/charge) (fig. 1,2,3).

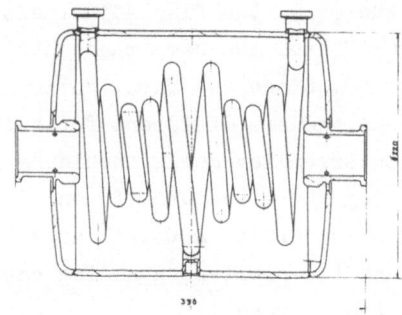

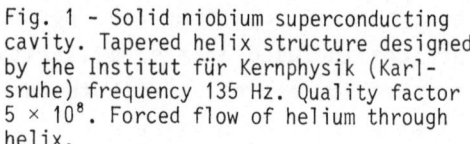

Fig. 1 - Solid niobium superconducting cavity. Tapered helix structure designed by the Institut für Kernphysik (Karlsruhe) frequency 135 Hz. Quality factor 5 × 10^8. Forced flow of helium through helix.

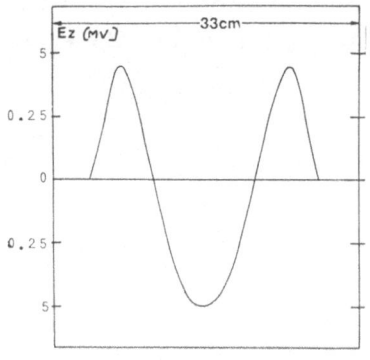

Fig. 2 - Accelerating field as obtained on axis by bead perturbation measurements. Maximum field at the surface of the helix 16 MV/m.

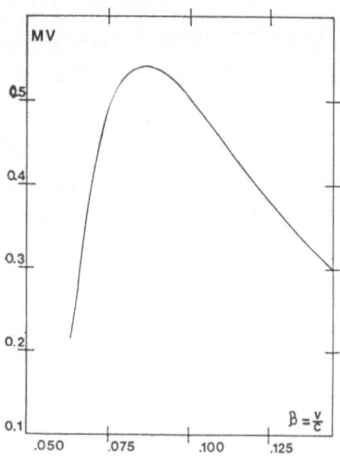

Fig. 3 - Accelerating voltage versus velocity. Maximum accelerating voltage : 560 KV (i.e. 2,2 MV/m)
 A single type of cavity will be used throughout the machine.

It is expected that the construction of the machine will be completed in 1986. In the present paper we discuss the optics and dynamics of this machine.

2. CALCULATIONS OF THE HEAVY ION TANDEM POSTACCELERATOR

2.1 General equations of electric and magnetic fields

In the following we shall assume that our cavity is running through a $TM_{01,n}$ mode, implying cylindrical symmetry around the optical axis Oz, and we shall treat the problem in cylindrical coordinates (r, θ, z).

The fields E and B, solutions of the Maxwell equations are given by :

$$E_z(r) = \cos(\omega t+\phi) \sum_{i=1}^{M} A_i I_0(K_i r) \sin\lambda_i z \qquad (1.a)$$

$$E_r(z) = -\cos(\omega t+\phi) \sum_{i=1}^{M} \frac{\lambda_i A_i}{K_i} I_1(K_i r) \cos\lambda_i z \qquad (1.b)$$

$$B_\theta(r,z) = \sin(\omega t+\phi) \frac{\omega}{C^2} \sum_{i=1}^{M} \frac{A_i}{K_i} I_1(K_i r) \sin\lambda_i z \qquad (1.c)$$

with : $\lambda_i = \frac{\pi i}{L}$ and $K_i = \sqrt{\lambda_i^2 - (\frac{\omega}{C})^2}$

where L is the electric length, ω the pulsation and ϕ the phase.

We can get by bead perturbation measurements the actual values of the field $E_z(r=0)$. These values are treated by least square fitting procedures. We then obtain the best values of the coefficients A_i (for a given number M of spatial harmonics).

2.2 Motion of a charged particle

Let us denote m, v, q respectively the mass, velocity and charge of a particle of the beam. The radial motion is given by the following second order differential equation

$$\frac{d^2 r}{dt^2} = \frac{q}{m} (E_r(z) - V_z B_\theta(r,z)) \qquad (2.a)$$

or approximately, with $r' = \frac{dr}{dz}$ by :

$$\frac{dr'}{dz} = \frac{q}{mV_z^2} (E_r(z) - V_z B_\theta(r,z)) \qquad (2.b)$$

The longitudinal velocity V_z can be obtained from the differential equation :

$$\frac{dv_z}{dz} = \frac{1}{mv_z} E_z(r) \qquad (3)$$

we shall obtain the motion of the particle inside each cavity from the simultaneous resolution of equations (2) and (3).

2.3 Resolution of motion equations

As velocity increases (or decreases) throughout the cavity, it is convenient to solve the differential system (2) and (3) by numerical integration, dividing the electric length L is small steps. Then we can assume that the longitudinal velocity remains constant along the length of one step.

Our objective is twofold :
- define the optical line and the beam optics,
- simulate the beam behaviour and determine the values of the parameters critical to buncher focussing and the values of the phase of the accelerating cavities.

In order to attain these two goals, two programs have been developped.

2.3.1 *Beam optics calculations*

The widely used transport program has been modified to include accelerating cavity matrices, Tandem matrix, and the angular straggling introduced by the high energy stripper. In particular, the velocity changes were included in the fitting procedures.

Let us examine the computation for an accelerating cavity. When $(k_i r)$ is small, the asymptotic expansion of the modified Bessel function $I_1(k_i r)$ is given by :

$$I_1(k_i r) = \frac{k_i r}{2} + \varepsilon(r^2)$$

In this case, we can rewrite (1.b) and (1.c) :

$$E_r(z) = r\, E_r^*(z) \quad \text{and} \quad B_\theta(r,z) = r\, B_\theta^*(z) \tag{4}$$

where $E_r^*(z)$ and $B_\theta^*(z)$ are independent of the radial coordinate r.

The evolution of the radial coordinate r from step i to step i+1 can be expressed in terms of current values of r_i and r_i' following the Euler method

$$r_{i+1} = r_i + h r_i'$$

where h is the length of a step. If we consider equation (2.b) we can write :

$$\int_{ih}^{(i+1)h} dr' = r_{i+1}' - r_i' = \frac{q\, r_i}{v_i^2} \int_{ih}^{(i+1)h} (E_r^* - v_i\, B_\theta^*(z))\, dz \tag{5}$$

We can then see immediately that the transfer matrix from step i to step i+1 in terms of current values of r_i, r_i' and v_i is given by

$$\begin{pmatrix} r_{i+1} \\ r'_{i+1} \end{pmatrix} = \begin{pmatrix} 1 & h \\ -\frac{1}{f} & 1 \end{pmatrix} \begin{pmatrix} r_i \\ r'_i \end{pmatrix} \qquad (6)$$

with

$$-\frac{1}{f} = \frac{q}{mv_i^2} \int_{ih}^{(i+1)h} ((E_r^* - v_i B_\theta^*(z))) \, dz$$

The longitudinal velocity at step i+1 is defined in terms of the current values of v_i and r_i by :

$$v_{i+1} = v_i + \frac{1}{mv_i} \int_{ih}^{(i+1)h} E_z(r_i) \, dz \qquad (7)$$

The relation (7) is directly deduced from equation (3). All the integrals are evaluated by Simpson's rule.

2.3.2 Beam dynamics program

A six dimensional ray tracing program has been developed using equations (2.b) and (3). These equations must be integrated to a very high degree of accuracy in order to insure that the errors do not grow too rapidly. The purpose of this section is to discuss in which way the perturbations due to the numerical approximations can be bounced.

Methods of calculations

First we will examine the calculation of the phase variation throughout a cavity. Let us suppose that the current longitudinal velocity v_i remains constant along the length of one step. In this case, the phase at step i+1 may be related to that at step i by :

$$\phi_{i+1} = \phi_i + \frac{h\omega}{v_i} \qquad (8)$$

As the velocity of the accelerated particles is small (particularly in the case of the heaviest ions) the phase varies very rapidly along the trajectory in the cavity and equation (8) introduces a truncature error. This error will be approximated for each step by :

$$\frac{1}{h\omega} |\Delta\phi_i| = \frac{|v_{i+1} - v_i|}{v_i^2} + \frac{(v_{i+1} - v_i)^2}{2 v_i^3} + \ldots \qquad (9)$$

The effect of the total amount of error introduced in phase calculations is discussed in detail in ref. 1.

To reduce that truncature error, we will make use of the higher accuracy semi empirical relation

$$\phi_{i+1} = \phi_i + \frac{3 \, h\omega}{v_{i+1} + v_i + v_{i-1}} + \frac{(v_{i+1} - v_{i-1})h\omega}{v_i^2} - \frac{1}{2} \frac{(v_{i+1} - v_{i-1})^2 \, h\omega}{v_i^3} \quad (10)$$

Integration of equations (2.b) and (3) are performed with the predictor-corrector formula due to Milne and originally derived from the Adam-Bashfort explicit multistep method. Let :

$$y' = f(x,y), \quad f_k = f(x_k, y_k) \quad \text{and} \quad h = x_{k+1} - x_k$$

The predictor-corrector formula which we used is the following :

Predictor : $\quad y_{k+1}^p = y_{k-3} + \frac{4h}{3}(2f_k - f_{k-1} + 2f_{k-2}) + \frac{14}{45} h^5 y^5(n_1)$

Corrector : $\quad y_{k+1}^c = y_{k-1} + \frac{h}{3}(f_{k+1}^p + 4f_k + f_{k-1}) - \frac{1}{90} h^5 y^5(n_2) \quad (11)$

where $\quad f_{k+1}^p = f(x_{k+1}, y_{k+1}^p) \quad$ and $\quad x_i < n_1, n_2 < x_{i+1}$

These two particular multistep equations have given us the best results in terms of absolute stability against step size and magnitude of the error.

Design of the beam dynamics program

The program computes the evolution of a beam burst (which can include up to 100 particles) as it travels along the machine. Each particle can be followed individually in the six dimensional phase space through the different optical lenses and dynamical components of the beam line (Tandem, stripper, bending magnet, quadrupole, solenoid, cavity). The beam envelope can also be observed anywhere in the machine. This envelope is described in the program by the concentration ellipse of the statistical distribution of individual particles in the 3 two-dimensional phase subspaces. In the energy-phase subspace the area of the concentration ellipse is given by

$$A = 4\pi \sqrt{\sum_i (\Delta w_i)^2 \, \sum_i (\Delta \phi_i)^2 - \left(\sum_i \Delta w_i \Delta \phi_i\right)^2} \quad (12)$$

where Δw_i and $\Delta \phi_i$ are the energy and phase displacement of each trajectory with respect to the center of gravity of the beam burst. Concerning classical lenses, several options included in the program allow the computation to proceed using either transfer matrices, either analytic field or map field measurements. For bending magnets, the program is able to provide computations of astigmatic and spherical aberrations.

3. RESULTS OF COMPUTATION

3.1 Beam optics

The new features implemented in the TRANSPORT program did not affect its versatility for beam optics calculations. We present here the machine lay-out computed with this program (fig. 4).

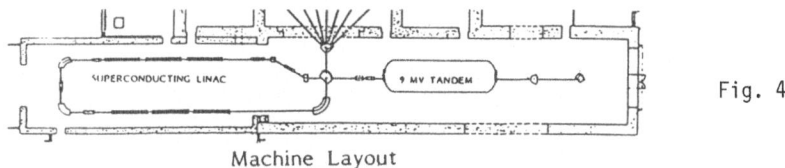

Fig. 4

Machine Layout

The particles outgoing the Tandem are injected together with second stripping and bunching through an achromatic 2 × 30° bend. Each accelerator module consists of a cryostat housing 8 cavities and 2 focussing solenoïds at 4°K. In order to house the machine in the available space and to switch the beam back to the existing experimental equipment, the six accelerator modules are separated by an isochronous U turn followed by a single cavity rebuncher. The quadrupole triplets preceding the two accelerator legs are used to focus the beam into a cylindrical waist (x=y=5mm) which matches the solenoïd soft focussing in order to minimize the radial oscillations of the beam envelope.

The calculations were performed with an overestimated emittance corresponding to the geometrical size of the Tandem terminal stripper canal. Actual measurements of 2 and 5.5 mm.mrd for 50 MeV ^{12}C and 69 MeV ^{56}Fe respectively are well below this safe limit.

3.2 Beam dynamics

At the present day the study of the beam dynamics is not fully completed. However, with the help of the ray tracing program, we already have obtained enough informations to define a dynamic matching mode for the machine. It showed us that the values of the buncher focussing parameters and the phases of the accelerating cavities could be roughly guessed from the theory developed for relativistic proton linacs. The full analysis of the beam burst concentration ellipses, obtained with the exact ray-tracing program, shows that by slightly modifying the above mentioned guess values, it is possible to keep down the beam size in the Energy-Time subspace (figs. 5-14).

REFERENCE

1. P.M. Lapostolle, Equations de la dynamique des particules dans un accélérateur linéaire à protons. AR/Int. SG/65-11 (31.5.1965).

CONCENTRATION ELLIPSES (SOLID LINE) IN THE PHASE SPACE AND INDIVIDUAL TRAJECTORIES (DOTS). ^{16}O Tandem energy: 51 MeV, charge 5+, post stripping charge 7+ (75 trajectories). *Emittance in Tandem stripper* $x = y = \phi = 0; \theta = 0$ mrd, ± 5 mrd; $dp/p = 0$, $\pm 2 \times 10^{-4}$, $\pm 5 \times 10^{-4}$; $\Delta t = 0$, $\pm 2 \times 10^{-10}$s, $\pm 5 \times 10^{-10}$s. *Matching mode*: cavities 1 to 5 $\phi_0 = -30$ deg, cavities 6 to 48 $\phi_0 = -15$ deg

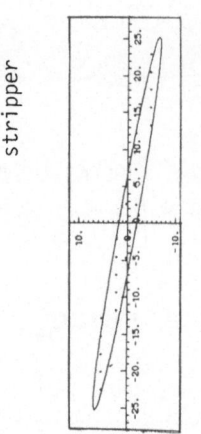

Fig. 6 - Tandem stripper

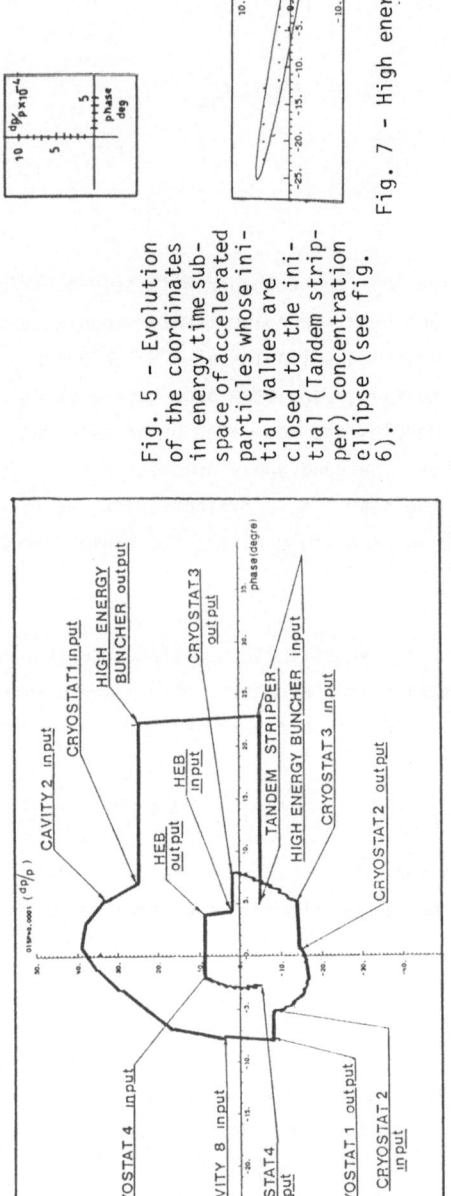

Fig. 5 - Evolution of the coordinates in energy time subspace of accelerated particles whose initial values are closed to the initial (Tandem stripper) concentration ellipse (see fig. 6).

Fig. 7 - High energy buncher input

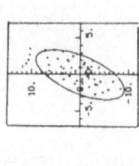

Fig. 14 - cryostat 3 output

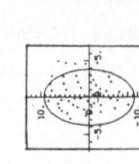

Fig. 13 - cryostat 3 input

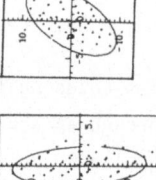

Fig. 12 - cryostat 2 output

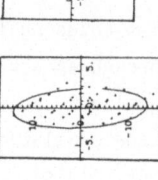

Fig. 11 - cryostat 2 input

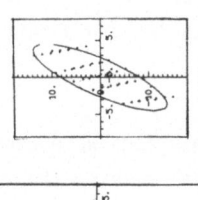

Fig. 10 - cryostat 1 output

Fig. 9 - cryostat 1 input

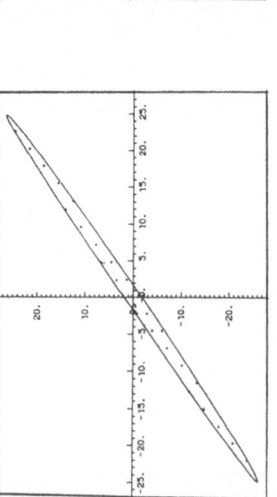

Fig. 8 - High energy buncher output

ELECTRON INJECTOR COMPUTER SIMULATIONS
D. TRONC
CGR MeV - 551 rue de la Minière - 78530 BUC France

Introduction

We present contributions for electron injector computation and design, describing a simple but complete simulation code implemented on a personal computer, giving the main design choices taken for the BCMN[1] and LEP[2,3] high intensity injectors and for the ORION self-focussing injector.

Electron dynamics are characterized by the predominant effect of the first "accelerating" cell, in contrast with proton dynamics[4]. In this region shorter than an RF half-wavelength the non-linear bunching and acceleration can only be simulated in a step-by-step procedure. Analytical "adiabatic" approach cannot help the designer but he can take advantage of non-repetitive features to obtain radial RF self-focussing together with longitudinal bunching[5].

The simulation code

An interactive code is the preliminary condition for design work. The 1973 ESPEL code[6] took only account of the preexistent fields E_z^p, B_z^p and of the space charge fields E_z^i, E_r^i (axis z, radius r, overscripts p for preexistent and i for space charge). A 1980 version written in a collaboration between CGR MeV and Orsay LAL[2,7] included E_r^p, B_r^p and B_θ^i. Personal computers availability led us to write a condensed interactive 1983 version. It includes B_θ^p, data are easier to use and computation speed is increased. Table 1 presents the structure of this version :

1) The longitudinal calculation is made in a space representation. The z-axis is divided into successive domains delimited by abcissa input values z, E_z^p, B_z^p and RF phase data corresponding usually to perturbation measurements. Each domain is subdivised into subdomains for precise axial and radial calculations with E_z^p and B_z^p linearly interpolated. In a subdomain length h, for peak field E_{oz}^p, for dephasing $\emptyset$ between the RF field and the electron of energy V and of relative speed $\beta = v/c$, iterative calculation is made with :

$$V_1 = V_0 + h(E_{oz}^p + E_z^i) \cos \frac{\emptyset_0 + \emptyset_1}{2}$$

$$\emptyset_1 = \emptyset_0 + 2 \frac{h}{\lambda_0} \frac{2}{\beta_0 + \beta_1} \quad , \quad \beta_i = f(V_i)$$

2) The radial calculation utilizes the field values given by :

$$E_r^p = -\frac{r}{2}\frac{\partial E_z}{\partial z} \quad , \quad B_r^p = -\frac{r}{2}\frac{dB_z}{dz} \quad , \quad B_\theta^p = \frac{r}{2}\frac{1}{c^2}\frac{\partial E_z}{\partial t}$$

$$E_r^i - \frac{dz}{dt} B_\theta^i = E_r^i (1 - \beta^2)$$

in the following equations integrated with help of a Mac-Laurin expansion limited to two terms :

$$\frac{d^2 r}{dt^2} = \frac{e}{m}(E_r^p + (1-\beta^2) E_z^i + r\frac{d\theta}{dt} B_z^p - \frac{dz}{dt} B_\theta^p) + r(\frac{d\theta}{dt})^2 - \frac{\beta}{1-\beta^2}\frac{d\beta}{dt}\frac{dz}{dt}$$

$$\frac{d^2\theta}{dt^2} = \frac{1}{r}\frac{e}{m}(\frac{dz}{dt} B_r^p - \frac{dr}{dt} B_z^p) - \frac{2}{r}\frac{dr}{dt}\frac{d\theta}{dt} - \frac{\beta}{1-\beta^2}\frac{d\beta}{dt}\frac{d\theta}{dt}$$

3) The space-charge calculation uses a predetermined cylinders geometry of radius r and of lengths r, 2r, 4r, 8r, centered at the abcissa of the "electron" under influence. r is the mean bunch radius and two successive cylinders delimite pair of cases between them. E_z^i calculation uses the differences of the cases populations for each pair, weighting predetermined characteristic coefficients of the geometry. E_r^i calculation uses the cylinder which includes at least half the population included in the model and takes account of the difference between r and the radius value for the "electron" under influence. Such a rough simulation of the beam envelope behaviour is valid when space-charge effect is not predominant.

<u>The LEP high intensity injector</u>

A total charge of $0.15 \ 10^{-6}$C (15 A current, 10^{-8}s RF pulse length) is bunched and accelerated for positron production. <u>Figure 1</u> shows the geometry of the critical beginning of the injector.

The heavy short pulse beam loading is the main cause for the beam energy spead. It decreases when RF energy stacked per m increases. The limit is set by field breakdown and thermal constraints. We choose a triperiodic S-band SW structure with thick rounded irises[8]. For $\omega L \omega = 60 \ 10^{12}$, $I = 15$ A , $\tau = 10^{-8}$s , $E(t=0) = 16 \ 10^6$ V/m (the on-axis peak field reach $30 \ 10^6$ V/m), the relations :

$$W_e = W_o - W_\tau \quad , \quad W_o = \frac{E_o^2}{\omega L \omega} \quad , \quad W_e = I\tau \ E_{\tau/2} \ ,$$

gives $W_o = 4.1$ J , $E_{\tau/2} = 14 \ 10^6$ V/m , $W_e = 2.0$ J , $W_\tau = 2.1$ J

The next constraint is the space charge expansion effect : for an elementary spherical bunch of diameter 0.01 m and charge $5\ 10^{-9}$ C, $E_r^i = 0.5\ 10^6$ V/m. It is easily compensated axially by large E_z^p but needs a strong B_z^p confinement field. All stray magnetic field on the cathode must be avoided and a "gap" of field after the gun crossover induces scalloping. The experience gained on BCMN injector led us to design solenoidal coils of increasing diameter ("a conical current distribution") and adequate iron shields. The reduced diameter of the first coil insures a very steep B_z^p rise at the crossover and limited stray field.

Dimension choices for the best dynamic behaviour can now be made :

1) The 90 kV, 25 A gun geometry was derived from the PIERCE design. Simulations showed how the nice gun diode beam laminar flow becomes a turbulent one when grid control is used in a triode configuration. For intense short pulses, grid potential cannot be set at its equipotential value calculated for the diode configuration. As its potential rises much higher the bending of electron trajectories depends on their proximity to the bars of the grid. In fact the beam emittance is determined by the first millimeters of its trajectory.

2) Prebunching increases the density inside the acceptance phase space of the following periodic structure. Beam radius increase must be limited by the choice of a short distance between the modulation cell and the first accelerating cell.

3) The length choices for the first cells of the periodic structure results from a compromise between axial compression and energy spread.

The ORION self-focussing injector

This S-band SW biperiodic structure of 0.22 m length delivers 0.2 A at 5 MeV. It is used for medical or industrial application. The first design simulations showed the disastrous effects of preexistent intense E_r^p. In the LEP case the use of strong B_z^p limited their effects as well as space-charge ones. Here lower beam energy injection makes things worse. The traditional approach of optimizing the longitudinal dynamics and then compensating for adverse radial behaviour led to a critical magnetic lens located near the gun, to the loss of many electrons along the structure and to a large (simulated) target spot. We discovered that radial control was indeed possible by choosing a greater than usual length for the first cell[4,9]. The beam behaviour can be analysed on figure 2 as follows :

1) Axial compression occurs at the very beginning of the cell, from A to B. One-third of the electrons are bunched in one-sixth of an RF period for low injection energy of 10 keV to 20 keV. They will constitute the accelerated bunch. At the same time a complex radial evolution occurs. First electrons increase their distance to

the axis ("radius") when they enter in a decelerating defocussing field (trajectory a). Central electrons see an average zero field (trajectory b). Last ones enter in an accelerating focussing field and decrease their distance to the axis (trajectory c). After some millimeters electrons are bunched in phase but dispersed in radius and in energies.

2) In the "central" part of the cell from B to C a tenfold increase in energy occurs. C would be at the exit of the cell in a conventional design and a radial defocussing would occurs (dotted trajectories).

3) The length increase C to D insures that the bunch will cross the cell exit when RF field is zero or focussing (and decelerating).

4) The bunch enters the second cell when the field is already large and focussing. Afterwards the high energy reduces the radial effects.

Conclusion

Electron injector designs must optimize axial and radial behaviour together. This is easy with the advent of personal computers and a code implemented on less than 12 koctets. This short presentation was limited to the design guidelines as detailed simulation results are available in references given below.

References

1. A. BENSUSSAN, J.M. SALOME, NIM 155 (1978) 11-23
2. D. TRONC, J.L. AZAN, C. PERRAUDIN, J.P. GEORGES, M. AMMAD, CGR report ST 8029 (1980)
3. J.P. GEORGES, CGR report ST 8225 (1980)
4. D. TRONC, NIM 105 (1972) 335-347
5. French patent n° 83 14090
6. D. TRONC, thesis, Orsay (1973) 175
7. B. MOUTON, SA-LAL notes, Orsay (1980)
8. A. BENSUSSAN, D.T. TRAN, D. TRONC, NIM 118 (1974) 349-355
9. S. HADDAB, thesis, Paris 6 (1983)

Table 1 : ESPEL83 code structure

Input beam and structure data, precalculations and printings.
For each domain :

    For each electron, space charge calculation :
        $\bar{r}$, $\bar{\beta}$, cylinders phases
        For each neighbouring electron : Population accounting by phase comparison tests : next neighbour.
        E_r^i cylinder selection, E_r^i and E_z^i calculations
    Next electron.

    For each subdomain : For each electron
        Iterative longitudinal calculation
        Radial calculation
    Next electron : Next subdomain.

    Results printing

Next domain.

Figure 1. L E P high intensity injector (beginning)

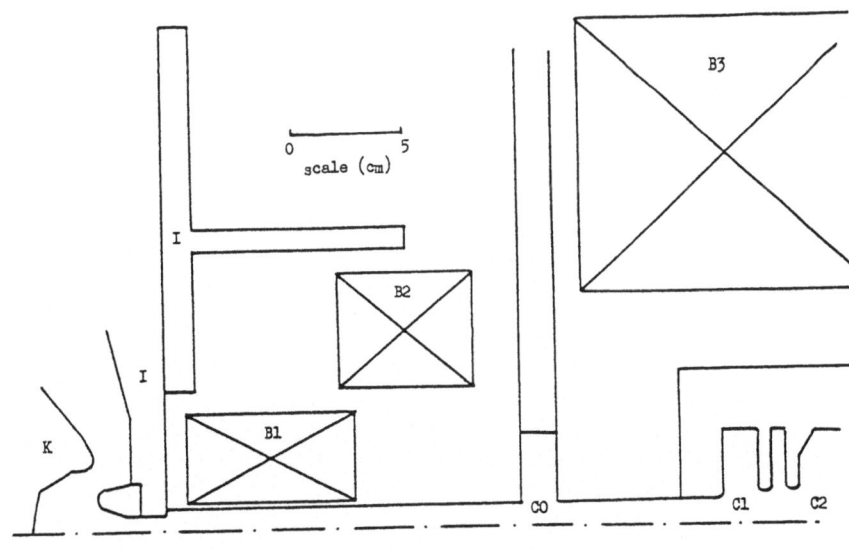

Figure 2. ORION self-focussing injector (beginning)

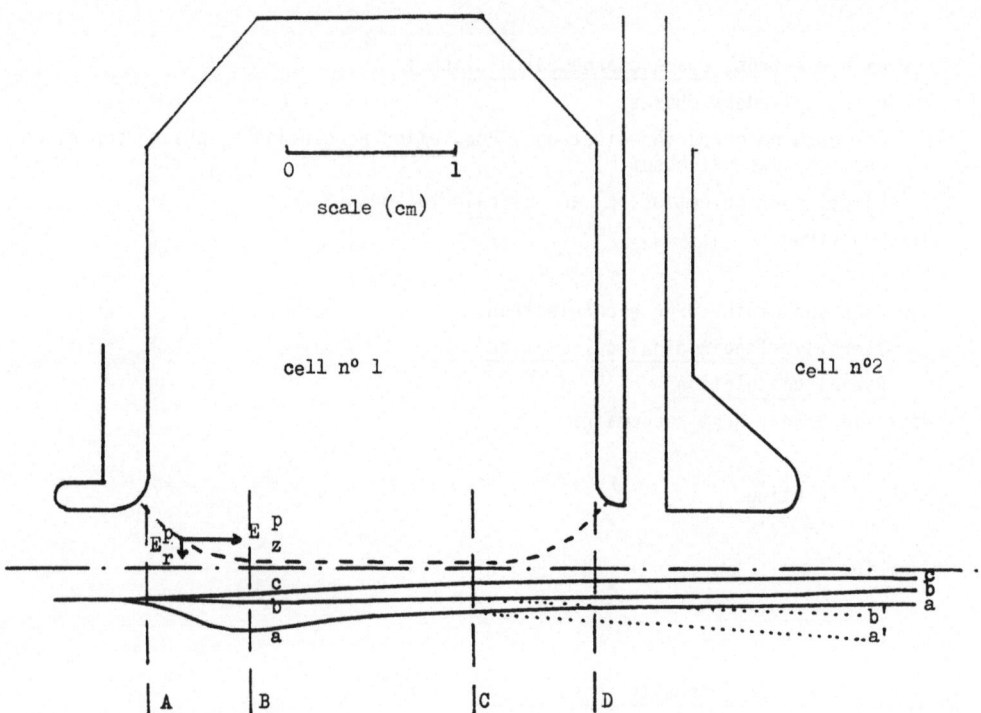

NUMERICAL SIMULATIONS OF ORBIT CORRECTION IN LARGE ELECTRON RINGS

G. Guignard, Y. Marti
CERN
1211 Geneva 23

1. INTRODUCTION

A computer program which simulates effects of possible error sources on the beam optics and the improvements due to orbit correction is a necessary tool for the study and design of large electon rings. Such a program called 'PETROS'[1] exists at the DESY laboratory. PETROS can work in two modes.

1. Uncoupled transverse motions are assumed and the usual three-dimensional matrices are used in each plane.

2. Coupled transverse motions are considered and five-dimensional matrices are used throughout.

It treats non-linear fields, effects of the radiation losses due to bending. It computes the linear transformation matrices of a ring structure, the corresponding betatron and dispersion functions, betatron and synchrotron frequencies. It calculates the five synchrotron radiation integrals, the damping partition numbers, the damping times, the length deviation of off-momentum orbits, beam emittances, bunch length, relative energy spread and synchrotron lifetime, the effect of prescribed or random distortions, taking into account the radiation losses due to bending. It simulates closed orbit corrections and gives the corresponding kick amplitudes. This program has been installed at CERN on the IBM computer by J. Kewisch and B. Zotter[2].

REFERENCES

1. K. Steffen, J. Kewisch, Study of integer difference resonances in distorted PETRA optics, DESY PET-76/09 (1976).
2. B. Zotter, A short guide for the use of program PETROS at CERN, LEP-70/37 (1978).

2. PETROC PROGRAM

Since the orbit correction methods were not well suited for large electron machines with variable phase advance, we improved at CERN the existing methods and we added new methods in the simulation program PETROC[3] (CERN version of PETROS). PETROC can correct the closed orbit in different ways according to the needs in the different parts of the machine. These correction methods are based either on local bumps or on a minimisation. In particular, a very powerful method MICADO[4] based on the least squares theory and using a small number of correctors has been installed. In order to exploit all the results, statistics on the corrector strengths and on the orbit deviations at the monitors, in the presence of reading or setting errors have been implemented.

2.1 Orbit Correction Methods

Four methods to correct the closed orbit are available in the program PETROC:

a. CORB: calculates the strengths of given orbit correctors by using a bump method. The closed orbit is corrected by successive local bumps. They are created by three correctors, i.e. the first, the second and the last one in a group of NK correcting dipoles. The remaining (NK - 3) dipoles are not used and NK can be 0, 3, 4, 5. (NK = 0 is used for skipping over correctors.) The number of pick-up monitors cannot exceed twice the number of correctors, since only two pick-ups can lie between two correctors. This arbitrary restriction does no harm, because increasing further the number of monitors would make the correction system too redundant.

```
Monitor number:
              11 12           13 14           15 16           17 18
─────────────x─|─|─x─────────|─|─────x───────|─|─────x───────|─|─x───
Corrector    C₁              C₂              C₃              C₄              C₅
number:

Kicks:        ε              εr₂              0               0              εr_NK
```

The ratio between the corrector strengths are given by

$$r_2 = C_2/C_1 = \sqrt{\frac{\beta_1}{\beta_2}} \cdot \frac{\sin(\mu_1 - \mu_{NK})}{\sin(\mu_{NK} - \mu_2)} \qquad (1)$$

$$r_{NK} = \frac{C_{NK}}{C_1} = \sqrt{\frac{\beta_1}{\beta_{NK}}} \frac{\sin(\mu_2 - \mu_1)}{\sin(\mu_{NK} - \mu_2)} \qquad (2)$$

$\mu_{NK} - \mu_2$ must be different from a multiple of π. That is why NK is a variable input parameter depending on the phase advance per cell. In addition the phase advance may differ from the regular one in some regions of a large machine and it is useful to have different NK values in the lattice and in these regions. Therefore two NK-values are given in the input: NK3 and NK4. The names for the orbit correctors may contain an F as 4th character, which is used as a flag to exchange the NK-values. If in spite of these facilities $\sin(\mu_{NK} - \mu_2)$ remains too small, the program tries to use C_3 instead of C_2. The bump can also be skipped. The orbit amplitudes due to the bumps are given by:

$$y_i = \varepsilon\sqrt{\beta_1\beta_i} \sin(\mu_i - \mu_1) \text{ at the pick-ups } i = 11 \text{ and } 12 \qquad (3)$$

$$y_i = \varepsilon\sqrt{\beta_i}[\sqrt{\beta_1} \sin(\mu_i - \mu_1) + r(2)\sqrt{\beta_2} \sin(\mu_i - \mu_2)] \text{ at pick-ups 13 to 18} \qquad (4)$$

and the bump strength characterized by the kick ε at the first corrector, minimizes the following function:

$$S = \sum_{i=11}^{18} (y_i + y_{mi})^2 + W^2\beta_1\beta_2 \sin^2(\mu_s - \mu_1)\varepsilon^2 \qquad (5)$$

with $\mu_s = \mu_2$ if $\mu_2 - \mu_1 < \pi/2$ and $\mu_s = \mu_1 + \pi/2$ if $\mu_2 - \mu_1 > \pi/2$. The quantities y_{mi} are the orbit distortions before correction and W is the weight attributed to the minimisation of the corrector strengths.

b. CORM: calculates the strengths of given orbit correctors by using a matrix inversion, a straightforward minimisation of the orbit distortion at all the monitors and of the corrector strengths with a weight attributed to them. With the matrix inversion routine (REQINV NAGLIB) it is possible to invert 450*450 matrices rather accurately but the method is time-consuming and should preferably be used for small machines, or for a preliminary correction of the effects of the insertions only using a small number of correctors.

c. CORL: The closed orbit is corrected by using a small number of magnets with the MICADO method based on the least squares theory and the Householder's transformations. The method is based on the minimisation of the Euclidian norm of the closed orbit distortion. On the first step, each magnet is tested singly and the one giving the minimum norm is retained. On the second step, pairs of magnets containing the one already chosen are tested and the pair giving the minimum norm is selected. The process is repeated until the peak amplitude of the residual orbit is smaller than some value given in advance or until a given number or the totality of the magnets has been used.

d. CORI: It is sometimes necessary to correct locally the effects of the insertion doublets which serve to create a low-beta value at the interaction point. These doublets are strong and are source of large orbit distortions. The unperturbed orbit patterns on both sides of the insertion are described by:

$$y(\mu < \mu_1) = \sqrt{\beta}(\mu) \cdot (a_1 \cos \mu + b_1 \sin \mu) \tag{6}$$

$$y(\mu > \mu_2) = \sqrt{\beta}(\mu) \cdot (a_2 \cos \mu + b_2 \sin \mu) \tag{7}$$

Measuring the orbit at a few well distributed positions at both sides of the insertion allows to calculate the coefficients a and b. Taking into account that the orbit for $\mu > \mu_2$ is the superposition of the orbit for $\mu < \mu_1$ and of the kick effect, we can write:

$$y(\mu > \mu_2) = \sqrt{\beta}(\mu)[a_1 \cos \mu + b_1 \sin \mu + \theta_1\sqrt{\beta_1} \sin(\mu-\mu_1) + \theta_2\sqrt{\beta_2} \sin(\mu-\mu_2)] \tag{8}$$

by equating (7) and (8) we get the kick amplitudes as function of a and b. It is then possible to introduce opposed kicks in correctors placed nearby the insertion doublets.

REFERENCES

3. G. Guignard, Y. Marti, PETROC users' guide, CERN/ISR-BOM-TH/81-32 (1981).
4. B. Autin, Y. Marti, Closed orbit correction of A.G. machines using a small number of magnets, CERN ISR-MA/73-17 (1973).

3. RESULTS AND CONCLUSIONS

First, we present results concerning the simulation of the orbit correction in the LEP-11[5]) machine with $Q_x = 58.34$ and $Q_z = 66.20$. The following r.m.s. values of positioning errors and field dispersion were used for the simulation: for bending magnets, horizontal displacement = 0.12 mm, tilt and twist = 0.24 mrad, field dispersion = 0.00025 and for quadrupoles, horizontal displacement = 0.12 mm, vertical displacement = 0.12 mm, tilt and asymmetry = 0.23 mrad, field dispersion 0.0005. In table 1, the CORL method was used and the number of correctors NC was varied.

Table 1: Variation of the closed orbit correction with respect to the number of correctors

NC	$\hat{x}$	$\hat{z}$	$\langle x \rangle$	$\langle z \rangle$	$\hat{\theta}_x$	$\hat{\theta}_z$	Computing time on IBM (sec)
0	35.06	22.55	7.92	5.91	—	—	21
10	5.37	5.99	1.06	1.40	0.611	1.38	52
20	3.26	3.71	0.84	0.88	0.508	1.06	57
50	3.11	3.08	0.69	0.69	0.363	0.708	73
100	2.58	2.56	0.66	0.79	0.353	0.813	97
200	2.45	2.42	0.63	0.72	0.363	0.757	135

where $\hat{x}, \hat{z}$ = maximum deviation of closed orbit in x,z-plane (mm)
$\langle x \rangle, \langle z \rangle$ = root meansquare deviation of closed orbit in x,z-plane (mm)
$\hat{\theta}_x, \hat{\theta}_z$ = maximum kick amplitude in x,z-plane (10^{-4} rad).

Table 2 shows the results of the simulation on 10 machines correcting first near the insertions (CORI) and then with the bumps' method (CORB).

The amplitude reduction due to the correction is 9 for $\hat{x}$, 12.5 for $\hat{z}$ and 4.8 for $\hat{D}_z$. Because of the decrease of the vertical dispersion and orbit, the average emittance ratio is 40, i.e. larger than the design value of 16.

Secondly, we present results related to the energy loss by radiation. The discontinuous replacement of the radiated energy induces opposite optical aberrations for electrons and positrons and miscrossings at the interaction points due to the combined effect of misalignments and aberrations[6]). The simulation of closed orbits before correction was done in the presence of radiation for positrons and electrons with Q_x = 90.34, Q_z = 94.20 and RF stations near I2 and I6 (voltage of 357.6 MV). Table 3 gives the ratio of the r.m.s. value of $\Delta x, \Delta z$ (difference of closed orbit for electrons and positrons at the interaction points) to the r.m.s. beam dimension σ_x^*, σ_z^* for the optically equivalent interaction point IP.

Table 2 Effects of closed orbit correction in LEP-11

Parameter	Unperturbed ring	Before orbit correction	After orbit correction
Closed orbit (mm)			
$\langle x \rangle$	0.0	6.83 ± 2.94	0.42 ± 0.14
$\langle z \rangle$	0.0	11.72 ± 4.23	0.49 ±
$\hat{x}$	0.0	26.87 ± 10.54	2.92 ± 1.56
$\hat{z}$	0.0	46.51 ± 17.01	3.64 ± 1.04
Dispersion			
$\langle D_x \rangle$	1565.5	1566.7 ± 13.8	1566.1 ± 0.6
$\langle D_z \rangle$	0.0	597.6 ± 225.6	145.7 ± 65.4
$\hat{D}_x$	2235.6	2927.3 ± 231.1	2321.6 ± 40.3
$\hat{D}_z$	0.0	2711.3 ± 997.2	558.1 ± 255.2
Short insertion			
$\langle \beta_x \rangle$	1.60	1.53 ± 0.18	1.58 ± 0.02
$\langle \beta_z \rangle$	0.10	0.16 ± 0.07	0.11 ± 0.02
$\langle D_x \rangle$	0.0	14.96 ± 1.12	9.21 ± 0.61
$\langle D_z \rangle$	0.0	1.06 ± 0.18	0.25 ± 0.03

Table 3 Ratio of the r.m.s. miscrossings to the r.m.s. beam sizes

IP	$\sigma(\Delta x)/\sigma_x^*$	$\sigma(\Delta z)/\sigma_z^*$
2/6	0.177	1.739
1/3/5/7	0.137	1.058
4/8	0.084	1.439

The largest miscrossings appear at the interaction points near the RF stations where the momentum is rapidly changing with the azimuthal coordinate.

These results show the high efficiency of the program and in particular of the CORL correcting method.

REFERENCES

5. A. Hutton, Parameter list for LEP version 11, LEP note 289 (1981).
6. M. Bassetti: Effects due to the discontinuous replacement of radiated energy in an electron storage ring, Proc. 1980 Particle Accelerator Conf., Geneva, p. 650.

SIMULATION OF POLARIZATION CORRECTION SCHEMES IN e^+e^- STORAGE RINGS

D.P. Barber, H.D. Bremer, J. Kewisch, H.C. Lewin, T. Limberg,
H. Mais, G. Ripken, R. Rossmanith and Ruediger Schmidt

Deutsches Elektronen-Synchrotron DESY
Notkestr. 85, D-2000 Hamburg 52, W. Germany

Abstract

Spin polarization in e^+e^- storage rings may be destroyed by various effects. Simulations based on the program SLIM have been performed so that ideas for correction of these depolarization effects could be tested. Application to studies of the consequences of closed orbit distortions and use of solenoids is discussed.

Introduction

It has long been recognised that electron and positron beams in storage rings can become vertically polarized as a result of emission of synchrotron radiation by the Sokolov-Ternov effect /1/. The maximum achievable polarization by this mechanism is 92.4 %.
In addition to the polarizing mechanism the spins also experience depolarizing perturbations as the electrons execute betatron and synchrotron oscillations with random amplitudes and phases in the fields of the machine magnets /2/. These effects are strongest when the following resonance conditions are satisfied

$$a\gamma = k \pm Q_I, \quad I = x, y, s \tag{1}$$

($a = (g-2)/2$, γ = Lorentz factor, k = integer)
and represent a severe limitation to attainable polarization at high energy.

Closed Orbit Distortion

One of the chief contributions to resonance depolarization arises when the ring is not completely flat.
The polarization points along the equilibrium spin axis, the so-called $\hat{n}$ axis /2/ which is a periodic unit vector whose direction can in general vary from point to point around the ring.
In a perfectly aligned (flat) storage ring particles travelling on the closed orbit feel vertical magnetic fields and the $\hat{n}$ axis is vertical.
In reality storage rings are never flat and the vertical profile of the closed orbit is "wavy". Since electrons on the closed orbit now feel the radial fields of the quadrupoles, the $\hat{n}$ axis is no longer exactly vertical.
Using the first order perturbation theory of SLIM /2/ it may then be shown that the spin vectors of individual particles which are executing horizontal betatron oscillations experience varying degrees of precession around the vertical depending on their amplitudes and phase. The ensemble of spin vectors thus becomes smeared or depolarized. These effects are strongest near the Q_x and Q_s resonances of Eqn. 1, and they increase with energy.
Clearly, the effect of horizontal motion can be suppressed if the $\hat{n}$ axis can be returned to vertical with the aid of the correction coils which control the vertical closed orbit so that horizontal and spin motion are decoupled. It may be shown that the angular deviation of the $\hat{n}$ axis from the vertical is proportional to /3/

$$[\oint B_r(s) \sin\varphi_s \, ds]^2 + [\oint B_r(s) \cos\varphi_s \, ds]^2 \tag{2}$$

where φ_s is the spin precession phase and $B_r(s)$ is the radial field on the closed orbit. The chief contributions to these integrals come from Fourier harmonics in the periodic $B_r(s)$ which are closest to the spin tune and which may comprise only a small part of B_r. If these harmonics can be empirically suppressed by selective excitation of vertical correction coils in a way that does not cause undue additional distortion of the orbit by excitation of other harmonics, then n̂ may again be brought close to vertical and the depolarization mechanism suppressed. These notions have been tested for PETRA and DORIS II /4/ using simulations based on the program SLIM. Results of a SLIM calculation for PETRA near 16.5 GeV (where the spin tune is 37.5) in which the correction coils generated an r.m.s. vertical closed orbit distortion of 1.2 mm are shown in Fig. 1 /3/. It is expected that the 37 and 38th sine and cosine harmonics of B_r are mainly responsible for the low predicted polarization. With the 4-fold symmetry of PETRA and its correction coils it is possible to choose various correction coil current distributions each of which excites only one of the dangerous harmonics. A Fourier analysis of the B_r on the closed orbit generated by SLIM indicates the strength of the 37 and 38th harmonics. Linear combinations of the 4 sets of coil currents are then applied to cancel them. The SLIM prediction for the polarization with the corrected closed orbit is also shown in Fig. 1 where the polarization is now seen to be much larger and the resonance effects reduced.

The effectiveness of this method has been verified experimentally /5/ at PETRA. Figs. 2a and 2b show the polarization versus the strength of the 38th sine and cosine harmonics at 16.5 GeV. Fig. 3 shows the polarization versus energy after optimizing the harmonics empirically. Without this optimization, measured polarizations were rarely above 30 % and not reproducible. However, Fig. 3 shows qualitative agreement with the result of Fig. 1 and shows that with closed orbit optimization, polarizations of above 70 % around 16.5 GeV may be achieved. The effect is reproducible and represents a big step forward in the simulation and control of depolarization effects caused by closed orbit distortions.

Detector Solenoids

With vertically polarized beams, uncompensated solenoids cause depolarization by tilting the n̂ axis around the beam direction by an angle /6/

$$\theta_s \simeq \frac{B_s L}{(\rho B)} \qquad (3)$$

(B_s = solenoid field, L = solenoid length).

The n̂ axis is then again no longer vertical in the arcs and interaction region quadrupoles so that, as before, depolarization occurs. In addition, an electron travelling at an angle x' to the axis in the horizontal plane has its spin axis tilted forward or backward by an angle /6/

$$\delta\theta_1 = a\gamma \, x' \, \theta_s \qquad (4)$$

in the longitudinal solenoid field and a total of:

$$\delta\theta_2 = (1 + a\gamma) \, x' \, \theta_s/2 \qquad (5)$$

in the radial end fields of the solenoid. The n̂ axis may be made vertical and the betatron depolarization also compensated by placing opposite polarity "antisolenoids" adjacent to the main solenoid as in Fig. 4. Normally, however, there is no space owing to the proximity of the focussing quadrupoles. The antisolenoids must then be placed further out in available space among the focussing quadrupoles as in Fig. 5. Since solenoids rotate the plane of the betatron oscillations by an angle $\theta_s/2$, the intervening quadrupoles must then be rotated by $\theta_s/2$ to avoid betatron coupling. The n̂ axis is now vertical in the arcs but still tilted by angle $\theta_s/2$ w.r.t. the quadrupoles and quasi-horizontal betatron motion will still cause depolarization if $x'_{sol} \neq x'_{antisol}$. A suitable optics must then be chosen and the antisolenoids positioned so that $x'_{sol} = x'_{antisol}$ ("the favoured position").

These ideas have been tested in simulations using SLIM. Fig. 6a shows the polarization prediction when antisolenoids are placed in the favoured position, whereas Fig. 6b shows the predicted polarization when the antisolenoids are placed so that $x'_{sol} \neq x'_{antisol}$ (the "unfavoured position"). In this latter case strong depolarizing resonance effects are seen, whereas these effects are strongly suppressed if the antisolenoids are correctly positioned.

Conclusion

The examples discussed show that the polarization simulation program SLIM provides a convenient and powerful means of testing ideas for improving polarization in electron-positron storage rings. It is now used routinely at DESY in this role.

References

/1/ A.A. Sokolov and I.M. Ternov, Sov. Phys. Dokl. 8:1203 (1964)
/2/ A.W. Chao, Nucl. Inst. Meth. 180:29 (1981)
   A.W. Chao, in Physics of High Energy Particle Accelerators,
         R.A. Carrigan ed. American Institute of Physics No. 87, New York
         (1982)
/3/ R. Schmidt, DESY M-82-22, Hamburg (1982)
/4/ H. Nesemann and K. Wille, DESY M-83-09, Hamburg (1983)
/5/ H.D. Bremer et. al., DESY 82-026, Hamburg (1982)
/6/ D.P. Barber et. al., DESY 82-076, Hamburg (1982).

Acknowledgements

We wish to thank Prof. Dr. G.-A. Voss for continued support and encouragement.

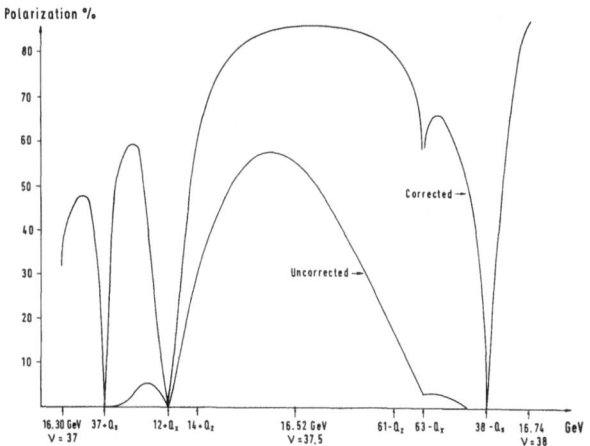

Figure 1. Polarization vs. beam energy for PETRA near 16.5GeV as simulated by SLIM with r.m.s. vertical closed orbit distortion of 1.2mm. The curves show the prediction before and after harmonic correction.

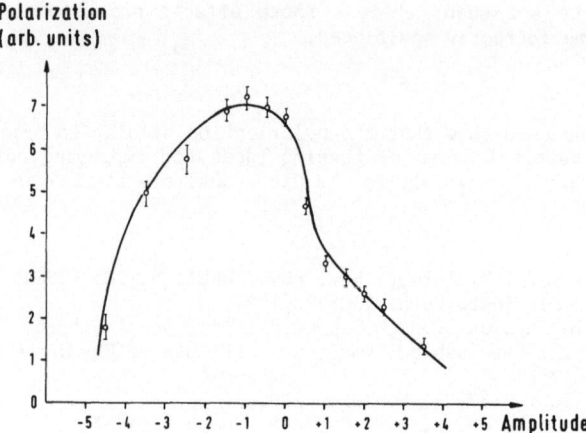

Figure 2a. Polarization measured at PETRA as a function of the strength of the 38th sine harmonic in the closed orbit. (arbitrary scale)

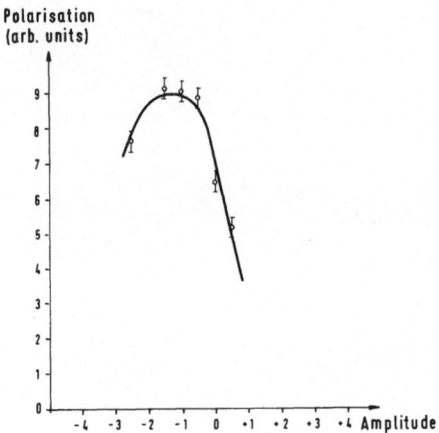

Figure 2b. As in Fig 2b. but for the 38th cosine harmonic.

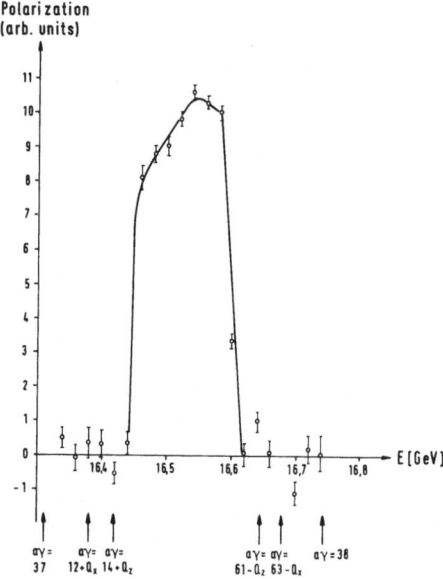

Figure 3. Polarization measured at PETRA vs. beam energy after optimising the 37th and 38th sine and cosine harmonics.

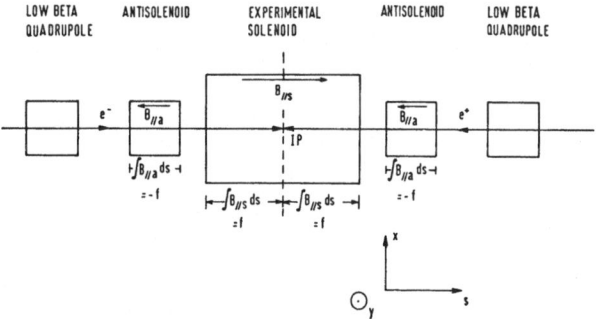

Figure 4. Ideal arrangement of detector solenoid and compensator solenoids: the compensators are adjacent to the main solenoid.

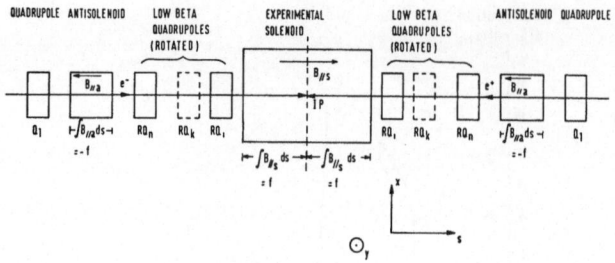

Figure 5. Solenoid compensation arrangement where the compensators are separated from the main solenoid by rotated low-beta quadrupoles.

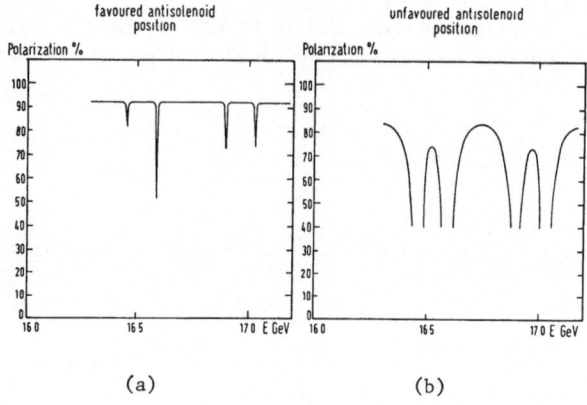

Figure 6. Polarizations predicted by a SLIM simulation for PETRA at 16.5GeV and for a typical detector solenoid. The curves show the polarization when the compensators are placed in the "favoured" position and the "unfavoured" position.

COMPUTATION OF ELECTRON SPIN POLARISATION IN STORAGE RINGS

Jörg Kewisch, Deutsches Elektronensynchrotron (DESY)
Notkestr.85
2000 Hamburg 52

1. Abstract

A new tracking program "SITROS" which enables simulation of polarising and depolarising effects in electron-positron storage rings is presented. The effectivness of the simulation is illustrated for the storage ring PETRA.

2. Introduction

Electrons or positrons circulating in a storage ring can become polarised during emission of synchrotron radiation by the Sokolov-Ternov effect /1/. The maximum possible degree of polarisation is 92.4%

Experimental studies on polarisation at PETRA /2/ have led to the achievement of a maximum measured degree of polarisation in PETRA of about 80%. This degree of polarisation could only be achieved when the uncompensated solenoidal detector fields were switched off and after a careful optimization of the vertical closed orbit.

The failure to reach the theoretical maximum degree of polarisation is caused by various depolarising effects. The first computer program capable calculating depolarising effects in electron-positron storage rings was the program "SLIM" developed by A. Chao /4/. SLIM calculates only the linear coupling between spin rotation and particle oscillations and is therefore restricted to the calculation of depolarising resonances of first order. The influence of higher order effects on polarisation driven by nonlinear forces (sextupoles, chromaticity, beam-beam force) is not taken into account. Since experimental work /2/ has indicated that nonlinear effects are important it has become necessary to develop a new program.

In contrast to SLIM which calculates the polarisation by a perturbation method, this new program is based on a tracking algorithm. Trajectories and spin vectors of a number of representative particles are traced over many revolutions and both linear and nonlinear effects and quantum emission are taken into account. The program is called SITROS.

3. Transformation Algorithm for a Section of the ring

The first task of SITROS is the calculation of the closed orbit. In an ideal machine the closed orbit goes through the center of the quadrupoles. In a real machine the closed orbit deviates from the ideal owing to magnet errors and the nonuniform distribution of cavities. All other trajectories will oscillate around this closed orbit. Thus the closed orbit defines the center of the bunch and can be taken as the average trajectory.

The motion of an electron is described by a 6 dimensional vector $\vec{X} = (x, x', z, z', s, \delta)$, where x, z, s are the horizontal, vertical and longitudinal displacements of the particle relative to the closed orbit particle and δ is the relative energy deviation.

In linear approximation the space coordinates are transformed from a point s_0 to an other point s_1 by a constant matrix M :

$$\vec{X}(s_1) = M \cdot \vec{X}(s_0)$$

SITROS includes 2nd order effects by using a transformation:

$$\vec{X}(s_1) = \vec{F}(x,x',z,z',s,\delta,x^2,xx',\ldots\ldots,s\delta,\delta^2)$$

which is applied not only for a single magnet but for bigger sections of the ring.

The spin direction of a particle is described by a vector . The rotation of the spin of a particle moving in an electromagnetic field is described classically by the well known BMT-equation /6/:

$$\frac{d\vec{S}}{dt} = \vec{S} \times \frac{e}{m\gamma} \left(\gamma a \left(\vec{B} - \frac{\gamma}{\gamma+1} (\vec{\beta}\cdot\vec{B})\cdot\vec{\beta} - \vec{\beta}\times\frac{\vec{E}}{c} \right) + \vec{B} - \frac{\gamma}{\gamma+1} \vec{\beta}\times\frac{\vec{E}}{c} \right)$$

A rotation is fully described by a vector $\vec{\Omega}$, whose direction is the rotation axis and it's length $|\vec{\Omega}|$ defines the rotation angle. The rotations in successive magnets in a section are combined into a single rotation vector /8/. This vector depends on the electromagnetic fields along the particle trajectory and can be written in the following way:

$$\vec{\Omega} = \vec{\Omega}_{c.o.} + \vec{\Omega}(\vec{X})$$

where $\vec{\Omega}_{c.o.}$ is the rotation vector on the closed orbit. The rotation vector $\vec{n} = \vec{\Omega}_{c.o.}$ for a whole revolution in an ideal flat machine is given by:

$$\vec{n} = (0, 2\pi\gamma a, 0); \quad \nu = \gamma a$$

The quantity ν is called the spin tune. If the spin rotation angle is not too close to a multiple of 2π, these quantities can be described in a good approximation by first and second order terms.

$$|\vec{\Omega}|, \Omega_x, \Omega_s = F(x,x',z,z',s,\delta,x^2,xx',\ldots\ldots,s\delta,\delta^2)$$

An electron loses energy stochastically by the emission of photons. Each photon emission excites synchrotron oscillations which are in general coupled to betatron oscillations. The stochastic character of the emission leads to a random phase and amplitude distribution of the oscillations, and together with the damping defines the beam size/10/. Coupling of the spin motion to these oscillations will result in depolarisation.

In order to save computing time the random energy jump in the SITROS program is only made in two bending magnets. This has the advantage that the transformations can be performed for a whole section containing many bending and quadrupole magnets, so that the computing time for the spin tracking is strongly reduced. The strength of these random jumps is chosen in such a way that the horizontal emittance of the beam is the same as in the PETROS program.

Since particles with different energies differ much less in their trajectories than in their spin behaviour, it is a good approximation to calculate the polarisation at different energy points by using the same space parameters at each energy so that in this version of SITROS 100 different energy points can be calculated in one job. This saves a factor of two in computing time.

4. Evaluation of the Degree of Polarisation

The SITROS program, as described so far, calculates the depolarisation rate of a perfectly polarised beam. The program starts with all particles on the closed orbit and with the spin vectors pointing in the direction of the so called n-axis /4/ which is the spin rotation axis on the closed orbit for one revolution. For particles remaining on the closed orbit this spin direction will be preserved.

Since the particles experience random energy kicks they will leave the closed orbit and the spins are kicked away from the direction of the n-axis by the quadrupole and sextupole fields.

The average deviation of the spin from the n-axis is zero far away from the depolarising resonances but accumulates to a non zero value near resonances.

We define a polarisation vector:

$$\vec{P} = \frac{1}{k} \sum_i \vec{S}_i \times \vec{n}$$

where k ist the number of tracked particles and $\vec{S}_i$ is the spin vector of the ith particle. The length P of the vector will decrease with the number of revolutions. In SITROS, it is assumed that the function P(t) can be described by the exponential function:

$$P(t) = P_o \exp\left(\frac{-t}{\tau_d}\right)$$

Taking two samples of this function at t_0 and t_1 the depolarisation time is:

$$\frac{1}{\tau_d} = \ln\frac{P(t_0)/P(t_1)}{t_1 - t_0}$$

τ_d converges after a few betatron damping times to the final value and the calculation can then be stopped.

The build up time of the polarisation is given by:

$$\tau_p = 98 \cdot \frac{R^3}{E^5} \frac{\langle R \rangle}{R}$$

where E is the beam energy in GeV, R is the magnet radius in meters and $\langle R \rangle$ is the average radius of the ring in meters.

From the values of τ_p and τ_d the equilibrium degree of polarisation can be calculated using /5/ :

$$P\infty = 92.4\ \% \ \frac{\tau d}{\tau p + \tau d}$$

5. Simulation Results for PETRA

Fig. 1 shows the degree of polarisation versus the spin tune for a M15 optics with a vertical closed orbit shape generated by a random distribution of kicks in the vertical correction coils. The rms amplitude of the closed orbit is 1.5 mm. The decrease of polarisation from the theoretical value of 92.4% can be explained by the occurence of various resonances. In general these resonances occur, when the condition

$$\nu + i\ Qx + j\ Qz + k\ Qs = m$$

is fulfilled, where i,j,k,m are integer. The resonances are expected to be strongest when the integers are small. The resonances $\nu \pm Qx = m$ and $\nu \pm Qs = m$ are most strong. Other strong resonances are $\nu \pm 2(Qx+Qs) = m$.

The resonances with $j \neq 0$ (Qz resonances) do not appear strongly in this calculations. Due to the fourfold symmetry, the driving terms of these resonances cancel each other in the different parts of the ring.

For comparison Fig. 2 shows the corresponding results of SLIM. The shape of the curves in Fig. 1 and 2 are roughly the same but SLIM can only detect the linear resonances $\nu \pm Qx = m$ and $\nu \pm Qs = m$. Due to the thin lens approximation of SLIM the Q-values are slightly different to those in Fig. 1.

It has been shown /7/ that the strength of depolarising resonances can be reduced by a special orbit correction scheme. The closed orbit is adjusted, so that harmonic components near the spin tune vanish.

Fig. 3 shows the polarisation of an optics where, with special symmetric kicks, the vertical closed orbit is distorted so that only the 4nth harmonic component exists. Thus the closed orbit is perfectly corrected with respect to the spin in the range between ν = 37 to 38.

The main resonances are seen in both the corrected and uncorrected optics, but in the optimized optics the resonances are smaller and the polarisation between them is higher. However, the correction scheme does not cure all problems as the corresponding SLIM calculation would suggest (Fig. 4).

Only two resonances can be seen in the SLIM result: $\nu - Qx = 14$ and $\nu + Qs = 38$. The resonances $\nu + Qx = 51$ and $\nu - Qs = 37$ are compensated by the symmetry of the machine.

6. Summary

This report introduces a new simulation program for calculating the degree of polarisation in electron positron storage rings and presents the first results achieved with this program. This program differs from the well known program SLIM in that it is a tracking program which allows computation of nonlinear and nonstationary effects.

Nevertheless, it is a common problem of all tracking programs that for exact calculations unlimited computing time is neccessary. This time is not available and therefore some simplifications and restrictions must be used. This report shows that despite these approximations and the classical treatment SITROS is a useful tool for the understanding of depolarising effects.

7. References

/1/ A.A. Sokolov and I.N. Ternov, Sov.Phys.Doklady 8 (1964) 1203.
/2/ H.D. Bremer et al. DESY 82-026 (1982)
/3/ H.D. Bremer et al. DESY M82-26 (1982)
/4/ A. Chao, SLAC-PUB.-2564
/5/ A. Chao, SLAC-PUB.-2781
/6/ for example J.D. Jackson "Classical electrodynamics",
    Wiley, New York 1975
/7/ R. Schmidt, DESY M82-20
/8/ R. Neumann and R. Rossmanith, Nuc.Instr.Meth. 204(1982) 29
/9/ J. Kewisch, Diplomarbeit, Univ. Hamburg 1978
/10/ M. Sands, SLAC-PUB.-121
/11/ J. Kewisch, DESY 83-032 (1983)

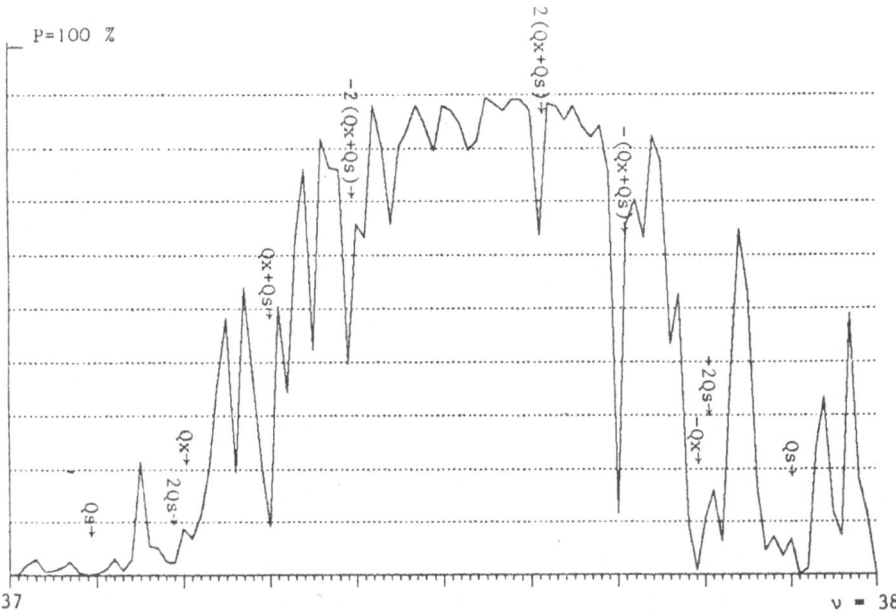

Fig. 1 : Polarisation versus spin tune of an optics with normal distributed vertical kicks calculated with SITROS.

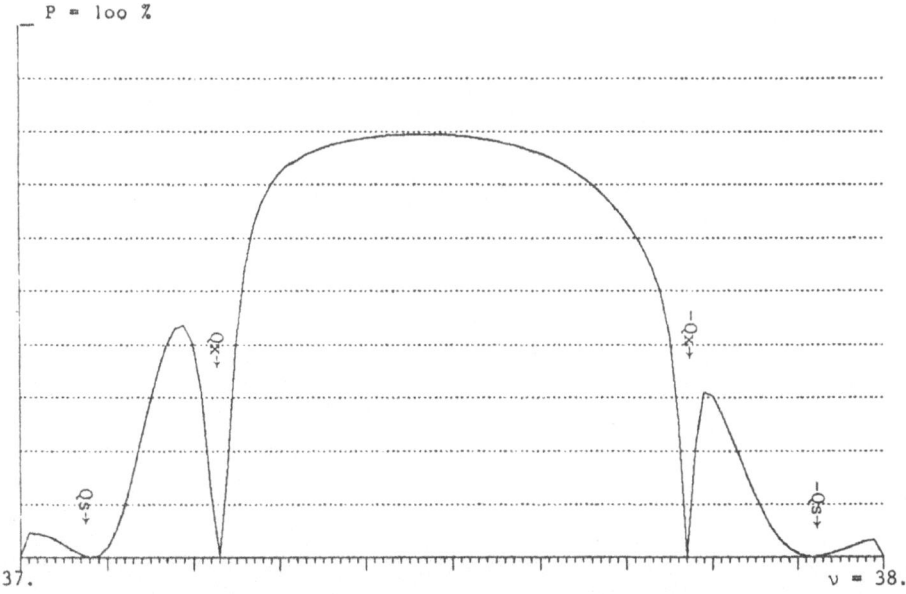

Fig. 2 : Polarisation versus spin tune of the optics of Fig.1 calculated with SLIM.

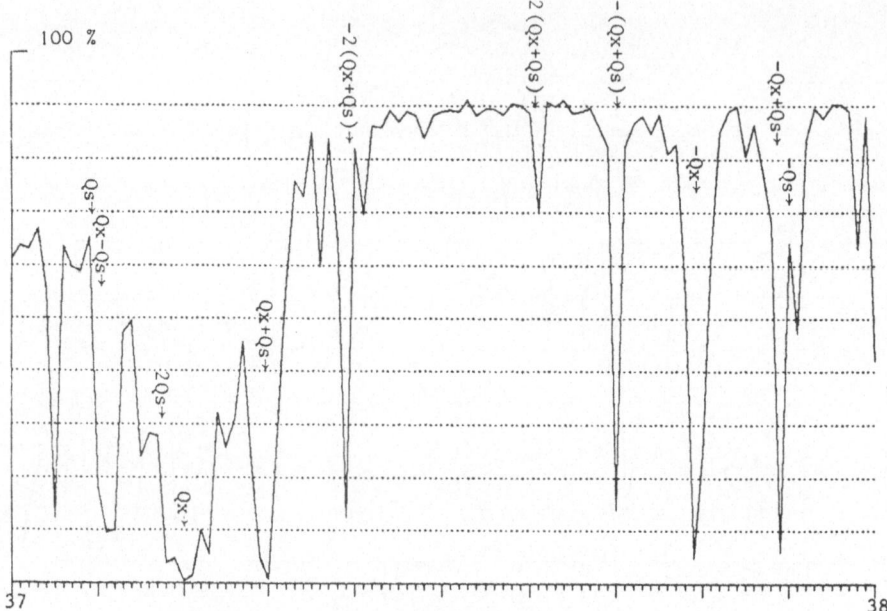

Fig. 3 : Polarisation versus spin tune of an optics with optimized closed orbit calculated with SITROS.

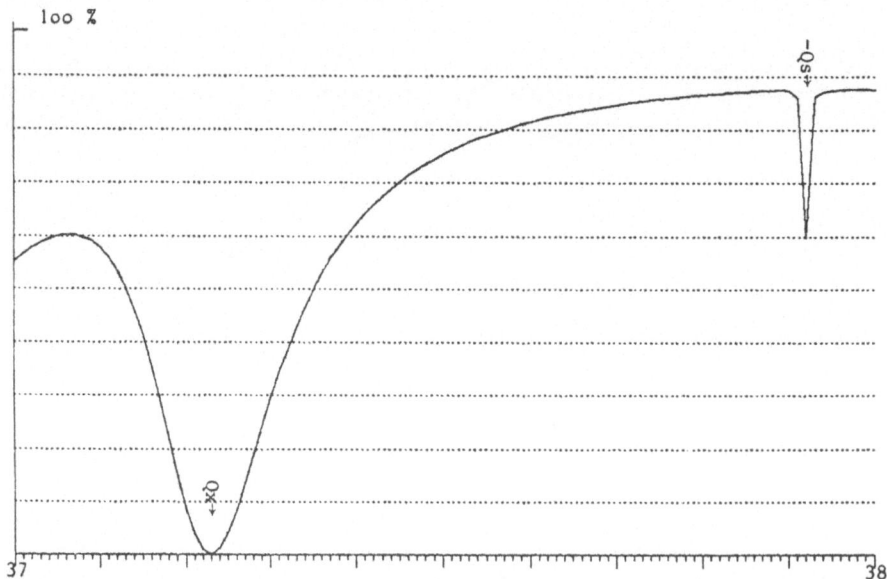

Fig. 4 : Polarisation versus spin tune of the optics of Fig. 3 calculated with SLIM.

ARCHSIM: A PROTON SYNCHROTRON TRACKING PROGRAM
INCLUDING LONGITUDINAL SPACE CHARGE*

Henry A. Thiessen and John L. Warren
Los Alamos National Laboratory
Los Alamos, NM 87545, USA

Summary

A particle-tracking program has been written for simulating the acceleration cycle of a rapid-cycling proton synchrotron. A lattice can consist of up to 100 cells and rf cavities. Transport of the beam in six dimensions includes all second-order optical terms. The rf field and proton velocity are treated exactly. Longitudinal space charge is handled in a self-consistent manner. The fluctuations due to the finite number of particles are handled by a Gaussian smoothing algorithm. The program runs on a VAX 11/780 and can track 100 particles without space charge through the full acceleration cycle from 0.8 to 32 GeV in 49 minutes. A thousand particles with space charge takes about ten hours of computer time.

Introduction

The motivation for writing this tracking program was the need to explore the effect of various rf accelerator-cavity parameters on the beam dynamics and stability of a proposed 32-GeV rapid-cycling synchrotron called LAMPF II.[1] LAMPF II is being designed for 100-µA average current. Initially there was some concern that space-charge forces might cause a strong blowup as the beam passed through the negative-mass phase transition at ~12-GeV. Simulation using ARCHSIM has demonstrated that probably there will not be a problem at transition.[2]

The reference design for the accelerator has a superperiodicity of 5. The straight sections are dispersion-free to reduce the possibility of synchrotron-betatron coupling effects. It is proposed that there be 60 rf cavities to provide a peak acceleration voltage of 14-MV/turn.

This paper will describe the organization of the program and the special features of the main subroutines. This will be followed by a discussion of runtime experience on the VAX 11/780.

Program Description

Figure 1 shows a simplified flow chart. INPUT reads in parameters characterizing the accelerator. It calls three subroutines whose purposes are to calculate the

---

*Work supported by the US Department of Energy.

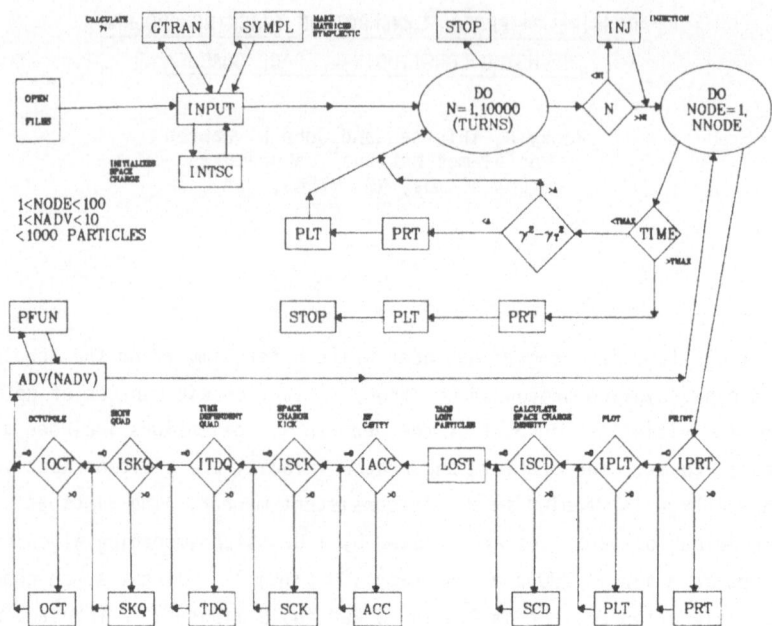

Fig. 1. Flow chart for ARCHSIM.

transition gamma γ_T, to initialize certain arrays for the space-charge calculation and to check (and if necessary correct) the transport matrices to ensure that they are equivalent to symplectic matrices. These transport matrices are obtained from programs such as DIMAT[3] or TRANSPORT[4] and were not intended to be used the 60 000 times necessary to simulate an acceleration cycle--small inaccuracies are magnified by such cycling. Making the matrices symplectic to the accuracy of the computer ensures that the phase-space volume of the bunches is not artificially increased during transport between rf cavities.

The main DO-loop on N is not intended to be terminated by the 10 000 turn maximum, but rather by the time, TMAX, that it takes to achieve the maximum energy. Particles are injected during the first NI turns. Particles are represented in the program by 9-D vectors. The first six components are the usual TRANSPORT variables $(x,\Theta,y,\varphi,\ell,\delta)$.[4] The other three components are the difference between the time taken by the particle to reach a certain location on the ring and the time that a synchronously accelerated particle would have taken; the difference in energy between the particle and the synchronous particle; and a tag (0 or 1) to determine if the particle has been lost, that is, if at some time its transverse (x,y) coordinates have exceeded apertures of the accelerator. This tag is changed in subroutine LOST.

The particles are transported from node to node around the ring by second-order transport matrices in subroutine ADV. This subroutine also calls subroutine PFUN, which changes the synchronous momentum. At each node several things may occur, depending on control parameters: The particles may have their longitudinal momentum and energy changed by a simulated rf cavity, in subroutine ACC. They may receive a linear change in transverse momentum in a thin-lens, time-dependent quadrupole (TDQ) or a skew quadrupole (SKQ). They may be given a nonlinear kick in an octupole (OCT). The longitudinal space-charge force is represented as a nonlinear kick (SCK) at the nodes. The size of the kick depends on the charge distribution, which is updated at a node (SCD). Because the longitudinal charge distribution changes slowly, it is not necessary to update it more than once per turn. At each node, one has the option of printing out various kinematic parameters such as synchronous momentum, rf-voltage and phase, synchrotron tune, relative changes in bucket area, Laslett tune shift, etc. One can also make plots of cross sections of phase space; for example, x-Θ, y-φ, x-y, ℓ-δ, etc. Because there is a special concern about the phase-space behavior near phase transition, special provision has been made for extra plots and printouts near γ_T. The width of the region of interest near transition is controlled by an input parameter DEL. Finally, printing and plotting occur at the end of the run when TIME $\geq$ TMAX.

INPUT. The input is organized into 13 sections dealing with (1) global parameters and universal constants, (2) injection, (3) longitudinal space charge, (4) machine apertures for subroutine LOST, (5) rf acceleration, (6) time-dependent quadrupoles, (7) skew-quadrupoles, (8) octupoles, (9) tranport matrices, (10) node parameters (IPRT, IPLT, ..., IOCT, NADV) that define the sequence of operations at nodes and the choice of transport sections between nodes, (11) synchronous momentum, (12) miscellaneous initializations, and (13) (for the future) restoration of arrays and variables in the restart mode. When running on a small time-share computer, it is impractical to simulate during one continuous run a complete acceleration cycle for 1000 pseudoparticles with space-charge effects. Soon the code will have a provision to store arrays and other parameters at TIME=TMAX for restart at a later time.

The need to check that transport matrices are equivalent to symplectic matrices has been mentioned above. First- and second-order matrices derived from TRANSPORT or DIMAT are not symplectic because transport variables are not canonical, that is, the variables Θ, φ, and δ are not the canonically conjugate momenta corresponding to x, y, and ℓ. For first-order transport, the symplectic conditions are equivalent to requiring that certain subdeterminants, for example, $R_{11}R_{22} - R_{12}R_{21}$, be unity. There are five such conditions in first order; second-order conditions are more complicated. The process of symplectification is described in a forthcoming Los Alamos internal report.[5]

INJ(ECTION). This subroutine has three sections. The first section generates a random point in a 6-D, tilted, phase-space ellipsoid to simulate a particle in the injected bunched beam. The emittance of the bunches from LAMPF are quite small compared to the desired emittance for LAMPF II. To avoid the transverse space-charge-force limit, steering magnets will be used to transversely offset injected bunches from the designed closed orbit. This is simulated in INJ by adding a sinusoidally varying offset to the particle's (x,Θ,y,φ)-coordinates, where x and y can have a different sweep frequency. The longitudinal phase-space enlargement is handled in two parts. The variable ℓ can be given a constant offset in INJ; then, the synchronous phase ϕ_{RF} can be varied sinusoidally in subroutine ACC during the injection period. The third section of INJ corrects δ for the fact that the synchronous momentum is changing from turn to turn, even during injecton.

ADV(ANCE). The main purpose of this subroutine is to multiply the first six components of the ray vectors describing particles by the first-order (6 by 6) and second-order (6 by 36) transport matrices. This is the most time-consuming part of the program, and special effort has gone into making the multiplication efficient. There are only 13 nonzero, nontrivial elements in the first-order matrix and 39 in the second-order matrix. They are put into nonarray variables to speed look-ups. Instead of doing matrix multiplications with DO-loops, explicit formulas are written out that use only the nonzero elements of the transport matrices. When desired, one can skip the second-order transport matrices for more speed. ADV calls PFUN, which updates the synchronous momentum p_{syn} and hence changes $\delta = (p - p_{syn})/p_{syn}$ for each particle. The variables Θ and φ also are renormalized by the change in synchronous momentum.

PFUN. The momentum rises as a function of time according to the formula PFUN = (PMAX + PMIN)/2. - (PMAX - PMIN) * COS (TWOPI * FREQ * TIME)/2., where FREQ is read in and corresponds to the inverse of the time for a complete acceleration cycle. Of course the purpose of ARCHSIM is to make changes in PFUN and ACC to see what effect they have on beam dynamics.

ACC(ELERATE). The current version of this routine attempts to keep the bucket area constant until the synchronous phase rises to $\sim 60°$. The change in synchronous energy required to keep up with the synchronous momentum change is

$$DW = \sqrt{(P^2 + M^2)} - \sqrt{(P_{LAST}^2 + M^2)} = V \sin(\phi) \quad , \tag{1}$$

where P is the synchronous momentum at the present node and P_{LAST} is the momentum at the previous node where ACC was called; V is the maximum voltage and ϕ the phase of the rf power. From Bovet et al.[6] the relative bucket area is

$$A/A_0 = \alpha(\phi)(\beta/\beta_0)\sqrt{\frac{V}{V_0}\frac{W}{W_0}\frac{\eta_0}{\eta}} \quad ,$$

where $\alpha(\phi)$ is a tabular function, W is the proton's energy, β is the usual relativistic proton velocity and

$$\eta = |\frac{1}{\gamma^2} - \frac{1}{\gamma_T^2}| \quad .$$

The subscript o means these quantities are evaluated at the beginning of the acceleration cycle. Setting $A/A_o = 1$, gives an expression for V as a function of ϕ that can be substituted into Eq. (1). This transcendental equation is solved numerically for ϕ. As γ approaches γ_T, ϕ will increase toward 90°. The rise is cut off at ~60° and ϕ is held constant until transition; after transition, ϕ is always changed to 180°-ϕ. V is allowed to rise to the maximum required (DW/sinϕ) and is not allowed to decrease. Toward the end of the cycle when DW gets small, this has the effect of bunching the beam in longitudinal space to produce the short pulses that will be required for some of the physics experiments.

SCK. The effect of longitudinal space-charge forces is included in the kick approximation. The energy (and hence δ) of each particle is changed at the nodes according to the formula

$$\Delta W = \frac{r_p mc^2}{\gamma^2} \left(\frac{h}{R}\right)^2 [1 + 2 \ln (b/a)] \frac{d\lambda(\phi)}{d\phi} L ,$$

where r_p is the classical proton radius, mc^2 the rest mass energy, $mc^2\gamma$ the proton energy, h the harmonic number, $2\pi R$ the machine circumference, b the beam-pipe radius, a the beam radius, $\lambda(\phi)$ the linear density of charge along the beam, and L the distance along the ring since the last kick.[7] The variable ϕ is proportional to the difference in the time the particle arrives at the node and the synchronous particle arrives at the node. The charge density and its derivative are calculated in subroutine SCD. With only 1000 particles, the density would have troublesome statistical fluctuations that would make $d\lambda/d\phi$ unrealistic. We get around this by making a Fourier transform of $\lambda(\phi)$, multiplying by a Gaussian smoothing function, taking the derivative of the function by multiplying the Fourier transform by the Fourier-transform variable, and then taking the inverse Fourier transform. The amount of Gaussian smoothing is controlled by an empirically adjusted parameter.

PRTPLT. The phase-space plots are done on the line printer using subroutines adapted from a high-energy physics program.[8] This makes the program more transportable but produces somewhat low-quality output as shown in Fig. 2. Our plans are to augment this plot routine with one that can produce plots for graphics terminals.

Run Time Experience

By varying the number of particles and choice of options for second-order transport and space charge, one can vary the run time from minutes to many hours. To examine characteristics of a certain rf-voltage and phase program, one can run using only one particle and no phase-space plots, except near transition and at the end. An acceleration cycle of 6700 turns with 5 nodes runs in about 2.5 min. One can get a better feel for the phase-space behavior by running with 100 particles and no space-charge effects. This takes about 49 min. A full 1000 particles with no space

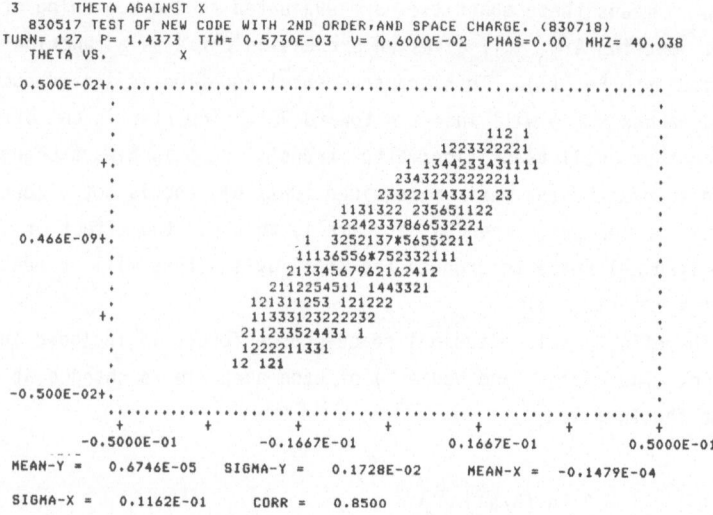

Fig. 2. Typical output of subroutine PRTPLT.

charge takes 7.4 hours. Space-charge calculations add considerably to the run time. At present, one can run 1000 particles through 2300 turns with space charge and full second-order transport in about 3.4 hours.

Acknowledgments

The authors would like to acknowledge important contributions to the code from R. K. Cooper, and useful discussions with E. P. Colton.

References

1. Henry A. Thiessen, submitted to Proc. 12th Int. Conf. on High-Energy Accelerators, Fermi Natl. Acc. Lab., Batavia, IL, USA, Aug. 11-16 (1983).
2. J. L. Warren and H. A. Thiessen, submitted to Proc. 12th Int. Conf. on High-Energy Accelerators, Fermi Natl. Acc. Lab., Batavia, IL, USA, Aug. 11-16 (1983).
3. R. Servranckx, Proc. 11th Int. Conf. on High-Energy Accelerators, Geneva, Switzerland, July 7-11, 1980, p. 556 (1980).
4. K. L. Brown, Adv. Particle Phys. 1, p. 71 (1971).
5. E. Forest and J. L. Warren, "Application of Symplectic Conditions to Second Order Transport Theory," Los Alamos National Laboratory Accelerator Technology Division, Group AT-6 internal report ATN-83-17.
6. C. Bovet, R. Gouiran, I. Gumowski, K. H. Reich, "A Selection of Formulae and Data Useful for the Design of A. G. Synchrotrons," CERN/MPS-SI/Int. DL/70/4 23 April, 1970.
7. A. Hoffmann, Proc of First Course of Int. School of Particle Accelerators, Erice, 10-22 November, 1976; published as CERN 77-13 (19 July 1977), p. 143.
8. R. Mischke, private communication.

A METHOD FOR DISTINGUISHING CHAOTIC FROM QUASI-PERIODIC

MOTIONS IN ORBIT TRACKING PROGRAMS

John M. Jowett
CERN, Geneva, Switzerland

In the design stage of an electron storage ring, its dynamic aperture (non-linear acceptance) is evaluated by means of orbit tracking programs. The closed orbit corresponds to an elliptic fixed point of the complete one-turn transfer map. Moving outwards from it, one generally observes a transition from regular stable motion to either unstable chaotic motion or quasi-periodic motion. For finite tracking times a band of sufficiently stable but chaotic motion is usually included in the notional dynamic aperture.

In order to distinguish quasi-periodic from chaotic motions one can calculate fractal dimensions of the computed power spectra of the motion. A method for provisionally assigning a dimension d to a sample of power spectra of an orbit is described and illustrated. A value of d sufficiently different from unity is the hallmark of chaos.

INTRODUCTION

In recent years there has been a degree of interest among storage ring physicists in the chaotic motion which is a universal feature of non-linear deterministic dynamical systems. Usually, this has been in the context of performance limitation by the beam-beam effect. However, even a single particle moving through the non-linear magnetic elements of a lattice can exhibit this kind of behaviour. A large and important part of the designer's job is to arrange for the motion to be regular ("integrable") in as big a volume of phase space as possible. This task has become more difficult as the size of storage rings has increased and their natural chromaticities have taken larger values, necessitating more complicated chromaticity correction schemes with several families of sextupoles. Guidance on the choice of sextupole strengths is available from programs like HARMON[1] but finding a good arrangement of sextupoles remains a difficult art.

Evaluation of the quality of a given scheme is the domain of orbit tracking programs. Usually these ignore the radiation effects on the motion of an electron so the system considered is Hamiltonian and periodic in azimuth s. In fact it is only in terms of this fictitious system that one can define the dynamic aperture of an electron ring. This is the largest connected part of the region of phase space in which the particle motion is not found to be unstable when tracked for a certain number, N_T, of turns around the machine. At present, computing time limits usually oblige one to make do with a scan of a two- or three-dimensional slice through this six-dimensional object.

SPECTRA OF CHAOTIC AND QUASI-PERIODIC ORBITS

The linear lattice optics is arranged so that an elliptic fixed point (corresponding to the closed orbit) of the complete one-turn transfer map lies somewhere in the middle of the dynamic aperture. In a neighbourhood of this point the motion is linear and, in an excellent approximation, contains only three frequencies (the linear betatron and synchrotron tunes).

Moving outwards from the fixed point, the motion may become either integrable (regular) or non-integrable (chaotic). It follows from Liouville's Theorem[2] on integrability that, in the integrable case, the particle is confined to a 3-dimensional surface in phase space, and that there exists a canonical transformation to action-angle variables J_i, ϕ_i (i = 1,2,3). The J_i label this surface and the ϕ_i change by an amount $2\pi Q_i(\underline{J})$ on each turn. Any linear combination, f, of phase-space coordinates is quasi-periodic. That is, its value on the kth turn may be written

$$f(k) = \sum_{\underline{n}} a(\underline{n}) \exp\left[2\pi i k \underline{n} \cdot \underline{Q}(\underline{J})\right] \qquad (1)$$

where $\underline{n} = (n_1, n_2, n_3)$ is an integer vector and $a(\underline{n})$ is a complex number satisfying

$$|a(\underline{n})| \leq F \exp\left[-\lambda(|n_1| + |n_2| + |n_3|)\right] \qquad (2)$$

for some positive numbers F and λ (independent of $\underline{n}$).

One may ask the question: does the region of regular (quasi-periodic) motion correspond to the dynamic aperture? The answer appears to be negative. The trajectories of some particles which get dubbed "stable" in a dynamic aperture scan appear wildly irregular and are presumably not quasi-periodic. One should therefore be wary of including them in the notional stable region because of their propensity to leak out on longer time-scales. Generally, the initial conditions for such particles are found near the edge of the dynamic aperture.

Among the methods available for distinguishing chaotic from quasi-periodic motion, a spectral analysis technique due to Blacher and Perdang[3] is conveniently applied to orbit tracking programs. We have adapted the well-known program PATRICIA[4] to illustrate it.

Although based on conjectures about difficult mathematical questions [the possible ways in which (1) and (2) may break down as f(k) becomes chaotic], their method has considerable (computer-)experimental evidence in support of it. To summarise just one of the arguments in Ref. 3, suppose that, in the chaotic case, (1) may be rewritten as a "Fourier integral"

$$f(k) = f_0(k) + \int_{-\infty}^{\infty} d\omega \, e^{-ik\omega} a(\omega) \quad (3)$$

with $f_0(k)$ quasi-periodic. Then $f(k)$ remains bounded and, if $a(\omega)$ were integrable, it would follow from the Riemann-Lebesgue lemma that $f(k) - f_0(k) \to 0$ as $k \to \infty$. Since this is not true one may conclude that $a(\omega)$ is not integrable. If it is bounded and well-behaved at $\pm\infty$, this means that $a(\omega)$ is discontinuous on a set of non-zero measure.

Indeed, the computed power spectra of apparently chaotic storage ring orbits exhibit broad jagged humps around low order resonances. As the resolution of the power spectrum increases, more and more structure is revealed, suggesting strongly that $|a(\omega)|^2$ is, in reality, a "fractal" curve[5]. That is, an object somewhere between an "ordinary" one-dimensional curve and a two-dimensional space-filling curve. The irregular sort of fractal, which we presume our power spectra to be, is characterised by a statistical form of internal homothety, and a dimension d. Several different definitions of d exist[3,5] but they all satisfy $1 < d < 2$ and differ from unity together. We adopt one based on the way in which the length of the graph of a given (discrete) numerical approximation to the power spectrum increases with frequency resoluton $\Delta\omega$.

Each approximate power spectrum is first renormalised so that its maximum value is unity. Then the lengths are fitted to an expression of the form

$$L(\Delta\omega) \sim (\text{const.})(\Delta\omega)^{1-d} \quad \text{as } \Delta\omega \to 0 \,. \quad (4)$$

Clearly d = 1 for a one-dimensional curve, all of whose detail can be resolved for some sufficiently small $\Delta\omega$. A power spectrum satisfying (1) and (2) also has d = 1. On the other hand all estimates of d > 1 must be provisional, pending better frequency resolution.

In Figs. 1 and 2 we show the power spectra of both horizontal and vertical betatron motion for two different initial conditions in a typical LEP lattice, with frequency resolution $\Delta\omega = 10^{-3}$. Figures 3 and 4 are logarithmic plots of the lengths of such curves for several different values of $\Delta\omega$ (obtained by tracking for different numbers of turns).

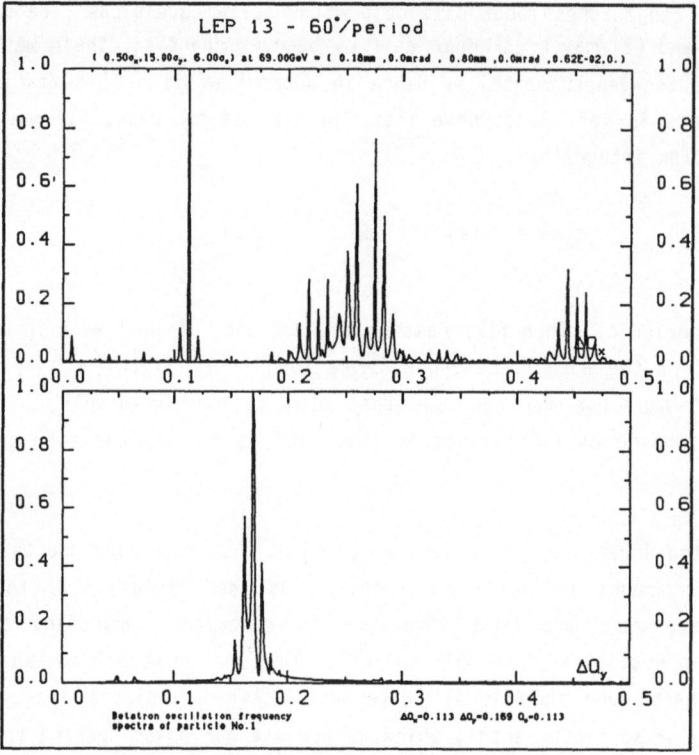

Fig. 1

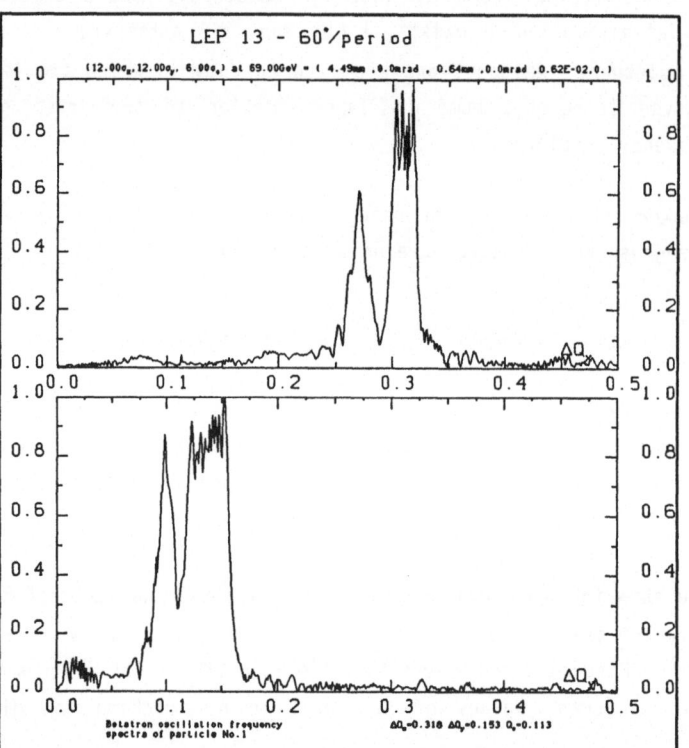

Fig. 2

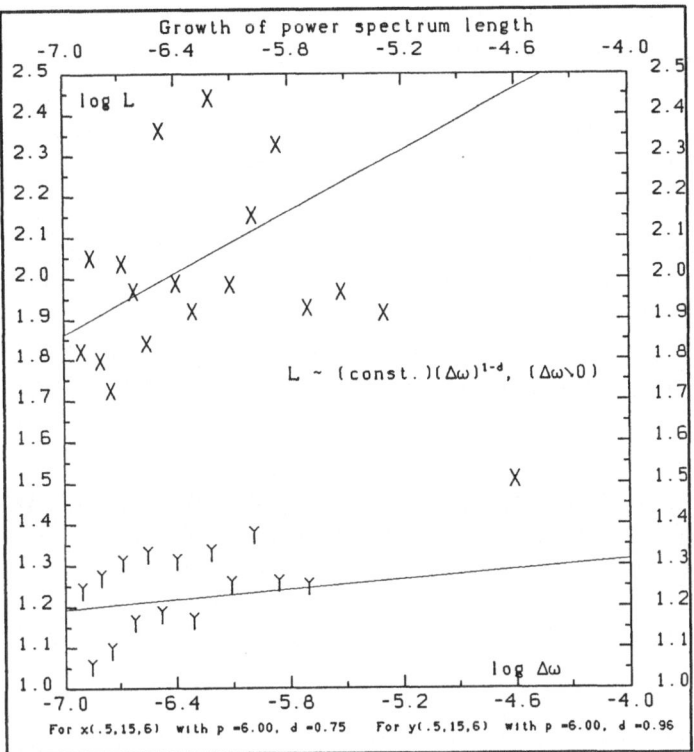

Fig. 3

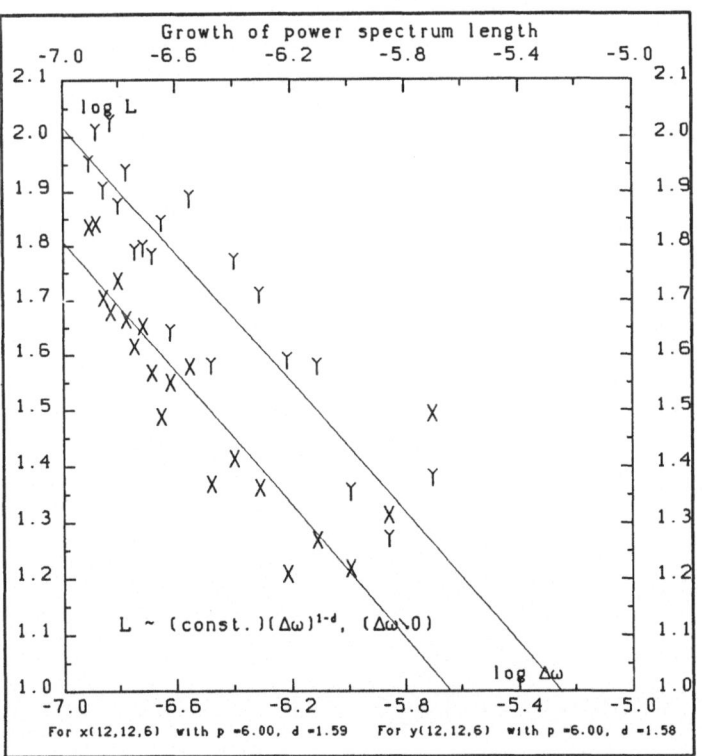

Fig. 4

[In the least squares fits of Figs. 3 and 4 a weighting factor of $(\log \Delta\omega)^p$, $p = 6$, is applied to emphasize the behaviour at small $\Delta\omega$]. The values of d are close to or less than 1 in the first case, which is identified as a regular orbit, and around 1.6 in the second. On a terminal screen, the latter particle is observed to jump about erratically, and we infer with some confidence that it lies in a chaotic region of phase space.

In the first case the fit is rather poor because $\Delta\omega$ does not get sufficiently small for (4) to be a very good approximation. Nevertheless it is difficult to mistake a quasi-periodic orbit for a chaotic one, because the fitted values of d tend to be < 1. On the other hand, the available resolution seems to be adequate to establish the validity of (4) in the chaotic case.

We conclude that the method of Blacher and Perdang may be a useful addition to an orbit tracking program. The Fast Fourier Transform algorithm provides an efficient way of generating the sample of spectra from just one tracked orbit.

REFERENCES

1. M. Donald and D. Schofield, CERN LEP Note 420 (1982).
2. V. Arnold, Les methodes mathématiques de la mécanique classique, Editions MIR, Moscow (1976).
3. S. Blacher and J. Perdang, Physica 3D, 512 (1981).
4. H. Wiedemann, PEP Note 220 and PEP Technical memo 230 (1976).
5. B. Mandelbrot, Les objets fractals, Flammarion, Paris (1975).

PATH - A LUMPED-ELEMENT BEAM-TRANSPORT SIMULATION PROGRAM WITH SPACE CHARGE*

John A. Farrell, Los Alamos National Laboratory
Los Alamos, New Mexico 87545, USA

Summary

PATH is a group of computer programs for simulating charged-particle beam-transport systems. It was developed for evaluating the effects of some aberrations without a time-consuming integration of trajectories through the system. The beam-transport portion of PATH is derived from the well-known program, DECAY TURTLE.[1] PATH contains all features available in DECAY TURTLE (including the input format) plus additional features such as a more flexible random-ray generator, longitudinal phase space, some additional beamline elements, and space-charge routines. One of the programs also provides a simulation of an Alvarez linear accelerator. The programs, originally written for a CDC 7600 computer system, also are available on a VAX-VMS system. All of the programs are interactive with input prompting for ease of use.

Program Descriptions

PATH consists of four programs: BEAM, TRAN, LINAC, and PLOT. The relationship between the programs is shown in Fig. 1. BEAM creates a file of randomly generated, 6-D particle coordinates. TRAN transports these particles through a beam-transport system and LINAC transports the particles through a linear accelerator. The input file for TRAN or LINAC can be from BEAM or from a previous TRAN or LINAC run; thus, the programs are very flexible. PLOT reads a file of particle coordinates from either BEAM, TRAN, or LINAC and produces one- or 2-D plots of the desired desired data. A more detailed description of the programs follows.

BEAM

The beam generating program contains four options for phase space distributions: uniform, Gaussian, Kapchinskii-Vladimirskii (K-V)[2] and binomial. The latter is characterized by a parameter m, and includes the most commonly used distributions.

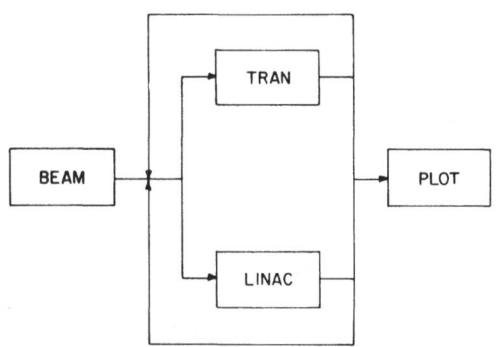

Fig.1. A flow chart showing the relationship between the four codes of the PATH group.

*Work supported by US Department of Energy.

A K-V distribution is obtained with m=o, uniform projections for m=0.5, and a "waterbag" for m=1. A Gaussian would be obtained for infinite m. The input beam parameters may be entered in either TRANSPORT notation[3] or Courant-Snyder notation[4]. BEAM contains a table of particle masses for convenience of input.

TRAN

TRAN is a modification of DECAY TURTLE. The input format is identical to TURTLE and TRANSPORT except that the title and initial beam-defining elements are not used. Calculations are done to second order in most elements; however, some elements include third-order calculations. It is anticipated that most elements will be extended to third order in the future. Longitudinal phase space has been included, also to second order, to permit the study of bunched beams and rf elements. The major addition to the program is a 3-D space-charge routine. This allows an element to be broken up into a specified number of segments with the space charge applied after each segment. The space charge is computed by summing forces between macroparticles to minimize artificial collision effects. The 2-D space-charge routines in r-z or x-y are also available. Generation of histograms is done slightly differently from TURTLE. The insertion of a 50; element causes the particle coordinates at that point to be written on a file for subsequent analysis by PLOT. The particle coordinates at the end of the system are automatically written to a file for use as an input file to TRAN or LINAC.

LINAC

LINAC simulates a conventional Alvarez linear accelerator. It is a modified version of the code PARMILA[5] and includes the capability for generating the drift-tube table. The space-charge routine is 2-D in r and z and normally is applied at the center of the rf gap where the beam is circular.

LINAC is relatively inflexible. It is also possible to simulate a linac with TRAN by individually putting in the quadrupoles and rf gaps. This simulation with TRAN may be preferable if one wishes to study the effects of misalignments, quadrupole harmonic errors, or perturbations to a single rf cell.

PLOT

PLOT is designed to produce graphic output on a Tektronix 4000-Series terminal. Because it is a separate program, one could write various versions to accommodate different output devices. The present version can produce 1-D histograms of any coordinate, either self-scaled or user scaled. It can also produce 2-D scatter plots or contour plots of any coordinate versus another. Another option is all six 1-D histograms, self-scaled on a single plot. Scatter plots of any phase-space projection can be made also. An encircled energy plot is available for convenience in analyzing the energy distribution of the beam on a target.

In addition to the graphic output, PLOT also produces a summary table giving the rms beam parameters at the end of the system and the number and locations of particles stopped by apertures.

EXAMPLES

Figure 2 shows a bending system consisting of a periodic system of four 90° phase-shift cells, each containing a dipole and two quadrupoles. This system automatically is achromatic to first order and can be made achromatic to second order by the addition of two sets of sextupoles located at the quadrupoles[6]. Figure 3 shows the results of a second-order PATH run through this system for a low-emittance, high-momentum spread beam with no sextupoles; Fig. 4 is the same system with the sextupoles turned on. The calculation was done using a uniform distribution.

Figure 5 shows a set of four, 60° bending magnets used to transport an electron beam. A slit is located at the center of the system to trim the momentum spread from 1 to 0.1%. The system was designed with TRANSPORT so that the monochromatic beam diameter was greater than or equal to the dispersion of a 0.1 % momentum-spread beam to ensure maximum transmission; Fig. 6 shows the result of the PATH simulation with a Gaussian beam. The summary table indicates that the transmission was 10%.

Figure 7 is a PATH calculation for a 20-keV proton beam accelerated to 100 keV by an electrostatic accelerator column. The beam current was assumed to be fully neutralized (0 current) on either side of the column and 0.1 A within the column. The calculation for the accelerator column was done to third order and a uniform distribution was used to enhance the visibility of the aberrations.

Figure 8 shows a longitudinal phase-space calculation for an rf buncher using a 750-keV proton beam. A plot was made every 0.5 m from the buncher to show the bunching action. Again a uniform distribution was used.

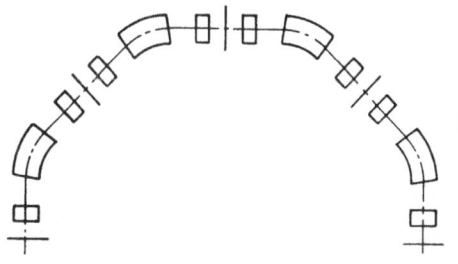

Fig. 2. The 180° achromatic bending system consisting of four identical cells Sextupole magnets are assumed to be incorporated within the quadrupoles.

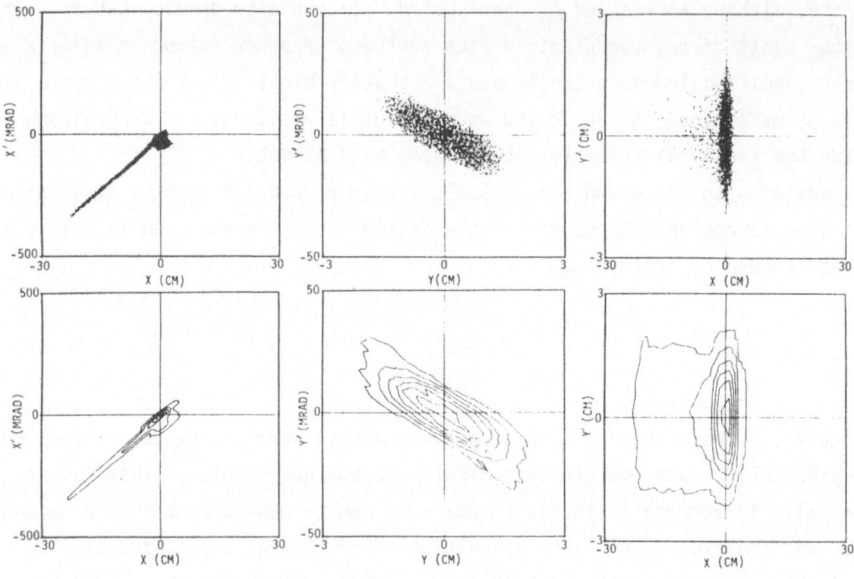

Fig. 3. A PATH calculation for the system shown in Fig. 2 with no sextupoles. Shown are scatter and contour plots for (X,X'), (Y,Y'), and (X,Y) plans.

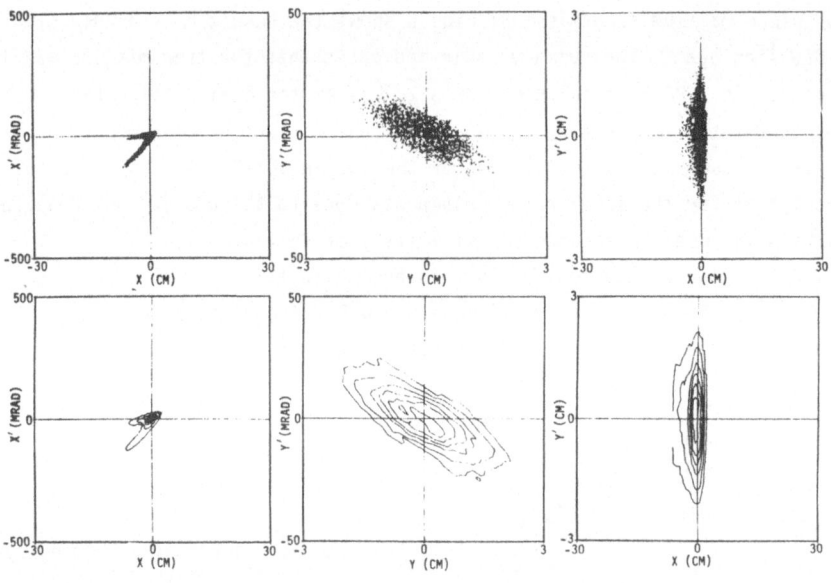

Fig. 4. A PATH calculation identical to Fig. 3 except with the sextupoles turned on.

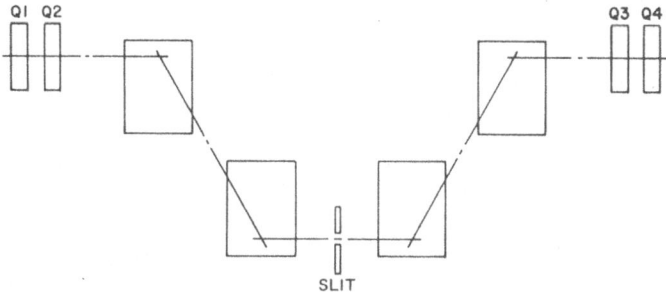

Fig. 5. An achromatic, four-bend momentum analyzing system.

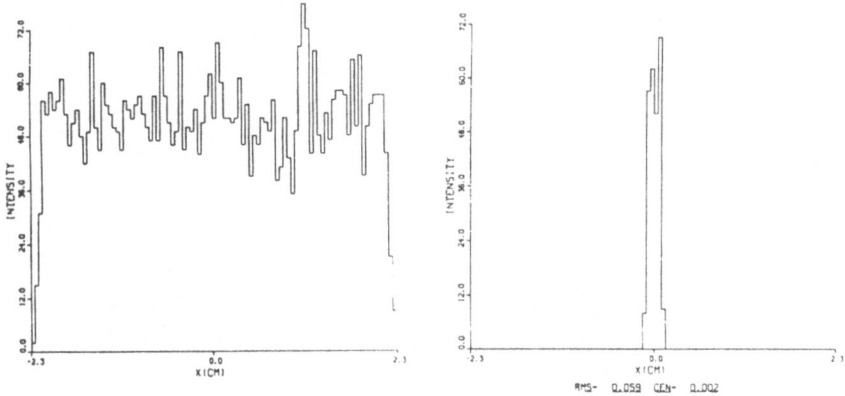

Fig. 6. A PATH plot of ΔP/P before and after the slit in the system of Fig. 5. A Gaussian input beam was used.

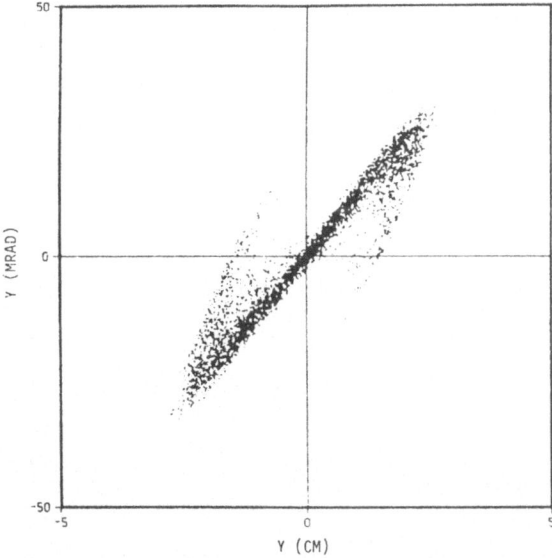

Fig. 7. A PATH simulation of a dc accelerating column that accelerates a 20-keV proton beam to 100 keV showing the effect of third-order aberrations.

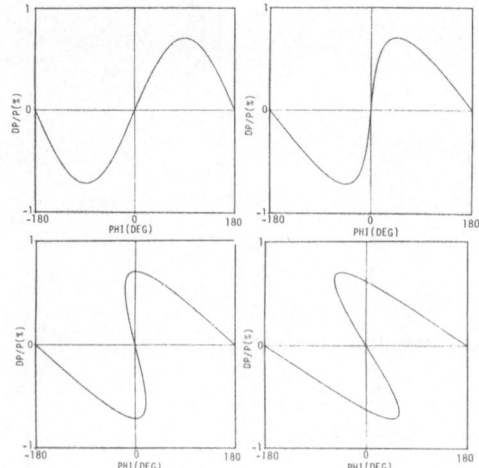

Fig. 8. Longitudinal phase-space plots showing the action of a single rf buncher at distances of 0.5, 1.0 and 1.5 m after the buncher.

Conclusions

PATH has proven to be a versatile and useful beam-transport simulation that has been applied to a wide variety of problems. The code is written in a modular fashion to facilitate inclusion of additional beam-transport elements as well as improved subroutines for existing elements. The compatibility with TRANSPORT and the interactive nature of the program make it easy to use.

Acknowledgment

The author is indebted to Ann Aldridge, Rob Ryne, and Dan Rusthoi for assisting in the development of the program.

References

1. Brown, K. L., and Iselin, Ch., CERN Report 74-2 (1974), DECAY TURTLE, A Computer Program for Simulating Charged Particle Beam Transport.
2. Kapchinskii, M. and Vladimirskii, V. CERN (1959), p.274, International Conference on High Energy Accelerators.
3. Brown, K.L., Carey, D. C., Iselin, Ch., and Rothacker, F., SLAC Report 91, Rev. 2 (1977), TRANSPORT, A Computer Program for Designing Charged Particle Beam Transport Systems.
4. Courant, E. D. and Snyder, H. S., Ann. Phys. 3, p.1 (1958), Theory of the Alternating Gradient Synchrotron.
5. Austin, B. et al, MURA Report 713 (1965), The Design of Proton Linear Accelerators for Energies up to 200 MeV.
6. Brown, K. L., IEEE Trans. Nucl. Sci. 26, p. 3490 (1979), A Second-Order Magnetic Optical Achromat.

WORKSHOP NO. 1. COMPUTER PROGRAMS FOR LATTICE CALCULATIONS

Convener: E. Keil, CERN

Notes by K.H. Reich, University of Dortmund

The aim of the workshop was to find out whether some standardisation could be achieved for future work in this field. A certain amount of useful information was unearthed, and desirable features of a "standard" program emerged. Progress is not expected to be breathtaking, although participants (practically from all interested US, Canadian and European accelerator laboratories) agreed that the mathematics of the existing programs is more or less the same.

Apart from the NIH (not invented here) effect, there is a - to quite some extent understandable - tendency to stay with a program one knows and to add to it if unavoidable rather than to start using a new one. Users of the well supported program TRANSPORT (designed for beam line calculations) would prefer to have it fully extended for lattice calculations (to some extent already possible now), while SYNCH users wish to see that program provided with a user-friendly input, rather than spending time and effort for mastering a new program.

Nevertheless there was a certain consensus about the desirable "new" features of such a program:

- "modern" input format,
- preferably interactive input and output,
- graphics routines not part of the program (but provision to use the graphics package of the user laboratory),
- hardcopy of interactive session (for analysis of improvements or failure),
- precision treatment of the effects of various magnet end fields,
- print-out of maximum values, wherever they occur,
(- active support of "authorised" version).

ANSI FORTRAN 77 was proposed as standard programming language but does not appear to be generally available. Also some feature or other desired by certain users (like an overlaid version) may not get enough support to be implemented early. Nevertheless, the prospects for a more general use of a single program (set) did not appear to be entirely negative.

As the main candidates MAD, SYNCH and TRANPORT stand out. Some of the attractive features of MAD, like the provision of linking the output to various specialised follow-up programs, was pointed out by a user. There was also a suggestion to consider MARYLIE as a candidate for the forthcoming tracking addition to MAD.

Clearly, the priority is in practice to add new features to existing programs rather than to improve them for easier use by newcomers. However, with active user support, even such improvements might not be totally out of the question, particularly if after this start interest continues. One could envisage to form a user community and to report on progress in this direction at the next conference.

DIGITAL CONTROL OF ACCELERATORS - THE FIRST TEN YEARS

H. Frese
Deutsches Elektronen Synchrotron (DESY)
2000 Hamburg 52, W. Germany

While both accelerators and digital computers have existed for more than forty years, it was not until the mid seventies that they were successfully united to form the computerized accelerators that are standard today.

When the first mini computers became available circa 1965 they were immediately employed at accelerator sites - for experiments. The physicists were very quick to put them to use in spite of their deficiencies: assembler programming, no operating system to speak of, and limited reliability. Controls people were sure that they did not need them and agreed that knobs and dials were the "natural" interface between men and machines anyhow. Also, a potentiometer setting was far cheaper and more reliable than a value in core storage plus its associated ADC.

Nevertheless, computers started creeping into control systems, took over the tedious tasks such as setting magnet currents and logging them, but always on a two-ported basis via an added connector guarded by that little switch that enabled good old manual control. Of course, computer breakdowns were entered into logbooks besides traditional obstacles such as water leaks or power supply failures.

Today we know that computers wouldn't become reliable until people decided to rely on them.

And the minis grew up, soon reaching the mature age of seven or eight, operating systems emerged, higher level languages, vector display screens - a powerful device was waiting for full scale employment when the demands came in the early seventies: FERMILAB and the CERN SPS, accelerators in the five km vs. the previous 500 m range which ruled out individual cabling of process equipment the to control room, complexities of acceleration and deceleration requiring the coordinated operation of various subsets of the equipment and injection and ejection switchyards just as difficult.

New tools for man/machine communication were emerging: the TV screen (possibly even in color), the touch panel, and the tracker ball. The computer console panel with its lights and toggle switches and the console teletype of old would leave the control room and be replaced by an integrated man/machine communication device designed to operate accelerators instead of computers. The hardware and software interface of the control computer would become double sided: software filters and translation programs were introduced between the operating elements and the control program proper as well as out towards the process equipment. This means that inside this filter all process variables, be they touch panel hits or position monitor values, can be treated in the same fashion by one common control language, and I think it is very appropiate to name a date here: it is year One of NODAL, the language with integrated equipment handling facilities developed for the CERN SPS. Two more points need to be observed: One mini is not enough, so use a school of them and interconnect them with a communication system of sufficient bandwidth. Also, provide a BASIC-like language interpreter for typed in user commands as well as complete operating programs. Would the ambitious scheme work? You all know the answer.

When PETRA and PEP were built four or five years later, a number of restraints were beginning to loose their significance:
- Operating systems were getting more comfortable - industry supplied multi-user systems would actually work.
- MOS memory started its logarithmic growth, storage cost ceased to be an obstacle both for program execution and for bit-map type raster scan displays. RGB screens could be used in numbers to provide a backdrop of constantly updated displays of various parts of the accelerator complex that the operator can switch to by simply turning his head.

Standardization of interface equipment was carried out in two alternative ways: While many people preferred the established CAMAC standard (which never could win on electrical or financial grounds), PETRA decided to integrate the formerly separate functions such as digital input/output, ADCs, and DACs in such a way as to provide a one-to-one correspondence between a piece of accelerator equipment and its 'responsible' control module.

After setting up the equipment and threading the first beams, people soon began to realize the intrinsic computer power of their control systems for coordinated operations such as controlling magnets in families to provide beam bumps. This, of course, was only the first step towards requesting a complicated operation, using a computer model of the accelerator to calculate the necessary ingredients and exspected effects, asking the operator for permission, and carrying it out. The execution provides the ultimate test of the validity of the model and probably will improve it for the next round. It should be noted that such a computer program acts as a digital model of the vast analog computer otherwise known as the accelerator, to which it is coupled through measuring and steering devices with only limited resolution and repeatability: real answers will only be provided by the real thing. Thus, machine study shifts will not disappear from accelerator schedules any too soon.

Control system designs for small accelerators were quickly following the examples of theis larger counterparts, both for new accelerator sites and for the tedious task of upgrading the outmoded control systems of existing accelerators to the new standard.

Meanwhile, back in Silicon Valley, micro computers started to grow, going through infancy much the same way as the minis had done ten years before: assembler programming by hand, no operating systems, but far more reliable than their ancestors. Today, a certain stage of maturity has been reached, but everybody agrees that at the current pace of development, he'll have the equivalent of a VAX/780 in his desk drawer in less than five years for less than 10000 deutschmarks or swiss francs.

Control system uses for micros are self-evident: they can replace the mini computers for operator console control, message transfer, and equipment control. Besides, they'll replace a lot of random TTL or CMOS logic in remote control modules and the equipment itself. In short, the micros are taking computer power into the process hardware itself. Because of their cheap price, sharing a CPU becomes a thing of the past. While elaborate multi-user systems are being designed by the industry, dedicated micros would not have much use for them. What they need is message transfer bandwidth and a system hierarchy that will keep long distance transmission of values of local interest to a minimum.

Also, while there are many micros already at work at the different laboratories for a vast variety of tasks, a coordinated, generalized, and systematic use of micro computers remains yet to be demonstrated.

These are the problems of the control system designers of today, with LEP, SLC, TRISTAN, and HERA under way and the DESERTRON entering serious discussions.

DISTRIBUTED DIGITAL CONTROL OF ACCELERATORS

M.C. Crowley-Milling
CERN

Geneva, Switzerland

1. INTRODUCTION

I must start by defining what I mean by distributed digital control, since the word "distributed" can be used in many different ways. I am going to consider the case of distributed intelligence - a system made up of a number of separate computers, each performing a significant number of tasks in a semi-autonomous way. This definition excludes the master-satellite type of system, if the satellites are just multiplexing data and control signals between them and the master, even if they are geographically distributed; on the other hand, it includes systems where the intelligence is distributed over a number of computers, even if they are gathered together in one room. The distributed intelligence also implies at least some distribution of the system data-base, if the inter-computer transactions are to be kept to a reasonable level.

I would also like to point out that there is a significant difference between a generalized distributed computer system and a distributed control system. In most of the distributed computer systems discussed in the literature, the aim has been to provide the user with a system where computations can be carried out on his behalf whenever spare capacity exists, and he has no interest in where this is, as long as the result is output to his local screen or printer. However, for distributed control systems, the hardware to be controlled is connected to specific computers, and so at least part of his programs must be executed at fixed locations.

I think the earliest example of a distributed control system, in the sense I have defined above, was that for the AGS at Brookhaven in the middle 1970's. Previously, a number of PDP-8 computers had been added to the control system to carry out specific and unconnected duties, but in about 1974 these were connected to a PDP-10, to coordinate their activities[1]). In contrast, the SLAC accelerator had a single PDP-9 computer, to which was added eight PDP-8 computers, distributed along the machine, but these only acted as remote multiplexors, and so this system, as it was then[2]), does not come into my definition of a distributed control system.

Distribution can be geographical, systematic, functional or mixed. Geographical distribution is used when an accelerator (and I include storage rings in the term accelerator) exceeds a certain size and the equipment to be controlled is concentrated at a limited number of points round the ring where access can be obtained, or, in the case of a Linac like SLAC, where the equipment repeats periodically along the accelerator. Then, a computer placed at each of these points can control a portion of each of the systems that extend all round the ring, power supplies, vacuum, beam monitoring, etc. Such general purpose computers can have almost identical software, which has to include the high level drivers for all the different types of equipment connected. One advantage of geographical distribution is that it can allow local operation of equipment consisting of a number of sub-systems, without connection to the rest of the network. For example, running the RF equipment can involve the vacuum system and power supply system.

Systematic distribution is more usual for smaller accelerators, where all the components of a given system, RF, vacuum, beam monitoring, etc., can be joined to and controlled by a single computer. In this case the software is different in each computer, but needs the high level drivers for only a restricted number of equipments. In a systematic distribution the systems can be subsystems of an accelerator, as in the examples given above, or complete accelerators in a complex of accelerators, as in the CPS, where the booster, the main ring, the antiproton accumulator, etc., each have a system computer[3].

Functional distribution is used to describe the situation where the duties carried out by a central computer are split up between a number of separate computers. An example of this is in the SPS system[4], where separate computers are used to drive the consoles, to provide a system library, to drive displays, to analyse alarms, for program development, etc.

The way control systems for past and present major accelerators have been distributed is shown in Table I. This has been compiled from published articles, and may not reflect the present situation, as control systems, like accelerators, evolve continuously.

In recent years, the development of microprocessors has had a considerable impact on distributed control systems. Up till now, these have mostly been incorporated in the interface system, usually in CAMAC modules, to carry out some task autonomously[5]. Now they are beginning to invade the equipment itself, taking over the duties of sequencing, function generation, surveillance, local servo loops, etc., which were previously carried out by the process computer or specialized hardware[6].

This tendency eases the process computers of some of their load, but it also introduces difficulties, as will be shown later.

The essential parts of a distributed control system are the computers, the data network to connect them together, the interface to the equipment, the software system to make them operate together, and the timing system to ensure synchronism Let us look at each of these in turn.

2. THE COMPUTERS

The same type of computers can be used for both centralized and distributed systems, but in general distributed systems use simpler computers but more of them, since each computer is performing fewer different tasks. Distribution is proposed to be carried out even further in the control system for LEP, where the usual multi-tasking mini-computers will be replaced by an assembly of micro-computers, each performing a single type of task[7]).

3. THE DATA NETWORK

In many of the early systems, special developments were necessary to interconnect the various processors, since few manufacturers could offer equipment for this purpose and even where such facilities were available, they were too slow for real-time applications. The first large network, at Fermilab, was of the single-master, multiple-slave type, using specially developed high-speed serial links[8]). The next large system, for the SPS, used a store-and-forward packet-switching system which treated all computers as equals, allowing any one to take temporary mastership[4]). More recently, the widespread development of Local Area Network (LAN) systems[9]) and attempts to get standards for them adopted could reduce the necessity for "in-house" development for future distributed control systems.

Data networks are usually classified by the topography; star, ring, highway, etc., and most of the networks for accelerator control so far have been of the star type. In this, each computer is joined by a separate line to a node which is responsible for routing the various messages. Large systems can have interconnected stars with several nodes (the SPS/CPS network presently has 5 nodes and about 60 computers, not all involved in controlling the accelerators). The switching at a node can be either circuit-switching or packet-switching. Circuit-switching operates in a similar manner to a telephone exchange; a connection is established between the two computers that want to communicate and this link is maintained until the transaction is complete. In a packet-switching system, messages are broken up into blocks or packets with headers giving address information and these packets are dynamically routed at the node, interspersed with packets from other sources for other destinations.

In store-and-forward systems, the node has buffer space to store temporarily the packets, so that a computer can send a packet at any time, without having to make certain that the recipient is ready for it.

Although some of the LANs being developed for general computer and terminal interconnection use the star topography, the majority are of the highway or ring type. In the highway LAN, all stations are connected to a single highway and when any one transmits, all receive the message, but it is only accepted by the one whose address corresponds to that in the destination header. There must be a mechanism to avoid errors if two stations try to send messages at the same time, and this can be done by allocation of time slots, at fixed or variable intervals, the latter being achieved by the circulation of a permit-to-transmit, known as a token. An alternative requires a station to listen and make sure that no-one else is transmitting before doing so. It must then continue to listen to check that no-one started to transmit simultaneously. In such a case, the station has to stop transmitting and wait a random time before trying again. This mechanism is known as Carrier Sense Multiple Access/Collision Detection (CSMA/CD). The best known system using this is ETHERNET[10], which has a 50-ohm coaxial cable as a highway, carrying signals at 10 Mbits/s. Such a system works well if lightly loaded, but is not very suitable for use in a distributed control system, since a maximum response time cannot be guaranteed, and there is no provision for priority access to the highway.

In the case of the ring type of LAN, the stations are connected to their immediate neighbours to form a closed ring round which messages or packets circulate in one direction. Each station, on receiving a packet, checks the address, copies it if the address coincides with the station address, and then repeats the packet to its downstream neighbour. When the originating station gets the packet back, it removes it from the ring. There are two main methods of providing controlled access to such a ring. In the first of these, packets are continuously circulated round the ring, either "empty" or containing a message. When it has a message to send, a station waits until it receives an empty packet, puts the message or part of it into the packet and sends it on. The Cambridge Ring is an example of this type[11]. In the second method, a token is circulated and a station must wait until it receives a free token before it can send a message. The latter method seems preferable for an accelerator control network, since it allows the use of a priority scheme for access to the ring[12].

Another division in LANs is between "baseband" and "broadband" systems. In baseband systems, the data stream is the only signal on the cable, and a bandwidth of not more than twice the data rate is required. A broadband system can be used for many services simultaneously, each being allocated a portion of the bandwidth.

In an attempt to avoid the chaos resulting from all the possible variations in these different types of LAN, an American IEEE working party has been drawing up proposals (P802) for standards in this field[13]. It recognizes that different configurations are required for different purposes, and so the proposed standards allow both ring and highway in a number of variations which means that their value is somewhat limited.

As far as I know, no "standard" commercial LANS are yet being used for accelerator control, but the LAN principles are being applied. For example, SLAC are using a CSMA/CD broad-band highway system of their own design[14]) and TRISTAN will use a slotted ring[15]). A token ring system is being considered for LEP.

Data transmission is usually carried out on coaxial or twisted pair cables, but the use of optical fibre cables is planned for a number of new machines, e.g. TRISTAN and LEP. One of the disadvantages of optical fibres is their low resistance to radiation, so that normally they cannot be used inside an accelerator tunnel.

4. THE INTERFACE

With the conventional minicomputer systems, there is no fundamental difference in the interface requirements between a centralized and a distributed system, although there may be detailed differences due to the distances involved, etc. The basic requirement is to transfer a number of digital or analogue values between the equipment and the computer, making the necessary transformations (D-A, A-D) on the way. This is normally carried out in a Command/Response manner, using a standard code, e.g. CNAF in CAMAC.

However, the situation changes when microprocessors are incorporated in the equipment. This primary interface is now between the equipment and its internal microprocessor, and a different kind of interface is needed between the equipment microprocessor and the process computer. Since there is intelligence in the equipment, communication does not need to be limited to Command/Response, and messages can be exchanged between the two parties. Instead of a series of CNAFs, the equipment can be told what to do in a message in the form of a string of characters, which can be meaningful to the user. This has the advantage that the equipment can be tested in isolation, using a simple terminal.

The physical links between the equipment microprocessor and the process computer have similar requirements to the intercomputer network, and a single type of LAN could be used for both purposes. However, the differences in distances, number of connections, etc., means that a single type that will cover all requirements will be more expensive than using two separate types, optimized for the application. For LEP, we intend to use the aircraft MIL/STD-1553B multidrop highway system[16]) to connect between the process computers and the equipment.

Although there are considerable advantages in putting microprocessors into the equipment, such as off-loading some of the work of the process computers, and making the equipment self-surveying and self-testing, there are also some disadvantages.

The main one is that the response time for simple actions is slowed down. With a conventional system, with equipment connected through CAMAC to a process computer, the latter can perform some action on the equipment in the time it takes to output a few CAMAC commands - of the order of 10-100 µs even with a slow serial loop.
However, with a microprocessor in the equipment, when the process computer sends a message, the microprocessor has to interrupt its present task, interpret the message, carry out the required action, compose a reply message and send it to the computer. This may take times in the range of 10-100 ms. In some cases, this increase in reaction time can be partly offset, if similar actions have to be taken on a number of equipments, by making provision for them to be taken in parallel, while a conventional system would have to take them sequentially.

A second disadvantage, more organizational than affecting performance, is that, since the primary interface between the equipment and the microprocessor becomes the responsibility of the equipment design group, and not the controls group, it is more difficult to obtain standardization of the hardware used.

5. THE SOFTWARE SYSTEM

The software system for a distributed control system has a number of requirements additional to those needed for a centralized system. They affect the applications language, the operating system and the database.

5.1 The Applications Language

While a distributed system could be programmed in any language, the applications programmers' task is made much easier if the language contains explicit provision for writing a program, different parts of which can be designated to run in different computers, with the necessary provisions for synchronization of returns and for exception handling. As far as is known, NODAL[17] was the first language to provide these facilities, through its EXECUTE, REMIT, WAIT and exception (!) commands.

Only recently have other languages become available with some of these features, the best known being ADA with its separate tasks and rendezvous mechanism. Even newer is OCCAM[18], which tries to provide the same facilities in a simpler package, aimed specifically towards process control.

5.2 The Operating System

A centralized system requires a computer with a very comprehensive operating system, since one computer has to perform so many tasks. Unfortunately, it seems that the more sophisticated an operating system becomes, the slower its response for real-time tasks, unless special tricks are used.

In a distributed system, the process computers controlling the equipment can have relatively simple multi-tasking executives, which can reside entirely in main memory, removing the need for discs, which is important for geographically distributed systems. One or more of the control computers must have an operating system with the full facilities, but this need not be involved in the fast real-time operations.

Of greatest importance is the communications package, which provides the means for the operating systems in the various computers to communicate in a homogeneous way. In most distributed systems in operation today, it has been necessary to write this specially for the system, as manufacturer's implementations, such as DECNET, are usually too slow for real-time control systems.

5.3 The Data Base

A distributed control system usually involves a distributed data base, since a central data base would require a computer to access this data base even when performing a local, autonomous action, which would increase the data network traffic appreciably, and slow the system down. Sometimes this difficulty is overcome by having a central data base containing all data, and then having subsets in the other computers, according to the duties they perform. This brings the usual problem of keeping up-to-date multiple copies of a data base.

In my opinion, a control system data base should be divided into a number of subsets, each subset being located where it is most appropriate, and any program requiring data should retrieve it from the computer holding the subset.

As an example, let us take the case of the extreme distribution, as proposed for LEP, where there are microprocessors incorporated into the equipment. Each of these microprocessors should hold the data table concerning that piece of equipment, such as, in the case of a power supply, demanded and measured values, maxima and minima, tolerances, conversion and calibration factors, etc. Although access to these data should be available from other computers, this is normally only required for fault-finding, etc. For most transactions, the equipment is sent a message to perform some action, or read or set a value in engineering units, the reply informing the requester if it is not able to do so.

At the next higher level, that of the process computer, the sub-data base takes the form of a directory, linking mnemonic names with the equipment addresses on the interface system, and holding information concerning interaction of one equipment with another on the same computer, etc.

Above the process computers, at the central control level, we need a data base that holds what I will call the machine physics parameters, the required values of Q, beta, chromaticity, etc., and the required settings of the various machine elements to give these values. This is the data base with which the control programs will mainly interact. The required settings will probably be obtained through some modelling or simulation programs, which in the past have necessarily been run off-line, but more and more frequently are being run on-line, as the computing power in the control system increases, and then access to a further data base is needed, that holding the machine optics parameters.

Other "central" data bases are those for the alarm system, allowing coded messages from the process computers to be analyzed and appropriate messages displayed to the operator, those containing "help" messages to guide the operator when in doubt, and those containing inventories of the equipment, with information on who to call when there is trouble, and where to find them.

From the above, which by no means includes all the data bases concerned with the operation and maintenance of an accelerator, it can be seen that the design and decomposition of the data base for a distributed control system is extremely important, and a badly designed system can result in continual frustration.

Allied with the subject of data bases is program storage. Some distributed systems use computers with sufficient local storage to hold all or most of the programs needed, but it is more usual to have some computers with a minimum of local storage and a central library from which they can load programs and run them as required.

6. The Timing System

In most accelerator systems there are three precisions of timing that are required; of the order of a second, of a millisecond and of a microsecond or less. For operations involving decisions by an operator, it is satisfactory if the system can respond in a time of the order of a tenth to one second. Most data transmission systems used with distributed control systems will allow responses on this time scale, even when several computers are involved, so no special provisions have to be made.

In the millisecond range, where, for example, synchronism between ramping power supplies controlled by different computers is required, or a closed orbit acquisition at a particular part of the cycle must be made, or a beam extraction started, it is unwise to rely on signals transmitted between computers, as to achieve a precision of even 5 ms requires special provisions. The problem can better be solved by providing a separate network carrying timing signals which can be distributed to the equipment with time critical requirements. Such timing signals can include a series of clock pulses, say at 1 ms intervals, with coded event markers interspersed. Then, rather than try to get a message through the system to say "do something NOW", one can pass a message at leasure, well beforehand, to say "do something on the nth clock pulse after event code X". With this system, it is possible to get synchronization, with a resolution depending on the clock pulse interval, to within the order of a few microseconds between hundreds of pieces of equipment, irrespective of the number of computers in the system.

Synchronism to precisions less than a microsecond usually are limited to the radio frequency or beam monitoring systems, and it is usual to make special separate provision for this, the computer system not being involved, except in setting programmable delays, phase shifts, etc.

7. Conclusion

The special properties, requirements and limitations of distributed control systems have been described. The main advantages of distributed over centralized systems can be summarized below, not necessarily in the order of importance.

(a) For geographically distributed systems, complete systems can be operated locally even when links to the centre and the central system are unavailable. Similarly, parts of the equipment can be tested and commissioned separately during the construction period.

(b) Distribution of tasks between a number of computers gives the possibility of parallel processing with a corresponding gain in speed and reduction in response time.

(c) Computer configurations and operating systems can be tailored to suit the application. New requirements can be catered for by the addition of one or more computers to the network without disturbance of the existing system.

(d) Local surveillance and testing can reduce the load at higher levels which only need to be informed if anything goes wrong. Diagnosis of faults can be easier, despite the increase in number of units, as they can be arranged to test each other. Maintenance can be easier in a modular system, by exchange of modules, and it is easier to provide redundancy in vital parts of the system.

(e) Different groups can work on different parts of the system with the minimum of interference.

(f) The needs for bandwidth in the communications system for a large machine are lower for a decentralized than for a centralized system.

(g) A distributed system can have economic advantages. It may require more hardware, which is becoming cheaper, but the software, which, if anything, is becoming more expensive, is generally simpler, and replicated.

References

1. "Use of a General-Purpose Time-Shared Computer in Accelerator Control", M.Q. Barton et al. Proc. 9th Int. Conf. on High Energy Accelerators, Stanford, May 1974.

2. "Initial Experience with a Multi-processor Control System", K.B. Mallory. Proc. 9th Int. Conf. on High Energy Accelerators, Stanford, May 1974.

3. "The Improvement Project for the CPS Controls", G. Baribaud et al. IEEE Trans. Nuc. Sci., NS-26, 3, June 1979.

4. "The Design of the Control System for the SPS", M.C. Crowley-Milling. CERN 75-20. Dec. 1975.

5. "A Versatile CAMAC Crate Controller and Computer", C. Guillaume et W. Heinze. Nuc. Instr. and Methods 177 (1980) 327-331.

6. "Integrating Local Intelligence into the LEP Power Supplies", J.G. Pett. Proc. 5th Int. Solid-State Power Conversion Conference, Geneva, Sept. 1982 (Intertec).

7. "Replacing Mini-computers by Multi-microprocessors for the LEP Control System", J. Altaber et al. IEEE Trans. Nuc. Sci., NS-30, 4, Aug. 1983.

8. "Intercomputer Communications in Real Time Control Systems", S.R. Smith et al. IEEE Trans. Nuc. Sci. NS-20, 3, June 1973.

9. "Untangling Local Area Networks", R. Parker and S.F. Shapiro. Computer Design, March 1983.

10. "Ethernet : Distributed Packet Switching for Local Computer Networks", R.M. Metcalf and D.R. Boggs. Communications of the ACM, Vol. 19, 7, July 1976.

11. "The Cambridge Digital Communications Ring", M.V. Wilkes and D.J. Wheeler. Proc. of Local Area Communications Networks Symposium, N.B.S. May 1979.

12. "A Token-ring Network for Local Data Communications", R.C. Dixon et al. IBM Systems Journal, Vol. 22 No. 1/2 1983.

13. "Local Area Networks : A Pair of Standards", M. Graube. IEEE Spectrum, June 1982.

14. "Wide-band Cable Systems at SLAC", W. Struven. IEEE Trans. Nuc. Sci. NS-30, 4, Aug. 1983.

15. "Design of the Control System for TRISTAN", H. Ikeda et al. IEEE Trans. Nuc. Sci. NS-28, 3, June 1981.

16. "Serial Digital Bus Heads for Industrial Systems". Electronic Design. 28, 19, Sept. 13. 1980.

17. "The NODAL System for the SPS", M.C. Crowley Milling and G. Shering. CERN 78-07.

18. "Process Oriented Language meets demands of Distributed Processing", R. Taylor and P. Wilson. Electronics. Nov. 1982.

Table

Publication date	Machine	Type of Distribution		
		Geographical	Functional	Systematic
1973	NAL	x		x
1973	SPS	x	x	x
1977	PETRA			x
1979	PEP	x		
1979	CPS		x	x
1979	CESR			x
1980	SNS (RHEL)	x		x
1980	ISABELLE	x		x
1981	SLAC	x		
1981	TEVATRON	x		x
1981	MEA (NIKHEV)	x	x	
1981	FMIT (LAMPF)			x
1981	Stonybrook			x
1981	TRISTAN		x	x
1983	LEP	x	x	x

Distributed Accelerator Control Systems

CENTRALIZED DIGITAL CONTROL OF ACCELERATORS

R. E. MELEN*
Stanford Linear Accelerator Center
Stanford University, Stanford, California 94305

Introduction

In contrasting the title of this paper with a second paper to be presented at this conference entitled "Distributed Digital Control of Accelerators,"[1] a potential reader might be led to believe that this paper will focus on systems whose computing intelligence is "centered" in one or more computers in a centralized location. Instead, this paper will describe the architectural evolution of SLAC's computer based accelerator control systems with respect to the "distribution" of their intelligence. However, the use of the word "centralized" in the title is appropriate because these systems are based on the use of centralized large and computationally powerful processors that are typically supported by networks of smaller distributed processors.

Linac and Beam Switchyard Computer Control System

Computers were first introduced into SLAC's accelerator control systems approximately 17 years ago when a SDS 925 was used in the beam switchyard area to provide monitoring and control for less than 50 power supplies and 1000 digital status bits.[2,3,4] In those times, the high cost of even simple computers combined with relatively primitive operations requirements made the use of computers in control systems a hotly debated subject. Several factions of SLAC felt that the advantages (or disadvantages) of computers did not justify their expense. Hence, several years elapsed before a Digital Equipment Corp. PDP 9 was installed in the main linac control system.[5,6]

The primary motive for using a computer in this system was for status monitoring. Another motivation was to provide automated reconfiguration of the linac after taking one of its 240 klystrons was taken off-line because of a failure and replaced with a spare unit. Ironically, this very complex problem remains today as one of the most demanding procedures that must be implemented in the new SLC control system.

Since the linac control system was originally designed as a manually operated system, the PDP 9 was essentially placed in series between the existing manual human interfaces and operators. The computer's intelligence was used to provide a more flexible man-machine interface and to extend the scope of the operators' observation abilities by providing extensive automated monitoring functions.

Before the first implementation of the PDP 9 computer system was complete, it became evident that a two control room approach was expensive and irrational and the switchyard and linac control rooms were merged into the switchyard control room, now known as Main Control Center (MCC).[7,8] Further, it soon became evident that the flow of information to and from the linac through the original manual interfaces was intolerably slow. Therefore, eight PDP 8s were placed along the 2-mile klystron gallery and used as intelligent data acquisition and distribution processors.[9,10]

The system was later expanded by the use of fixed program micro-processor controllers interfaced to the PDP 8s via serial links to provide specialized dedicated controllers to implement beam guidance,[11] phasing,[12] and triggering[13] functions.

With the exception of updating the system by replacing the PDP 9 and SDS 925 computers with PDP 11 computers, the basic concepts and architecture of the system remain relatively unchanged today.[14] It is serving the intended purpose as a very efficient "look-and-adjust" system.[15,16,17]

---

*Work supported by the Department of Energy, contract number DE-AC03-76SF00515.

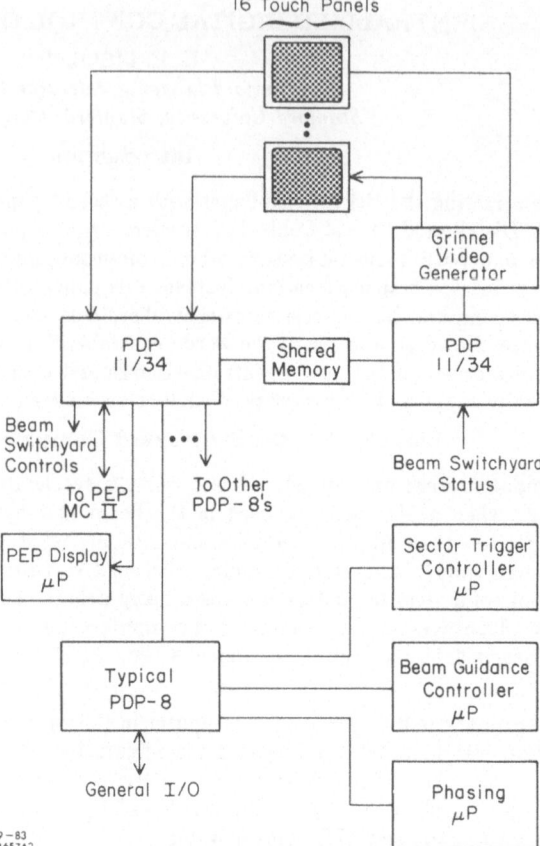

Fig. 1. The linac/switchyard computer control system.

SPEAR

Two considerations dominated the choice of computer architecture for the SPEAR computer control system[18] (Figure 2), which was implemented in 1970-72. First, a very tight budget and short construction cycle dictated a minimum cost and manpower effort. Secondly, there was a strong desire to provide a system that would provide extensive support of machine physics efforts via the use of real-time machine modeling. This concept would allow the machine operators to specify desired operating conditions such as tunes, beta values, dispersion, and energy and have the computer automatically calculated and set up the appropriate magnet and RF settings.

The second requirement exceeded the capabilities of the then current mini-computers. So it was decided that all of the objectives could be met by combining the control system computational needs with the needs of the two experimental areas, and a XDS Sigma-5 "timesharing" system was purchased.

Although the original computer configuration was woefully inadequate in terms of memory and disk space, it was expanded and both the control system and one experiment coexisted peacefully for several years. This sharing worked successfully primarily because the storage ring required very little "tuning" when the experimenters were taking data and vice versa.

The use of a large and computationally powerful processor in the centralized SPEAR control system had many advantages. The sophisticated CP-5 operating system, when combined with an advanced FORTRAN compiler, provided an excellent software development environment. Secondly, the resources available through the use of the Sigma-5 eliminated the need for a distributed system to acquire and process data from the physically small, 720 foot circumference facility, which greatly simplified the software task. These advantages allowed the applications software to be developed quickly with a small staff.

Probably the most significant weakness of this architecture was the fact that the CP-5 operating system had a relatively slow time-shared response-time. However, this weakness never really impacted the system performance since a storage ring has little need for faster than human response times (1-2 seconds). Further, the few needs for a fast response time were accomplished by providing carefully written "real-time' foreground tasks.

Through the succeeding years after its initial implementation, there has been very little need or interest in expanding the architecture of the system. New additions or changes have been easily implemented within the existing structure. However, maintenance support concerns for both the outdated hardware and the long neglected control programs finally forced a decision to replace the Sigma-5 and its in-house designed interfaces with a dedicated VAX 11/750 and CAMAC based data acquisition hardware. The new software system and human interfaces are based on the PEP control system while the CAMAC hardware uses elements from PEP, SLC, and SLAC experimental systems. This new system will go on-line within the next ten days.

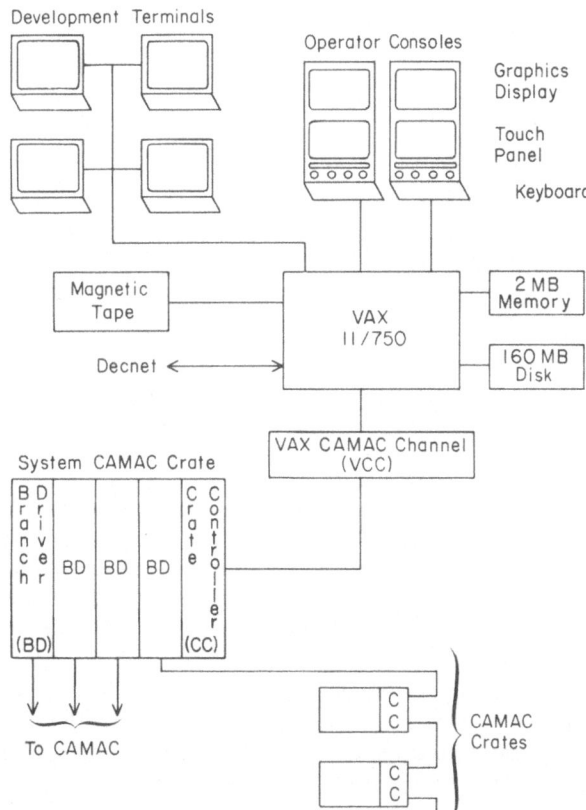

Fig. 2. The SPEAR computer control system.

PEP

The large physical size together with the large number of I/O points associated with PEP necessitated the use of a distributed network of computers to serve as intelligent data acquisition and distribution processors for a central computing complex.[19,20] As shown in Figure 3, the network contains ten ModComp computers and one VAX 11/780. The central MCIV computer is attached to a single operator console, and is connected via 500 kiloband serial links to nine MCII remote computers which are in turn interfaced to approximately 50 CAMAC crates via one megabaud serial SDLC links.[21,22,23] The MVIV central control computer is interfaced to the VAX via a similar link.

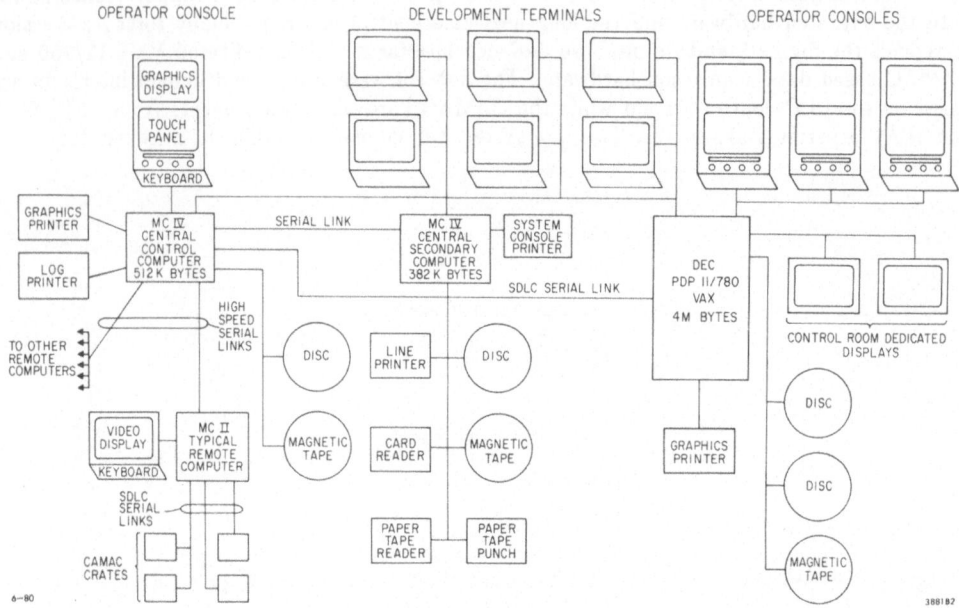

Fig. 3. The PEP computer control system.

In normal operation, the remote MCII computers continuously collect data from their respective CAMAC crates and then forward the refreshed data to the central MCIV computer at a rate of approximately seven refreshes/second. The central MCIV computer receives refreshed data from all of the remote processors and maintains a copy of the latest data for each signal in its RAM. This refreshed data is then sent as a block to the VAX at a rate of three blocks/second. Thus, continuously refreshed data can be accessed by application programs either in the MCIV or VAX. The only input data not read by this process is data collected by a slow digital volt-meter attached to each MCII. This data is delivered upon the request of a program in either processor. Output CAMAC data commands may originate in either processor. They are sent to the CAMAC crates via the MCIV and the relevant MCII.

In order the keep the data flowing at an acceptable rate, only the raw data, consisting of approximately 1200 points per MCII, is refreshed. Conversion to engineering units and other signal specific processing is performed only at the application task level on an "as needed' basis. Similarly, the output routines only process raw data.

The only "applications dependent' processing that occurs in the MCII's relate to the position monitor system which requires a complex readout procedure requiring several time delays. All other significant applications dependent programs, including limit checking and alarms, reside in either the central MCIV or the VAX.

The VAX executes nearly all of the higher level applications and modeling programs.[24,25,26] However, the MCIV does execute some display programs requiring fast refresh rates as well as a minimal set of programs for control and monitoring functions that can be employed in the event of a VAX hardware failure.

The addition of the VAX to the system architecture was a deviation from the original system design. This addition was necessitated by a general concern that although the hardware was adequate, weaknesses in the MCIV software prevented it from serving effectively as a stand-along central processor in the system. The software weaknesses existed in three areas.

First, the general quality of the operating system in terms of software development support left a lot to be desired. The general level and quality of support features were substantially below those provided by the Sigma-5 system at SPEAR and eventually by the VAX. This weakness had a significant impact upon the time required to develop and maintain applications software.

Secondly, the operating system lacked "robustness' in terms of its ability to continue running without crashing when it was asked to execute more programs than it could store in its RAM. This weakness was a significant problem because many of the modeling tasks required large blocks of memory.

Thirdly, the operating system was not "bullet-proof' in that relatively insignificant applications programming errors could cause a system crash or could degrade the system response to uselessness.

In contrast The VAX VMS operating system provides a stable, user-friendly, relatively bullet-proof system on which applications programs may be developed and run. The symbolic debugger is an especially valuable tool for debugging and maintaining real-time software. The relatively slow response time of the VAX is overcome by the use of the MCIV which serves as a dedicated front-end processor and generally handles procedures requiring faster than human response times. Further, acceptable system response times are achieved because PEP is a single purpose accelerator, and hence typically must support only three-four operating stations controlled by one or two operators.

SLC

In its final state, the SLC computer control system[27] will be an order of magnitude larger and more complex than any of SLAC's other accelerator control systems. In addition to modernizing and streamlining the operation of the present linac/beam switchyard system, the SLC system must provide a system based on machine modeling[28,34] to support the extensive accelerator development efforts required to develop an accelerating system meeting the tight SLC beam requirements.

In its final configuration, the SLC computer system will provide a combination of two VAX 11/780 central processors networked to 70-100 powerful micro-processor clusters, as shown in the block diagram in Figure 4. The micro-processor clusters interface with the equipment to be monitored and controlled through the use of CAMAC. These clusters will be located in each of the 30 linac sector alcoves and near the damping ring, electron and positron sources, and the SLC arcs and final focus.

The dual-VAX complex will serve to provide a centralized human interface for the machine operators and will be used to provide the on-line execution of the large modeling programs. In addition, these computers will serve to provide an environment for fast, efficient program development and maintenance for both the VAX and micro-processor clusters.

The distributed micro-processor clusters are based on the Intel Multibus architecture.[29] This architecture provides support for and arbitrary number of single-board computers (SBC) which communicate with each other through the use of shared memory and interrupts. The micro-processor clusters contain an Intel 86/30 SBC, 768 kilobytes of RAM and 8 kilobytes of EPROM. Various benchmark tests have shown that each micro-processor cluster has somewhere between 1/10 and 1/7 the processing power of the VAX 11/780. The micro-processor clusters are interfaced to CAMAC through a high-speed direct memory access (DMA) device based on the use of a bit-sliced micro-processor.

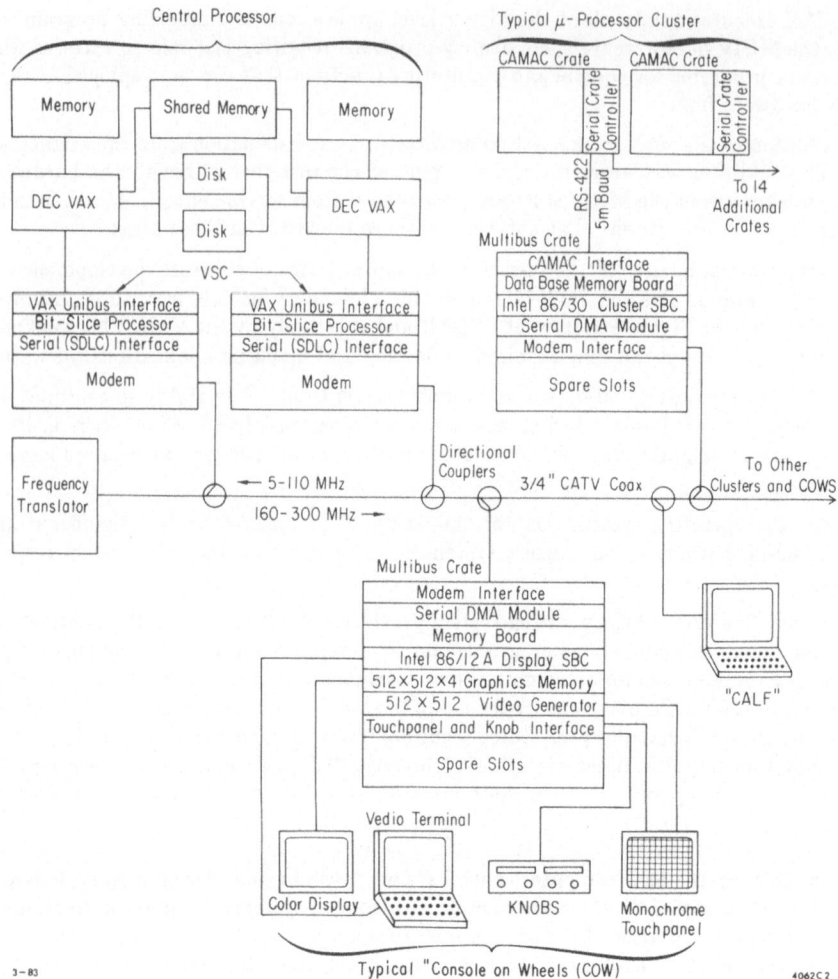

Fig. 4. The SLC computer control system.

Intelligence for the system is also distributed into the CAMAC crates via the use of dedicated controllers. One of these devices, the Smart Analog Monitor (SAM), is a Zilog Z80 based CAMAC module that continuously scans 32 analog channels and provides their floating point voltage values in either VAX or 8086 formats.

A second device, the Parallel I/O Processor (PIOP) CAMAC module, is a general purpose processor based on the Intel 8088 micro-processor chip and presents a standardized interface to the CAMAC data highway. This module provides a front panel port which is a differential transmitter/receiver version of the micro-processor's bus structure. This port provides a simple and straight forward method for interfacing special purpose "heads' that interface to specific devices or processes. So far this module has found use in the monitoring of the 270 linac klystrons' phase and amplitude[30] and also for their general monitoring and control.[31]

Programs for the PIOP may be developed by use of a cross-compiler or cross-assembler on the VAX. The compiled or assembled code may then be downloaded to the PIOP or it can be "burned' into EPROM's to provide a non-volatile program source.

There are two types of operator consoles in the system. The primary type has been given the name of console-on-wheels (COW) because it is a fully portable unit which may be connected at any point of the system's communications backbone. The second type is called a CALF and consists of an Ann Arbor Ambassador terminal with a modem which also allows it to be plugged into the communications backbone. Both the COW and the CALF communicate directly with the VAX. Software in the VAX allows the CALF to emulate a subset of the COW functions. The number of COW's and CALF's that can be supported by the system is limited only by the system's processing power.

The communications backbone for the system consists of a broadband (5-300 MHz) Cable Television (CATV) system that has the capability to support several hundreds of frequency-modulated signals on a single cable.[32]

Several sub-systems currently use the cable for communication. A high-speed, one Megabaud, polled network has been developed at SLAC to interconnect the micro-processors with the VAX's. A bit-sliced micro-processor is used to direct the sequential polling on the system and serve as a DMA channel to the VAX. This unit provides a maximum poll rate of 1000 polls/second. The use of this network structure is a departure from our earlier plans. We originally contracted with a commercial firm to provide a Carrier-Sense Multiple-Access Collision-Detect (CSMA/CD) network similar to that specified by Ethernet.[33] However, the development effort for this system turned out not to match our schedule needs, so an in-house solution was developed.

The CATV cable also supports terminal/VAX communications with equipment using protocols similar to Ethernet. The cable has a capacity for several hundred terminals. The same cable is also used to support television channels, voice channels, and point-to-point two megabaud data channels.

As previously mentioned, essentially all of the SLC software development is performed through the use of the VAX. Wherever possible, FORTRAN 77 is used for applications programming in both the VAX and the micro-processor clusters. FORTRAN 77 was chosen as the standard language because of its extensive support in the VAX, and because it is the most universally understood language. Although alternative languages could be used for the micro-processors, the consistency provided by standardizing on FORTRAN 77 has been a great advantage for both the development and support of applications programs.

A significant effort has been expended by SLAC to create an efficient and user-friendly environment for the development of micro-processor software. In collaboration with the Intel Corp., FORTRAN 77 and PLM 86 cross-compilers, a cross-assembler, and a cross-linker have been developed to support the 8086/8088 series of micro-processors. Further, a symbolic debugger has been developed to allow the remote debugging of micro-processor based programs.

Applications tasks executed in the VAX are written as structured subroutines which are attached to a VAX process that provides interface routines to the operator console, and to a structured database. This process also provides a scheduling service for the subroutines.

The micro-processor clusters provide local control algorithms for the operation of the technical equipment. In general, the micro-processors receive an operational configuration in engineering units for their equipment from the VAX. The micro-processors then insure that the equipment is set to that configuration and will only report back to the VAX when it is unable to achieve or maintain the desired configuration. The micro-processor clusters also provide monitoring information in engineering units to the VAX upon request. They also support a "pass-thru' mode for I/O commands from the VAX. The commands may originate from a VAX applications process or from a system user via individual, or a file of, interpretive commands. Micro-processor systems will be used in the future to implement time-sensitive digital control loops wherever required.

The first phase of the SLC control system implementation was brought on-line beginning August 1. This phase provides a single VAX 11/780 computer and micro-processor clusters in the first 10 sectors of the linac, the injector, and the Damping Ring. Several of the sub-systems are now operational and we are currently working hard to bring additional subsystems on-line.

It is much too early in the shakedown process to make any generalizations regarding the eventual performance of the system. However, two observations seem to be in order. First, the importance of on-line debugging and diagnostic aids cannot be overemphasized. In a complex system, it is extremely important to be able to trace problems efficiently in the system's operating environment. Although an extensive effort has been applied to providing these tools, we will continue to direct significant resources in this area. Secondly, the number of simultaneous users of the system has been overwhelming at times. It is not unusual to have 5 COW's, 5-7 CALF's, and 5-7 program development terminals simultaneously active. This concern has been partly alleviated by shifting some of the program development efforts to a second VAX. Though current response-times for the system may be tolerable even under heavily loaded conditions, with our current configuration it is evident that we will have to do battle with a response-time problem as the system expands to its fully-implemented state.

Summary

Upon careful examination of the architecture of SLAC's computer control systems described above, it becomes evident that the distribution of the systems' intelligence generally falls into three tree-like layers.

The first layer typically consists of a central computer complex incorporating one or more relatively large and powerful processors. The more modern systems use state-of-the-art 32-bit processors with several megabytes of RAM and several hundreds of megabytes of disk memory. Further, they support extensive user-friendly operating systems and program development facilities.

The second layer typically consists of several smaller processors which are downloaded from the central complex and whose primary task is to provide data acquisition and distribution. The more modern systems are 16-bit processors with several hundred kilobytes of RAM and no disk memory.

The third layer typically consists of several tens or hundreds of micro-processors, each dedicated to a single device. The micro-processors for these "dedicated intelligent controllers' are small and inexpensive and typically require less than 32 kilobytes of RAM or EPROM memory. Their hardware may be general purpose in nature or may be built into the architecture of the device itself. Figure 5 illustrates several of the relevant features of each of these layers.

This paper serves to illustrate that "for better or for worse," SLAC is committed to the centralized digital control of its accelerators.

Acknowledgements

The list of key people contributing to SLAC's control systems over the last 15-20 years is too extensive to single out individuals. Instead we have tried to reference publications that reflect the many hundreds of man-years of effort that has been applied to the design, implementation and support of these systems.

Central Computer Complex	Remote Data Acquisitions and Distribution Processors	Dedicated Intelligent Controllers
• Provides centralized human interface for operating personnel.	• Collects and distributes I/O data for the central computer complex.	• Dedicated to a fixed well-defined task for a single device.
• Is sufficiently powerful and contains sufficient resources to support the real-time execution of computationally demanding and physically large machine modeling programs.	• Performs tasks requiring greater I/O rates or faster response times than can be provided by the central computer complex.	• Provides high I/O rates and fast response times.
• Maintains both volatile real-time and non-volatile disk-base centralized data bases.	• Monitors device stations and reports important changes to the central complex.	• Provides a simple software interface to the device.
• Serves to synchronize the over-all system operation.	• Initializes front-end devices and controllers.	• May be downloaded from remote processor or may use EPROM memory.
• Provides flexible and efficient operating systems and program development facilities.	• Maintains a local general-use database.	• Does not use an operating system.
• Provides cross-compilers, cross-assemblers and remote debugging facilities for development of remote processor programs.	• Supports multi-tasking real-time operating system.	• Does not support extensive on-line debugging capabilities.
• Maintains current program images for all processors in the system.	• Supports remote debugging features to allow efficient on-line checkout through the central computer complex.	
• Provides diagnostic support for all processors in the system.	• Does not support mass storage.	

Fig. 5. Comparison of typical features for the three layers of intelligence.

References

1. M. Crowley-Milling, "Distributed Digital Control of Accelerators", paper to be presented at lecture session LS-B1 of this conference.

2. S. K. Howry, R. Scholl, E. J. Seppi, M. Hu, D. Neet, "The SLAC beam switchyard control computer," IEEE Trans. on Nuclear Science, NS-14,3, 1066 (1967).

3. S. K. Howry, SLAC Report CGTM 10, "BSY Control Computer System Language," (1966).

4. S. K. Howry, SLAC PUB-248, "A Concise On-Line Control System," (1966).

5. K. B. Mallory, "The Control System for the Stanford Linear Accelerator Center," IEEE Trans. on Nucl. Sci., 1022-1029, (1967).

6. K. B. Mallory, "Some Effects of (Not Having) Computer Control for the Stanford Linear Accelerator Center," IEEE Trans. Nucl. Sci., NS-20 (1973).

7. K. Breymeyer, et al., "SLAC Control Room Consolidation Using Linked Computers," IEEE Trans. Nucl. Sci., NS-18 (1971).

8. S. Howry, R. Johnson, J. Piccioni, and V. Waithman, "SLAC Control Room Consolidation-Software Aspects," IEEE Trans. Nucl. Sci., NS-18, 403-303 (1971).

9. K. B. Mallory, "Initial Experience with a Multi-Processor Control System," Proceedings 9th International Accelerator Conference on High Energy Accelerators, (1975).

10. K. B. Mallory, "Control Through a System of Small Computers," IEEE Trans. Nucl. Sci., NS-22, 1086-1087 (1975).

11. W. C. Struven, K. B. Mallory, "Two Micro-computer Controller Applications at SLAC," IEEE Trans. Nucl. Sci., NS-24, (1977).

12. S. K. Howry, A. Wilmunder, "A Micro-procesor Controller for Phasing the Accelerator," IEEE Trans. Nucl. Sci., NS-24, 1804-1806 (1977).

13. S. K. Howry, "Trigger Pattern Generation by Computer," SLAC TN-75-5 (1975).

14. V. Davidson and R. Johnson, "Present SLAC Accelerator Computer Control System Features," IEEE Trans. Nucl. Sci., NS-28 (1981).

15. D. Fryberger and R. Johnson, "An Innovation in Control Panels for Large Computer Control Systems," IEEE Trans. Nucl. Sci., NS-18, (1971).

16. K. Crook, "CRT Touch Panels Provide Maximum Flexibility in Computer Interaction," Control Engineering Magazine, 33-34, July 1976.

17. K. Crook and R. Johnson, "A Touch Panel System for Control Applications," SLAC PUB-1861 (1976).

18. A. M. Boyarski, A. S. King, M. J. Lee, J. R. Rees, and N. Spencer, "Automatic Control Program for SPEAR," IEEE Trans. Nucl. Sci., NS-20, 580-583, (1973).

19. A. Chao et al., "PEP Computer Control System," IEEE Trans. Nucl. Sci., NS-26, 3268-3271 (1979).

20. R. Melen, "The PEP Instrumentation and Control System," Proceedings of the 11th Internation Conference on High-Energy Accelerators, pp. 408-420 (1980).

21. J. D. Fox, E. Linstadt, and R. Melen, "Applications of Local Area Networks to Accelerator Control Systems at the Stanford Linear Accelerator Center," IEEE Trans. Nucl. Sci., NS-30 (1983).

22. J. R. Kersey, "Synchronous Data Link Control", Data Communications, McGraw-Hill Publications, 49-60 (May/June 1974).

23. A. Altmann, R. Belshe, R. Dwinell, J. Fox and N. Spencer, "CAMAC Micro-processor Crate Controller, Revision J," PEP internal document.

24. M. H. R. Donald, P. L. Morton and H. Wiedemann, "Chromaticity Correction in Large Storage Rings," IEEE Trans. Nucl. Sci., NS-24, 1200-1202 (1977).
25. E. Close, M. Cornacchia, A. S. King and M. J. Lee, "A Proposed Orbit and Vertical Correction System for PEP," IEEE Trans. Nucl. Sci., NS-26, 3502-3504 (1979).
26. M. Donald et al., "Some Schemes for On-Line Correction of the Closed Orbit, Dispersion and Beta Function Errors in PEP," IEEE Trans. Nucl. Sci., NS-28, (1981).
27. R. Melen, "A New Generation Control System at SLAC," IEEE Trans. Nucl. Sci., NS-28 (1981).
28. M. Lee et al., "Mathematical Models for the Control Program of SLAC Linear Collider," IEEE Trans. Nucl. Sci., NS-28, (1981).
29. Intel Multibus Specification Manual 98006832-02, Intel Corporation (1979).
30. J. D. Fox and H. D. Schwarz, "Phase and Amplitude Detector System for the Stanford Linear Accelerator Center," IEEE Trans. Nucl. Sci., NS-30, (1983).
31. R. Keith Jobe, "A New Control System for the SLAC Accelerator Klystrons for SLC," IEEE Trans. Nucl. Sci., NS-30 (1983).
32. W. Struven, "Wide-Band Cable System at SLAC," IEEE Trans. Nucl. Sci., NS-30 (1983).
33. The Ethernet, Digital Equipment Corporation, Intel Corporation, and Xerox Corporation (1980).
34. M. J. Lee et al., "Models and Simulation," paper to be presented at lecture session LS-C2, this conference.

CONCURRENT CONTROL OF INTERACTING ACCELERATORS
WITH PARTICLE BEAMS OF VARYING FORMAT AND KIND

P.P. Heymans and B. Kuiper
for the PS Controls Group.

CERN, 1211 Geneva 23, Switzerland

1 Introduction

Accelerator complexes in some form or another exist at a number of laboratories, such as CERN, DESY, FNAL, SLAC, KEK, GANIL, NSLS and others. Firstly, there is the trivial case of tandem machines, in which sub-ranges of the total energy sweep are covered by different accelerators best suited for that range. Typical is the example of the linac/synchrotron tandem. Secondly, in a number of cases the implementation of the next generation of machine at a particular laboratory could be made more economically by using the previous generation accelerator as an injector or, more recently, as a partner device (DESY, FNAL, SLAC, CERN). Thirdly, certain installations have been conceived from the beginning as multi-purpose facilities and use consciously a modular approach to synthesise the required functions while at the same time allowing for a stepwise implementation (GANIL, NSLS, KEK).

Accelerator complexes may be ranked by the intricacy and time density of their beam transactions and by the pressure of their operations. Intricacy increases with the number of machines and transfer channels which the beam traverses and with the number of transformations in time structure and geometry (beam gymnastics) which it undergoes up to its destination. Time density increases with decreasing intervals between transactions and with the number of concurrent ones. Operational pressure increases with the total number of hours run in the year, hence - but far more than proportionally so - with the brevity of start-ups or changes of operational mode and with the diversity and frequency of those changes.

It is obvious that this scale of values must somehow map onto the complexity of the controls problems and this proves to happen in a strongly progressive way. When going upward in this scale one inevitably reaches a point where more explicit provisions have to be made in the controls in order to cope with the complexity aspect of the accelerator facility.

These provisions may be divided into two broad classes: (i) coordination and synchronisation problems between accelerators, and (ii) provisions in order to cope with many data, short intervals and high concurrency (real-time problems).

2 Controls issues in accelerator complexes

The pressure of the numerous physics groups around some of the present accelerator complexes can be high and one can respond to this need by the technique of CYCLE SHARING. In this scheme the various physics groups are concurrently supplied with beams, each group with the particles and beam properties of their choice. These are produced in interleaved cycle sequences which are periodically repeated in a so-called SUPERCYCLE. Each physics group has the impression of being the sole user of the accelerator complex since they receive only their specified beam in periodic bursts at the location of their experiment. They may only be aware of the cycle sharing through the interval between beam bursts, which may match or exceed the time necessary for processing their previous particle burst's data. Control-wise, the problem is complicated by the fact that the beam may traverse several accelerators and different ones for different users. The supercycle therefore looks different for each accelerator in the complex. A situation as described creates a strong pressure for short acceleration cycles and at the CERN PS accelerator complex their duration can be as short as one second. The cited facts have a number of consequences in terms of controls :

(a) In the relevant accelerators of the complex it becomes necessary to change beam properties (intensity, time structure, geometry, kind of particles) from cycle to cycle. This implies cycle-to-cycle refreshment of the process hardware's working registers (if one excludes multiplication of the latter) and cycle-to-cycle processing of acquisitions and refreshment of displays. In the CERN PS complex this technique is in use since 1975 and is called PULSE-TO-PULSE MODULATION (PPM).

(b) The pressure of operations makes that, during agitated times (change of mode, start-up, fault chasing), several operators have to interact with one and the same accelerator. If one excludes multiple dedicated consoles for each accelerator in the complex, this requirement points to general purpose consoles. Furthermore, the quick succession of different cycles imposes temporary assignment of each console to one chosen cycle type, excluding information from other cycles. The operator thus gets the impression of working alone on one accelerator with only one beam, while in reality sharing the same physical accelerator with other consoles working on different beams. At the CERN PS complex this VIRTUAL ACCELERATOR facility is in operation since 1980.

c) Although it would be conceivable to set up the cycle sequences and relevant beam exchanges separately in each of the cooperating accelerators, this would be a highly mistake prone approach since the intricacy of beam transactions makes it near to impossible to oversee the overall context. As changes of supercycles may happen daily or even several times a day, the above approach would reduce the stable operation time to a small portion. It is a much safer approach to compose an overall programme by choice of compatible cycle sequences and their operations from finite sets of previously tried options. This may be done by interaction on synoptic displays that signal or even forbid incompatible or incoherent combinations (intelligent editor). Programs using the edited input must also assure the cycle-to-cycle concertation of operations in the accelerators. In the CERN PS complex this coordinating task is performed by a computer assisted system called the PROGRAMME LINES SEQUENCER (PLS), in use since 1975.

d) Several consoles may interfere by acting on the same parameters within the same cycle. This danger can be eliminated by PARAMETER RESERVATION for the first calling console; the second one can then only acquire but not control that parameter. Parameter reservation has sometime been controversial since costing extra programming effort to implement, some extra interaction response time and an emergency release for hangups. In simple accelerators like the CERN SPS it could be avoided by personal contact of operators. In highly intricate complexes like the PS, operators may lose track of individual parameters of working-sets and more so when interacting by more abstract parameters such as SIN or COS orbit deformations, which is a growing trend. For these reasons parameter reservation is in use at CERN PS since 1980.

e) In each accelerator, one or more sophisticated beam instrumentation systems must be addressed from and return relevant acquisitions to different consoles working on different contexts, often on the same cycle. This problem is further complicated by the fact that each of the contexts may require different options of the same instrumentation system. Potential conflicts must therefore be prevented and arbitrated. In the CERN PS complex this is done by the facility MULTIPLE TRIGGER for MEASUREMENTS (MTIM), in use since 1982;

f) The pressure for strong parallelism requires that the console hardware and software support a substantial number of SIMULTANEOUS interactive programs and displays on the same console.

g) The pressure for short cycle durations creates stringent real-time problems. In accelerator complexes these are severely aggravated by the need of substantially more parallelism in displays in order to keep adequate overview of the intricate context. A number of REAL TIME DEVICES may then become necessary in order to help the application software meet its goals.

3 The PS Accelerator Complex

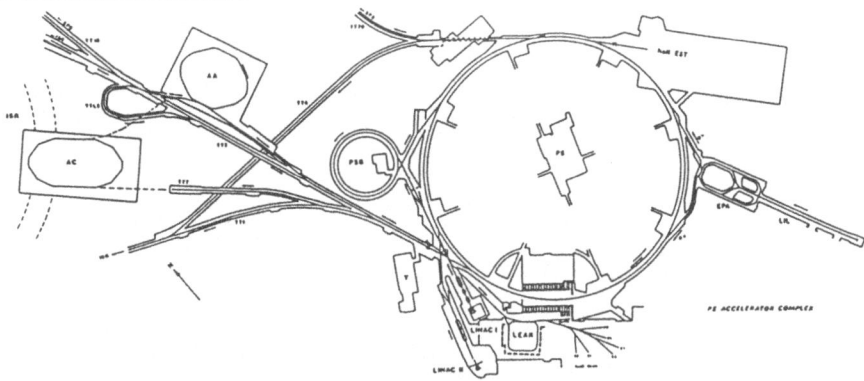

Fig.1 : The PS accelerator complex.

We limit ourselves to a citation from the recent "Woods Hole" conference report [1] : "... °The Proton Synchrotron (PS) ... at CERN, is an immensely versatile machine. It accelerates protons, antiprotons, deuterons, and alpha particles more or less on demand, serving as injector to CERN's higher energy accelerators and storage rings, as well as providing beams for its own fixed-target program. It is being modified to serve as the injector of electrons and positrons for LEP...". A detailed coverage may be found in References [2] [3] [4].

3.1 Beam Transactions

They are best summarised by the synoptic diagram (Fig.2). It represents a fictive supercycle in which all the above mentioned beam transactions are shown together; although fictive on the operationnal side, this supercycle could be implemented controls-wise.

The PS can <u>receive</u> particles from three sources : (i) protons, deuterons or alphas from one of the two Linacs, (ii) antiprotons from the Antiproton Accumulator (AA), and after 1985, (iii) electrons and positrons from the LIL and EPA complex.

It can <u>send</u> these particles to different users : (i) protons to its own 25 GeV physics experiment area (EAST zone), (ii) any particle to one of its beam dumps (internal or external), (iii) protons to the neutrino experiment line, (iv) protons to the Antiproton Accumulator line where they are converted to antiprotons on a target and then accumulated in AA, (v) protons (and/or antiprotons) to one of the two rings of the Intersecting Storage Rings (ISR), (vi) protons (and/or antiprotons) to the SPS, (vii) antiprotons to the Low Energy Accumulator Ring (LEAR), and after 1986, (viii) electrons and positrons to SPS for LEP.

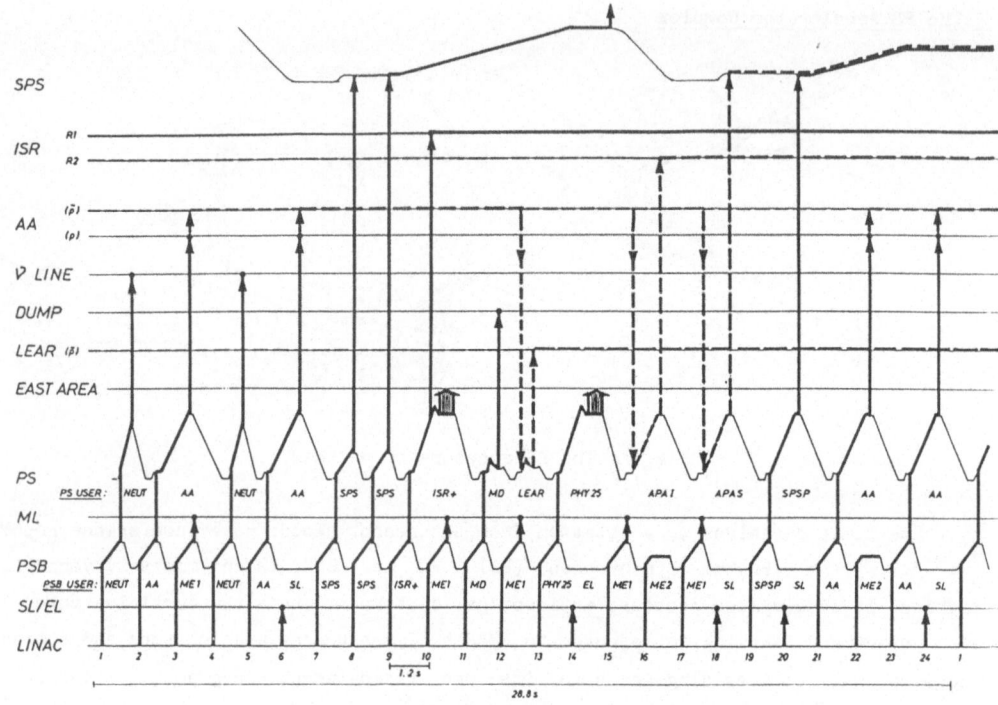

Fig.2 : Beam Transactions in a "Fictive" Supercycle.

In addition, for parasitic Machine Experiments (ME), intermediate Linac pulses may be sent to its own Spectrometer (SL) or Emittance (EL) Lines or to the PS Booster (PSB) which may send it to its Measurement Line (ML) for ME's.

3.2 Controls System Layout

Main operator consoles work on process hardware through a network of ND-10 and ND-100 minicomputers[5]. Communication is via an SPS type packet switching system, using a central store-and-forward message handling computer (MHC) in a star configuration.

The controls system has thus a process oriented part and a common, i.e. operations and systems oriented part. The structure of the former follows the one of the process and there are no general purpose computers like in SPS. The use of serial CAMAC makes this possible since computers do not have to be geographically distributed. Consequently the consoles predominantly communicate with one process computer at a time although there may be a growing number of exceptions to this as the accelerator complex grows more intricate.

The short cycle times exclude massive program file transfers over the network and require that a large fraction of the application programs be compiled. Thus there is no use for a library computer like in SPS and program files are stored on the relevant computers' discs. There is further decentralisation of files and processing into the microprocessor based Auxiliary Crate Controllers (ACC) in near to every CAMAC crate. They handle time critical transactions such as cycle-to-cycle refreshing of process parameters and buffering and preprocessing data bursts from the beam instrumentation.

Synchronisation is twofold : by computer settable preset counters on pulse trains from clocks and stepping integrators of the magnetic field, and by so-called program lines, i.e. serial telegrams containing information about the imminent cycle and the next one. The serial telegrams are distributed by the Program Lines Sequencer (PLS) to the interface and computers.

On the common side, the CONSole computers for the main operator consoles, assisted by microcomputers for displays. The TREES computer manages the data-bases for the touch-panel trees as well as for the so-called working-sets. These are sets of process parameters relevant to typical operational contexts and reserved when calling the latter. The computer also handles alarm messages and drives the analog Signal Observation System (SOS) through CAMAC. The MCR computer drives the logging printers, the graphics plotter and handles tasks like the link with the CERN main computing centre.

As mentioned in section 2, the new PS control system includes a number of devices to cope with the specific problems of interacting accelerator complexes. Below these are discussed under two headings : coordination and synchronisation on the one hand, and software devices on the other.

3.3 Coordination and Synchronisation

The coordination of the accelerators and their operations is done by a dedicated master programmer, called Program Lines Sequencer (PLS) [6]. It determines what control values the relevant pieces of process hardware will use in each cycle. This includes the special case of simple gating. The PLS does this by referring to preprogrammed options (coherent sets of values) through messages encoded in the so-called PLS TELEGRAM, which is distributed to relevant process hardware, interfacing hardware and computers.

Synchronisation between processes is by exchange of pulses, derived from pulse trains via presettable counters started by key pulses from the process. The values set in these counters are in turn subject to the PLS telegram's message.

Due to the important number of concurrent physics experiments and the frequent changes of beam sharing between them, including unscheduled ones in response to temporary inability of one experiment or beam facility, the coordination system must be extremely flexible. The coordination includes also a safety aspect handled by including so-called external conditions which, if not fulfilled, may stop or modify the programme through modification in real time of the PLS telegram. SPARE USERS are preprogrammed and may be then given beam automatically instead of the normal user if the latter is unable to receive it.

A PLS telegram is a serial message of 256 bits, containing information about the imminent cycle and also about the next one (for equipment needing advance notice). The telegram contains all necessary information for selection of the relevant sets of parameters in the PPM scheme (see sec. 3.4), i.e. magnetic field cycle, intensity level, RF harmonic number, beam destination, etc. The PLS telegram is derived from data on normal and spare user, entered interactively at the console. It is generated by a set of real-time tasks in the PLS computer, on receipt of an interrupt, every Linac pulse and every supercycle.

The telegram is subject to "external conditions" from (i) the process (e.g. a switching magnet ON or not), (ii) the security chains, or (iii) "no beam request" from a user. Conditions are interrogated before each cycle and if not fulfilled, the PLS will try to set up a telegram for a pre-programmed spare user, or otherwise dump the beam.

Creation and modification of sequence is done interactively from the Main Consoles. It is possible, through the main tree touch-panel, to set any "global" information (see below) about the accelerators' behaviour, to create a new supercycle or to modify an existing one; to store it into an archive or to retrieve it, possibly to modify the archive, and to send it to the PLS working data-tables. One may also read and modify the influence of external conditions affecting the PLS telegrams.

Firstly, the accelerators to participate in the supercycles to be created, and how they participate, is entered interactively on the graphics sreen. Secondly, the operational characteristics of each different user for each accelerator are set interactively on the colour screen (Fig.3). When doing this on a running sequence, the changes will be sent also to the on-line data-tables.

In order to create a supercycle, the beam users are then entered successively, the Linac pulse counter being automatically incremented according to the corresponding PS cycle length. If the operational mode corresponds to a PSB injection at 800 MeV into the PS, then the PS user is automatically copied to the corresponding PSB cycle, and the PSB cycles not used by PS are available for PSB Machine Experiments (ME1 or ME2) or the Linac beam is routed to its own Spectrometer Line (SL) or Emittance Line (EL): see Figure 2.

USER MODIFICATION - PS 1983/SEP/18-09:12:13

	SPS	ISR	AA	APTST	MD	PHY25	APA	LEAR
CYCLE TYPE...	A	B	C	C	B	B	C	D
HARMONIC NB..	420	20	19-21	6	20	20	6	10
LE WORK POINT	LEHIG	LELOW	LEINT	LELOW	LELOW	LELOW	LELOW	LEDEC
HE WORK POINT	HEA	HEB	HEC	HEC	HEMD	HEB	HEC	NOHE
INJECTION....	800MV	800MV	800MV	800MV	800MV	800MV	800MV	800MV
LINAC DEST...	PSB	PSB	PSB	PSB	PSB	PSB	.....	.....
SIM:IP=0(Y/N)	N	N	N	N	N	N	N	N
PBAR INTENS	-	-	-	-	-	-	-	-
INTERNAL DUMP	-	-	-	-	D93	-	-	-
EJ16 DEST	FTS	FTI	FTA	ATPP	-	-	ATPA	ATPA
MISC OP 1	-	-	-	-	-	-	-	-
MISC OP 2	-	-	-	-	-	-	-	[illegible]
OPERATIONS	-	FE16I	FE16A	[illegible]	-	[illegible]	[illegible]	-
	-	-	-	[illegible]	-	-	[illegible]	-
	-	-	-	-	-	-	-	[illegible]
	[illegible]	-	-	-	-	-	[illegible]	[illegible]
DMP 93 TIMING	0	0	0	0	0	0	0	0
DMP 97 TIMING	0	0	0	0	0	0	0	0

Fig.3 : User characteristics.

```
PS - CYCLE MODIF.(800MEV)        1983/SEP/18-09:09:02
PS CYCLE NO:   5   SAME AS >              LINAC PULSE NO:  7
NORMAL PROGRAMME:                  SPARE PROGRAMME:
         USER >ISR                          >PHY25
     H.E. OPS ->FE16I                        SE62
              -/-                            FEKIL
              -:-                            -
              -:-                            -
     MISC OP1 -:-                            -
     MISC OP2 ->-                            -
    EJ16 DEST ->FTI                          -
    PBAR INT  ->-                            -
     INT  DUMP->D93                          -
  DUMP 93 TIM->  660
```

Fig.4 : On-line Modification of the PS supercycle.

Modifications of the sequence (Fig.4) can be made on-line any time during operation : the user, the spare user, or some of the operations involved by these users (Fig.3) may be changed interactively after reading the actual situation.

A repetitive display of the PLS on-line conditions, refreshed at Linac pulse rate, is available on request on the colour-TV screen and on the Video distribution network. This gives immediate information about elements in bad state, leading to user substitution (spare user) or dumping the beam.

There exist programs for so-called global operations on the PLS functionning : (i) SET-UP reloads from disc all the data required by the sequencing (data-tables, running of the PLS real-time tasks, PLS-decoder settings, specific hardware, clock delays, etc.), (ii) PAUSE makes the corresponding accelerator refuse the beam, (iii) FULLSTOP stops the beam in all accelerators; (iv) another facility permits quick exchanges of 2 sets of supercycle data between 3 disc buffers and the working buffer.

A number of specialists programs allow : (i) changing the PLS information names or logic; (ii) changing the external condition matrices; (iii) displaying the internal PLS tables driving the supercycles; (iv) archiving actual PLS supercycles, displaying chosen archives, modifying them, or loading the PLS computer with a previously archived PS supercycle.

Two safety devices may be mentioned : (i) any modification made to any of the PLS information is first saved on disc before being sent to the computer memory. This feature allows for recovery in case of a computer crash or power-fail; (ii) secondly, a special PLS reservation scheme, as distinct from the working-set reservation mentioned in sec.2-d, exists to prohibit two operators at two different consoles to interfere and modify concurrently any bit of the PLS sequence or condition.

3.4 Software Issues

<u>Pulse-to-Pulse Modulation.</u> (PPM) Microprocessor based Auxiliary Crate Controllers (ACC's) in nearly every CAMAC crate contain data-tables relevant to each cycle of the supercycle. Each table contains the control values of the operations programmed in that cycle as well as locations for the acquired values of the same. The control values of each table are transferred to the working registers in the interfacing CAMAC modules at specific instances, upon receipt of interrupts by key timing pulses in the cycles. The table is for each cycle chosen according to the PLS telegram received. The latter is decoded in a CAMAC module PLS-RECEIVER which is interrogated by the ACC before the transfer. Acquisitions are made and placed in the table, following other interrupts, relevant to the process in question. Control values, labelled with one of the PLS conditions, are entered asynchronously into the Equipment Modules (EM)[7] in the relevant minicomputer. Acquired values are <u>likewise</u> requested through the EMs. The PPM scheme is described in detail in[8].

<u>Virtual Accelerators.</u> The cycle to which the console will be locked is first chosen as an option on the main tree touch-panel. This results in an automatic labelling of all control values set from the console and of the acquired values requested. Since the cycle type and not the cycle number is set as label, the console will set all cycles of the same type and receive refreshed data from all cycles of that type, as long as there is no overload. The operator is protected by the parameter reservation scheme against interference by other consoles in the chosen cycle.

<u>Parameter reservation</u>[9,10]. Parameter reservation in PS is per so-called Working-Set, i.e. a set of process parameters whose use has proved typical in a particular operational context. When going down the touch-panel tree, the context and thereby the working-set is chosen at a level below which the relevant parameters are looked up in a data-base located in the TREES computer, whereupon control but not acquisition of these parameters is blocked for other consoles. In order to also protect against interference from terminals directly connected to the process computer, the blocking is done at the Equipment Module level. When going up the TREE, the parameters of the working-set are released. There is a separate emergency release for the case that the regular one fails.

<u>The MTIM facility</u>[11]. As a further refinement of the virtual accelerator scheme, one may work with several consoles close to independently on one and the same beam measurement system inside the same accelerator cycle. This scheme allows to select and reserve different sets of measurement instants of one and the same instrument while the respective measured data are returned to the requesting console as if each

of them were the sole user of the instrument. A condition is that the interval between the first and last chosen instant for one console does not overlap with the corresponding interval in the other consoles. The facility, which solves some stringent real-time and arbitration problems, is structured as a general purpose framework so that it may be used for a variety of beam measurement instruments. It is successfully in use for ten different instruments.

Concurrent Interaction and Display [12]. The systems software in the console supports up to 5 interactive programs concurrently. It also supports the possibility of sharing the same physical screens between several independent application programs. The colour-TV and graphics screens are divided in a number of zones which the programmer may combine or use separately. There are independent and protected software access channels, which allow an application to run on part of screen without perturbation of concurrent programs, even interactive, sharing that screen.

Some real-time devices [13]. Transmission of the intense data flow for concurrent refreshed displays at cycle times of the order of one second has problems of its own. The PS Control system has dedicated software mechanisms for limiting the transmission load. This mechanism exists at two levels : in the process computers and in the console computers for image generation.

4 Conclusions

Complexes of interacting accelerators, if used intensely with frequently changing programmes, have controls problems not normally occuring in simple accelerators or tandems. These relate to : (i) coordination in real time of operations in the cooperating accelerators, (ii) fast and reliable composition of overall operational programmes, (iii) efficient switchover from one such programme to another, and (iv) strong concurrency hence real-time problems. Such complexes therefore benefit from controls structures deliberately laid out to cope with their particular requirements.

Amongst the world's accelerator complexes, the CERN PS occupies a singular place by its high versatility and intricate beam transactions and by the pressure of its operations. The problems (i), (ii) and (iii) are solved by the PLS system. The problem (iv) is dealt with by powerful console hardware and software, by distributing intelligence in the process interface and by sophisticated real-time software devices.

The PPM scheme with its centralised coordination by the PLS makes control of the complex homogeneous. Central interactive supercycle composition allows to oversee the whole context. The PLS application programs give great flexibility for the many changes in the operational schedule and make switchover efficient. This centralised coordination approach also works for reliability. The strong parallelism of the consoles and the concurrency of the applications software are indispensable for change of context and trouble shooting.

In the coming adaptation of the PS accelerator complex to its function of an electron/positron injector to LEP, the PLS system might be extended to at least 16 users, some historical compromises will be corrected and some improvements suggested by operational experience will be included. More powerful microprocessors will be introduced at the process side and at the displays, and the existing software will be further decentralised towards the microprocessors in the interface. The overall solutions as described above would, however, all be conserved.

References

1) US Department of Energy. Report of the 1983 HEPAP subpanel on New Facilities for the US High Energy Physics Program. Washington D.C. 20545 July 1983.

2) The PS staff, presented by R.Billinge, The CERN PS Complex : a multipurpose Particle Source. 12th Int.Conf.High En.Accel. 1983 Fermilab. (CERN/PS/83-26).

3) R.Billinge and E.Jones, The CERN Antiproton Source, 12th Int.Conf.High En.Acc. 1983, Fermilab (CERN/PS-AA/83-25).

4) LEP Injector Study Group, The Chain of LEP Injectors, IEEE, Trans. Nucl. Sci., Vol.NS-30 nr.4, p.2022 (1983).

5) G.Baribaud, S.Battisti, G.P.Benincasa et al. The Improvement Project for the CPS Controls. IEEE, Trans. Nucl. Sci., Vol.NS-26 nr.3, p.3272 (1979).

6) J.Boillot, G.Daems, P.Heymans, M.S.Overington, Pulse-to-pulse Modulation of the Beam Characteristics and Utilization in the CERN PS Accelerator Complex. IEEE, Trans. Nucl. Sci., Vol.NS-28 nr.3, p.2195 (1981).

7) A.Daneels, E.Malandain, M.Martini, and P.Skarek, Standard Software Modules for Equipment and Composite Variable Control. PS/CO/WP/83-77 submitted to IFAC 84.

8) G.P.Benincasa, F.Giudici, and P.Skarek, Fast Synchronous Beam Property Modulation using a large Distributed Microprocessor System. IEEE, Trans. Nucl. Sci., Vol.NS-28 nr.3, p.2192 (1981).

9) J.Boillot, M.Boutheon, D.Dekkers et al. Operation Oriented Computer Controls for the CERN PS Complex. IEEE, Trans. Nucl. Sci., Vol.NS-28 nr.3, p.2261 (1981).

10) D.Heagerthy, P.Heymans, J.Kenaghan, Ch.Serre, Interactive Control of the CERN Proton Synchrotron Complex. PS/CO/WP/83-79 17/6/83 submitted to IFAC 84

11) G.P.Benincasa, F.Giudici, and N.Vogt-Nilsen, A Multi-user Microprocessor-based Meas. System for the CERN PS Accel.Complex, PS/CO/WP/83-76, submitted to IFAC 84

12) F.Perriollat et al. Les Consoles Centrales du Nouveau Systeme de Controle du PS. CERN/PS/CCI/Note/77-28 1977.

13) L.Merard, Th.pettersson, Ch.Serre, Concurrent Execution of Real-Time Displays. PS/CO/WP/83-78 17/6/83 submitted to IFAC 84

INTEGRATED CONTROL AND DATA ACQUISITION OF
EXPERIMENTAL FACILITIES

F. Bombi, JET Joint Undertaking, Abingdon (Oxon, UK)

Introduction.

JET is the largest single project of the co-ordinated nuclear fusion research programme of the European Atomic Energy Community (EURATOM) aimed at proving the feasibility of nuclear fusion as a new energy source [1]. The JET machine was successfully started in June of this year after the conclusion of a five year construction phase [2]. The experiment is using the tokamak magnetic field configuration and will have a greater performance capability than any other machine of this type in the world. Size and complexity of the machine, its large pulsed power supplies, additional heating systems and sophisticated diagnostic measurement systems require a comprehensive and high performance control and data acquisition sytem.

This paper describes the architecture of the distributed computer system and of the interface electronics used to connect the computer with the experimental facility. The essential software components and their engineering are also described, and the solution given to the problem of safety in the complex distributed system is discussed. Finally cost, manpower and implementation timescale are analysed.

1. Objective and Basic Design Features.

The main objectives of the JET Control and Data Acquisition System (CODAS) can be summarised in the following points:
  - centralised control and monitoring of all actions to be performed during normal operation;
  - capable of handling a large number of data, up to $10^{**}5$ engineering data per pulse and in excess of $10^{**}6$ plasma diagnostic data per pulse;
  - availability during commissioning of the apparatus with the capability of independent operation of each subsystem;
  - extendibility with regard to future demands;
  - safe and reliable in the sense that pulsing should be prevented in the presence of faults or operator errors and automatic shut down should be initiated by a fault in an essential part of the plant.
To meet these demands CODAS was conceived with the following main features:
  - use of computers of the same family for control, monitoring, data acquisition, storage and analysis;
  - distributed system with a high degree of modularity and capability of autonomous operation;
  - use of CAMAC as a computer-independent standard interface to the process and to the operators´ consoles;
  - integrated operation through three communication paths: computer communication system, Central Interlock and Safety System, Central Timing System;
  - integrated acquisition and archiving system to file and retrieve all acquired data which are stored together with the machine settings;
  - connection to the computer facilities of the Host Laboratory for off-line data storage and analysis of experimental results.

The system is logically organized alongside the hierarchical structure depicted in Fig. 1. At the top level (Level 1 or Supervisory Level) a number of computers are in charge of the overall experiment supervision, co-ordinating the activities of a number of other computers, each in charge of the data acquisition and the control of a functionally homogeneous part of the experiment called a subsytem. The collection of all the subsystems constitute the next layer in the system hierarchy (Level 2 or Subsystems Level). Each subsytem is, in turn, composed of a number of local units at the lower level in the hierarchy (Level 3 or Local Units Level).

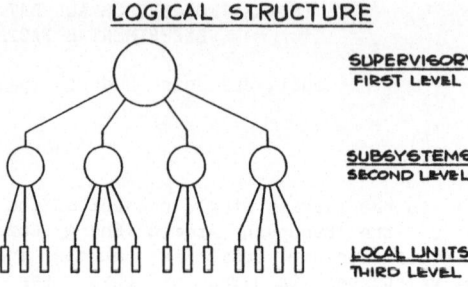

Fig. 1

Communication in the system follows the hierarchy in the sense that interaction between local units in a system is achieved through the subsytem computer and co-ordinated operation of the subsystem is the responsibilty of the supervisory level.

This architecture has the advantage of being simple and easy to understand. It is open ended in the sense that more local units can be easily added to a subsystem and more subsystems can be added to the complete system. It is particularly well suited for a step-wise integration. In the case of JET this has been achieved starting with the commissioning of the local units followed by their integration in the various subsystems. Each subsystem has been in turn integrated into the supervisory level obtaining a complete working environment. A top down integration could have also been proposed from the beginning if sufficiently clear goals for the overall operation had been defined early enough, a top down approach will certainly be used in the future for the addition of new subsytems while the experiment is in operation.

The major disadvantage of the hierarchical architecture is related to the inherently unpredictable response speed to the commands sent through the computer communication and the insufficient reliability inevitable in a high complexity distributed system. Both problems were clear at the design phase and have been overcome by the use of separate paths to distribute timing signals (Central Timing System) and of an independent high reliability interlock (Central Interlock and Safety System).

2. Computer network

Although the system is functionally and logically organized in a hierarchical structure, a symmetrical double-star network is used to provide communication paths between the computers (Fig.2). The computers at the centre of the network (hubs) have the sole role of switching messages to the proper destination. The initial design of the network included two hubs in consideration of throughput and reliability factors, but the use of two computers proved also convenient from the operation point of view, as it allows the splitting of the network in two independent

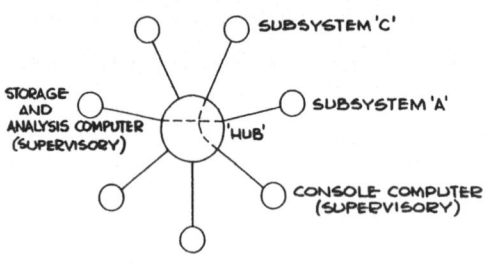

Fig. 2

halves - one on-line and the second used for software and hardware development.
Links within the network are implemented using modified HDLC boards (MEGALINK) [3] capable of operating at the speed of 1 Mbit/s in full duplex. Off-site communication with the Host Laboratory IBM-CRAY relies on a standard HDLC link operating at 300 kbit/s over a 2.048 Mbit/s British Telecom digital line. The network uses standard Norsk Data software (XMSG-M) which provides task to task communication between remote machines. The basic vendor-supplied software is supplemented by a JET developed layer which provides fully transparent communication services to the user programs. The system can handle short high priority messages (used for control purposes) and long messages (used for high throughput data acquisition). The communication software is protected by a comprehensive end-to-end protocol which prevents global crashes in the event of network overload or localized computer failures.

The time required to send a short message between two tasks active in remote machines is of the order of 10 ms, typical throughputs of the order of 40 kByte/s can be achieved during file transfer, which involve three computers (source, destination and one hub).

3. Subsystem structure

Subsystems are designed in order to be able to operate either independently or connected to the supervisory level. Each subsystem is equipped with a computer and a CAMAC serial loop. Two different configurations are used: for the control subsytems 16 bit ND-100 computers are used with .5 MByte of main memory and one 75 Mbyte disk drive, 32 bit ND 500 computers with 1.25 Mbyte of main memory and two 75 MByte disk drives are used for the diagnostic subsystems.

Between 3 and 12 crates are used in each subsystem in accordance to its complexity. The highway, which connects the crates, is driven by a fast driver board housed directly into the computer bus. The driver can provide concurrent DMA and PIO transfers, in DMA the driver can fully exploit the highway speed operating in pipelined block transfer at 5 MHz. The serial highway driver in conjunction with the handler implements most of the error detection and correction procedures recommended by the ESONE Committee [4].

To overcome long distances and the high electromagnetic noise environment typical of the JET experiment, the serial highways are implemented with fiberoptic links through the use of U-port adaptors. The adaptors provide facilities to bypass each crate, to select the main or the back-up loop and to loop collapse partially incomplete or faulty loops. Each U-port adaptor is equipped with an auxiliary D-port normally used to drive a mobile operator console during commissioning runs.

CAMAC crates are housed in cubicles of standard configuration, each cubicle providing interface to one or more local units in accordance to their geographical distribution. Plant cables are terminated into screw terminal strips and then routed to signal conditioning modules built in Eurocard mechanics. Signals are finally connected to the CAMAC modules (Fig. 3).

The majority of the circuits carry on/off control and monitoring signals which are conditioned by the Line Surveyor Driver (LSD) modules. An auxiliary HL Logic bus drives the LSD modules which provide isolation, fan in and fan out to a standard CAMAC module. Each LSD module appears as a sub-address of the driving CAMAC module.

A number of TMS 9900 microprocessor based auxiliary controllers are used as programmable function generators, to drive large relay multiplexers, to provide fast real time control actions and to interface non standard equipment.

4. Basic Software structure

The software used in the system is based on the standard Norsk Data operating system Sintran III (the release H is used at present) with minor "patches". The most noticeable addition to the system is the CAMAC serial driver handler, developed by Nork Data UK against JET specifications. On top of the basic operating system the application programs rely upon the use of a real-time database and a set of CAMAC drivers. The real-time database holds in a table called hardware tree, the description of the crates connected to each computer. In conjunction with the appropriate driver routines it allows access, using a symbolic name,

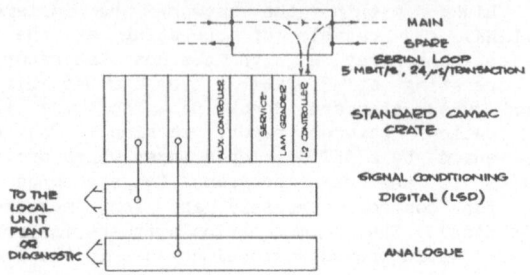

Fig. 3

to an input or output line (a point) and it defines access rights, calibration factors, etc. The hardware tree is converted, at load time, into a memory resident table used to translate, with the help of a disk resident point names dictionary, symbolic names into physical addresses. Point names are unique through the system and application programs can access with the same type of calls both local and remote points.

The next layer of software can be divided into two categories, namely, the programs used to operate the experiment facility and the auxiliary plant on a continuous basis, and those used to manage each experimental pulse. In the first group we can name: MIMIC, the console package, the Plant Status Image program, Continuous Monitoring Program and the Alarm Handling package. MIMIC provides graphic colour pictures dynamically updated as a function of the status or value of a defined set of points. Together with the touch panel and the trackerball handler they constitute the main operator/machine interface. The Plant Status Image program manages points in the computer memory, either the image of plant variables or the result of a software action, eg limit checks. The Continuous Monitoring System links dynamically actions to specific events, in conjunction with the other products named allows the majority of control requirements to be resolved without the need for special purpose software.

The pulse related software is constituted by the suite: General Acquisition Program (GAP), Pulse File Manager (PFM) and Immediate Pulse File Program (IPFP). GAP and PFM provide a generalised facility for the setting up of an experimental pulse and for the subsequent data acquisition. The package is driven by a database which describes the actions to be performed before and after each pulse. Before the pulse, to prepare the files required to store the experimental data and to initialize the CAMAC interface to the conditions required for the pulse, and after the pulse to retrieve the experimental results from the interface memories and accumulate them in a disk file (JET Pulse File).

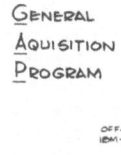

The creation of a central JET Pulse File is achieved in two stages (Fig. 4). First the experimental data are collected in a file at the subsystem level. In a second phase the subsystem files are transferred in a global file at the supervisory level. Two identical copies of the file are created simultaneously, one in the Storage and Analysis computer and the second in the remote IBM-CRAY mainframe used for off-line data analysis and reduction.

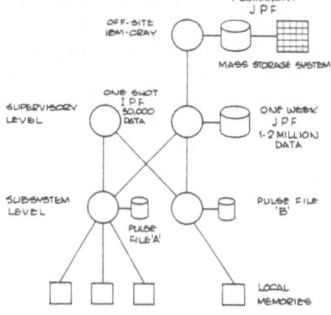

Fig. 4

The Author aknowledges the high professional skill and dedication of the CODAS Division staff and of the contractors working at the design and implementation of the system who made possible its timely start in June 1983.

----

[1] EUR 5791e (EUR-JET-R8): The JET Project - Scientific and Technical Development 1966, Published by The Commission of the European Communities (1977).

[2] P.H. Rebut and B.J.Green: Status and Programme of JET, to be published on Plasma Physics (Jan 1984).

[3] ND-12.018: HDLC - High Level Data Link Control Interface (Nov 1978).

[4] ESONE/SD?02: Recommendations for CAMAC Serial Highway Drivers and LAM Graders for SCC-L2.Commission of the European Communities (1977).

SOFTWARE ENGINEERING TOOLS

Roman Zelazny
RCCAE CYFRONET, 05-400 Otwock-Swierk, Poland

>...The tools we are trying to use and the language or notation we are using to express or record our thoughts are the major factors determining what we can think or express at all!
>
>The Humble Programmer
>E.W. Dijkstra

INTRODUCTION: LIFE-CYCLE MODEL AND ITS TOOLS

The invention and usage of tools created a man, made him "a professional". Every profession has its tools of trade, which are the products of long development and evolution.
Such situation exists also in the field of software development. After the initial period of enchantment and magic spell caused by the first successes of programming there came a period of reflections and criticism. Some of the confronted problems were connected with the following phenomena plagueing the computer applications [1]:

cost/schedule overruns,   maintenance difficulties,
poor visibility into development status,   inconclusive verification,
unreliability,   inadequate or nonexistent documentation

Complete change of goals and demands with respect to programmers and software developers have transformed the field into a discipline of software engineering, which undergoes the process of continuous development and refinement. A more systematic analysis of the software development and the maintenance process disclosed that this activity can be divided into phases with reasonably well formulated inputs and outputs. Although there are slightly different formulations of such a division due to different authors, however to big extent, the common consensus casts the software development and maintenace process into the following life-cycle phases:

1. Requirement analysis and definition   4. Implementation (coding)
2. Architectural (global) design   5. Validation and verification
3. Detail design   6. Operation and maintenance

Generally speaking the first phase should, on the basis of end-user requirement formulation and analysis, end up in the requirement documentation (specification), specifying the end-user contractual wishes with respect to the ordered software system (its functional and performance capabilities).The next phase, the global architectural design, should analyse and produce the global functional structure of the systems (model), which is to fullfil (achieve) the requirements (specifications). The detail design phase should be devoted to the analysis and specification of functional objectives of components (modules) together with their interface characteristics. The next phase, coding or implementation, should, on the basis of the former phase documentation (specification), produce source code implementation of components (modules) and the whole system. The validation and verification phase checks the consistency of individual modules and the whole system with the design and requirement specifications from the point of view of functionality and performance criteria. Output of this phase is the adequately documented system, ready for operation and maintenance.

It is again a common consensus that the careful management throughout the life-cycle is critical to the success of any software project. Although teams working on software projects should be as small as possible and consist of the best professionals contemporary software, due to its size and complexity, is being produced collectively. Being an intangible and collectively performed product software must be carefully managed. Unfortunately software mangement is not yet well developed and understood. Again software management requires the tools and visibility. Part of this is to be accomplished by imposing discipline, standards, documentation and verification procedures. To assist in all those activities during the life-cycle phases and management of software projects a big amount of various tools has been created and developed. They can be categorized as cognitive, augmentive and notational tools. Cognitive tools enhance the intellectual capabilities of software developers providing problem-solving techniques, such as hierarchical decomposition, information hiding, structured coding etc. Augmentive tools increase the practitioner's "power". Usually they are software systems themselves and include such implementation-oriented tools as compilers, text editors, debugging packages as well as tools for the pre-implementation (requirement definition and design) and post-implementation (verification, testing and maintenance) phases. Notational tools, i.e. languages, modelling formalisms, provide media of expressing and communicating ideas, concepts, structures, processes, relations. The lists of such tools used in various phases of the software development life-cycle and management supervision can be found in [2].

Soon it was realized that the variety of tools and methods needs an integration in a methodology, it means creation of software development environments or software development systems. There is a number of requirements or desirable characteristics, which are to be met by a methodology , if it is to enhance the productivity and effectiveness of the software development activities. Let us formulate after [3] some of them:

1. The methodology should cover the entire software development life-cycle.
2. The methodology should facilitate transitions between all phases of the cycle (both forward and backward).
3. The methodology must support achievement and determination of systems correctness throughout the development cycle from the point of view of functionality and performance of the system.
4. The methodology must support the software development team, organization and management, enhancing communication within a team and with the external world, creating the visibility in the extent of progress, remaining tasks, various system characterisics, even those which were not explictely provided in the course of development.
5. The methodology must be able to cover a large class of software projects.
6. The methodology must be user friendly minimizing training (learning efforts) for all parties involved.
7. The methodology must be supported by automated tools, which are integrated in a uniform, compatible and flexible environment.

To make these concepts feasible, there is again a common consensus, that the basis of the software development system (environment) must be a data base with a proper data management system, sometimes called software engineering data base. In this data base all relevant information concerning the whole software development phases is performed with the assistance of respective methods and tools using effectively the content of data base.

The possible architecture of such a software development system is presented schematically in the Fig. 1. It is based on the notion of four processors or front-end "machines" corresponding to the distinct four phases of the development process: requirement definition, design, coding and verification. These processors allow the developers to formulate their concepts and plans in a more or less formalized language and to perform a suitable analysis to end-up with a more or less formalized specification of requirements, global design architecture, system components (modules) and codes (programs) which are finally the basis for validation and verification process.

According to other views [1] the process of verification and validation should permeate all activities of the development process. Particularly transitions from one phase to the other require the verification and testing (validation) to check whether the considered phase fullfilled the objectives formulated not only by the previous phase with its respect but also by the former ones. The visibility of achieved objectives and testing of performance constraints imposed on the problem solution is in all software projects, particularly large ones, the utmost vital goal of management procedures. This point of view considers, that validation and verification cannot be treated as a separate life-cycle phase, but should be a part of each of the life-cycle phases, performed during the transition to the next phases. These concepts are illustrated by Fig. 2 and Fig. 3.

PROGRAMMING SUPPORT ENVIRONMENTS (PSE)

Let us now discuss problems connected with integration of tools into an environment or a system and the ideas and examples of realizations which encompass the whole life-cycle model of the software development process.

Historically speaking the first attempts to create the environment of development tools were based on the implementation (coding) phase concepts. Using operating system and programming language capabilities with a couple of such tools as text editor (formatter) one can control the source code versions and produce relevant system documentation. A notable example is the UNIX-system based software environment, which is better called programming support environment, as well as the Programmer's Workbench based on the UNIX-system [4]. The great advantage of the UNIX-system is its capability to hook processes together using the pipe mechanism combined with the notion of standard input and standard output. The operating system manages the buffering and control necessary to feed the standard output of the first process as a standard input to the second process. This capability allows to link many existing or newly prepared tools (programs) to create a larger system in an easy and convenient way. Such philosophy to "think small" is a basis of enormous power and effectiveness of such environments. One must however admit, that the UNIX-system has been created rather for experienced and highly sophisticated programmers, which are able to use it very effectively for well conceived technical tasks. The easiness of creating, on the basis of the UNIX-system, a more "integrated" programming and documentation environment is again demonstrated by SOLID, a SYSTEM FOR ON-LINE INFORMATION DEVELOPMENT [5]. It allows archiving all the source for programs and documents, executable programs and formatted documents and translating via procedure-generation all types of source into final products. The pre-implementation activities based on different methodologies and metaphores must be performed manually and incorporated in a way dependent on personal preferences, experiences and views. Analogous plans based on the much higher level language ADA are well known under the cryptonime STONEMAN [6]. The ADA program support environment (APSE) is to be developped for embedded computers. This fact is of importance for contemporary physical experiments on the beam of accelerator particles and accelerator control systems, where computer systems are also embedded in larger physical environments.
The APSE must provide a well-coordinated set of useful tools with uniform inter-tool interfaces and with communication through a common data base, which acts as the information source and product repository for all tools. Those tools should be composable, user selectable and able to communicate through the common data base.

The environment should facilitate the development and integration of new tools and the improvement, updating and replacement of tools. The system shall provide a helpful user interface, satisfying the human engineering requirements, both for interactive and batch users. Communication between users and tools shall be according to uniform protocol conventions. Whenever possible the concepts of the ADA language should be used in APSE allowing to use it as a command language. This is to be achieved through lower levels within the APSE; the kernel (KAPSE) and the Minimal toolset (MAPSE), Fig. 4. Those levels can be characterized, following [6], as follows:

Level 0: Hardware and host software.
Level 1: Kernel ADA Program Support Environment (KAPSE), which provides data base, communication and runtime support functions to enable the execution of an ADA program (including MAPSE) and which present a machine-independent portability interface
Level 2: Minimal ADA Program Support Environment (MAPSE), which provides a minimal set of tools, which are both necessary and sufficient for the development and continuing support of ADA programs. These tools will be written in ADA and supported by KAPSE.
Level 3: ADA Program Support Environments (APSEs), which are constructed by extentions of MAPSE to provide fuller support of particular applications or methodologies.

This model provides a consistent user interface through the KAPSE, which defines the host system interface. Additional tools, written in ADA, can easily be added and subsequently transported to another APSE. Thus the KAPSE may be viewed as virtual machine for ADA programs, including tools written in ADA. The presented model emphasizes the support of the implementation phase of the software project, whereas the support to earlier phases is mostly textual. The power of a language is not utilized to the full extent. However, the concept of a package or module and the distinction between its specification and implementation (body) allows the developer to make at least some conceptual steps towards detail and global design. The possibility to generate the structural elements of a system creates new capabilities, which can be used to develop new tools, as for example the System Composition and Generation Tool of the Gandalf system [9]. However, more is to be expected. An analogous approach of creating a program development environment centered around one single language, this time the system implementation language CDL2, is the so-called CDL2 Laboratory [10], developed at the Technical University of Berlin and commercialized by Epsilon. The language and Laboratory have the following properties, which are relevant for our discussion:

- CDL2 programs are created with the usage of the hierarchy PROGRAM/MODULE/LAYER/SECTION/PROCEDURE/CALL/PARAMETER/OBJECT.
- The EXPORT/IMPORT interface at the module level and the EXTENSION/ABSTRACTION interfaces at the section level.
- The same language serves as a command language (more complex commands may be written as CDL2 procedures).
- Logging, filing, editing, compiling and debugging facilities are merged into a uniform development process, controlled by one simple, high level language.
- The information available in the data base forms a hierarchy according to authorship and language contructs.
- It deals with the source text and interfaces as well as annotations to the text. Input retrieval and manipulation of information is supported by a general selection mechanism and by options. The mechanism is entirely in terms of the language and the software.
- There is a provision for separate compilation of modules and integral compilation of programs as well as any intermediate forms.
- Reanalysis of modules and programs can be based on incremental analysis of sections.

The Fig. 5 gives an overview of the components of the laboratory. All presented examples attacked the problem from the implementation (coding) concepts and tried to extent their capabilities toward design phase (and possibly towards testing and verification activities). The higher level programming languages constructs make

FROM THE REQUIREMENT FORMULATION

There are views that the whole problem should be attacked in a more systematic way starting from the requirement analysis and definition, which should be machine-readable and allows for automated analysis. Informal formulation of the requirements in natural language introduces ambiguities and misinterpretation. It is worth while to mention only the early attempt [11] to create a system, called SAFE, which could transform informal specifications into formal ones through context match with interactive participation of the user. Further results of this attempt are not known to the author. The general approach is then to formulate the requirements in a more or less formalized language with specific syntax and semantics. Thus to formulate "what" is to be achieved by the information system some kind of modelling formalism must be applied, which allows us to discribe this part of the "real world" or the "universum of discourses" which is of interest. Basic components are "things", "processes" (entities), relations among them, and their "properties" (attributes). Thus we have arrived at the ERA (entity, relationship, atribute) model [7]. This model is a basis of one of the most succesful and pioneering approaches to the problem of requirement definition and analysis due to D. Teichroew and his group at the Michigan University [8], known as the ISDOS PROJECT. Fundamental concepts of this approach are PROBLEM STATEMENT LANGUAGE (PSL), PROBLEM STATEMENT ANALYZER (PSA) and DATA BASE.

The PROBLEM STATEMENT LANGUAGE is constituted from such elements (types) as "objects", "relationship" and "properties". Dependent on the number of types of those three constituents a specific version of PSL can be created automatically by the META SYSTEM (sometimes this version is called SXL, System Description Language for Methodology X). Those versions or methodologies are to be applied to specific application domains (various parts of the "real world"). One can create as many as necessary but fixed amount of object types such as : input, output, process, processor, set, group, entity, element etc. The user gives his/her names to particular instances of those types. Those objects may be connected via different relationships. There must be again a fixed amount of such relations. They describe various aspects of information processing systems such as: system boundaries (RECEIVES, GENERATES), system structure (PART OF), data structure (CONTAINED IN, IDENTIFIES), data derivation (USES, DERIVES, UPDATES), size (VALUES) and dynamics (TRIOGGRES) etc. The relationships permitted between objects depend on their types; consequently many inconsistencies may be detected automatically. Moreover most relationships appear in both directions (binary relationships). The statement that the object one uses the object two is exactly equivalent to the statement that the object two is used by the object one. This makes it possible to produce on demand a "where-used list" for any object. This gives also the user a list of all objects which may be affected, in particular which object is to be modified. Each object can be described by assigning values for various properties (attributes). These include synonyms, key words and descriptions (any text desired to describe the object). New properties may be also defined.

The PROBLEM STATEMENT ANALYZER operates on the PSL statemnents recorded in the data base to produce a number of useful summaries and analyses, such as :

        Formatted problem statement
        Directories and keywords indices
        Hierarchical structures reports
        Graphical summaries of flow and relationships
        Statistical summaries.

Some of these capabilities are well suited to supporting system design and analysis activities, going beyond the requirement definition and analysis. At any time PSA can be asked to produce reports on all or any selected part of the data. The data presented in any given report may have been entered by different analysts at different times. This is very important from the point of view of the collective usage of the PSL/PSA resources by a team of developers. PSA uses a specified set of commands with the help of which different queries can be formulated to produce reports, analyses and summaries useful for the system developers or supervision of the management. Summing up, PSL/PSA gives the possibility of formalized statement of the functional properties of the system development with the very powerful analysis tools and reports. The user interface requires training and skills but the results are rewarding. The missing dimension is the performance description. This is at the moment the most difficult aspect of the information systems to deal with. The ISDOS PROJECT proposes the method to tackle this problem to some extent. On the basis of the content of the data base the automatic generator generates the simulation package, which can produce the required performance figures to be confronted with demands. A version of PSL used for this purpose is sometimes called DSL (DYNAMIC STATEMENT LANGUAGE)[12]. PSL/PSA have many followers. One of them was DAS, which has been transformed into AIDES, used by Hughes Aircraft Company [13]. Another, RDL (REQUIREMENTS AND DEVELOPMENT LANGUAGE) developed and used by Sperry Univac [14], has been generated by means of the quoted META SYSTEM of the ISDOS PROJECT. The peculiarity of the latter language (RDL) is the existence of objects, defined distinctly for other phases of the software life-cycle. These characteristics allow more systematic tracing of objectives, concepts, implementation, tests for management and verification purposes.

More known and better described in the literature is the SOFTWARE REQUIREMENT ENGINEERING PROGRAM (SREP) or METHODOLOGY (SREM) [15,16,17,18] developed by the US Army Ballistic Missile Defensive Advanced Technology Center. It owes much to the concepts of the ISDOS PROJECT and certain parts are strongly derivatives of PSL/PSA. Its language and analyzer is called REQUIREMENTS STATEMENT LANGUAGE (RSL) and REQUIREMENTS EVALUATION AND VALIDATION SYSTEM (REVS), its data base is called ABSTRACT SYSTEM SEMANTIC MODEL (ASSM). The main new aspects of the SREM can be briefly characterized as follows. RSL is an extensible language in the sense that some primitive type concepts have been initially built in, which can in turn be used to define additional necessary language concepts. Those primitives types are: elements, attributes, relationships and structures. The first three primitives correspond to PSL primitives. The basic difference is that one can define new types of these primitives by the extension capability of RSL. The permanent non-extensible primitives of RSL are structures, the representations of two-dimensional flow graphs, so called requirements networks, or R-NET's. They consist of nodes, which specify processing operations, and of the arcs which connect them. The basic nodes include ALPHA's, which are the specification of functional processing steps, and SUBNET's, which are specifications of processing flows at a lower level in the hierarchy. The SUBNET is an ALPHA, which is expanded to include internal details of the processing. More complex processing flow situations are expressible in RSL by the use of structured nodes, which fan-in and fan-out to specify different processing paths. The structured nodes are the AND, OR and FOR EACH. The syntax of structures in R-NET's is similar to the syntax of many structured programming languages and enforces a discipline on the user through the use of a fixed set of flow primitives (they are not extensible). Such approach allows to express not only functional but also performance requirements in terms of constraints on processing paths (flows). Statements in RSL are being translated by the RSL translator and entered into ASSM. The design of ASSM is different from the one used in PSL/PSA. It provides a decoupling between the input language, RSL, and the analysis tools, REVS. This decoupling permits for the extension capability of RSL and great freedom in the design of REVS. The tools merely access the ASSM and are in no way dependent on the RSL syntax. They can be modified and added with the evolution of SREM. Among the variety of tools in REVS one should mention the interactive R-NET generation tool, which provides graphic capabilities through which the requirement engineer may input, modify or display R-NET's. The graphic or RSL language representation of R-NET's are completely interchangeable. Another important REVS tool is the automatic simulator generator, which takes the ASSM representation of requirements of the data processing system and generates from it simulators of the system. They are of the discrete event type and are driven by externally generated stimuli. There are two distinct types of simulators. The first one uses functional models of the processing steps. This type of simulation serves as a means to validate the required flow of processing against higher level system requirements. The other type of simulator uses analytical models. It is used to define a set of algorithms for the system, which have the desired accuracy and stability. Both types of simulators are used to check dynamic system interactions and performance criteria. Simulation codes are written in PASCAL. Of course, there exists also a group of tools which check statistically (whithout simulation) for completeness and consistency in the requirement specifications. They detect deficiencies in flow processing and data manipulations, stated in the requirements.

In summary SREM appears to have met most of its goals of producing unambiguous, testable requirements for data processing functions and performance. Within the framework of this research the various decomposition alternatives were considered: the verification graphs method, the Petri Net method and Finite Machine approach. The latter approach uses the graph model of decomposition as a relation between function and composition (i.e. a graph) of subfunctions. An important requirement is the computability of funtion performance indices from the performance indices of the composition of subfunctions. The automated aids (REVS) were transported to CDC host computers and run times reduced by a factor of one hundred to yield a cost effective engineering tool. It is beccoming at present a proven engineering tool for defining and validating software requirements in a stand-alone fashion. The initial research objectives of achieving a smooth transition from software requirements into process design is still a research topic of current interest. Adding the PROCESS DESIGN ENGINEERING phase to SREM makes from it a SYSTEM DEVELOPMENT SYSTEM (SDS). The objective of the PDE part of SDS is to provide decomposition techniques for mapping the processing onto a software architecture described by a process design language, to provide the tools for verification of the design, for evolving the design into the complete code and for preventing or detecting errors as soon as possible. The resulting PROCESS DESIGN METHODOLOGY (PDM) [19] and PROCESS DESIGN LANGUAGE (PDL) [20] provide an integrated approach to the outlined problems. The methodology provides procedures for partitioning requirements into tasks and defining scheduling techniques for meeting response time requirements. Top-level designs specify processor assignments, task interfaces, and control characteristics. Process design tools to support the methodology include software library management and configuration control aids, process construction aids for composing functional models or algorithm level simulations from a library, aids for automatic recompilation of modified code and specification of data collection techniques.

PDL2 is based on PASCAL. In addition to the PASCAL features of strong data typing, block structuring and scope to variable it contains primitives for task synchronization, assertions, variable lenght arrays, and escape statements. Experience gained on PDL2 was an input to the requirements of ADA.

The development of the SDS resulted in advancing the state-of-the-art of software engineering in several ways [18]:

1. Formal languages, tools, and methodologies were developed for all phases of the software development process proving the feasibility of their ideas and setting directions of further developments.
2. SDS demonstrated that errors can be caught early in the software development process to improve reliability and reduce software costs and software risk.
3. Both SREM and PDM demonstrated that simulations can be produced from requirements (design specifications) with the help of automated tools reducing the risk of programmers misinterpretation.
4. Both SREM and PDM demonstrated that a methodology can be defined in terms of objective milestones. Automated tools can be used to verify the satisfaction of those criteria.

Experience with SDS has led to the identification of a number of issues, which are important for software engineering environments:

1. The software engineering environments must start at the top system level, and address the decomposition and allocation of functional and performance requirements of the system. Without this foundation one runs the risk "of soving the wrong problem".
2. The utility of methodologies is greatly enhanced by the existence of smooth transitions between system requirements, software requirements, software design and implementation. The unified model of software requirements, design and code is necessary in order to achieve full integration of methodologies. Until this problem is solved, gross inefficiencies will occur at the interfaces between methodologies.

MORE CONTEMPORARY APPROACHES

Research and feasibility demonstrations of SDE's were carried out by many groups and institutions. Let us mention briefly other efforts and lessons learned on their experience and achievements, based on some more contemporary approaches. Extremely important new concepts of the decade were the concepts of abstractions. They originated from pioneering works of Parnas on the concepts of mudularization [21] and on specification techniques [22], as well as from class concepts in SIMULA [23]. These kinds were later on developed by Guttag, Zilles and others [24, 25, 26, 27, 28] and form a basis of recent developments in software engineering.

One of good examples of such developments is the HIERARCHICAL DEVELOPMENT METHODOLOGY (HDM) originating at the Stanford Research Institute [29, 30, 31, 32]. HDM requires that a system be structured horizontally and vertically. A system is first of all decomposed into separate levels or abstract machines. Each level provides a set of facilities to the next higher level in the hierarchy. The facilities provided by the top level are those available to the user. The lowest level is called the primitive machine, upon which the entire hierarchy rests. The primitive machine provides all the facilities the designer regards as given, whether it be a hardware machine, the abstract machine presented by the operating system, or a hypothesized machine provided by a higher-level language processor (e.g. a PASCAL machine). An abstract machine in HDM consists of a set of internal data structures, which define its state and a set of operations, which can access and modify the state. An operation invocation causes a state transition, and a program invocation causes a sequence of state transitions. Besides the vertical decomposition into levels of abstract machines, any stem of any size must be also decomposed horizontally within each level into distinct units or modules. One way to modularize an abstract machine is by facility: each facility is encapsulated in a distinct module. Then the effects of likely future changes can be minimized. It isalready a commonly accepted fact, that each module's external behaviour must be separable from its internal details. This is the precise role played by the module's specifications. It defines external behaviour of a module without revealing the module implementation. Independent specification allows independent implementation and proof. Thus the system developmennt process, as well as the system structure, can be modular, and can proceed in a step-by-step manner. Central to HDM is its specification language, SPECIAL (SPECIFICATION AND ASSERTIONLANGUAGE). A substantial subset of SPECIAL has been formalized within the framework of the theory of Boyer-Moore [33, 34]. This formalized subset serves as the basis for the SRI verfication systems. An HDM module is composed of data structures and operations. The state defining data structures are characterized by their functional behaviour,they are referred to as state-functions. The specification of a state-function provides the state-function's signature and constraints on its initial value. A state change is described by relating the post-invocation values of state-functions to their pre-invocation values. A returned value is defined by constraints it must satisfy. An exception return occurs when an exception conditionassociated with the operation is satisfied. An exception return precludes a state change. For historical reasons the constructs provided by SPECIAL do not describe operations and state-functions but rather V-functions, O-functions and OV-functions derived from Parna's terminolgy [22].

Besides the module specification language SPECIAL, which is also used to specify mapping functions, HDM uses other languages as well. HSL (HIERARCHY SPECIFICATION LANGUAGE) is used to describe the structuring of modules into abstract machines and of machines into systems. ILPL (INTERMEDIATE LEVEL PROGRAMMING LANGUAGE) is the abstract machine programming language, used to record module implementation decisions. In addition an executable language (e.g. PASCAL OR MODULA) is used for final implementation code.

A very strong feature of HDM is its verification capability, due to its formal basis. There are two forms of verification possible, the design proofs and implementation proofs. Design proofs show that the system specification possesses certain properties. Implementation proofs are to show that the systems implementation meets its specification. The specification checkers include the usual parsers, type checkers, and prettyprinters for SPECIAL, HSL and ILPL specifications. They also enforce various external consistency criteria, concerned with how the individual specification fits into the entire system specification. The multilevel security verifier is used to verify that the design, as given by a set of SPECIAL module specifications, satisfies a particular security model. It uses the augmented version of the Boyer-Moore automatic theorem prover. The HDM approach to implementation proofs is based on the Floyd-Hoare method, in which a program is proved correct with respect to entry and exit assertions. A full description of the method of hierarchical proofs is given in [35]. There are also implemented MODULA and PASCAL verification systems.

On the basis of the demonstrated, evident appropriateness of HDM particularly for module design, specification and verification the HDM "second generation" methodology is under way. SPECIAL is to be extended to keep up with other current language designs and is to put data abstraction on a par with procedure abstraction. There are plans to integrate ADA with HDM. The ADA-oriented tools should allow to check mechanically properties of HDM modules implemented in ADA. The specification is to be oriented not only toward design goals but also to much bigger extent toward requirement presentation. There are tendencies to use SREM capabilities, like R-NETs, in SPECIAL, as well as other SREM approaches, like the advanced development data base concept for recording design decisions, maintaining versions and configurations and coordinating the diverse designs and analysis tools of the environment. It is to be studied how the concept of a library of modules can help the user in coordinating requirements, specifications, implementation and documentation (data base module specifications). It will be evidently very interesting to observe the further HDM development, oriented toward more uniform Software Development Environment, using a kind of wide-spectrum language, being a symbiosis of specification and implementation language, with requirement specification possibilities and powerful computer assisted verification capabilities.

Another interesting approach is proposed by Higher Order Software, Inc. and known under the name of HOS [36,37,38]. The HOS theory is based on a set of six axioms that describe the properties of hierarchical system structures, called controllers or modules, which execute functions. The modules exist at the node, just immediately higher on the tree relative to the functions it controls. A module has the responsibility to perform a function. Every function receives input from and produces output for its controller either directly or indirectly. Systems are decomposed into HOS control maps. This is done via AXES, a specification language of HOS. It is a complete and well-defined language capable of being analyzed by a computer. It provides mechanisms to define data types (in order to identify objects), functions (in order to relate objects of types) and structures (in order to relate functions). From the axioms of HOS a set of three primitive control structures has been derived. The primitive control structures identify control schemata on sets of objects. A mechanism for defining an algebra for each distinct set of objects is provided in AXES. To form a system new control structures are defined in terms of primitive structures or in terms of other nonprimitive control structures. Once we have a library of control structures, data types, operations and derived operations we are ready to form a particular AXES definition using these mechanisms. Those definitions are entered to the computer interactively, either graphically or textually. The machine-readable form enables the ANALYZER to automatically check logical completeness, detect interface errors, timing or data conflict errors, data definition errors, and control errors. Detected errors are displayed to the analyst, who can correct them using a graphical editor. AXES specifications proven to be logically consistent by the ANALYZER are automatically passed to the RESOURCE ALLOCATION TOOL (RAT). RAT is to map the proven functional specifications into the target environment generating code for the target machine. It is an automatic programmer converting specifications received from the ANALYZER directly into source or object code for the selected machine. Those programs are guaranted to be consistent with the proven specification. They can be converted to FORTRAN, ADA, COBOL etc. The HOSUSE. It provides means of executing programs at any stage of development. Partially completed programs can be executed by simulating uncompleted branches of the program structure. The system prompts the analyst to enter data for the unimplemented branches. This simulation mode gives the analyst and the user a chance to modify and refine the functional and performance requirements of the system, Fig. 6. Thus the HOS-methodology provides a totally automated implementation of any definable system from machine-readable, logically verifiable, correct specifications. The specifications may utilize any syntax that is automatically converted into the AXES specification language via an interactive graphic editor. No human intervention is required to convert the specification into code and system documentation. HOS evidently displays capabilities which complement many existing methodologies and proves that automation of implementations is feasible and practical.

This review shows that the idea of software development environments is not far from practical implementation. Many concepts and activities, tools for requirement definitions, specifications, various stages of design, implementation, verification and maintenance have been elaborated and tested in experimental setups proving their practical capabilities. We are not too far from the implementation of such systems as for example COMPASS, encompassing almost all of the existing state-of-the art concepts of software engineering [39].
But what about a possibility of still another approach?

Interesting views and proposals of other concepts appear in the literature in continous way. See, for example, the paper of Winograd [40] on Higher Level Programming Systems and of Wasserman and Gutz [41] on the future of programming. There is going on much in the field of algebraic specifications. Let me particularly quote the DFG-project ACT on algebraic specification techniques in Berlin [42, with references herein] developing ideas of ADJ group. "But this is a material for another story".

CONCLUSIONS

The software development systems are being constructed and applied in many branches of research and industry with success, improving software products as concerns costs, reliability, maintenance problems, effectiveness.

If this is the case the question arises, which of contemporary and future software engineering tools and systems could be adapted and applied in the field of physics, particularly in the field of physical experiments on the beam of accelerator particles (both high and intermediate energy). This class of experiments seems to be at the moment well defined, with sharply determined profile, which has a tendency to stability over a reasonably long period of time ahead. Those experiments are heavily based on intensive computer applications. These applications have been and are to be developed still in a more or less spontanic, non-correlated, non-coordinated fashion. It is a fact of life, that on the other hand, those experiments on the beam of accelerator particles are being carried out by larger and larger teams consisting of people from many laboratories. This tendency will build up converging with the tendencies of building larger and more unical accelerators, using international support. The access to those accelerator-monstres will be then more difficult, the experiments themselves will be more complicated, carried out by more people, from a larger number of laboratories. There appears then a question: could the usage of accelerators and organization of experiments on them be optimized by the more effective and refined usage of computers both for accelerator control, beam supervision and administration and experiment data acquisition and processing? Many efforts in this direction have been already undertaken. Some of them were presented in the framework of this conference. More concerted and organized effort is necessary. One of the ways is to consider the development of Software or System Development Environments fitted to particular fields, applications. Evidently it can be conceived to consider the digital control of accelerators and the beam administration as well, contained, specific branches, for which modelling and specification concepts could be well defined and necessary tools and environment contructed to make the system development more effective and useful.

The same can be told about the experiments on the beams of accelerator particles. Some concepts and thoughts about so-called Experiment Information Systems have been proposed recently by the author of this review and his collaborators [43]. They can be very briefly summarized, that software development environment envisaged for those experiments could be the basis not only for the development of necessary software but also through the distributed computing facilities could give possibilities of more coordinated, concerted usage of experimental data, data processing programs and methods, logical experiments, reevaluation of published results etc. Such research, possibly organized within the framework of the LEP-Project, is worthwhile of careful consideration and is convergent with other efforts already undertaken by ECFA and CERN, as well as by other European and American institutions and organizations.

This paper is devoted to the 10-th anniversary of the creation and starting of computational activities of the Regional Computing Center CYFRONET of the former Institute of Nuclear Research, Otwock-Swierk, Poland. On this occasion I would like to thank all my coworkers and friends for their active participation in and contribution to the conceptual, organizational scientific, technical activities of the Center. Particular thanks and gratitude are due to them for their support, acts and words of appreciation and farewell expressed by them after my dismissal from the post of the Director of the RCC CYFRONET on 18-th of July 1983. This dismissal has been due to the differences in opinion between me and the authorities of the National Atomic Agency on the policy directions concerning the further development of the CYFRONET Center.

REFERENCES

1. L.J. Osterweil, A Software Lifecycle Methodology and Tool Support, in W.E. Riddle and R.E. Failey/Ed/, Software Development Tools, Springer Verlag, Berlin, Heidelberg, 1980
2. W.E. Howden, Contemporary Software Development Environments, Com. of the ACM, 25, 318,1982
3. A.I. Wasserman, Automated Tools in the Information System Development Environment, in H.J. Schneider and A.I.Wasserman/Ed/,Automated Tools for Information Systems Design, North-Holland, Amsterdam,New York,Oxford, 1981
4. R.W. Mitze, THe UNIX-System as a Software Engineering Environment, in H.Hünke/Ed/, Software Engineering Environments,North-Holland, Amsterdam, New York, Oxford,1981
5. M.H. Bianchi, R.J. Glushko, J.R. Mashey, A Software-Documentation Environment Built from the UNIX Toolkit, in H.J. Schneider and A.I. Wasserman/Ed/, Automated Tools for Information System Design, North-Holland,Amsterdam, New York, Oxford, 1982

6. J.N. Baxton, L.E. Druffel, Requirements for an ADA Programming Support Environment: rationale for STONEMAN, in H. Hünke/Ed/, Software Engineering Environments, North-Holland, Amsterdam, New York, Oxford, 1981
7. P.P. Chen, The Entity-Relationship Model - Toward a Unified View of Data, ACM Transactions on Database Systems, 1,9/1976
8. D. Teichroew, E.A. Hershey, PSL/PSA: A Computer-Aided Technique for Structured Documentation and Analysis of Information Processing Systems, in IEEE Trans. on Software Eng.,Vol.SE-3, 41/1977
9. A.N. Haberman, D.E. Perry, System Composition and Version Control for ADA, in H. Hünke/Ed/, Software Engineering Environments, North-Holland, Amsterdam, New York, Oxford, 1981
10. M. Bayer, B. Böhringer, J.P. Dehottay, H. Feuerhahn, J.Jasper, C.H.A. Koster, U. Schmiedecke, Software Development in the CDL2 Laboratory, in H.Hünke/Ed/,Software Engineering Environments, North-Holland, Amsterdam, New York, Oxford, 1981
11. R. Balzer, N. Goldman, D. Wile, Informality in Program Specifications, IEEE Trans. on Software Engineering, SE-4, 94/1978
12. D. Teichroew, S. Spewak, E.A. Hershey III, Y.Yamamoto, C. Starner, Computer-Aided Modelling of Information Systems, Proc. COMPSAC 79
13. R.R. Willis, AIDES: Computer-Aided Design of Software Systems-II, in H.Hünke/Ed/, Software Engineering Environments, North-Holland, Amsterdam, New York, Oxford, 1981
14. H.C. Heacox, RDL: A Language for Software Development, ACM Sigplan Notices, 14, 71/1979
15. C.G. Davis, C.R. Vick, The Software Development System, IEEE Trans. on Software Eng., SE-3, 69/1977
16. M.W. Alford, A Requirements Engineering Methodology for Realtime Processing Requirements, IEEE Trans. on Software Eng.,SE-3, 60/1977
17. T.E. Bell, D.C. Bixler, M.E. Dyer, An Extendable Approach to Computer-Aided Software Requirements Engineering, IEEE Trans. on Software Eng., SE-3, 49/1977
18. M.W. Alford, C.G. Davis, Experience with the Software Developmment System, in H.Hünke/Ed/, Software Engieering Enviroments, North-Holland, Amsterdam, New York, Oxford, 1981
19. S.N. Gaulding, J.D. Lawson, Process Design Engineering - a Methodology for Real-time Software Requirements, Proceedings 2nd International Software Engineering Conference, 1976
20. R. Kopang, Process Design System - An Integrated Set of Software Development Tools, Proceedings 2nd International Software Engineering Conference, 1976
21. D.L. Parnas, On the Criteria to be used in Decomposing Systems into Modules, CACM, 15, 1053/1972
22. D.L. Parnas, A Technique for Software Module Specification with Examples, CACM, 15, 330/1972
23. O.J. Dahl, C.A.R. Hoare, Hierarchical Program Structures, in O.J. Dahl, E.W. Dijkstra, C.A.R. Hoare, Structured Programming, Academic Press, 1972
24. J.V. Guttag, J.J.Jorning, The Algebraic Specification of Abstract Data Types, Acta Informatica, 10, 27/1978
25. J.V. Guttag, Abstract Data Types and the Developmnent of Data Structures, CACM, 20, 396/1977
26. B. Liskow, S.Zilles, Specification Techniques for Data Abstractions, IEEE Trans. on Software Eng., SE-1, 7/1975
27. B. Liskow, Modular Program Construction Using Abstractions, in D. Björner/Ed/, Abstract Software Specifications, Lecture Notes in Computer Science nr 86, Springer, Berlin, Heidelberg, New York, 1980
28. S.N. Zilles, Algebraic Specification of Data Types, Project MAC Progress Report, MIT, Cambridge, 1974
29. K.N. Levitt, L. Robinson, B.A. Silverberg, The HDM Handbook, Volumes I-III, Computer Science Laboratory, SRI International, June 1979
30. J. Goldberg, Hierarchical System Development, Final Report, Computer Science Laboratory, SRI International, June 1978

31. B.A. Silverberg, On Overview of the SRI Hierarchical Development Methodology, in H.Hünke/Ed/, Software Engineering Environments, North-Holland, Amsterdam, New York, Oxford, 1981
32. W.D. Elliot, B.A.Silverberg, K.N. Levitt, A Critique of HDM, Technical Report CSL-131, Computer Science Laboratory, SRI International, November 1981
33. R.S. Boyer, J.S. Moore, A Formal Semantics of SRI Hierarchical Program Design Methodology, Computer Science Laboratory, SRI International, November 1978
34. R.S. Boyer, J.S. Moore, A Computational Logic, Academic Press, 1979
35. L. Robinson, K.N. Levitt, Proof Techniques for Hierarchically Structured Programs, CACM 20, 271/1977
36. M. Hamilton, S. Zeldin, Higher Order Software - a Methodology for Defining Software, IEEE Trans. on Softw. Eng. SE-2, 9/1976
37. M. Hamilton, S. Zeldin, The Relationship between Design and Verification, The Journal of Systems and Software, 1, 29/1979
38. M. Hamilton, S. Zeldin, The Functional Life Cycle Model and its Automation: USE.IT, Technical Report Nr. 36, Higher Order Software, Inc. Cambridge, MA, December 28, 1982
39. H.J. Schneider, Techniques and Formal Tools for Design, Realization and Evaluation of Evolutionary Information Systems, Interner CIS Bericht 8/81, Technical University Berlin, Institut für Angewandte Informatik, 1981, to be published in: I.Hawgood/Ed/, Proc. IFIP TC-8
Working Conference on Evolutionary Information Systems, North-Holland, Amsterdam, New York, Oxford
40. T. Winograd, Beyond Programming Languages, CACM,22, 391,1979
41. A.I. Wasserman, S. Gutz, The Future of Programming, CACM, 25, 196,1982
42. H. Ehrig, W.Fey, A Method for the Specification of Software Systems: From Formal Requirements to Algebraic Design Specifications, Preprint May 1982, Technical University Berlin, Institut für Software und Theoretische Informatik. Short version of this paper is published in: W. Bauer/Ed/,Informatik-Fachberichte 50, Springer Verlag, 1981
43. T. Czosnyka, J.Grabowski, P. Strzalkowski, R. Zelazny, Experiment Information Systems/a Proposal/, Preprint of the Institute of Nuclear Research, Otwock-Swierk,October 1981

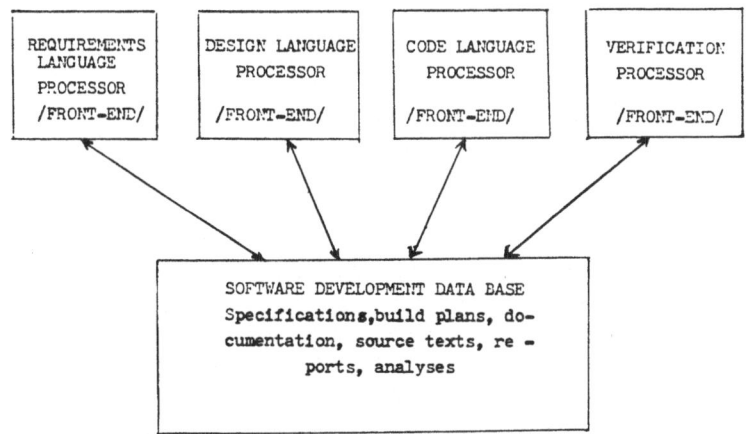

Fig. 1

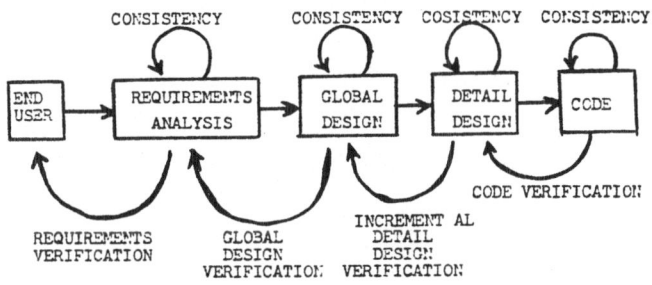

LIFE-CYCLE VERIFICATION

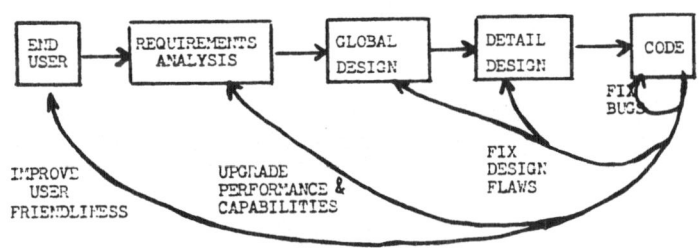

LIFE-CYCLE VALIDATION

Fig. 2

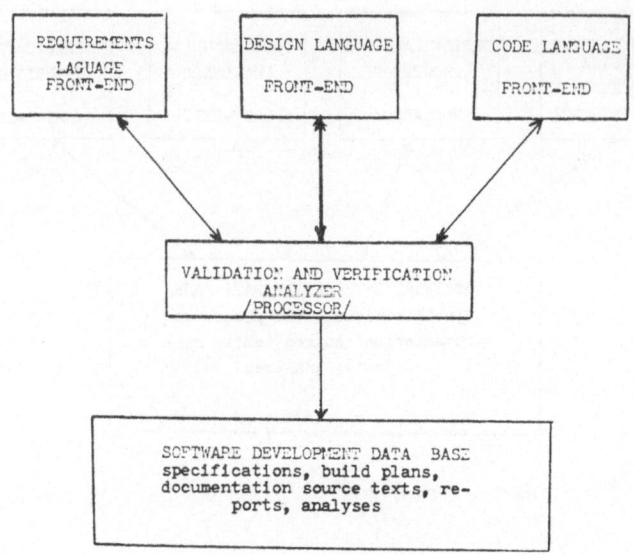

Fig. 3

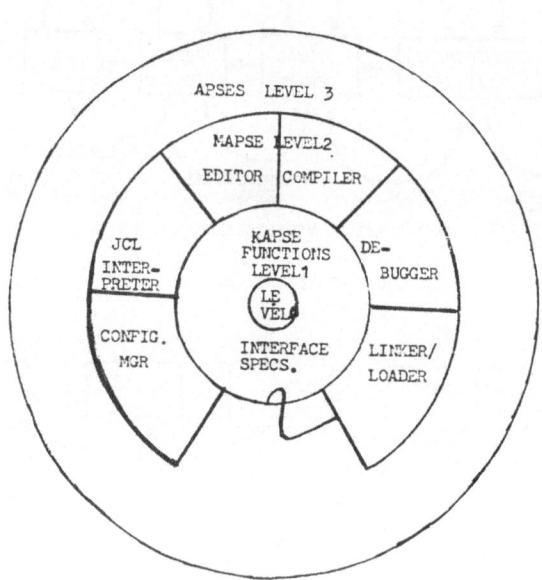

Fig. 4

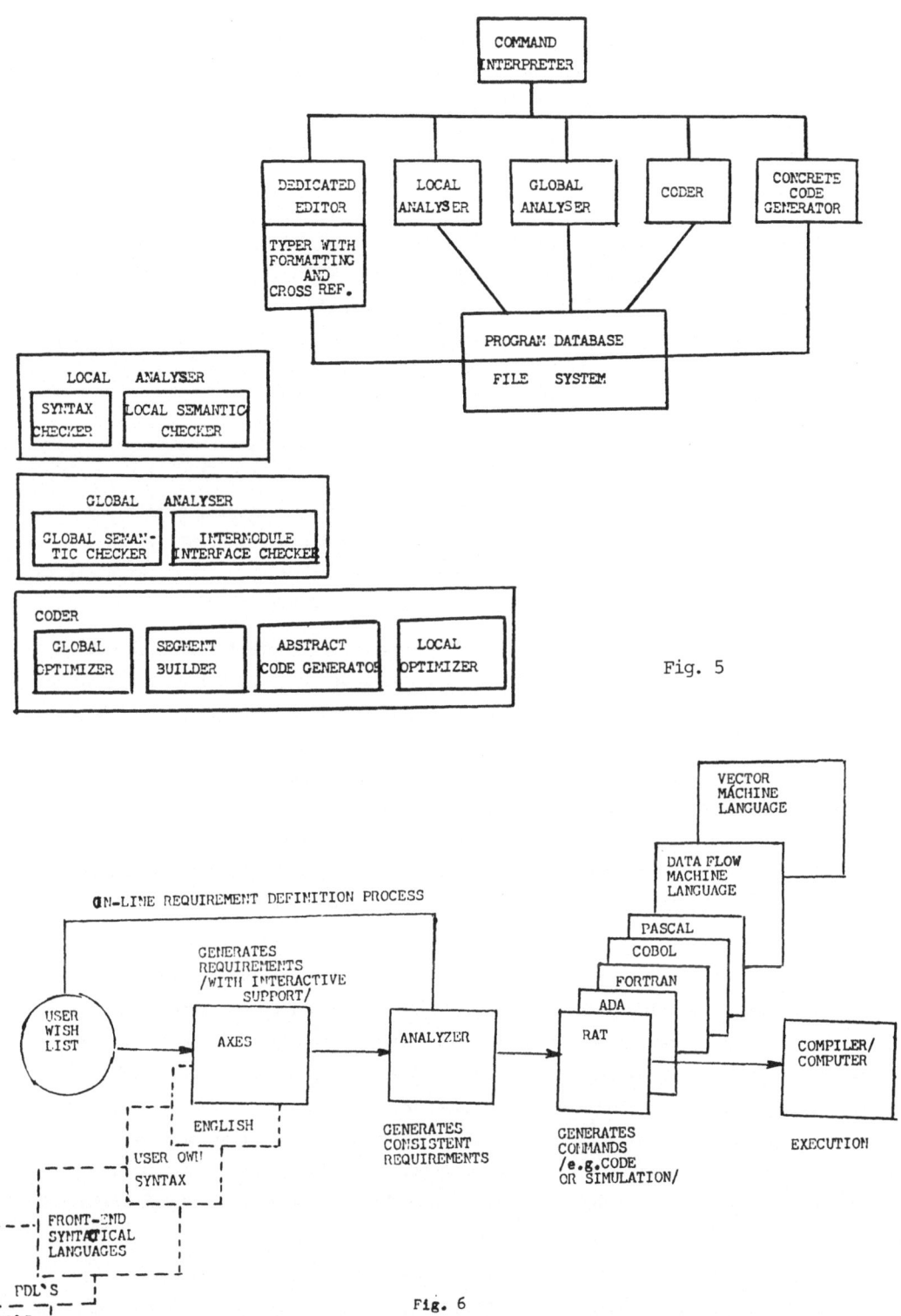

Fig. 5

Fig. 6

Centralization and Decentralization
in the TRIUMF Control System
D.A. Dohan and D.P. Gurd
TRIUMF, 4004 Wesbrook Mall, Vancouver, B.C., Canada V6T 2A3

1.0 INTRODUCTION

The increased demands of an expanding accelerator laboratory have made it timely to consider strategies for expansion of the TRIUMF Control System. These requirements have led to reflections on one of the major themes of this conference – centralized vs distributed digital control systems for accelerators. This paper discusses the way in which the TRIUMF system successfully combines elements of both approaches.

2.0 CENTRALIZED AND DISTRIBUTED CONTROL

The phrase "distributed processing" does not have any single, simple and precise meaning. We have identified four aspects of digital control systems; hardware, software, data bases, and operator control stations, all of which may be distributed to varying degrees in different system configurations.

2.1 Hardware. Multiple distributed processors economically provide more CPU and Input/Output cycles per unit time, as well as the possibility of hardware redundancy. The degree of centralization, however, is not determined by the power of the processors, nor by their geographic distribution.

What most clearly characterizes such a system is the topology of the communication system linking its processors. Some topologies may require all messages to pass through a single computer node, whereas others may permit any processor to communicate directly with any other. Topology and software together should be designed to minimize the interprocessor communication required.

2.2 Software. The software analogue to the distribution of processors is the distribution of "tasks," which may be grouped according to geography, equipment type or by function. The amount and level of task grouping determines the selection of an operating system, which may range from multi-user multi-task to dedicated single task systems. One advantage of the former is the ease of intertask and application-driver communication. However considerable overhead results from task and process switching and the possibility exists of one task or user interfering with another. A potential disadvantage of the latter case is the overhead of increased communication, which can be minimized by designing tasks which are as independent as possible.

2.3 Data Base. An element of any control system which may be more or less distributed is its data bases. Several data base types, each of which may be distributed to a different extent, are discussed below.

Firstly, there is a fixed data base frequently referred to as the "device tables," which describes machine parameters such as device name, access method, and operational limits. Closely associated with this data is the software required to use it. These tables and associated software might exist in one place only or could be distributed for easy access by those processors requiring them.

A second conceptually distinct fixed data base describes console devices, such as buttons, knobs, and displays. Again, this requires fixed, descriptive data, and programs which refer to it.

A third data base describes complete machine parameterizations required to achieve specific operating conditions, such as energies and timing structures. In situations where the operating mode requires rapid or frequent changes, this must be immediately accessible to the processors using it.

The final type of data base is the "live" or variable data which represents the current state of the accelerator. This includes raw single parameters, such as magnet fields or beam current, as well as calculated parameters, such as beam polarization or emittance.

2.4 Consoles. The widespread use of redundant, fully assignable operator consoles imposes constraints on the distribution of hardware, software, and data bases. A completely general console must have access to all accelerator processors, processes, and data. This may be achieved in several ways, ranging from one processor responsible for all consoles, to a number of interconnected processors, each responsible for a separate part of the console process. Some degree of local control is also required for many subsystems. This may be achieved by truly local control stations having access only to the hardware, software, and data required; or by the "soft" dedication of an otherwise fully assignable console.

Regardless of how multiple control stations are implemented, the possibility of conflicts between different stations requesting access to the same parameter must be resolved. The complexity of the resulting protocols will depend largely upon strategies adopted for the distribution of processors, tasks and data bases.

3.0 THE TRIUMF CASE

The TRIUMF control system is considered to have three levels: a CAMAC level, an executive level, and an applications level. Each level is discussed with a view to showing how its design has incorporated characteristics of both "centralized" and "distributed" digital control. Figure 2 is a schematic representation of the complete system, showing the rather arbitrary division between levels.

3.1 The CAMAC level. The lowest, or CAMAC, level is a conventional parallel branch system having seven branches and 43 crates. No attempt is made to centrally store the distributed "live" data base of raw machine parameters contained within this system.

Many crates house TRIUMF designed intelligent controllers known as TRIMAC (2), which has recently been upgraded to accommodate 5K of RAM and 32K of EPROM. Figure 1 shows three sample applications of TRIMAC processors in the CAMAC layer, where distributed architecture is most clearly represented at TRIUMF. Each interacts with the CAMAC system and executive layer in a somewhat different fashion, and represents differing degrees of distributed control. One (fig. 1b) is a stand-alone system, passing data to a dual-ported memory in a central system crate. Another (fig. 1c) is an independent control system using a PDP 11/34 and serial CAMAC highway, with TRIMACs for local control.

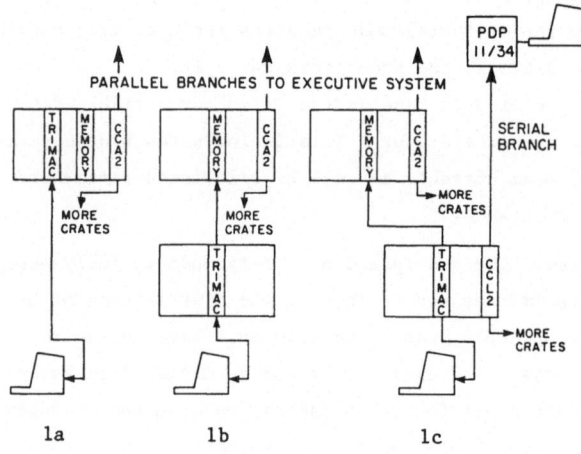

Figure 1
Typical TRIMAC System Configurations

Figure 1a represents a dedicated TRIMAC based beam steering system designed to centre and align the low intensity polarized proton beam being delivered to a sensitive asymmetry measuring experiment. Signals read from secondary emission plates (SEMs) are averaged, misalignments determined, and a matrix inversion used to calculate corrections to four beam steering elements.

Control is passed from the operators to this loop by the setting of a flag, matrix elements and other loop parameters in the shared memory. In this case the "local" control console is located in the experimental data acquisition room. The data is also made available in machine readable form to the data acquisition system. All relevant loop parameters are directly available to the TRIMAC, except for one steering magnet whose controller is housed in another crate, and whose TRIMAC calculated set point must be moved by an executive level computer. The overall system configuration makes this extremely easy to do, but it is annoying nonetheless. Another bus structure, such as FASTBUS, would eliminate this limitation which is imposed by the CAMAC system structure.

3.2 The Executive Layer. Elements of both centralization and decentralization are to be found in the "executive" layer, which is discussed under the headings used to characterize distributed processing: hardware, software, data base, and consoles.

3.2.1 Hardware. The distinguishing feature of the executive level hardware is the use of multiple mini-computers, all centrally located in the main control room, and of three independent interprocessor channels: the Multi-processor Communications Adaptor or "MCA"; the Multiport memory or "MPM": and CAMAC (See figure 2).

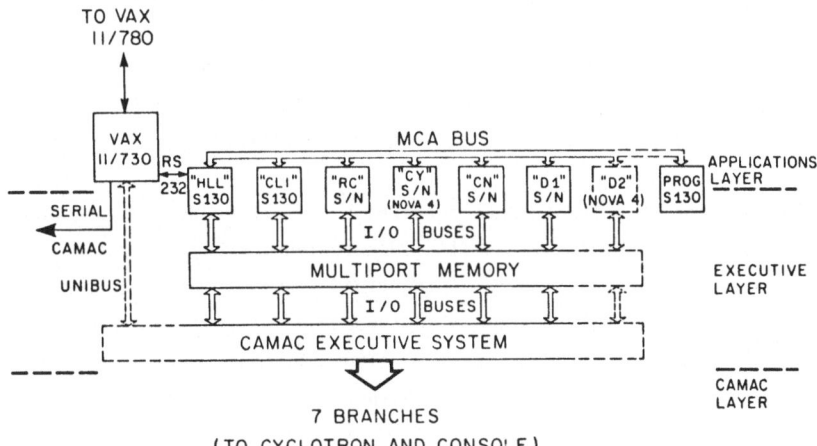

Figure 2
The TRIUMF Control System

The MCA is a vendor supplied parallel direct memory access bus which links the executive level computers, and allows several processors to work together in truly distributed fashion to perform a single function. For example the "CY" computer collects data which is passed to "CN" for display on the main console. The link also allows programs to be loaded quickly and easily directly from the program development computer.

The MPM is a TRIUMF designed multi-ported memory, housed in a CAMAC crate (2). In principle, ports adapted to any computer bus could be designed compatibly. Programmed I/O, DMA, and message passing modes are all available.

The TRIUMF CAMAC interface is a multi-sourced system based upon the GEC Elliott "Executive Crate" architecture, which provides a versatile multi-processor data flow node. The present system of six computers is soon to be expanded to eight by the addition of two active executive crate extenders, and of a third crate to accommodate new source interfaces. This extension will permit direct interfacing of the 32 bit VAX 11/730. Further expansion is limited by bandwidth (now only 15 - 20% used) to a maximum of 30 sources.

3.2.2 Software. Executive level software consists of a small, real-time multi-tasking scheduler as well as the common subroutines in the Supernovas. These routines, which together occupy about 14 K words in each 32 K word computer, give access to the MCA, MPM and CAMAC. TRIUMF has examples of three types of operating systems: multi-user, multi-task in the VAX and Eclipses; single-user multi-task in the Novas and some TRIMACs; and single task in other TRIMACs. Program development is carefully isolated in a separate system.

3.2.3 **Data Base.** The TRIUMF executive layer provides access to the various data base types described in section 2.3, and it is here that the blend of centralized and distributed concepts is most apparent. Device tables are located in the fully centralized MPM, yet each central system processor has direct access to them. The interpretation of these tables, however, is by common subroutines repeated in each computer.

There exists at present no data base in the TRIUMF system to describe console devices - each program contains its own dedicated subroutines for that purpose. This deficiency is characteristic of neither centralized nor distributed data bases - it is simply poor design.

Finally, the 'live' data base describing current machine parameters is distributed throughout the CAMAC system (1). Nonetheless, every parameter is directly accessible to any computer source on the CAMAC system. This includes calculated parameters which may be evaluated by any processor at any level and then stored either at a CAMAC address or in the MPM. Thus all machine parameters will be directly available to the VAX when it is interfaced to the executive crate, and any complex parameter which it might calculate, such as beam emittance, will be readily available to all central system computers, with no need for any communications protocol or overhead. This example demonstrates the value of an appropriate mix of centralized and distributed philosophies in the TRIUMF design.

3.2.4 **Consoles.** The implementation of several categories of operator consoles at TRIUMF provides another example of the mixture of centralized and distributed ideas at TRIUMF.

The unique main console is largely dedicated, requiring specialized software, not universally accessible, to interface to it. By interfacing the central console to a CAMAC branch, it becomes directly available to all executive and applications level computers. This permits a high level application to attach a complex control algorithm to a main console knob directly.

The REMCON consoles provide unsophisticated readback and control of all accelerator variables (1). They are perhaps the earliest examples of fully assignable, redundant consoles. Interfaced via CAMAC, but distributed geographically around the laboratory the panels are all serviced by the same computer, labled "RC" in figure 2. They offer truly distributed control as a result of their access to both the distributed (CAMAC) and centralized (MPM) data bases.

The TRIMAC-serviced operator consoles in the CAMAC layer are truly local, having access only to that data collected within the local subsystem.

Although the TRIUMF configuration would make a protocol to eliminate conflicts easy to implement, no such precaution has been taken. Conflicts seldom arise, and they are most efficiently resolved by direct oral-aural inter-operator communication on the voice band.

3.3 **The Applications Layer.** The applications layer currently consists of a DEC VAX and the six Data General Novas, linked to the executive level as shown in figure 2.

Application level software in the Novas is distributed by function, rather than by accelerator subsystem, and the symmetry of executive level interfacing allows any processor to run any program. A consequence is that if one computer should fail, some degree of control is maintained over all cyclotron subsystems.

The VAX has been added to introduce a more sophisticated level of accelerator physics into cyclotron development and operation. For these purposes a powerful CPU capable of executing complex fitting and analysis codes was required. A 32 bit VAX 11/730 was selected, in part because a large amount of relevant software, developed at other accelerator laboratories, was available. The VAX communicates with one Central Control System computer, known as "HLL," by an RS232 link. This allows HLL to act as a "front end" to the VAX for some processes, and provides an indirect connection to the MPM tables. The VAX has an independent serial CAMAC highway, and will shortly be interfaced directly to the central CAMAC system. In addition, it has a direct RS232 DECNET link with the TRIUMF Data Analysis Centre VAX 11/780.

Considered on its own, the VAX is a highly centralized system using a multi-user, multi-task operating system (VMS) to perform all operations in a single processor. It operates on its own higher level data bases incorporating logical beam dynamic parameters made up of combinations of single parameters.

The use of HLL as a "front end" and MPM port introduces some decentralization, as will the planned interface to the distributed CAMAC data base, which will also give the VAX access to the main control console, and make possible the development of a powerful graphics console on the VAX VT640 terminal.

Finally, a planned Ethernet local area network will expand the applications layer upward and make possible the distribution of a variety of applications throughout a network of processors large and small.

4.0 CONCLUSIONS

Distributed control is characterized by a number of different concepts relating to hardware, software, data bases, and control stations. Although some control system designs are more centralized than others, all contain elements of both approaches. In particular, the TRIUMF system contains a unique blend of centralized and distributed attributes, deriving primarily from the multi-sourced CAMAC and memory systems at its executive node.

REFERENCES
(1) D.P. Gurd, D.R. Heywood, and R.R. Johnson. Proceedings of the Seventh International Conference on Cyclotrons and their Applications (Birkhauser, Basel 1975) p.561.
(2) D.P. Gurd, D.R. Heywood, and J.V. Cresswell. Proceedings of the Ninth International Conference on Cyclotrons and their Applications (Les Editions de Physique, Les Ulis, 1981) p.565.

THE FERMILAB ACCELERATOR CONTROLS SYSTEM

D. Bogert and S. Segler

Fermi National Accelerator Laboratory*

P.O. Box 500

Batavia, Illinois 60510

U.S.A.

The Fermilab Accelerator Controls System has been substantially increased in scale and capability to support the new superconducting accelerator. New controls have been designed specifically for the superconducting system, and these controls involve more than 500 micro-processors distributed around the accelerator complex. In addition, the accelerator's original central control computer system has been replaced by a new network of two DEC VAX 11/780's and 21 DEC PDP-11/34's and the Linac has been reinstrumented.

All the components for the Fermilab accelerator complex (200 MeV/c Linac, 8 GeV/c Booster, Main Ring, Tevatron, future P-bar source, Fixed Target Switchyard, Collider Interaction Region) are controlled from one central "Main Control Room". An overview of the entire controls network is found in Figure 1. There are fourteen identical control consoles. Each console is supported by a DEC PDP-11/34. A partial list of the salient features is:

1) RSX 11 operating system supporting up to four simultaneous user tasks.

2) All console equipment is supported from a CAMAC crate connected to the PDP-11/34 via a serial link. This permits the physical location of the console to be arbitrarily far from the supporting PDP-11/34.

3) A keyboard, touchpanel, "track ball," and interrupt button for operator interaction.

4) Two Hitachi color Monitors each supported by an in-house designed hardware driver, a Lexidata Inc. high precision colorgraphics monitor, and Tektronix 613 monochrome storage scope with monochrome hardcopier unit.

Programs of two types, "Primary" and "Secondary" Applications "pages" (PA's and SA's) permit the Consoles to serve as the human interfaces to the controls system. The PA's have access to the complete hardware resources of the console system. One example of a PA is the "parameter page" which is a general purpose program for presentation of data in real time for any desired set of devices known to the system data base, including analog readback, setting, digital status, and alarm limits for each device. This set of devices can then be saved for future perusal. Other specific PA's include the refrigerator loop control page, the orbit control page which calls up the orbit smoothing program on the VAX, and the accelerator clock page which allows one to set up the timing of all the various accelerator clock events. A SA can run as a background task at a given console independent of what PA may be running at that console. The SA controls a given hardware resource only until a PA

*Operated by Universities Research Association Inc., under contract with the U.S. Department of Energy.

requires that resource. Should that occur, the SA is automatically terminated. Examples of useful SA's are those which control alarm annunciation, slow and fast time plotting, as well as ring-wide plots of vacuum readings or refrigerator temperatures.

There are two VAX 11/780 main frame processors linked via a shared memory, one designated the "Operational Vax" and the other the "Development Vax." The Operational Vax supports a large central data base which contains addressing information for all "control points" in the accelerator system. It also supports certain "central applications programs" not run from the consoles. The Development Vax is used to support software development for computers throughout the control system. A system called the Applications Program Librarian (APL) has been implemented for the development of applications programs. It serves as a librarian for maintaining source code for existing applications programs. It is also possible to develop software independent of APL on the development VAX and to try it out in the debugging phase as a temporary application. This allows for faster turn around times during the debugging phase. Examples of "Central Applications" are the logging of data for later analysis, the task of sending alarm messages on to the console computers, and the job of gathering data from the beam position processors as input to do orbit smoothing.

Seven "Front End" PDP-11/34 computers are used to interface the "Central Host" (Consoles and Vaxes) to the various discreet accelerator components. A "Front End" attempts to hide from the Host computer the individual differences of the various

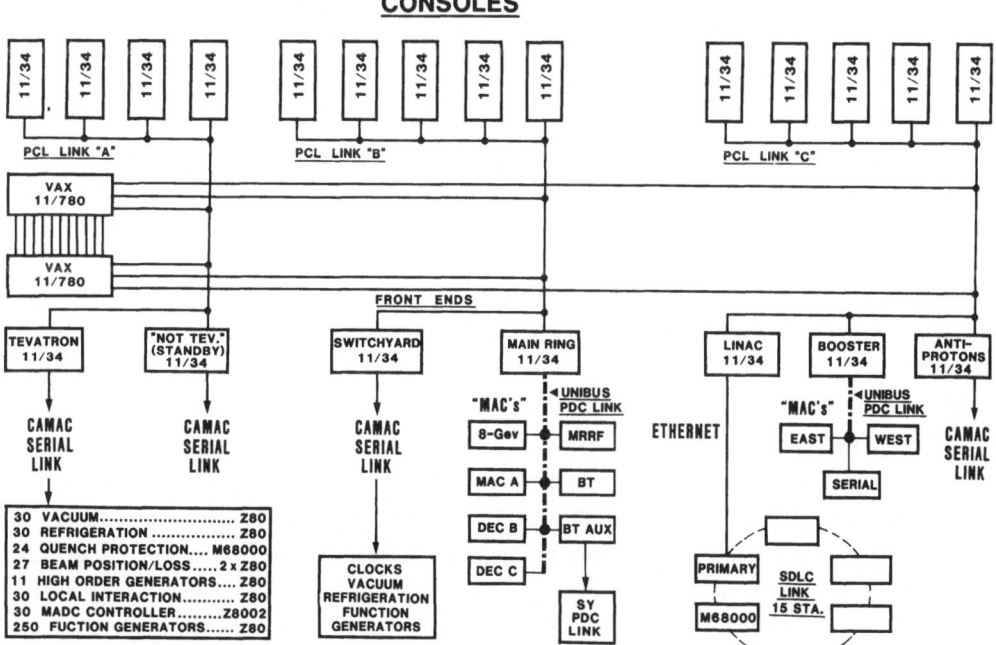

Figure 1: The Fermilab Accelerator Controls Computer Network

subsystems. This allows the "Host" to treat all requests for receiving or sending data identically. Individual software drivers such as that for CAMAC and those for communicating with the "smart" subsystems are located in the appropriate Front End. The Front End also has the task of scanning for any alarms and then alerting the Host as to devices which are at values outside their nominal limits. This includes the unsolicited messages generated by the "smart" subsystems as well as LAM's which can be generated by "dumb" CAMAC modules such as power supply controllers.

A variety of bus and computer configurations are used to connect the various accelerator components to the several Front Ends. These include:

1) The revised Linac system, which is a set of 15 distributed Motorola 68000 based multibus microcomputer systems interconnected by a fiber-optic SDLC link. One 68000 microcomputer, designated the "primary station," is interfaced via an Ethernet link to the Linac Front End.

2) The original 8 GeV/c Booster system, interfaced to three Lockheed MAC 16 minicomputers via an in-house designed Unibus to MAC-16 PDC crate link.

3) The original Main Ring System is also interfaced to Lockheed MAC-16 crates.

4) A number of independent serial CAMAC systems, each of which is driven by a Front End with essentially identical driver code. CAMAC is used for the Tevatron, the Anti-proton Source, and new switchyard equipment.

In some of the systems, but most significantly the Tevatron Control system, a very extensive use of distributed processing has been made. As indicated, in the case of the Tevatron, several hundred processors are involved. All 21 PDP-11/34's and the two VAXes are interconnected with DEC PCL (parallel communications link) buses. Three such buses are used since the fixed time slice architecture does not permit a very high multiplicity on any one bus.

The Tevatron Clock System incorporates a clock encoder which is controlled via an 11/34 Front End. The encoder serves to set up a 10 bit encoded message with a start bit and parity bit on a 50 mhz modulated rf serial link with 100 nsec precision in real time. A clock decoder chip was developed at Fermilab which can be programmed to "listen" selectively for any eight of the 256 possible encoded events. This system makes possible time synchronization of the various systems with the beam. There are two classes of clock events. There are events which occur regularly in time cycle after cycle and can serve as triggers for predet timers. Then there are events which are encoded and sent in real time based on some specific requirement having been met.

The former accelerator control system consisted in total of approximately 6000 control points. The introduction of the distributed processor systems to support the cryogenic refrigeration plant, quench protection, extensive beam position/beam loss monitors, correction function generation, and similar tasks has raised the potential number of control points and the words of information for read back to over 100,000. It is therefore no longer possible to update every piece of information from the entire accelerator complex into a central data pool at 15 hz as was done for the original 6000 control points. The new system is based upon the philosophy of "selective data acquisition" of only those items currently required at the Host system level. Although an individual refrigeration control microcomputer may gather several hundred pieces of information constantly, and use this information for the closed loop control of a satelite refrigerator, only a small fraction of this data may ever be returned to the host. Within a given console computer there is one data pool area shared by all of the possible four simultaneously running applications programs for the return of selective data. To acquire a piece of data using the data gathering services of the control system, an applications program first makes a request to the "Data Pool Manager" task resident in that particular console PDP 11/34. This request is usually quite minimal; the "device name" and a "frequency time descriptor" are sufficient. The frequency time descriptor is a specification of how frequently the data is to be read: examples might be 15 hz, 1/2 hz, once only, or " on every event 'i'." Frequencies greater than 15 hz are supported, but with the intervention of intermediate processors which may buffer the data and provide an ordered (or even time stamped) list as a vector packet delivered to the console at 15 hz. Very slow repetitive data acquisition (e.g once a minute) may use the "once only" acquisition mode, wherein the request for data is actually cancelled and recreated once a minute to conserve buffer space. Data gathering "on event 'i'" permits one to specify particular events on the Tevatron serial clock as that point in time when one wishes a reading to be made. This is particularly useful when studying behavior at times when voltages and currents are being rapidly ramped, such as for injection or extraction "bumps."

The "data pool manager" then makes a request across the PCL network to the central data base for the complete set of addressing information for the particular "property" (such as 'reading' in this example) for the device name specified. This process is repeated for all devices specified to the data pool manager by the applications program. All requests for services for devices connected to a particular front end are transmitted as a "list" in a single network request to that front end. The information transmitted to the front end consists of the rest of the addressing instructions which it will need to service the data request. This information will vary markedly for different types of front ends and for different devices within a given front end. For example, in the case of the Tevatron there are both "intelligent" and "dumb" modules distributed around the serial CAMAC system. A request for a periodic reading of a "dumb" device at 15 hz is serviced by the Tevatron front end itself. A request for the periodic reading of data from an "intelligent" subsystem, such as an ADC reading from a refrigerator microcomputer, is passed on as part of a list to the refrigerator computer system. The addressing is only distributed to the subsystem once at list set-up time; from then on the front end is permitted to request the return of data (readings) in the list at the appropriate frequency using only a list identifier for description. The subsystem has the obligation to have gathered the data at the appropriate frequency and to have it available when so requested. The Front End then assembles a response list of the appropriate data at the appropriate frequency and transmits the data to the requesting console system. The console's data pool manager stores the received data in the data pool and notifies the applications program of the arrival of new data. The applications program may then plot, display, or calculate with the newly arrived data as desired.

Additional examples of the variations in data acquisition in different Front Ends are the Linac and Main Ring systems. In the case of the Main Ring, the older MAC-16 subsystems are programmed to deliver all data to the central host in a fixed order list at 15 hz. The Main Ring Front End PDP-11/34 accepts all data (i.e. it is not selective) but then only sends data to the applications programs in lists in response to requests made to it. This approach is possible because of the relatively minimal number of data points collected, so sufficient storage is available in the Main Ring Front End to store all of the data. The Linac is slightly different. All

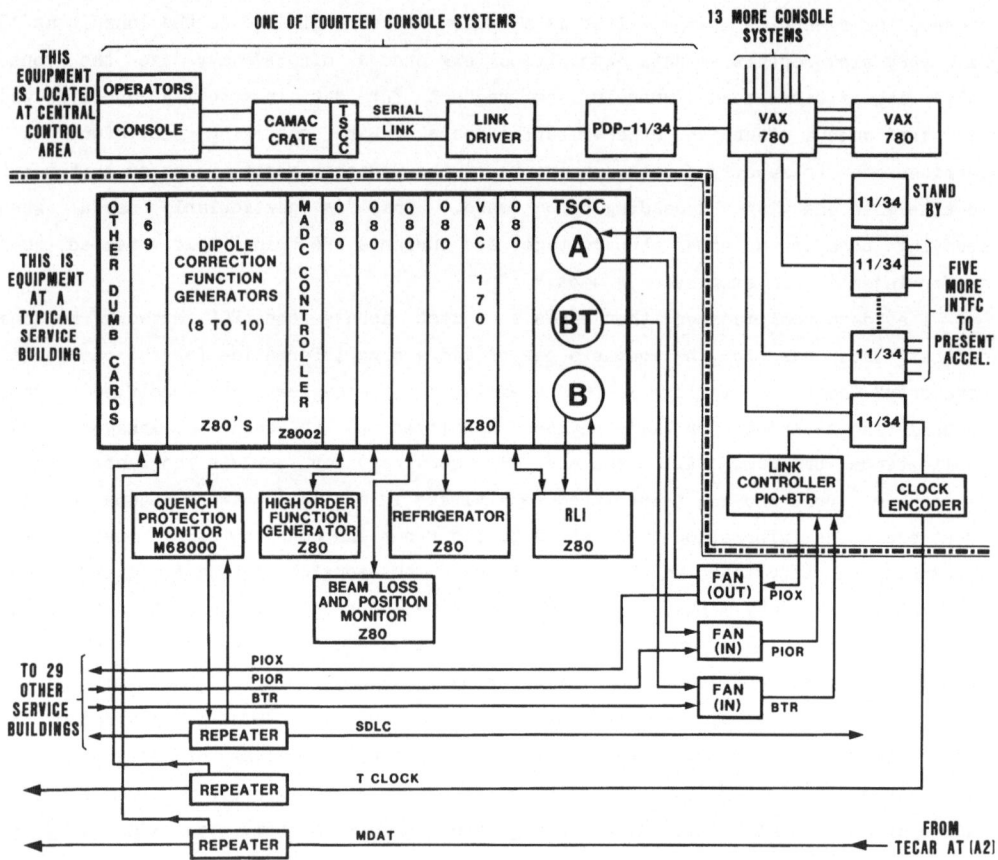

Figure 2: The Tevatron Controls System showing Serial Link, CAMAC crate, Serial Crate Controller (TSSC), distributed microprocessors, Local Interaction (RLI), and Clock

the data is collected at 15 hz in the distributed M68000 microcomputer systems, but only passed to the primary and the Linac Front End by lists as requested.

Certain other simplifications of data gathering specification have been introduced to minimize network traffic and to conserve list specification buffer space. An especially useful example is the "wild card device." This device permits one to invoke the great symmetry of the Tevatron control system. For example, one may request the reading of a single ADC channel in a refrigerator microcomputer system to be returned to the application from all such systems, and in a logically or geographically sensible order (i.e. in order around the ring.) Thus, one may recieve an ordered reading of a given device at 30 locations with just one request. In addition, ordered arrays of information from a particular processor may be specified, and these arrays may be gathered as "wild cards."

The distributed processors provided for the Tevatron control system are indicated in Figure 2. Microcomputers have been provided for vacuum scanning, refrigerator and compressor control, magnet protection, beam position/intensity/loss monitors, real time function generation, power supply excitation, quadrupole extraction resonance control, and fast time plotting of ADC channels. The larger systems could not be supported on sufficiently few CAMAC cards so as to be sensibly organized. Systems such as the refrigerator microcomputer require eight standard multibus cards, and are therefore housed in an individual multibus crate. An inter-computer protocol for handling Front End to microcomputer subsystem communication is used to communicate to all the intelligent subsystems capable of a variable set of readings and settings. This protocol has been coded for the varying processors used in these subsystems so that the Front End has need of only one driver for all these systems. Some hardware options have been included so that even byte ordering differences in two-byte long words are handled when speaking to Zilog Z80's as compared to Motorola 68000's.

It has by now been possible to gain considerable experience in utilizing the controls system for the various parts of the accelerator complex. The controls for the 24 satellite refrigerators and the six sets of compressors have run dependably over many months of testing. The controls system has been used "on-line" to do orbit smoothing. This application utilizes the beam position processors to gather data on beam position. The Operational VAX then calculates corrections to the existing orbit and sets up new waveforms for the 216 dipole correction elements around the ring. This application utilizes almost the entire controls system and has worked very well in the initial tuning of the Saver. For the Tevatron RF, control is required for supplies as well as curve generators which drive the RF and monitor its stability. The RF system has performed well from the first time it was turned on, successfully captured the Saver beam, and accelerated it. At present, beam has been accelerated to 600 Gev/c at an intensity of $2\times10^{**}12$ and to 700 Gev/c at lower intensity. Beam has been successfully captured and circulated at 512 Gev/c for 800 seconds with a loss of only a few percent.

THE CONTROL SYSTEM FOR THE DARESBURY SYNCHROTRON RADIATION SOURCE

D.E. Poole, W.R. Rawlinson and V.R. Atkins[*]
Science & Engineering Research Council
Daresbury Laboratory
Warrington WA4 4AD
England

*Now with Shell Research Ltd.

1. INTRODUCTION

The Daresbury Synchrotron Radiation Source [1] is the world's first electron storage ring purpose built to provide intense beams of VUV and x-rays for research embracing a wide range of scientific disciplines. The source comprises three accelerators, a 12 MeV linac, a 600 MeV synchrotron and the 2 GeV storage ring itself. The control system uses a two level computer network which is linked to the Daresbury site central computer, the latter being used for software preparation and applications requiring bulk data storage. This paper describes the design of the system hardware and software, and the interfaces to the plant and the operators. The facility has been fully operational for over two years and operational experience and proposed future developments are discussed.

2. COMPUTER SYSTEM

The system is based on the use of a two level network of computers, as shown in Figure 1. One minicomputer system is dedicated to each of the three accelerators, with a fourth system providing controls for the experimental beam lines. The main function of these machines is control of the plant interface. The main computer of the control system provides network communication, utilities for maintaining the system and the operator interface.

Hardware

The computers are all from the Perkin-Elmer (Interdata) range. The four minicomputers are model 7/16s and 8/16s with 64 Kb memory. They have no mechanical peripherals and all interfacing is via CAMAC. All programmes are resident together with a

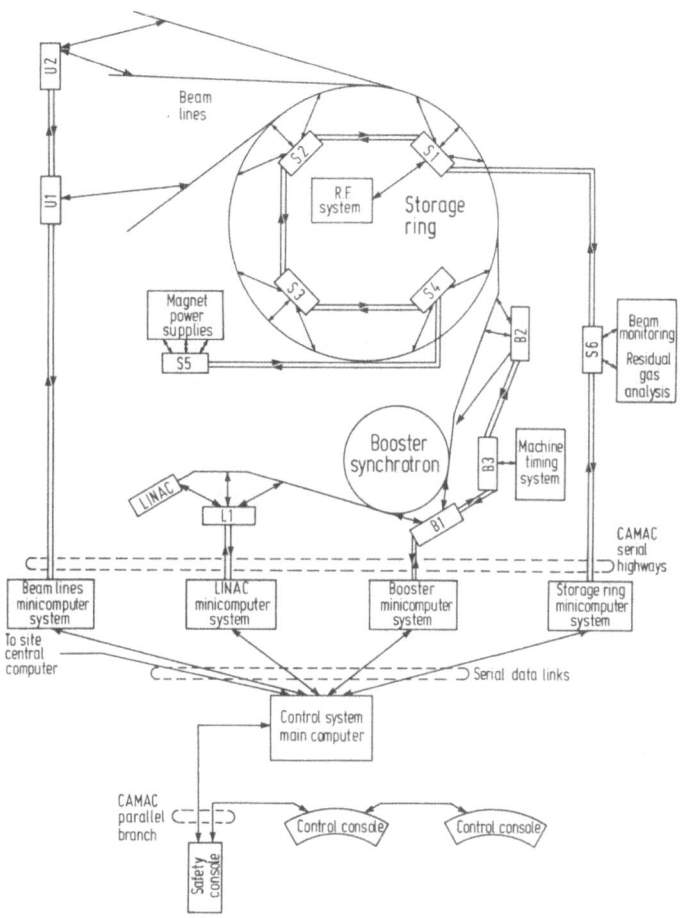

Figure 1. SRS Control system block diagram.

subset of the control system database. The emphasis is on very high reliability since re-initialisation at this level could lose a stored beam.

The main computer is a model 7/32 with 320 Kb of memory, two 10 Mb disc drives and a line printer. The operator control consoles are interfaced via CAMAC together with several VDU's and printer terminals. Re-initialisation of this system causes no change in the state of the plant.

System Software

The 16 bit machines use the Perkin-Elmer operating system OS/16-MT which has been enhanced to provide support for CAMAC and network communications. The 32 bit machine uses the manufacturer's multi-tasking operating system OS/32-MT. This has good support for real time applications, but to provide an efficient control system, three levels of enhancements were added.

Support for CAMAC is in the form of an additional supervisor call enabling tasks to access CAMAC modules, and also enabling special operating system device drivers to be written so that peripherals interfaced via CAMAC are treated as standard. User tasks are prevented from accessing CAMAC modules for which a system driver is present.

The network communications package uses a protocol developed as a site standard for Daresbury Laboratory. CAMAC to CAMAC serial data links operating at 614 kbits/sec are used, with all the line level protocol being handled within a special operating system device driver. An extra supervisor call handles the network protocol at the call level, and thus an efficient communication system is available to user tasks.

The third extra supervisor call provides access to plant parameters [2]. A user task may request access to a table of parameters which may be selected at random from the full database, and may then monitor or control the analogue settings or the status of those parameters. Only one user task is allowed to control a parameter at any time, but any number may monitor it. An important feature of the process control package is that it allows parameters to be defined which compute their values from groups of other parameters. These "virtual parameters" may be selected and controlled from user tasks as normal parameters. Examples are the storage ring average vacuum pressure and the booster energy. The storage ring contains 16 multipole magnets [3] which can be programmed to produce any correction field up to duodecapole. The virtual parameter facility makes it possible to present these to the operator as if they were normal magnets.

Application Software

The high level language selected for application level programming is RTL2 [4]. A cross compiler which runs on the central computer and a self compiler for the Perkin Elmer machines have been produced at Daresbury together with a comprehensive library of support procedures. A large number of application tasks exist, but for routine operation, only about eight are required. These provide for:-

1) Execution of a sequence of control operations defined in command files.
2) Random control of any parameters directly by the operator.
3) Selection of "Fill storage ring" mode by the timing system.
4) Selection of "Stored Beam" mode.
5) Ramping of the storage ring energy.
6) Application of correction field patterns on the multipole magnets.

7) Plotting graphs of any parameters against other parameters or against time.
8) Supervision of the personnel safety system during search and lock-up of the machine areas.

There is a suite of tasks which are scheduled for periodic execution at varying intervals, which together provide a comprehensive data logging facility, including an archived operational history of the machine over its life. A further group of tasks of a rather specialist nature assist with accelerator physics investigations into machine behaviour, and there is the usual collection of utilities and diagnostics programmes for maintenance and fault finding on the control system itself.

3. OPERATOR INTERFACE

It seemed logical to interface the control consoles to the main computer of the control system, because the criterion of high reliability which justifies the minicomputers interfacing the plant does not apply, and interposing another minicomputer would simply slow down data transfer to and from the consoles. It is also more difficult to interface an operator than a plant item, and the more powerful 32 bit computer with better facilities for asynchronous I/O performs this duty more effectively. The two main operator consoles are equipped with:

a) A keyboard which has an additional function key pad seen as a separate device by the operating system, and five special keys used to initiate execution of the most frequently used tasks;
b) A colour display of which the screen is divided to provide two logical devices. The lower portion, 5 lines with scrolling, echoes input from the keyboard and is accessible for displaying messages written by any user task. The upper portion of 26 lines can be operated in page mode or scroll mode and is normally used exclusively by one task at a time.
c) A tracker ball, used to manipulate a cursor on the colour display.
d) A light pen which can be used interactively with the colour display.
e) A control knob which can be used to adjust any analogue parameter on the system.
f) A high resolution refreshed vector graphics display.

For plant commissioning and fault finding small portable control consoles are available, interfaced at the 16 bit minicomputer level which can be plugged in local to the plant.

4. PLANT INTERFACE

The CAMAC Serial Highway was chosen as the means of bridging the distance between the computer room and the accelerators. The storage ring uses a highway 700 m long operated in byte mode. One or more serial CAMAC crates form the nucleus of a plant control station (Figure 2). Positions for control stations are chosen to minimise problems of cabling between them and the plant. The linac has one, the booster three and the storage ring six.

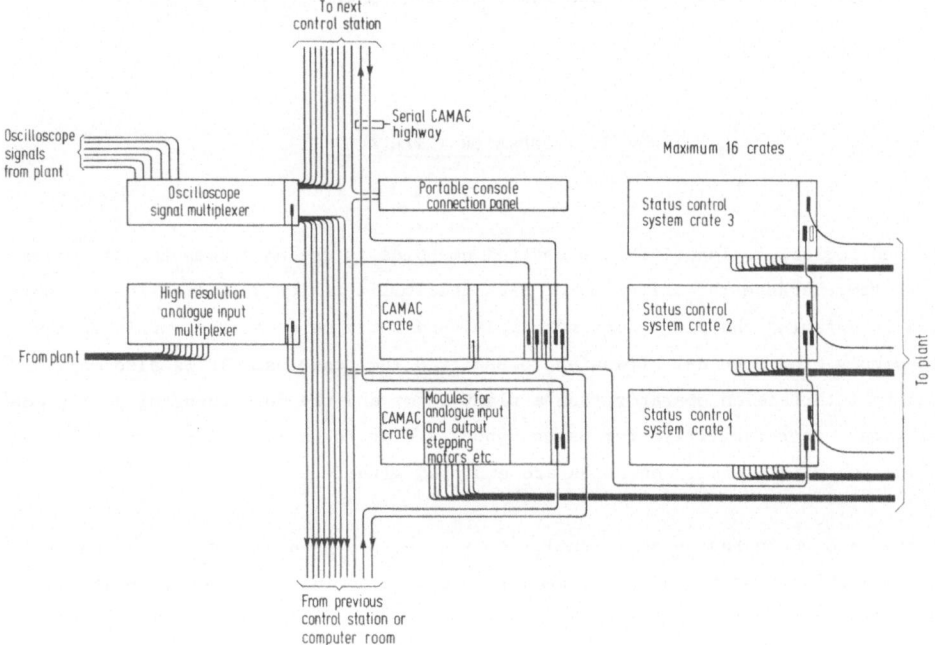

Figure 2. Components of a typical control station

Analogue control is achieved by commercially available CAMAC modules providing either two channels at 15 bits + sign resolution or sixteen channels at 11 bits + sign resolution. Analogue monitoring of the higher resolution signals is via reed relay multiplexers and for the lower resolution signals, CAMAC modules providing 32 channels are used.

For controlling and monitoring of plant status, a specially designed multiplex system external to CAMAC is used. This combines in a modular system the hardware to test the state of contacts in the plant via a pulsed highway 20 times per second, the conversion from electronic logic levels to potential free contacts able to switch relay circuits within the plant, and all interlock protection functions. Periodic self-checking of the protection logic is incorporated to ensure that this system is at least as fail-safe as conventional relay logic.

A further special multiplex system is used to convey analogue signals for oscilloscope monitoring in the control room. Eight co-axial cables link all the control stations. Four of these terminate on each of the two control consoles and the operators may request any signal which is interfaced to the system to be fed into any of the four cables serving his console.

5. OPERATIONAL EXPERIENCE AND FUTURE DEVELOPMENT

The system has been in use since 1978 for commissioning of the accelerators and since early in 1981 for scheduled operation of the facility. The machine system has been expanded to cater for control of the beam outlet ports as new beam lines have been constructed, and the additional 16 bit computer for controlling beam line hardware has been added. The reliability of the computer hardware has been excellent, and the 16 bit systems routinely run untouched for a complete machine cycle of six to eight weeks. The policy of keeping the function of the 16 bit systems as simple as possible without any dependence on mechanical peripherals has proved to be justified. The mean time between failures of the 32 bit systems is typically a few days but can fluctuate considerably, improving as errors are found and eliminated, deteriorating after the installation of a major modification.

Response to operator commands when several application programmes were executing was initially unsatisfactory. The decision to provide access to the whole accelerator complex from any task means that requests to access the process control package are queued and the speed of transmission of data packets over the networks was found to be the principal cause of delays. Features originally specified, but not available at the time of machine start up have subsequently been installed, doubling the speed of network operations. These include hardware vectoring of interrupts from CAMAC, multichannel direct memory access for the six data links on the 32 bit computer and rewriting the call handler for the 16 bit computers. For speed of implementation, this was originally produced as application tasks written in RTL/2, but the new version is modelled closely on the more efficient call handler in the 32 bit machine.

Once the facility had become fully operational, the amount of time available for operating system development has been negligible, even during machine shutdowns the control system is still needed. Since considerable effort is involved in incorporating the special enhancements into any new release of the operating system, the policy adopted was to move to a new release only if it offered additional features of relevance to the control system. The 32 bit operating system currently in use is 3 revision levels out of date. It is now important to bring it up to date in order to avoid the problem of obsolescence which several accelerators have already encountered. It is quite common for the construction period and useful life of an accelerator to exceed the time for which it is economical to maintain the hardware of one computer generation. It is expected that the SRS computers will be replaced in 1985 by which time they will be 11 years old, and that instead of four 16 bit minicomputers there will be two 32 bit machines, one for the linac and booster, and one for the storage ring and beam lines. These will be linked to a larger 32 bit machine as at present. The new hardware requires the latest release of the operating system, and one machine from the Perkin-Elmer 3200 series has been purchased in order to carry out the necessary development work to prepare for the changeover.

The status control and monitoring system has proved the versatile and reliable facility that it was intended to be and the number of crates in service is now 30. The fail-safe features incorporated in the logic have successfully ensured that the interlock protection has never failed in a dangerous situation. However, a significant number of intermittent problems in the electronics have caused difficulties in diagnosis, and it is felt that a new design incorporating a microprocessor could provide a more intelligent checking process, and more versatile control for certain special items and it is planned to develop this soon.

REFERENCES

1. SRS Design Study Team. Daresbury Laboratory Report DL/SRF/R2 (1975).
2. Daresbury Laboratory Technical Memorandum DL/SCI/TM19A.
3. R.P. Walker, Proceedings 7th Int. Conf. on Magnet Tchnology, Karlsruhe, June 1981.
4. J.G.P. Barnes and R.J. Long, Proceedings, IEE Conference on Software for Control, Warwick, July 1973, P75.

THE MICROPROCESSOR-BASED CONTROL SYSTEM
FOR THE MILAN SUPERCONDUCTING CYCLOTRON

F.Aghion, S.Diquattro, A.Paccalini
E.Panzeri, G.Rivoltella

Università degli Studi di Milano
Istituto Nazionale di Fisica Nucleare

1. INTRODUCTION

The present status of the design of the computer control system for the Superconducting Cyclotron,under construction since two years at the University of Milan,is here discussed.
We may briefly recall that this new machine has been designed to have a K_{FQC}= 200 and a K= 800, enabling energies of 100 MeV/n for fully stripped light ions and 20 MeV/n for uranium; the Cyclotron will allow operation with both internal ion source or injection from a Tandem; eventually axial injection from an external ions source could be implemented. The magnetic field will span between 22 and 48 Kgauss and the R.F. range will be between 15 and 48 MHz., for 100 KV dee voltage, and harmonic operation for h=1 to h=4.
Two years ago,an outline of the main ideas in the design of the control system was presented at the International Conference on Cyclotrons held in Caen. (1)
Since then, first achievements as well as the continuous progress in large scale integrated circuits have strengthen our decisions.
Not only microprocessors are becoming more and more reliable,powerful, and cheap but also analogue-to-digital converters,digital-to-analogue converters and high-speed communication components are becoming convinient giving us the chance to integrate intelligence in every part of the equipment we think useful, thus reducing interaction between the latter and the control system to a high-level flow of informations.

2. THE CONTROL NETWORK

Designing distributed control systems, several solutions are attractive, but two main schemes are usually taken into account: CSMA/CD configurations and token-passing rings. (2-5)
Usually CSMA/CD techniques are popular in transaction oriented environments, such as computer-to-computer communications where access delays can be tolerated.
In real-time applications where access must take place before a fixed time for an event to occur, token-passing schemes are generally preferred.
A deeper look at the most up-to-date solutions proposed by control companies and laboratories, reveals an interesting point: the more the control stations become powerful the more indeterminate access methods such as CSMA/CD are becoming popular also in time critical processes.
In fact it turns out that real-time applications are assured by the control stations that would be still operable in case of failure of the communication system and this latter becomes a computer-to-computer not time critical link.

Advantages of a distributed architecture can be briefly underlined:
- control stations can be designed and tested,for a large percent, as stand-alone controllers
- communication on the network can be in a high-level message form, and for instance, ASCII characters easy to be viewed on a terminal or printed, can be used
- each control station can be detached,for maintenance or improvements implementations, from the control system and operated by means of a local terminal or a portable console.

3. HARDWARE ARCHITECTURE

An optical bus with a star topology will connect through an Ethernet network a PDP 11/44 with up to fifteen microprocessor-based peripheral control stations. (fig.1)
The mini-computer provides large programs execution, interfaces with storage disks, magtapes and printers, and will act as a supervisor in the system.
The control stations,directly interfaced with the machine, are based on a crate using the proposed IEEE 796 standard bus (Multibus) architecture. Such a crate supports an arbitrary number of single boards computers, thus allowing to control contemporaneous processes in parallel.

3.1 Inside the control stations

We have specified four different functions inside each programmable control station :
- interface to the analogue world
- processing capability
- communication
- display capabilities

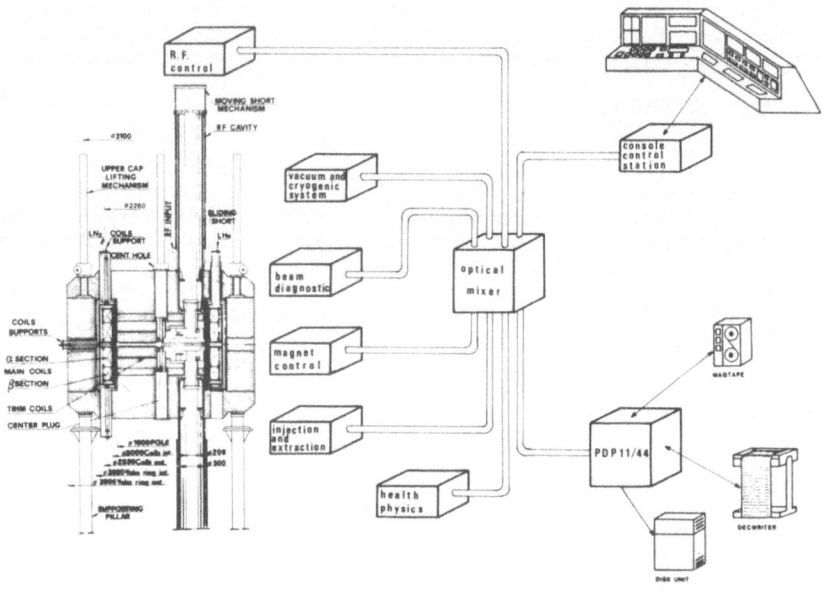

Fig.1 - Lay-out of the control system

Each function has a dedicated processor and Multibus[6] is used for data exchange.
Intel SBC 86/14 has been chosen as our standard microcomputer. It is based on a 16-bits microprocessor (iAPX86) and supports anumeric data processor (iSBC337), up to 64 Kbytes of RAM and up to 64 Kbytes of on-board EPROM.
A small number of standard boards for controlling the Ethernet network, for A/D and D/A conversion and for graphics, has been selected from the market. Limiting the range of different boards allows an easier design and simplified maintenance.
Nethertheless we are convinced that the choice of a standard bus like Multibus gives us the chance to remain open to meaningful improvements in technology, with a minimum effort for upgrading.

3.2 The optical bus

Interest in designing a local area network like Ethernet, in which cables are replaced by optical fibers, essentially derives from the following technical advantages that seem quite interesting in accelerating machine environments :
- inherent ground isolation
- no crosstalk due to signal radiation
- immunity to electromagnetic fields interference
- frequency-independent attenuation

Choice of a network topology with optical fibers is limited by physical reasons; thus a complete likeness with Ethernet is not possible.
In fact a linear bus system with optical fibers should require as many T-couplers as many drops are considered in the design.
It is easy to calculate that only a very small number of stations can be connected to the bus, owing to the high attenuation introduced by each T-coupler (about 3 dB).
On the other hand a star topology allows for a constant attenuation value that can be considered in the flux budget calculations.
At present, transceivers must operate with a maximum power attenuation of 27 dB, in a network with sixteen stations appended to the bus. This value takes into account losses introduced by connectors, the star coupler, and cable allowing 500 meters spacing between each node.
Some investigation is actually in progress on nuclear-radiation effects on fibers ordered from different manufacturers, in a simulation of a cyclotron environment and results will be meaningful before the end of this year.

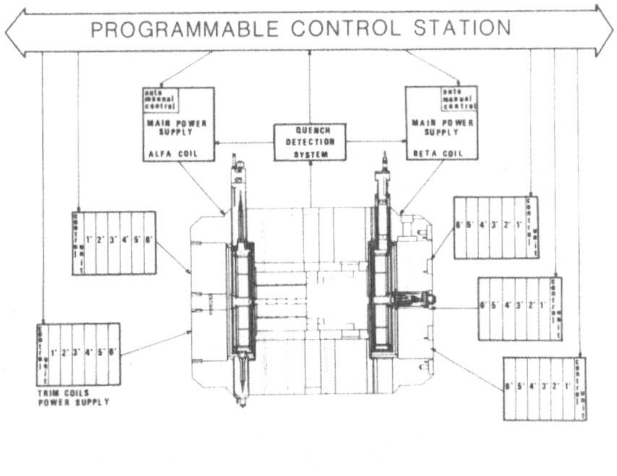

Fig.2 - The power-supply control

4. SOFTWARE

It is well known that software has become the most important item in a man-hours and costs budget for a control system.
At present we are investigating on a software design methodology using silicon software components as a potential solution to this kind of costs and efforts.
No standardisation will be done on languages even if we espect that PLM/86 will be extensively used at board level, FORTRAN for cal-

culations and Assembler for time critical routines.

5. THE POWER SUPPLIES CONTROL STATION

Two home-made power supplies will feed the main coils of the machine with a maximum current of 2000A, while 30 commercial power supplies will be connected to the trim-coils.
Closed -loops, regulations,sequencies,ramps and status acquisitions are performed by the programmable control station, while A/D and D/A converters plus alocal control unit for manual operation are fitted in each power supply.
At present we are debugging this control stationthat will be fully operating for the first magnetic measurements on the machine.

REFERENCES

1. F.Aghion, G.Rivoltella, L.Troiano - Outline of the control system for the Milan Superconducting Cyclotron - Ninth Int. Conf. on Cyclotron and their applications - Caen 1981
2. W.Busse - Advantages and constraints of modern accelerator control systems - Ninth Int. Conference on Cyclotrons and their applications - Caen 1981
3. H.D.Lancaster, S.B.Magyary, J.Glatz, F.B.Selpha M.P.Fahmie, A.L.Ritchie, S.R.Keith, G.R.Stover and L.J.Besse - A Microcomputer Control System for the Superhilac third injector - Proc. of the 1979 Linear acc. Conf. - Montauk N.Y.
4. R.Melen - A new generation Control System at SLAC - Proc. of the 1981 Part. Acc. Conf. IEEE Trans. Nucl. Sci. NS-28 N.3 part 1
5. M.C. Crowley-Milling - The Control System for LEP - Proc. of the 1983 Part. Acc. Conf. IEEE Trans. Nucl. Sci. NS-30 N.4 part 1
6. Multibus is a trademark of Intel Corp.

THE ELSA CONTROL SYSTEM HARDWARE

Ch. Nietzel, M. Schillo, H.J. Welt, C. Wermelskirchen
Physikalisches Institut der Universität Bonn
Nußallee 12, D-5300 Bonn 1

Introduction

ELSA is an Electron Stretcher and Accelerator ring[1,2] fed by the Bonn 2.5 GeV Electron Synchrotron and has been designed to provide electron and bremsstrahlung beams with high duty cycle.

In stretcher mode operation electron pulses from the synchrotron are injected into ELSA with a maximum rate of 50 Hz. The electrons are then ejected from ELSA at a constant rate within 20 msec or more. The duty cycle will be of the order of 95%.

When used as a post accelerator to yield up to 3.5 GeV electrons ELSA is fed with 1.75 GeV electrons from the synchrotron. Times for ramping up and down are both fixed to 150 msec. With a maximum length of the high energy flat top of 500 msec and a 20 msec injection plateau a duty cycle of up to 60% will be achieved.

ELSA is planned to operate in the stretcher mode at the end of 1985 and as a post accelerator about one year later.

Requirements

A suitable computer based control system is required to operate the accelerator in its different modes.

In the stretcher mode fast electron transfers (< 1 µsec) from the synchrotron to ELSA and subsequent continuous extraction with constant external beam intensity have to be handled.

In the post accelerator mode synchronous control of a multitude of different magnets (dipoles, quadrupoles, sextupoles, kickers, septa) and the RF system in predetermined sequences must be achieved. To provide individual shapes of the ramps for these elements is one of the major tasks of the control system.

Human interaction is to be done through central operating consoles. Here particular emphasis has to be given to the aspect that the new accelerator will not be operated alone by trained staff but also by the experimentalists themselves.

All the support equipment (vacuum pumps, valves, cooling, and interlock) has also to be controlled by the system.

Furthermore we have to take into account the future integration of the 2.5 GeV synchrotron. Its present manual control will be partially discarded and its elements connected to the new control system.

Components

The overall structure of the projected control system was considerably influenced by related systems already in existence at CERN and DESY. There are three main sectors (fig. 1).

a) <u>Computer Configuration</u>

The central computing resources will be provided by two VAX-11/750 computers from Digital Equipment. One of these is in continuous use for the operation of ELSA (Control processor). The other one will be utilized for development of new control software. It will also serve as a standby for possible failures of the first processor, being able to take over its tasks at short notice.

Later on the control of the synchrotron will be a further task for the standby processor.

Both computers will be linked via DECnet (a serial data link provided by Digital Equipment). It will be mainly used to copy the actual set of machine parameters to the standby processor at regular intervals for backup purposes and archiving on its more extensive set of peripherals.

The control processor forms the link between process peripherals and console elements. Here the main tasks are:
- maintenance of a database for machine parameters
- display of the actual parameters for the operator
- distribution of operator commands to the microprocessors of the process control
- supervision of the total control system and reaction to alarm messages
- logging of accelerator operations and equipment status
- calculation of correlated parameters (ramps, beam lines).

b) <u>Operating Consoles</u>

The consoles are to display the machine status in a comprehensive manner. Thus the application of colour graphic video devices is an essential requirement.

Because of financial and staff limitations we were fortunate to be allowed to use the PADAC[3] console system interface standard developed by DESY.

ELSA is controlled by two equal priority operating consoles. Both contain two 19" colour TV monitors, an alphanumeric keyboard, and several trackballs.
They can be used together with displayed cross-hair cursors as menu selectors or to input analogue data (replacing dials and knobs). The raster-scan colour monitors offer a resolution of 256*512 pixels with 7 possible colours. Space is provided for additional equipment, eg. up to three 12" monitors, oscilloscopes

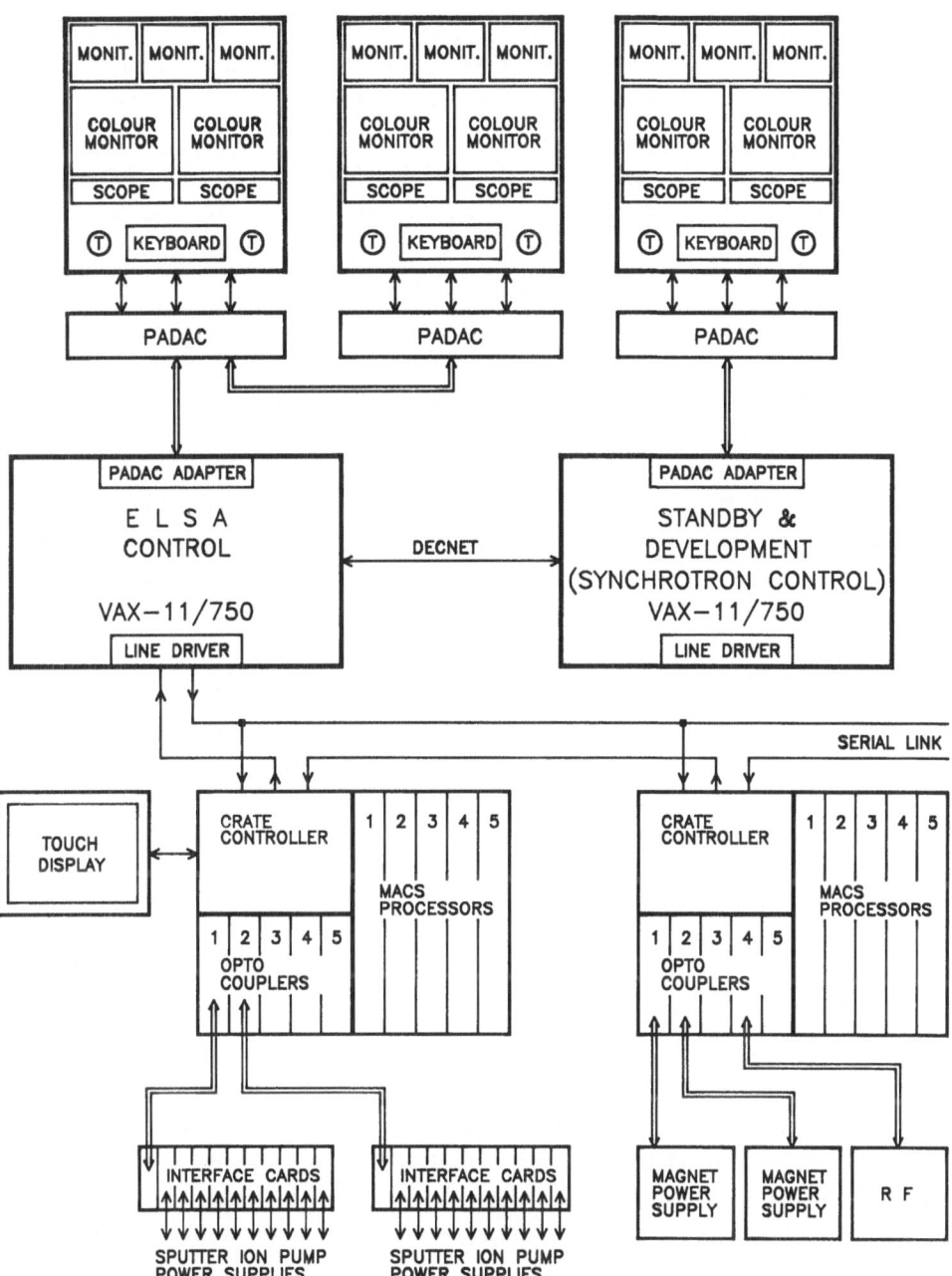

Fig. 1 : Control system with sample ELSA equipment

etc. An extra operating console attached to the standby processor will prove to be of great convenience for the development of control programs and their graphical output in the initial stage and will be used later on for the control of the synchrotron and its injector (LINAC).

Through the extensive sacrifice of bulky manual controls all the operating equipment for ELSA can be placed in the existing synchrotron and linac control room.

c) Process Interface

The interfacing of a variety of different equipment (eg. magnets, RF, vacuum, valves, cooling, and interlock) forms a vital aspect in the design of the control system. The use of CAMAC was excluded by its prohibitively high costs. Similarly the utilization of the DESY built SEDAC[4] system required the development of a multitude of different specialized modules, ruled out for reasons of the limited manpower available. The serial process interface MACS (Microprocessor-Aided Accelerator Control System) was designed to be the solution. It resembles SEDAC's structure, but simplifies the hardware development through the use of standardized microprocessor modules. These modules will also relieve the host computer of most of the simple monitoring tasks.

The following elements make up the system (fig. 1):
- line driver (UNIBUS module)
- crate controller for the processor crates
- microprocessor modules
- opto coupler modules
- interface cards in subcrates.

The line driver transmits serialized data and commands from the control computer to the processor crates and receives messages from the microprocessors. Data transfer between control computer and processor modules is achieved by a DMA method. The serial link is planned to obey the HDLC[5] protocol. Blocks of data up to 16 kbyte can be sent in a single "frame" with 1.25 Mbit/sec. Transmission error detection and recovery is according to the HDLC protocol. This way it is possible to transfer the different ramps given by up to 1500 data items directly to the memory of the processor modules employed.

A different competitive serial technique is the MIL-STD-1553 specification. A decision in favour of either has yet to be made.

MACS uses the SEDAC topology: a star network for the down transfer from the control computer to the MACS crate controllers and a daisy-chain for the way up.

Hence no collisions can occur in the down transfer of data and commands. Response messages are always synchronized by the daisy-chain mechanism and the automatic polling of the line driver.

The double-height eurocard processor modules reside in the processor crates together with opto coupler modules and the crate controller. Bytewise data transfer takes place between processor modules and the crate controller, which serializes/parallelizes the data and monitors the HDLC protocol.

A terminal connection is available on every crate controller. This allows testing and maintenance as commands to individual modules can be given locally and even without the use of the control computer.

The identical processor modules (fig. 2) form the lowest intelligent level in the whole control and monitoring system. They are adapted to the particular equipment controlled through the use of dedicated software. In the development stage that software will reside in the processor modules' RAM and after thorough testing in EPROM. Through these especially taylored programs the individual microprocessors know the characteristics of "their" equipment enabling its autonomous control. Under ordinary operating conditions only messages concerning faults are transmitted up to the control computer.

The different types of equipment are differently interfaced to the system. All devices requiring short reaction times (magnets, RF, beam transfer) have their individual processor modules directly connected through device dependent opto coupler cards.

Equipment used for the control and monitoring of vacuum, temperature, and interlock, where hundreds of parameters are to be handled, utilizes processor modules in conjunction with subcrates containing the necessary interface electronics like ADCs, DACs, switches, inputs. The reduced speed of their readout matters little as independent circuits for the safety of personnel and equipment are employed. Acquisition is used here mainly to obtain data for comprehensive status displays and diagnostics at the operating consoles.

An example for the subcrate architecture is the following structure of the vacuum control: The power supplies for the 50 sputter ion pumps are installed in 5 racks each containing 10 power supplies and a MACS subcrate. Each power supply is connected to an individual eurocard module of identical type containing optically isolated inputs and output latches and a 12 bit ADC. One processor module serves each power supply rack (subcrate) to which it is connected through opto couplers. All the processor modules for the power supplies reside in one processor crate.

For the local control of the vacuum system a touch display connected to this processor crate is used to directly monitoring and switching the individual vacuum pumps.

Status

The mentioned subcrates for the sputter ion pumps have already been built and are fully tested. The first version of a MACS processor module exists and runs special diagnostic software. The design of the turbomolecularpump (TMP) interface is finished and the magnet power supply connections have been defined.

References

1) Vorschlag für den Bau eines Stretcherringes am 2.5-GeV-Elektronensynchrotron der Universität Bonn
   Internal Reports BONN-IR-79-31 (Oct 79)
   BONN-IR-82-17 (May 82)

2) D. Husmann, Invited Talk for the US Part. Acc. Conf., Santa Fe 1983
   Internal Report BONN-IR-83-6 (March 1983)

3) G. Hochweller, H. Frese IEEE NS-$\underline{26}$, 3382 (1979)

4) H. Frese, G. Hochweller IEEE NS-$\underline{26}$, 3385 (1979)

5) ISO 3309, Entwurf DIN 66221

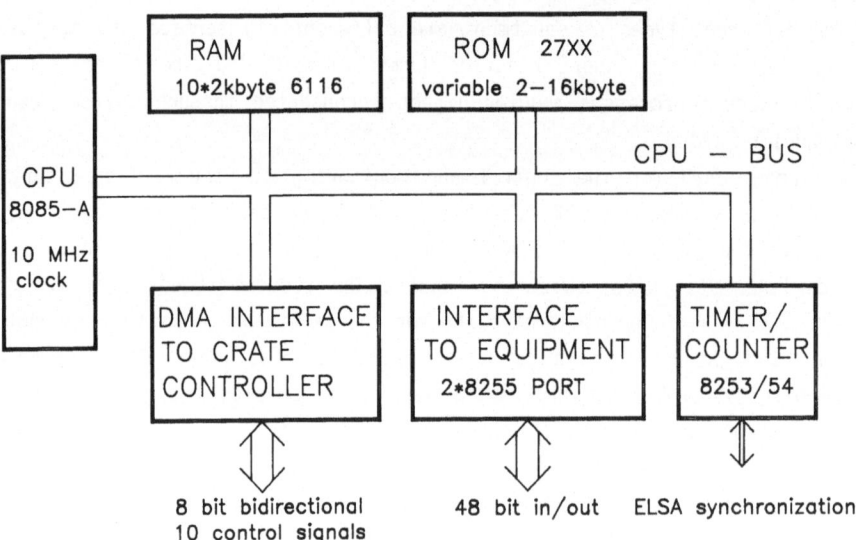

Fig. 2: MACS processor block diagramm

COMPUTER CONTROL SYSTEM OF POLARIZED ION SOURCE
AND BEAM TRANSPORT LINE AT KEK

J. Kishiro, Z. Igarashi, K. Ikegami, K. Ishii, T. Kubota,
A. Takagi, E. Takasaki, Y. Mori and S. Hukumoto

National Laboratory for High Energy Physics, Japan

1. Introduction

A new project to accelerate the polarized proton up to 12GeV is now in progress at National Laboratory for High Energy Physics in Japan (abbreviated KEK)[1,2]. The project includes the construction of an optical pumping polarized H^- ion source, a new 750keV preinjector and a beam transport line with the length of about 40m. The ion source and the beam line are consisted of so many magnets and power supplies, about a handred in whole, that we developed a computer control system with the use of the serial data transfer linkages. Because of the inclusion of the equipments with the different electrical potential in this system, the optical fiber cables are used for the serial highways to ensure the high noise margin and electrical isolation.

2. Layout of the preinjector and the beam line

When the project of polarized proton acceleration was authorized, there was no enough space around the present preaccelerator building. Thus, the new building was decided to be built behind the present one and the accelerated polarized H^- ion would be transported by a 40m beam line to the injector linac.

The polarized H^- ion source is placed in a 750kV high voltage terminal and consists of 32 power supplies for its equipments[3]. The beam transport line, called LEPBT, connects between new preaccelerator and the present injector linac and consists of 54 quadrapole magnets, 5 bending magnets and a solenoid magnet.

All of these many magnets and power supplies are controlled at the local control station beside the new preaccelerator building. A small computer is placed at another side of that building because of the cleanest condition at there.

3. Power supply control

We made standards for the control procedures of the power supply and other equipments as follows;
 1) ON/OFF function should be carried out by a power relay which should be driven by a small relay controlled by a TTL circuit,
 2) Out put current of each power supply should be controlled by a low level reference voltage ranging from 0 to 10V,

3) The monitoring of the out put current should be carried out by observing the induced low level voltage in a shunt resistance and it should be ranging from 0 to 10V,
4) The break down function at emergent condition of each power supply or magnet should be carried out by a relay logic circuit and the status of such condition could be transfered to an interface circuit and
5) It should be electrically isolated between the interface circuit and the control one.

4. Layout of the control system

The control system is consisted of a sixteen bits small computer with its standard peripheral equipments. Two CAMAC crates are installed in this system. The first one is connected to the buss-line of the CPU with the use of a dedicate crate controller. In this first crate, a CAMAC serial driver (SD) is installed in order to extend the CAMAC serial highway. The second crate is connected to this CAMAC serial highway with the use of a serial crate controller (SCC).

This configulation of the crates enables us;
1) The extention of the system will be accomodating in future and
2) The system can be easily supported at the break down period of the main CPU with the use of a single auxiliary crate controller which will be installed in the first crate.

From those CAMAC crates, the four power supply serial highways (PS SHW) are extended to the power stations. These serial highways are consisted of the optical fiber cables. The optical fiber cables enable us to connect many equipments with different electrical potential in a single system and also make us easy to place those in the much noisy surroundings.

At the power station, the power supply interface modules which were developed at KEK are connected to the power supply serial highways by the so-called multi-drop style. The system configuration is shown in Fig. 3.

The control station is placed beside the new preaccelerator building. At this station, one color CRT graphic display and two blak/white CRT character display units are placed in order to monitor the system equipments. There are three manual controllers and one key-boad terminal at the conrol station which enable us to control the power supplies according to the standards mentioned in previous section (Fig. 1).

The manual controllers are connected to one of the CAMAC crates in order to control the power supplies manually. However, in the usual case, the out put current and the status of each power supply are observed by a software program and the long term deviation of each current are compensated by this software program automatically.

Fig. 1 Control station

5. Module description

We developed three modules at KEK.
1) Power supply interface modules,
2) CAMAC serial tranceiver modules and
3) Electrical/optical signal converters.

The power supply interface circuit is constructed in a double width NIM module case. It contains both the TTL logic circuit which acts as the interface circuit between MIL-1553B signal and the power supply function signal and the small relay logic circuit which delivers the interlack functions (Fig. 2).

The interface circuit contains the serial/parallel converter, address decoder and status register. And also it contains a digital/analogue converter which delivers the low level reference voltage corresponding to the out put current of the power supply and an analogue/digital conveter which enables us to observe the out put current. Both of these converters are electrically isolated in order to ensure the high noise margin of each circuit and to sustain a possibility of connencting to the power supply with floating elecrical potential.

The CAMAC serial tranceiver acts as a serial/parallel converter between the CAMAC parallel signal and MIN-1553B pallel one.

Fig. 2 Power supply interface module.

6. Software program

The control program was developed at KEK. It consists eight tasks and the significant tasks are as follows;

    1) Scanning task

        The status and the out put current of all the equipments are observed every thirty seconds by this task. And the out put current deviation is compensated sumiltaneously.

    2) Interlock task

        When a power supply or a magnet is suffered from an emergent condition the interlock circuit cuts off its out put power without software program concerned. At the same time, the interface module transferes the status word indicating which emergent condition be caused. This task is triggered by this status word and tried to recover that condition. If this recovery action has failed the task indicats it on the color CRT display and awakes an operator to it by ringing a buzzer.

    3) Manual control task

        This task is triggered when a manual controller is operated. This task interprets the bottom operation and transfers it to the power supply interface module concerned.

The control program should be executed on the fore-ground. And the back-ground is reserved to the application programs. This configulation enables us to develop an application program independently to the control one. The access from an application program to one of the system equipments is enabled by the fourth task;

    4) Fore-gound/back-ground communication task

        This task provides twenty-four subroutines which should be called by supervisor mode from user application programs.

7. Conclusion

The test operation were started from this January. During this operation some debugings of the software program and hardware circuit were carried out and the construction of this system was successfully completed by this April.

We took much cares of the electrical isolation between equipments. Namely, we used the optical fiber cables to extend the serial highways. And in a power supply interface module, a part of the interface circuit connected directly to the power supply is isolated electrically from other control circuit. This isolation is very efficient to construct a system in a noisy surroundings and also to sustain a possibility of connecting many equipments with different electrical potential in a single system. Especially, our system contains a very high voltage equipment, Cockcroft preaccelerator with 750kV, and sometimes it causes much noises by the discharge phenomenon. However, our system has not suffered from those large noises.

References

1. S. Hukumoto et.al.   Proc. of Particle Accelerator Conf.
       Washington, D.C., U.S.A. 1981.
2. S. Hiramatsu et.al. Proc. of 5-th Lin. Sympo. on High Energy Spin Phys.
       BNL, 1982.
3. Y. Mori et.al.   Proc. of 5-th Int. Sympo. on High Energy Spin Phys.
       BNL, 1982

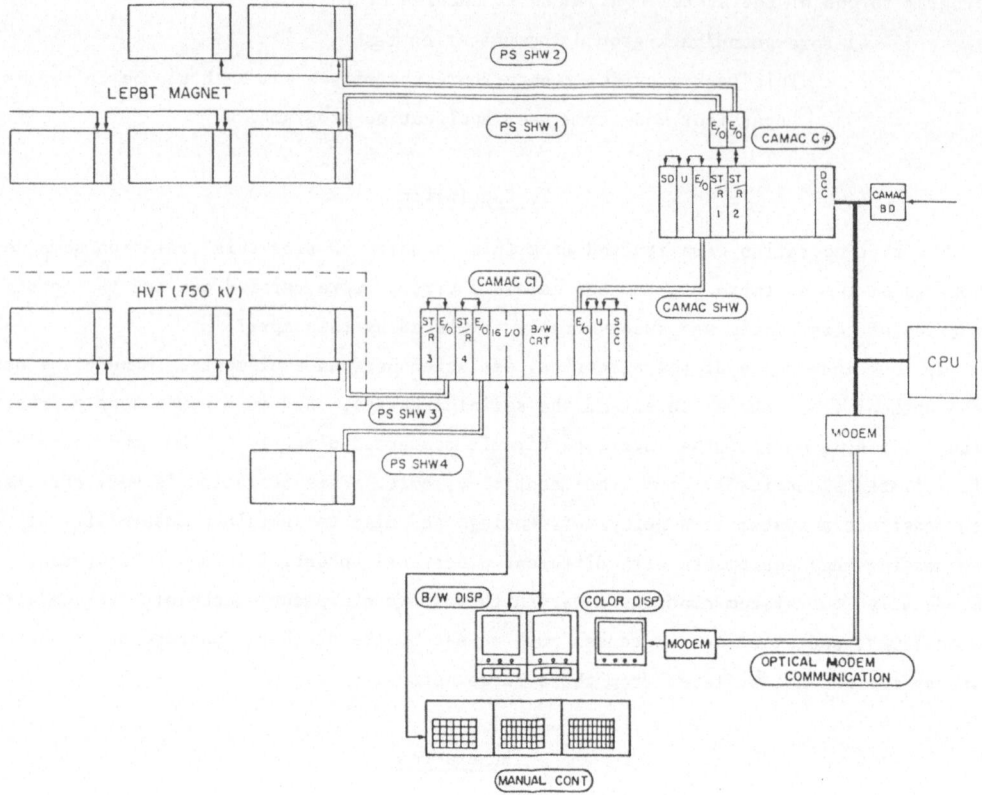

Fig. 3 System block diagram.

COMPUTER CONTROL SYSTEM OF TRISTAN

A. Akiyama, K. Ishii, E. Kadokura, T. Katoh, E. Kikutani
Y. Kimura, I. Komada, K. Kudo, S. Kurokawa, K. Oide
S. Takeda and K. Uchino

National Laboratory for High Energy Physics
Oho-machi, Tsukuba-gun, Ibaraki-ken, 305, JAPAN

Abstract

The 8 GeV accumulation ring and the 30 GeV x 30 GeV main ring of TRISTAN, an accelerator-storage ring complex at KEK, are controlled by a single computer system. About twenty minicomputers (Hitachi HIDIC 80-E's) are linked to each other by optical fiber cables to form an N-to-N token-passing ring network of 10 Mbps transmission speed. The software system is based on the NODAL interpreter developed at CERN SPS. The KEK version of NODAL uses the compiler-interpreter method to increase its execution speed. In addition to it, a multi-computer file system, a screen editor, and dynamic linkage of datamodules and functions are the characteristics of KEK NODAL.

Introduction

TRISTAN is an electron-positron colliding beam facility now being constructed at KEK [1]. It consists of three cascade accelerators: a 2.5 GeV linac, an 8 GeV accumulation ring (AR) and a 30 GeV x 30 GeV main ring (MR). Last two are controlled by a single system, namely TRISTAN control system.

The first beam injection to AR is scheduled in November 1983, while MR is now under construction and the target date for the first collide is 1986.

At present, nine minicomputers form a network for AR control.

System architecture

The complexity and the size of TRISTAN compel us to adopt a distributed computer control system. About twenty 16-bit minicomputers (Hitachi HIDIC 80-E's) are distributed around the accelerators. These computers are linked together by optical fiber cables to form an N-to-N token-passing ring network. A token is circulating along the network. If a node wants to transmit a message to another node, it must catch a free token before it sends a message as a packet; the destination node copies the message, while the packet on the ring travels to the source node, where the packet is destroyed and a free token is sent again to the ring. The function of message switching is distributed among the receiving nodes; each node recognizes the message addressed to it. The transmission speed on the optical fiber cables is 10 Mbps and the overall transmission capacity is approximately 600 kbytes/sec. Figure 1 shows the layout of the TRISTAN control system.

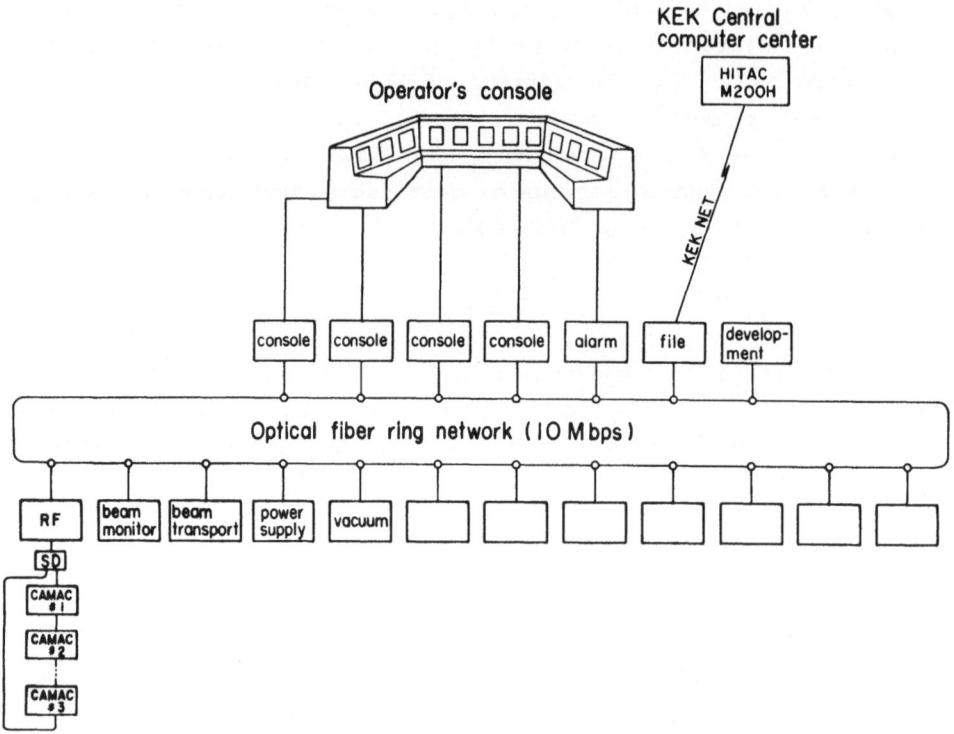

Fig. 1   Schematic layout of the TRISTAN control system.

The minicomputers are classified into two groups: the system computers and the device-control computers. The system computers are located in a central control room. Each system computer processes one of the control center functions such as the operator's console function, the alarm and logging function, the program library and database function, and the program development function. These computers as a whole make a control center of the system. The device-control computers control the hardware equipments such as magnets, power supplies, radio-frequency systems, etc. They are installed in the site buildings around the accelerators.

From each computer, a 2.5 Mbps bit-serial CAMAC highway extends to the equipments. The number of CAMAC crates for AR control is 40, while that for MR is 140. The advantages of CAMAC and CAMAC serial highway are: (1) we can easily make a test bench with a low-cost microcomputer, since CAMAC is computer independent, (2) many kinds of CAMAC modules are commercially available from many venders, (3) CAMAC serial highway is one of the most powerful, commercially available standards for the long-distance process control data-highway, and (4) the bypass and the loop-collapse functions of CAMAC serial highway are useful for maintenance.

Tasks that are not manageable with the minicomputers are done by the KEK central computer (Hitachi M200H). One minicomputer in the network and the central computer are connected by KEKNET [2], an in-house high-speed network at KEK.

NODAL system

The software system of the TRISTAN control is based on NODAL developed at CERN SPS [3]. This language is an interpreter and enables us to develop programs interactively. Moreover the multi-computer programming facility of NODAL is effective for the unification of a distributed computer system.

The KEK version of NODAL has the following enhancements over the SPS NODAL: (1) the fast execution speed due to the compiler-interpreter method, (2) the multi-computer file system, (3) the screen editing facility, and (4) the dynamic linkage of datamodules and functions.

To overcome the slow execution speed of NODAL due to the interpreting scheme, we adopt the compiler-interpreter method in KEK NODAL. When a programmer writes one line of a source code and presses the return key, the compiler part of NODAL translates the source code to an intermediate code. The interpreter part of NODAL interprets the intermediate codes at run time. This method can save time for changing a constant in ASCII string form into a floating point form, time for rearrangement of the order of operators and operands, etc. at the interpreting phase. As a result, two to three times speedup is achieved. Table 1 summarizes the result of a benchmark test for KEK NODAL. It shows the necessary time for a command to be executed in msec.

In SPS NODAL the user can save and load NODAL program files freely among minicomputers. In KEK NODAL data files also can be read/written without any restrictions among computers. Moreover, the user can freely input/output from/to line printers, magnetic tapes and typewriters attached to the different computers.

To ease the programming, the screen-editing facility is implemented in KEK NODAL. The user can make and change NODAL programs by moving a cursor on the CRT screen to the desired position to change, insert or delete characters.

Datamodules and functions are written in PCL, a FORTRAN-like compiler language for process control programming on HIDIC 80-E. The KEK NODAL system has a screen editor for PCL programming, too. The specifications of this editor are same as those of the NODAL editor.

Each datamodule or function is compiled, linked and loaded on memory or disk. The KEK NODAL interpreter searches the location of a datamodule or a function at run time and links it dynamically. There is no need for re-linking a datamodule or function to NODAL interpreter when a new datamodule or function is added to the system; in the static linkage method, time-consuming re-linking is necessary.

Acknowledgements

We wish to thank Professors T. Nishikawa, S. Ozaki, T. Kamei and S. Shibata for their supports and discussions. We also wish to thank Dr. H. Ikeda for his contribution to the early stage of this work.

Table I   Result of benchmark test

FOR LOOP	0.319 msec
SET A=1	0.463 msec
SET A=B	0.668 msec
SET A=A+1	1.066 msec
SET A=A-1	1.063 msec
SET A=A*0.9999	1.066 msec
SET A=A/0.9999	1.059 msec
SET A=A^0.9999	1.453 msec
SET D1(4)=1	1.150 msec
SET D2(4,4)=1	1.414 msec
SET D3(4,4,4)=1	1.673 msec
IF B>1;	0.771 msec
WHILE B<=1;	0.769 msec
DO 33.20; 33.20 RET	0.612 msec
DO 54 ; 54.10 GOTO 54.20; RET	1.152 msec
$SET C="A"	0.879 msec
$SET D="A" "B"	1.683 msec
$SET DS(4)="A"	1.134 msec
$IF C="X";	1.113 msec
$SET C=1	3.664 msec
$DO "S A=1"	265.306 msec
SET A=SIN(1.001)	1.569 msec
SET A=SQRT(1.001)	1.522 msec
SET A=DEF(1.001)	3.173 msec

* DEF is the name of a defined function, which consists of only one line: 4.10 VAL X.

References

1. T. Nishikawa and G. Horikoshi, contributed paper to 1983 Particle Accelerator Conference held at Santa Fe.

2. Y. Asano et al., Nucl. Instrum. Methods 159 (1979) 7.

3. M.C. Crowley-Milling and G.C. Shering, CERN 78-07 (1978).

THE SYSTEM FOR PROCESS CONTROL AND DATA ANALYSIS BASED ON MICRO-COMPUTER AND CAMAC EQUIPMENT IN THE LAE 13/9 LINEAR ELECTRON ACCELERATOR

Z.Zimek, J.R.Zablotny

Institute of Nuclear Chemistry and Engineering

Dorodna 16, PL-03-195, Warszawa, Poland

The linear electron accelerator LAE 13/9 had been installed in Departament of Radiation Chemistry and Technology, Institute of Nuclear Chemistry and Engineering in 1972. It has been in constant use since than. Technical and running parameters of LAE 13/9 allow to apply this machine both for research within radiation chemistry and for the radiation processing at the semi-technical scale. LAE 13/9 is a linear waveguide accelerator in which electrons are accelerated in the field a traveling electro-magnetic wave. This wave propagating itself within two sections diaphragmatic waveguide powered by a single klystron. The basic parameters of the linac are as follow

energy	5 -13 MeV
energy spread	$\pm$ 6 %
average beam power	9 kW
pulse current	0.5 - 1.5 A
pulse duration	0.5, 2.5, 5.5 μs
two exit of the beam:	
"direct"	$\varnothing$ 10 mm
"bent and scanned"	25 x 500 mm
UHF energy	1818 MHz

No computing equipment had been originally installed in accelerator.

Further exploitation of LAE 13/9 had showed that computer could be used on two ways: in a data handling system for pulse radiolysis experiments where fast kinetics have been recorded and analysed and for digital control of accelerator. According to that two partly independent systems have been developed.

The system for acquisition and analysis the data consist of Mera 60-30 minicomputer /PDP 11/03 equivalent/ with peripherals, transient recorder and CAMAC units. A time-resolved transient absorption or emission produced by a single pulses of radiation is measured. The basic scheme of the system is showed on Fig. 1.

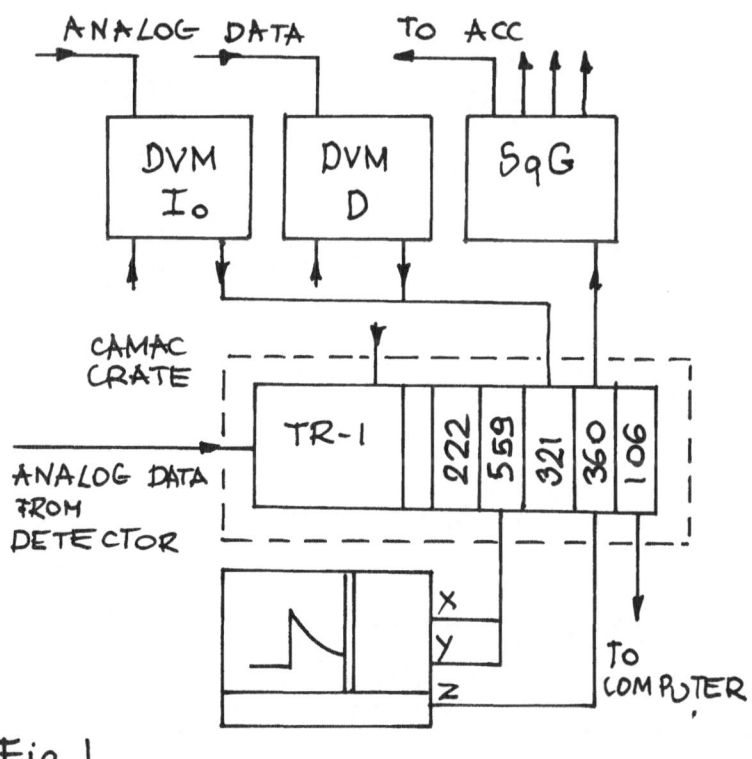

Fig. 1.

The Mera 60-30 computer has 28 k memory with word size 16 bits. It has been equiped with: linear printer, display, tape reader and puncher, dual floppy disk, dual mag tape and CAMAC interface. It will be operating on-line with experiment. Aflow diagram for the computer processing of the pulse radiolysis data is showed on Fig. 2.

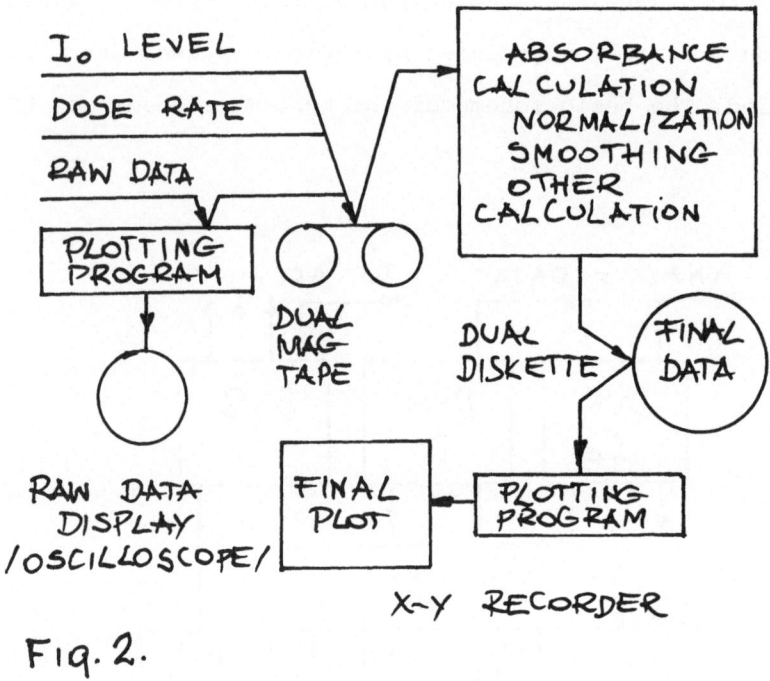

Fig. 2.

The system will be programmed in FORTRAN or BASIC with diskettes running under RSX-11. Interfacing to the experiment is done by means of CAMAC. Some DVM as the analogue-digital converters have been used for gathering auxiliary data of pulse radiolysis experiments.

The digital control of electron beam parameters can be also important part of experiment. That program has been started from the control of the conditions of large scale irradiation processes. The system showed on Fig. 3. allows to keep the constant dose rate despite of accelerator instabilities. It has first class importance for the most of radiation processes.

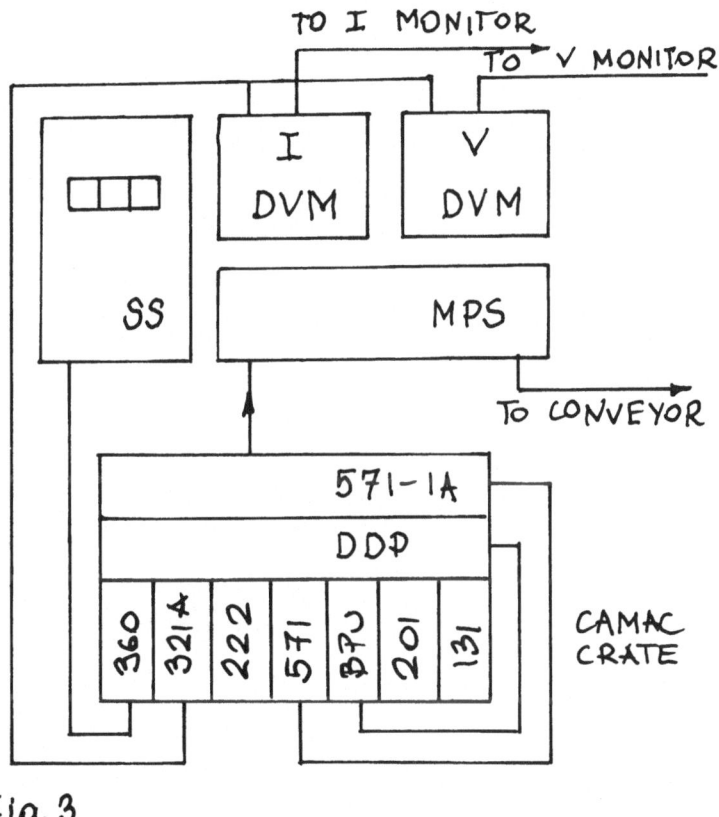

Fig. 3.

The CAMAC units and equipment belonged to accelerator have been used to fulfil all necessary functions like: reading true value of beam current and spead of conveyor, programming the constant dose rate for the certain process, adjustment of speed of conveyor according beam current instabilities, signalize to high and to low level of the beam current and the stop the process when the proper dose rate can not be reach.

The other parameters will be stabilized by the system in future. One the of profit of this approach is that system can be installed part by part. The full project has been developed by staff of accelerator division at a minimum cost, both in money and also in manpower.

to Fig. 1.

TR-1 - transient recorder

360 - driver register

222 - programmable ROM

I_o - analysing light level

106 - CAMAC interface

559 - display driver

SqG - sequence generator

D - dose rate monitor

to Fig. 3.

SS - signalize system

DVM - digital voltmeter

I - beam current

V - spead of conveyor

MPS - motor power driver

DDP - digital display and programmer

131 - processor

201 - memory

BPU - by-pass unit

222 - programmable ROM

360 - driver register

321A - input gate

SOME FEATURES OF THE COMPUTER CONTROL SYSTEM FOR THE SPALLATION NEUTRON SOURCE (SNS)
OF THE RUTHERFORD APPLETON LABORATORY

T. R. M. Edwards
Rutherford Appleton Laboratory
Chilton, Didcot, Oxon OX11 0QX, ENGLAND

1. Introduction

The Spallation Neutron Source (SNS) is a high intensity pulsed neutron source currently under construction at the Rutherford Appleton Laboratory, Chilton, England. It is based on an 800 MeV proton synchrotron, which has a pulse repetition rate of 50 Hz. From a control point of view, the SNS consists of several hundred operating parameters and over a thousand status indications which must be set and monitored by the control system. This is achieved by a hierarchy of computers consisting of GEC 4070 minicomputers at the higher level and microprosessors at the lower level.

2. Aim and Philosophy of the Control System

The main aim of the control system is to provide the means for centralising both acquisition of data from the accelerator equipment and control over that equipment. The accelerator can then be operated in an integrated manner from a central control room. As a consequence of the design, local control of major subsystems of the accelerator, eg. the 70 MeV Injector, is also achieved. Indeed, this philosophy of 'local' individual control is also extended to certain individual items of equipment.

A further aim is to provide a system with sufficient flexibility to allow, on the one hand, direct interactive development of sophisticated control algorithms and on the other, semi-automatic or push-button operation. In addition, it has to be simple to use, so that these functions can be programmed by the physicist directly without the need for computer experts.

From a functional point of view, the two factors that impinge most on the design of the control system are the high beam intensity and the pulse rate of 50 cycles/sec. The former is such, that the loss of two consecutive pulses through the beam pipe cannot be tolerated, while the latter allows, at most, 20msecs computation time between pulses. The phiolology is, therefore, that all equipment protection and personnel safety interlocks are hardwired and that the computers take no <u>active</u> part in their operation. In general, the control over equipment is supervisory and the system acts in a set point and monitoring mode.

It is also intended that wherever possible, dedicated control/monitoring functions be off-loaded onto microprocessors, which become part of the equipment. Their operation need only be supervised by the higher levels of the control system, thus reducing data traffic to the main computers.

3. Overall System

Figure 1 shows the main consituents of the computer control system. It is based on a distributed star network of GEC 4070 computers. Each computer is configured with 256 kbytes of main memory, 9.6 Mbytes of disc storage plus floppy discs. The computers run under the SNS Control Operating System (SCOS) which has been developed by the Controls Group. The console computer acts as the node and is connected to 3 satellite computers. A second console computer will be incorporated at a later stage. The program development computer is linked into the network and is used for offline system software development. Each satellite supervises the operation of a major subsystem of the accelerator viz Injector, Synchrotron, Extracted Proton Beam and Target. The console computer interfaces the main control desk from which the operation of the accelerator is coordinated and supervised. Local control consoles are also provided at each satellite.

The primary interface to the computer is CAMAC and the devices and facilities on the control consoles are driven through this. The accelerator equipment itself is interfaced via the 'general purpose multiplex' system (GPMPX). This is a system developed for the SNS providing a set of standard interface modules which are located in crates near the equipment.

4. The Network

The SNS control network provides complete interconnection between the GEC 4070s of the control system. Via this network, operators/physicists are able to access information from any item of accelerator equipment by the use of simple commands. The network is packet switching, with a layered protocol. At the physical level, the interconnection is made via CAMAC to CAMAC links. The Message Transfer Unit (MTU) has been designed as the interface and this provides a full duplex, 8 bit parallel data link. The unit has 2 x 1024 byte buffers and handshakes data at both byte and complete packet level. The unit performs error checking by appending CRC words to the data stream and automatically re-tries in the event of failure. The transmission rate over the physical link is 250 kbytes/sec.

The protocol adopted at the packet level is a subset of level 3 of X25. Each packet carries source/destination address and control information. The packet size is variable up to a maximum of 1024 bytes.

5. The Control Language GRACES

One of the most fundamental decisions in the design of the Control System was the adoption of an interpreter as the command/control language. This philosophy had proved very successful at the SPS, CERN with NODAL[1]. An interpreter is a program which executes instructions in a high level language directly rather than translating them into machine code and then executing the compiled program. The chief advantages are ease of learning by a 'non-expert' and the rapid debugging loop possible. A disadvantage is the slower speed of execution; however, since the SNS control system is supervisory in nature, this is not a drawback.

The interpreter developed for SNS is GRACES[2] (GEC-Rutherford Accelerator Control Executive System). All control/monitoring of the SNS will be performed via the GRACES language. In fact, GRACES is not a pure interpreter like NODAL, but a semi-compiling interpreter. This means that statements are compiled into a convenient internal form and this is then interpreted. This approach means that GRACES runs about 10 times faster than NODAL and about 10 times slower than FORTRAN. In addition to acting as a programming language, GRACES serves as the operating system command language providing commands for handling disc files, displaying data files, editing programs etc.

5.1 Interrupt Handling

GRACES provides a simple and consistent way of handling interrupts from three sources - interval timer (TIME), CAMAC (LAMs), SNS multiplex system interrupts (MINTS). The commands available are:-

SET interrput	to identify the interrupt expected
WAIT interrupt	for synchronous interrupt handling
ON interrupt GOTO	for asynchronous interrupt handling
RESUME	to return and continue from where the program was interrupted
NORESUME	to ignore return
CANCEL interrupt	to cancel receipt of the interrupt

These features have proved very useful in the writing of control programs to cope with external stimuli.

5.2 Scheduling

In addition to the ability to switch control within a running program in response to an interrupt, the GRACES programmer can also request that a completely independent program be run in response to an interrupt. This is termed scheduling and the command is of the form:-

    RUN 'filename' ON interrupt.

The interrupt is defined as in 5.1. The program 'filename' is then run on a different copy of GRACES (see section 6). Alarm handling and regular surveillance programs are scheduled in this way.

5.3 Network Commands

A very important requirement in a distributed computer control system is the ability for programs to communicate and interact across the network. On the SNS four categories of exchange are catered for:-

1. Remote Filing
Complete files can be transferred from one computer to another. These can be GRACES programs or data files. The transfers can be from file store to file store or file store to device, eg. line printer, etc. This feature is used to downline load new system generation files from the program development computer. An example of the syntax is:

    COPY file 1 TO [remote computer] file 2

this copies file 1 from the local computer to file 2 in the remote computer.

2. Remote Job Submission
All or part of a GRACES program may be sent across the network for immediate execution on a remote computer. Data may also be exchanged. This feature is very necessary for the supervisory and coordinating activity of the console computer. Global information can be gathered on the accelerator by sending routines for execution, in parallel, on the satellites with subsequent remission of data. An example of the command is:

    EXEC [remote computer] n, A

where n constitutes the group of GRACES line numbers to be executed on the remote computer and A is the data array exchanged.

3. Remote Terminal Access

A user at the console computer can connect his terminal directly to any of the computers on the network. This facilitates 'immediate mode' communication with a GRACES on the remote computer. The command is simply:

    REMOTE [computer no]

4. Remote Scheduling

An extension of the scheduling commands mentioned in 5.2 allows programs to be scheduled remotely. This means that an interrupt on a satellite computer can cause a program to be run locally and also schedule a program on the console computer, say. The GRACES command to instigate this takes the form:-

    RUN [remote computer] 'filename' ON interrupt

This will cause the program 'filename' to be run on the remote computer on receipt of the interrupt.

5.4 FORTRAN Interface

Normal use of GRACES does not demand much floating point arithmetic and indeed for extended numerical calculation it is slow. To overcome this problem, a GRACES/FORTRAN interface has been provided. Users can write time consuming routines in FORTRAN, compile/link them and then call them from a GRACES program. GRACES data arrays can be exchanged. Entry to the FORTRAN is by a simple call:-

    CALL FSUB (A,B,...) A,B are arrays

6. The GRACES Environment

The 'Environment' is the overall software system on a given computer, incorporating GRACES, which provides all the facilities necessary for operators/physicists to control the accelerator. Various copies of GRACES are provided on each computer to satisfy the differing requirements (see fig 2).

Scheduling (LGRAC, SGRAC)
One of the most important requirements is the capability to run background monitoring/surveillance tasks at regular intervals and/or run programs in response to hardware generated interrupts eg. alarms. LGRAC together with a scheduling process (SCHED) provides for this. The GRACES programs are stored as files on the disc storage medium of the computer. A later extension will be the addition of a 'short term scheduled GRACES', SGRAC, to run certain high priority programs at a higher repetition rate. These will be short programs held in the online memory.

Inter-Computer Programming (XGRAC, RGRAC)
As discussed above, two situations are catered for - remote submission of GRACES programs with remission of data and remote terminal access. XGRAC receives and executes routines sent from remote computers, while RGRAC acts as an interactive GRACES to a remote terminal.

Operator Interaction (IGRAC, QGRAC)
Any operator/programmer situated at a control desk will communicate directly or indirectly with the interactive GRACES, IGRAC. Normal operator initiated access to the accelerator is done through this. A second operator linked GRACES is also provided at the control desks for use in conjunction with IGRAC - this is the 'Quickie', QGRAC. This allows operators to suspend IGRAC activity eg. during lengthy test loops and execute independent commands. On completion of these, IGRAC resumes execution.

Roving Terminal (TGRAC)
This GRACES caters for terminal input from various sites around the accelerator. It is used for simple equipment testing or program development without the sophisticated input/output of the control desks.

In addition to the various GRACES, the environment also provides the means of accessing the accelerator hardware. This is based on the Data Module System developed at the SPS[3]. An important feature of the software system is that it is identical on all computers, except for the hardware specific Data Modules. As a result of this, system changes need only be done on the offline program development computer and the system files subsequently transferred across the network. In addition, a method of dynamic linking has been developed for the Data Modules such that these can also be changed and loaded separately without stopping the computers.

7. Microprocessors

Microprocessors are used on the SNS where subsystems have complex acquisition or control requirements. As mentioned earlier, the high pulse rate means that pulse by pulse monitoring is not practical through GRACES on the main computers. Rather this is achieved by interfacing microprocessors directly to the equipment. The main control computers then supervise and direct the microprocessors' control and acquisition function. Many of these set-ups exist on the SNS and a standard microprocessor interfacing module (type E) has been included in the GPMPX system, driven by a standard Data Module, to cater for this. It has been used to interface Motrola 6800, RCA 1802 and INTEL 8080A systems.

Microprocessors have also been installed in the control desk to reduce the load on the console computer. The Analogue Waveform Selection (AWS) system is driven in this way as well as the Operator Selection Tree (OST).

A single board microprocessor based on an Intel 8080 has been developed as a standard for SNS applications. The user friendly philosophy has been extended to this and a high level language, microGRACES[4], has been provided. This forms a subset of GRACES, with the range of commands limited so that the complete editor/compiler is held in ROM on a single board. A GPMPX interface is also provided so that stand alone microGRACES systems can be configured to control major items of equipment.

8. Summary

The system described has been used to date to commission the 70 MeV Injector from its local console and has achieved the design aims of user-friendliness and subsystem control. The next phase will be overall control of the SNS from the main console.

REFERENCES

1. M C Crowley-Milling, G C Shering, 'The NODAL System for the SPS', CERN 78-07 (1978).
2. R Brewer, P P Haskell, 'GRACES Reference Manual', Rutherford Report, RL-79-31 (1979).
3. M C Crowley-Milling, 'The Design of the Control System for the SPS', CERN 75-20, (1975).
4. D David, G G Hicks, R H C Morgan, 'Microprocessors in the SNS', Proceedings 11th International Conference on High Energy Accelerators, (1980) p444.

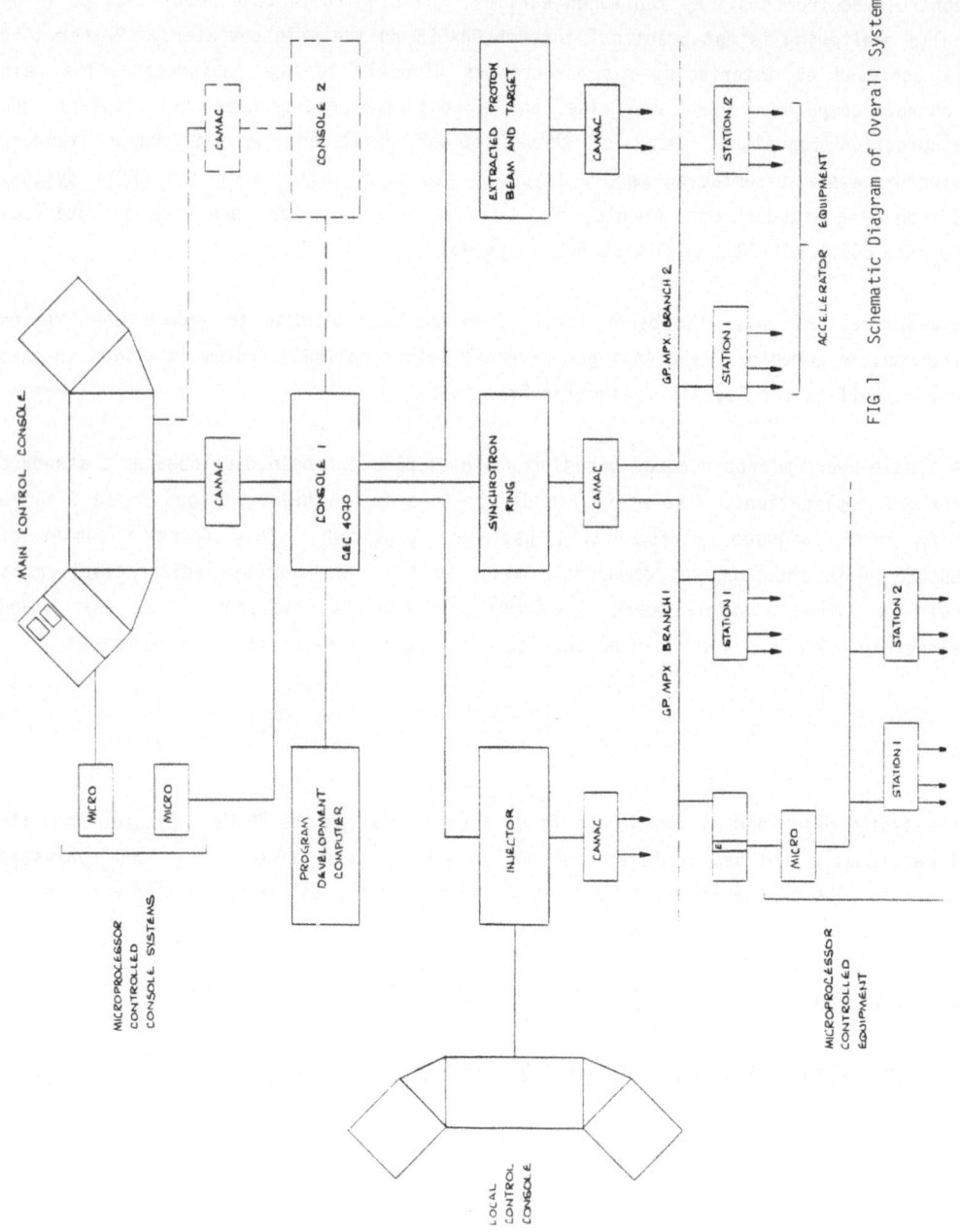

FIG 1 Schematic Diagram of Overall System

385

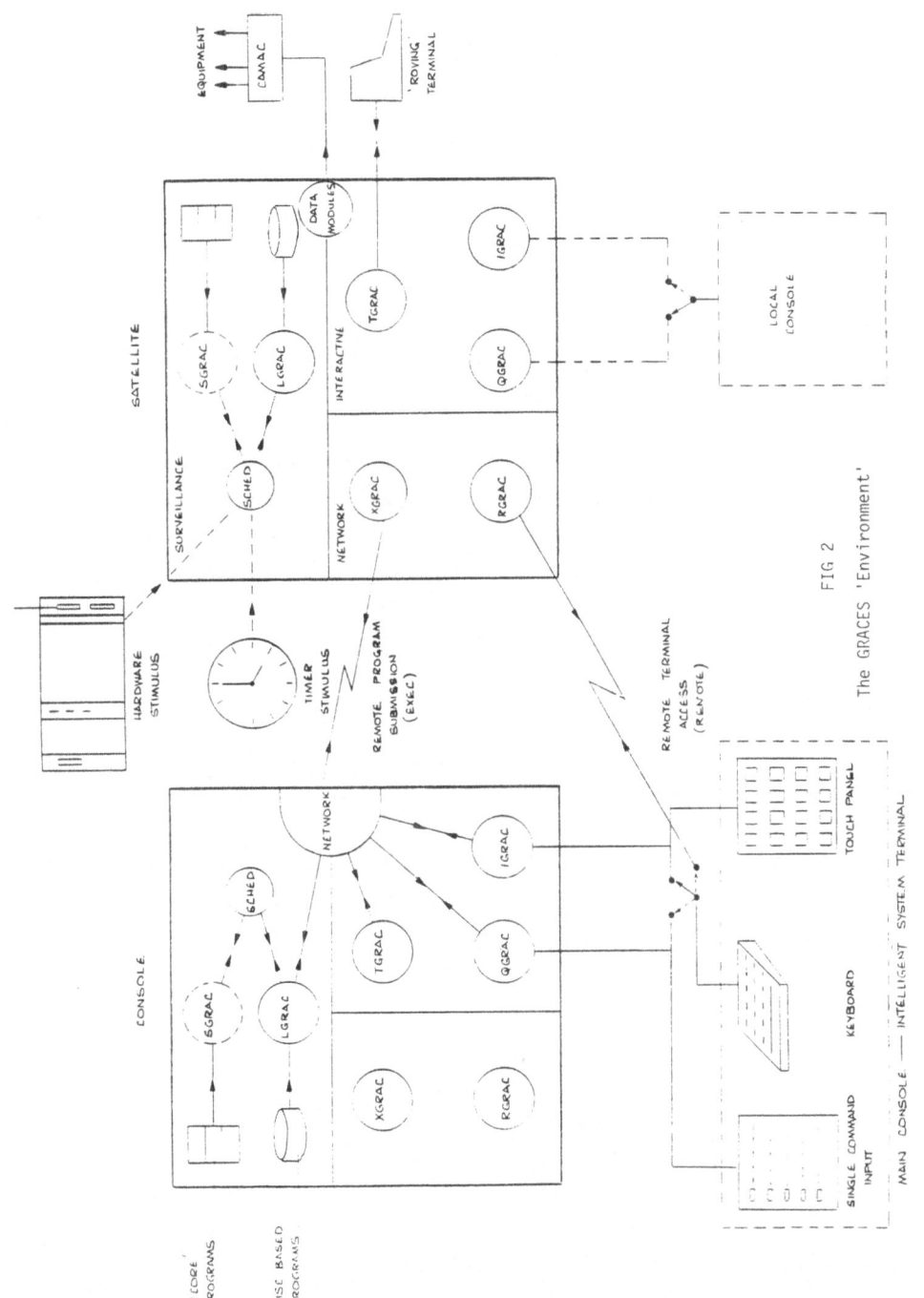

FIG 2 The GRACES 'Environment'

Design Criteria for the Operation of Accelerators Under Computer Control

P.D. Eversheim and P. von Rossen

Institut für Strahlen und Kernphysik
der Universität Bonn
Nussallee 14-16

D - 5300 Bonn
West Germany

Abstract

The control of technical systems is achieved by means of software and/or hardware. Some features of control systems are discussed like flexibility, reliability, effectiveness, speed, and transparency that lead to criteria how to balance hardware against software and how to design soft and hardware. Examples are given in terms of:

    i) A computer controlled energy variation of the Bonn Isochronous Cyclotron, beam handling system, data aquisition, and kinematic compensation

   ii) A computer controlled polarized ion source.

1. Introduction

The design and the set-up of a computer control for accelerators in general turns out to be a little bit frustrating for various reasons. Since usually the work is split up among several colleagues each of whom solving problems in his special field there is often something the authors like to call "pointing moment". This is the moment when something does not work out as it was intended to and everyone is pointing at each other with the words:" It's your fault".

To avoid this situation it is very important that every coworker understands to a high degree the problems that are involved with the handling of the accelerator, the connection of the accelerator to the hardware, the interfaces, and the software. Then there is a good chance that an "optimum" solution to a problem is found. Often there are competing solutions which involve soft- or hardware only. The best solution lies usually in between a pure soft- or hardware solution with respect to costs (efficiency), reliability, flexibility , speed, and transparency.

The authors found that giving account of these just mentioned terms for even detail problems clarified the way to an appropriate solution. Therefore, these terms are discussed in more detail in the following, whereby it is shown that there are some criteria which these terms have in common.

2. Discussion of the terms

Reliability is perhaps one of the most important (and most difficult) properties that has to be achieved. It is clear that adding a sophisticated computer system to a complex accelerator each of which having its own limited reliability does not lead in general to a system with a better overall reliability. Certainly some controlling circuits can be added but each of these circuits has a limited reliability too. It is an interesting question to find for a given system the relation between added control circuits and improved reliability.

In the practical approach one tries to make each system as reliable as necessary right away. Critical in terms of reliability are compared to integrated circuits[1] moving mechanical parts like relais , connectors etc.. For this reason fast serial ports are desirable since only a few integrated circuits are necessary and one coax cable to establish the communication to different stations in the accelerator area and the main computer. Thus only a few parts are used and little hardware which as such improves reliability.

Moreover caring for redundancy of critical parts of the system increases reliability too. This may even lead to an odd number of main computers which run parallel. In case of a malfunction the proper decision is made on the base of the majority.

Flexibility in this sense is a property which has to do with the future. Since every accelerator facility has to meet the developments in science there is a steady (and necessary) process of changes. A computer control for an accelerator puts a corset on it with respect to cables, special interfaces, and software. The software for instance usually is developed for a certain set of applications and has an underlying concept. If a new requirement has to be met which breaks this concept the software gets "spoiled" and complicated. If this happens it shows up in extensive programming time. Therefore only a "simple" concept is able to meet even future requirements. Such a concept is a structured, segmented tree like block concept, where the connection from one program block to the next is done by well defined and for all blocks alike parameter sets.

Now a new application for the software involves only the exchange or addition of some program blocks. With the above mentioned serial ports it makes no difficulty to add a new parameter or to change its destination. Only new cables have to be pulled from the probably new interface in the next nearest station.

Effectiveness deals with the costs of the system with respect to the hours spent to design and set up the system, the hardware costs and the maintenance. Software costs are held down if the above mentioned block concept is realized. Programmers can work parallel and since block connections are well defined and alike this helps to have the whole program run without major difficulties. To keep the software costs low the main computer should not be involved in closed loop controls for parts of the system, since this is rather time consuming to program. Such closed loop controls should be rendered to autonomous subsystems[2] which get via a station their leading value. To have these autonomous systems cuts the maintenance costs too since these systems have well defined tasks that can easily (without support of the main computer) be checked upon.

The speed of the computer control system is kept high if it is not -as mentioned above- involved in a closed loop control for a part of the system. If such a closed loop control has to be digital a decentralized slave processor can do the job.

The serial ports for in- and output should be "intelligent" and transfer from and to certain memory areas of the main computer the data. Thus the serial ports look from the point of view of the main computer like normal memory locations. The computer memory should be large enough to keep the whole program.

The concept for the main computer should aim at rendering as many tasks as possible to decentralized autonomous systems and to have it handle solely the algorithm for the control of the system. The output to the stations are leading values only. The output to the operator is prepared by a special processor that has access in the above mentioned way to all relevant memory areas of the main computer.

By transparency[2] the authors understand all aspects of a system that enable an easy and fast judgement on the state of the system. Since reliability is limited there will always be some malfunction. To find the faulty part or logic structure in a complex system is often very difficult and time consuming. Autonomous systems with well defined tasks and a clear block concept for the software help to diagnose and overcome the malfunction.

Interrupts, however, usually are not "transparent". Especially in order to trace accidental interrupts caused by spurious spikes it helps to have a special interrupt processor. This processor sorts out, verifies, handles minor interrupts (like informations to the operator etc.), and keeps track of these interrupts in a non volatile memory (first in first out). Proper verification procedures guarantee that the main computer is alarmed only if there is a particular system state that has to be taken care of.

3. Examples

Figure 1 shows the set up to a computer controlled energy variation of the Bonn Isochronous Cyclotron. The mini computer

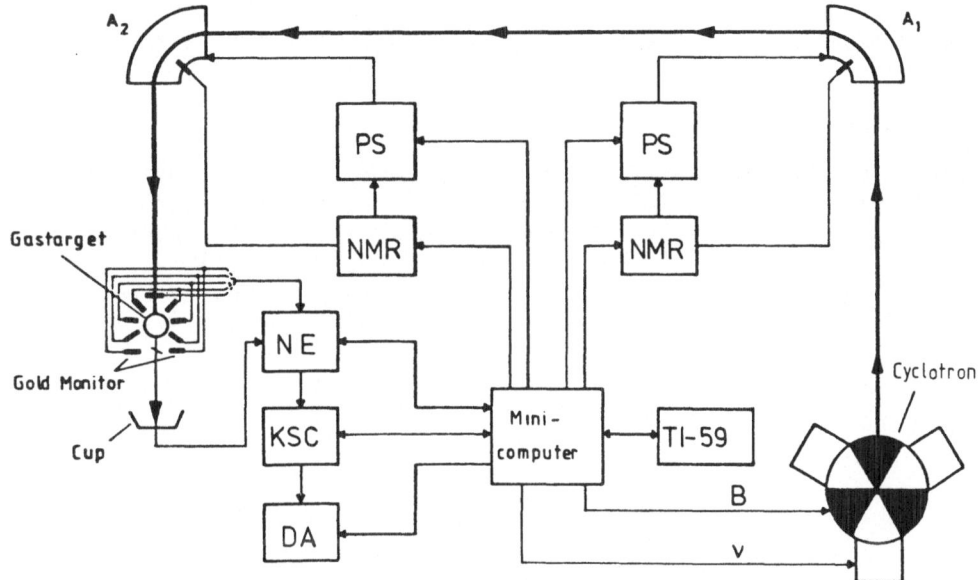

Fig. 1 Schematic of a set-up for a computer controlled energy variation. PS - Power supply, NMR - Nuclear magnetic resonance, NE - Nuclear electronic, KSC - Kinematic spectrum compensation, DA - Data aquisition, TI-59 - pocket calculator, A1/2 - Bending magnets, B and ν - leading values for the magnetic field and frequency of the cyclotron.

provides only leading values for the NMR's (nuclear magnetic resonance), data aquisition, kinematic spectrum compensation, nuclear electronic , and the cyclotron (effectiveness, transparency). There are decentralized closed loop controls (autonomous systems) in the NMR's for the bending magnets A1/2 , magnet power supply for the cyclotron, oscillator frequency, and spectra stabilization. The mini computer is supported by an independent pocket calculator which performs as a slave processor all necessary calculations.

In 3) we describe a rather specialized inexpensive (efficient) stand alone system of a digital control that easily can be connected to other computers (flexibility, speed, transparency).

4. Conclusions

A dominant concept in all these considerations turned out to be segmentation. Segmentation on the hardware level, segmentation on the software level. The main computer should be relieved from as many tasks as possible. These tasks should then be rendered to autonomous systems. A fast serial link cares for the communication between the main computer and the subsystems. A computer control fulfilling these criteria will be an effective means in the operation of accelerators.

Literature

1) Winfried Görke, Zuverlässigkeitsprobleme elektronischer Schaltungen, Bibliographisches Institut, 820/820a (1968)

2) P.D. Eversheim, P. von Rossen, B. Schüller, F. Hinterberger, and, K. Euler, Nucl. Instr. Meth. 157 (1978) 311-314

3) N.W. He, P. von Rossen, P.D. Eversheim, and R. Büsch, Computer Aided Control of the Bonn Penning Polarized Ion Source, This Proceeding

Computer Aided Control of the Bonn Penning Polarized Ion Source

N.W. He[+], P. von Rossen, P.D. Eversheim, and R. Büsch

Institut für Strahlen und Kernphysik der Universität Bonn
Nussallee 14-16
D - 5300 Bonn
West Germany

[+]Tianjin University Department of Basic Science
Tianjin
Peoples Republic of China

Abstract

A CBM computer system is described which has been set up to control the Bonn Polarized Ion Source. The controlling program, besides setting and logging parameters, performs an optimization of the ion source output. A free definable figure of merit, being composed of the current of the ionizer and its variance, has proven to be an effective means in directing the source optimization. The performance that has been reached during the first successful tests is reported.

1. Introduction

A substantial improvement in the operation of the new superconducting penning ionizer of the Bonn Polarized Ion Source[1] has been obtained through the use of a CBM-4032 desk-top computer. The purpose of the computer control was not only to speed up the setting and logging of the source parameters, but to do an optimization of the source's output parameters under program control. Such a feature was especially desirable, as the optimal setting of the source's parameter tends to drift due to varying vacuum and surface conditions inside the ionizer.

2. Hardware Configuration

The computer system is arranged of a CBM-4032 desk-top computer with a floppy disk as mass storage for programs and data and a matrix printer for hard copies of records[2]. The schematic layout is shown in figure 1. An added PC-board with two versatile interface adapter IC's (6522) gives additional 40 I/O channels for the communication with peripheral devices. The system is designed to allow future extension not only to the axial injection line but to the cyclotron as well. Especially the possibility of using other low cost computer units and interfacing them to the host computer eases the effort of adding even complex structures to the system.

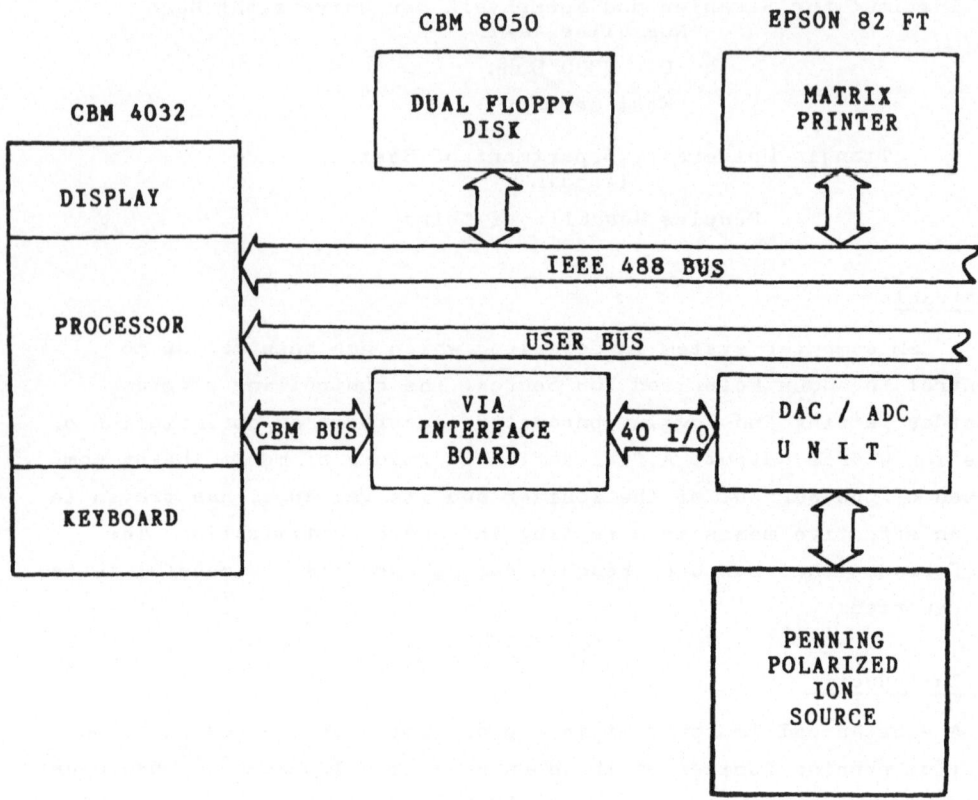

Fig. 1 Schematic of the Ion Source Control

The Penning discharge mode of the ionizer is controlled by eight electrodes E1 through E8, the main electrodes for the discharge mode being E1 through E6 (cf. Fig. 2). The entire ionizer can be put to a high voltage thus giving the proper energy of the extracted ions. The schematic of the controlling high voltage power supplies is shown in figure 3. The eight power supplies for the electrodes E1 - E8 are mounted inside a "hot-rack". With the power supply "beam potential" a high voltage is applied to the "hot-rack". The control signals for the supplies in the "hot-rack" are fed via optical links to the control unit ensuring thus the necessary isolation towards ground. The power supply connected to the einzellens matches the ion beam to the acceptance of the axial injection line.

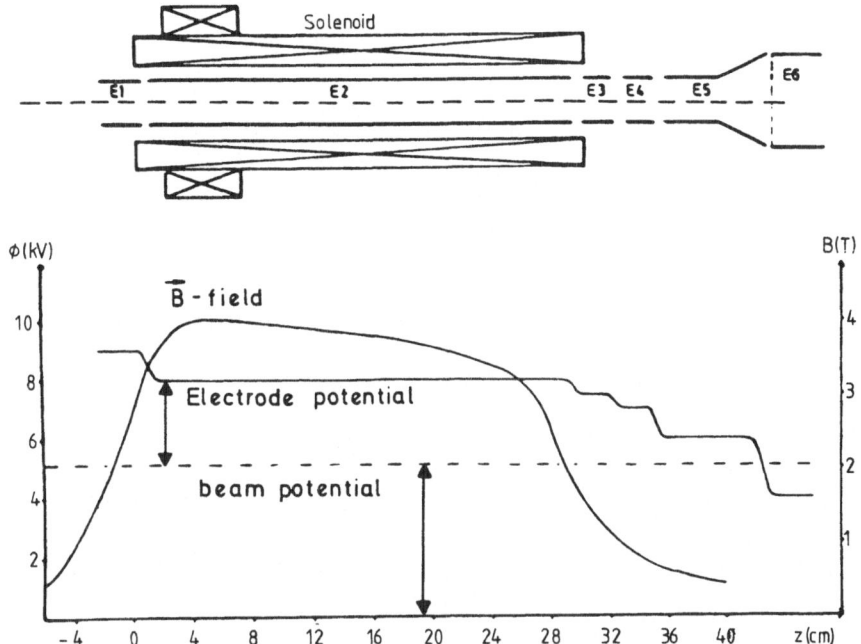

Fig. 2 Detailed view of the ionizer, with electrodes E1 through E6. Below, with the same scale a typical voltage configuration is shown together with the magnetic field B of the solenoid.

3. System Software

The difficulty in laying out the program structure results from transforming the knowledge of an experienced operator into a workable program code. A theoretical approach would be to describe the output current IC as a function of the parameters E1 - E8 and time t.

$$IC = f(E1, E2, \ldots, E8, t)$$

Unfortunately it contains a dependence of the time t explicitly. Besides this, the hypersurface spanned by E1 - E8 is not known analytically. Additionally the previous experience showed that the maximum ion current alone would not be the best quantity of optimization. Therefore, we have created a figure of merit (FGM) which took into account the stability of the discharge mode. The measurement of the current is done in a repetitious way to reduce the influence of accidental fluctuations. By having for instance five current values available for a specific parameter setting it is possible to compute a mean and its variance, which allows a judgement on the stability of the discharge mode.

We used for our purposes a FGM defined as FGM = $(IC)^A/(\sigma^2)^B$. IC is the measured ion current and σ^2 is its variance. The exponents A and B

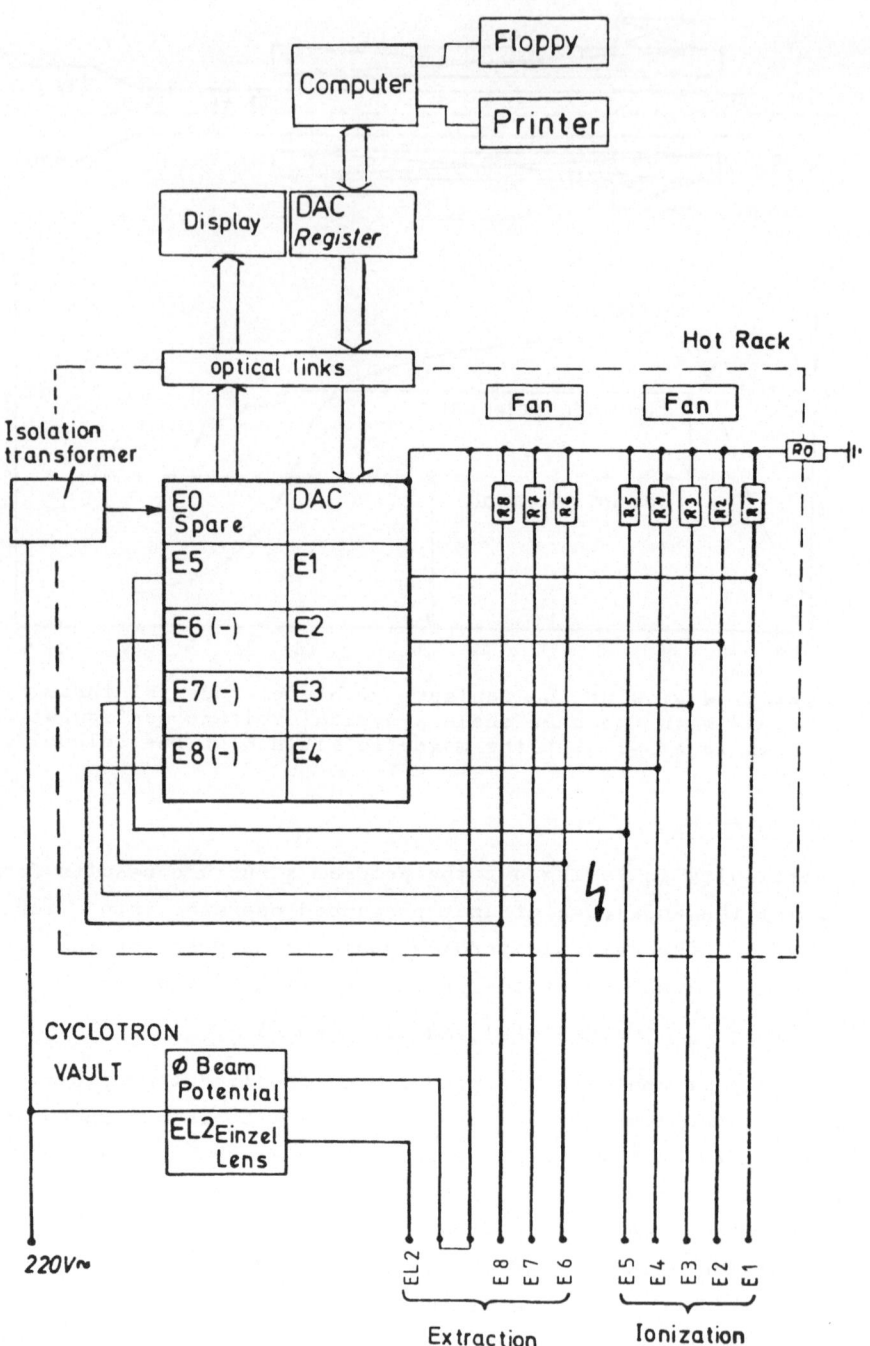

Fig. 3 High voltage power supplies E1 - E8 and their connection to the electrodes and the computer.

can be chosen for best results in current respectively stability, according to the needs of the experiment. This sensitivity to strong fluctuations helps also to avoid being trapped by instable modes.

The optimum is searched for by a modified grid search[3]. A parameter is varied until the maximum is exceeded. With the last three points, a parabula is calculated and the parameter is set to the optimum. Then the next parameter is optimized and so on until it starts with the first parameter again. Since it turned out that the parameters are relatively independend from each other this method works out well, though provisions are made to define pseudoparameters by which the problem of coupled parameters could be overcome. A pseudoparameter is defined by being a function of the setting of some other parameters i.e. E1...E8.

In order to guarantee fastest data handling all essential parts of the code are programmed using an assembler language[4] besides multi tasking is provided to control the ion source in the hold mode, when not much CPU time is necessary. A detailed documentation of the software is given in [5].

4. Practical Tests of the Optimizing Software

As stated before no exact knowledge of the dependence of the FGM concerning the eight parameters exists. Therefore, it was important to experience whether the developed algorithm would establish a stable digital closed loop control system. Fig. 4 shows copies of a chart recorder plot with the total ion current (Y-axis) versus time (X-axis).

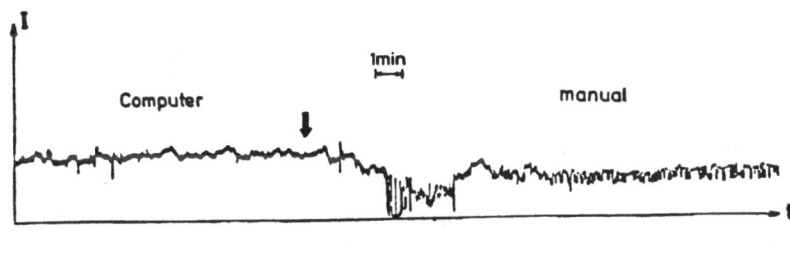

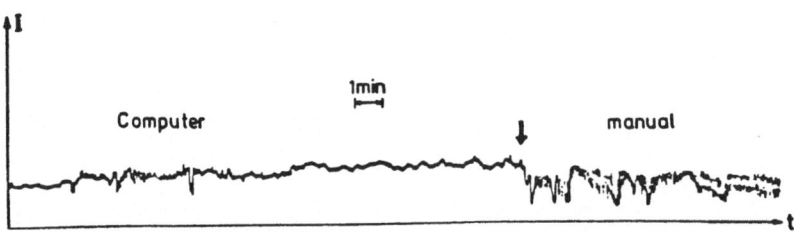

Fig. 4 Ion current with different tuning procedures.

The charts show the ionizer under computer control on the left side. Only the ion current was taken as FGM. Beginning at the time marked with an arrow it was tried manually by various operators to improve the current. As can be seen the manual attempts resulted in a poorer performance in both cases.

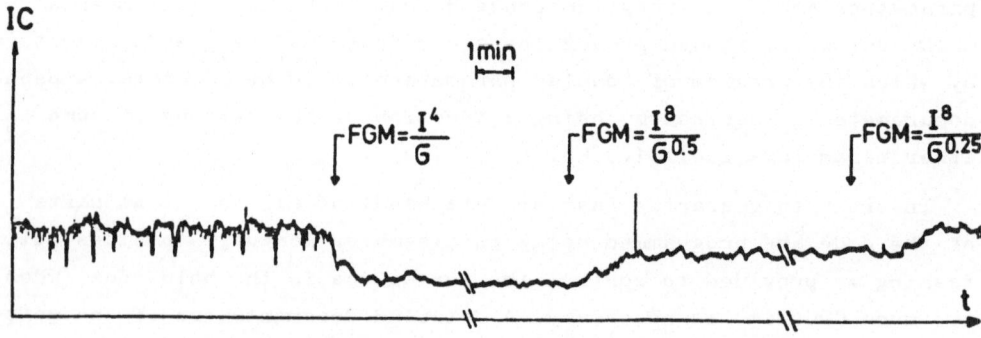

Fig 5   Ion Current under different Figures of Merit

Fig. 5 shows a study with different definitions of the FGM. On the far left only the current signal IC is optimized. As soon as the FGM is changed to $IC^4/(\sigma^2)^{.5}$ the settings of the ionizer are altered by the computer to give lower but more steady current. Weighing the current stronger leads to an improvement in current until with the definition $IC^8/(\sigma^2)^{.125}$ nearly the old current value is reached.

Another example that demonstrates the behaviour of the control loop is shown in Fig. 6.

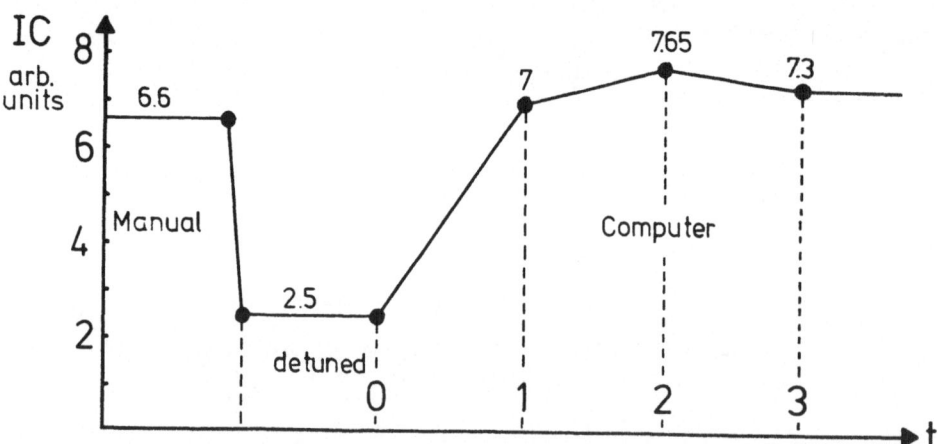

Fig. 6   Ion current versus iteration cycle.

The ion source, after having been optimized manually, is deliberately
detuned and control is rendered to the computer. The numbers on the
X-axis show the iteration cycles. Already after one iteration cycle
of all parameters a current value is obtained that exceeds the old
maximum. The next cycle gives additional improvement of the current
and reaches nearly the steady state.

5. Conclusion

This work has shown that by means of a low cost desk-top computer
system it is possible to optimize even such a complex apparatus like
a polarized ion source. The difficulties result not that much from
the hardware and the interfaces but from the establishment of a
reliable program code which incorporates the operation experience
thus far obtained in manual tuning.

The algorithm used has proven to be a potent means for finding
an optimum under various conditions of the ion source. The successful
test showed clearly the superiority of the computer tuning versus
manual tuning procedures.

Literature

1) H.G. Mathews, Dissertation, IAP Bonn, 1979
2) R. Büsch, Diplomarbeit, ISKP Bonn, 1983
3) P.R. Bevington, Data Reduction and Error Analysis for Physical Sciences, New York, 1969
4) J. Scanlon, 6502 Software-Design, Indianapolis, 1980
5) Nai Wen He, Computer Optimization of the Bonn Polarized Ion Source, ISKP Report, 1983

TREATMENT AND DISPLAY OF TRANSIENT SIGNALS

IN THE CERN ANTIPROTON ACCUMULATOR

T. Dorenbos
European Organization for Nuclear Research (CERN)

1211 Geneva 23, Switzerland

Introduction

The CERN Antiproton Accumulator (AA) is a storage ring for antiprotons. These are produced by a high energy proton beam striking a tungsten or copper target. The antiprotons are focused with the aid of special devices like a magnetic horn and pulsed electromagnetic quadrupoles, and, with the aid of fast pulsing kicker magnets, put onto an orbit in the AA ring. Here the antiprotons are cooled stochastically (reduction of momentum spread) during about 2 seconds, after which they are slightly decelerated and stacked on orbits where they receive a further cooling treatment and where they stay stored. During normal production runs, this process is repeated every 2.4 seconds, and antiproton beams have been stored for over a month. If the antiprotons are needed for experiments, the stack, or part of it, is accelerated slightly, and the particles put onto the ejection orbit. When the experiments are ready to receive antiprotons, first three bunches of protons are accelerated to 270 GeV/c through the complex via the proton synchrotron (PS) and super proton synchrotron (SPS). Then three bunches of antiprotons are ejected from the AA and accelerated also to 270 GeV/c but, of course, are stored in the SPS in a manner such that they are circulating in the opposite direction to the protons. The experiments are carried out at points where the counter rotating bunches of protons and antiprotons collide (Fig. 1).

Although the AA is a DC machine, the injection and ejection procedures are very much time-related to the PS. Therefore, a good display of injection and ejection parameters, together with their time relationship, is of great help to the operators who can then, based on this observation, take proper action to improve the efficiencies of the various processes.

In the past these signals were mainly displayed on blurry storage oscilloscopes with very often maximum intensity, shortening the tube life to a minimum. Now, with the advent of microprocessors, fast digitizers and cheap memory, things can be done in a much more convenient was, having the signals available nearly full time on relative cheap oscilloscopes using normal settings.

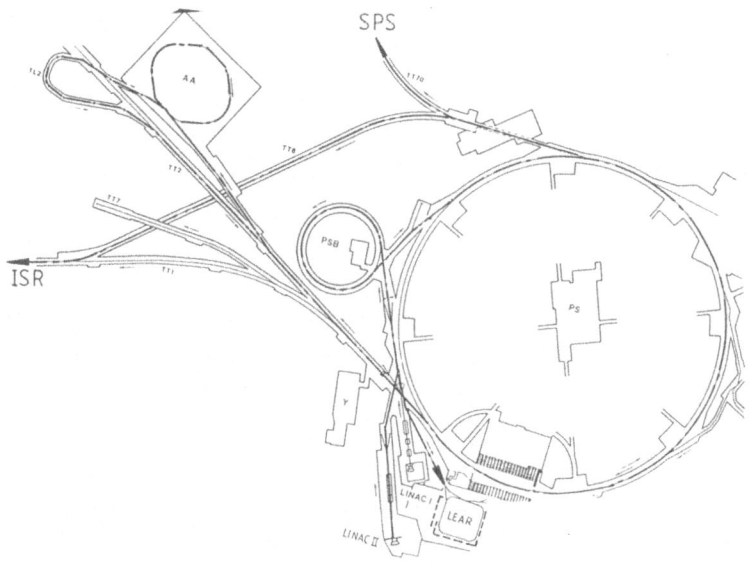

Figure 1 - CERN p̄ Source.

Requirements

The requirements for the analog signals to be displayed, were specified by the PS Operation Group responsible for day-to-day operation. They are the following (Fig. 2):

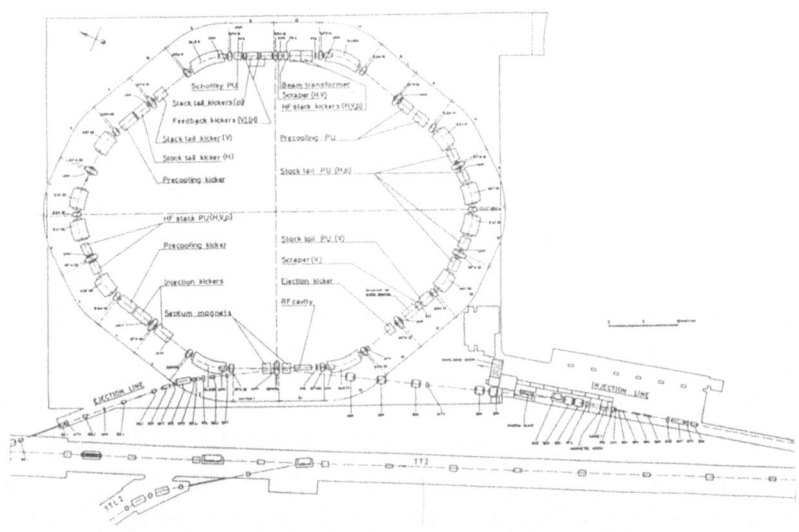

Figure 2 - Antiproton Accumulator: General Layout.

1. The signal of the beam current transformer placed just in front of the target; it represents the proton beam current hitting the target (length: ~ 0.5 µs).

2. The gate of the electronic equipment, integrating the above signal; this is to check whether the timing of this equipment is correct with respect to the beam signal (1.5 µs).

3. The current of the magnetic horn. The magnetic horn is a thin walled (0.5 mm) cylindrical device installed just after the target. During injection a current of about 150 kA is fed through it, creating a very strong focusing field. The horn current has about a half sinewave shape with a length of 40-50 µs, and is created by a capacitor discharge. The timing of the discharge should be such that the top of the discharge curve coincides with the beam hitting the target in order to provide maximum antiproton collection.

4. The signal representing the sum of the currents through the injection kickers (kicksum). The injection kicker is a pulsed deflection magnet, which kicks the injected antiprotons onto their orbit in the AA-ring. It consists of 10 modules, each one pulsed at about 4000 A with a pulse width of about 700 ns. Timing of this device is also very improtant and has to take into account the time of flight of the antiprotons from the target to the injection point in the AA-machine.

5. The signal of the tachometer connected to the shutters. The shutter consists of many slabs of ferrite, closing the gap of the precooling kickers, thereby eliminating the influence of the kicker signal on the stack. Once the injected beam is pre-cooled, the shutter is opened and the beam is displaced onto the stack orbit. The signal of this tachometer shows the proper working of the shutter mechanism.

6. The signal of the voltage program for the radio-frequency (RF) cavity. This signal shows the modulation function of the RF amplitude, for the accelerating cavity, when the injected beam is decelerated to be put on the stack.

All the above signals are related to injection. For ejection the following signals must be available:

7. The signal of the ejection kicker. This is the same signal as mentioned under 4, however coming from the ejection kicker which kicks the beam out of the AA-ring into the transfer channel.

8. The signal of the beam current transformer placed in the transfer line about 12 m downstream of the ejection point of the machine. It shows the ejected bunch of antiprotons ejected towards the PS.

9. The gate of the integrator for the above signal. This is the same signal as mentioned under 2, however with a different timing.

10. The signal of the injection kicker for the PS. The timing of this signal takes into account the time of flight from the AA to the PS.

11. The signal of an electrostatic pick-up station which normally shows the bunch of antiprotons on the ejection orbit. The fact that it is located near the ejection kicker has the advantage that it also picks up a spurious signal when this kicker is pulsed, that is when the beam is ejected. The signal then shows clearly the position of the ejected bunch within the kicker pulse.

Hardware

In designing the system, use has been made of standard commercially available equipment. The digitizers are made by LeCroy, USA. Most of the signals can be handled by the same model, which is the TR 8837F. This is a 32 Msample/second, 8 bit transient recorder with 8 kbytes of built in memory. It has 16 MHz analog bandwidth and a DAC is built in for reconstruction of the signal on an oscilloscope. It is housed in a one-wide CAMAC module (Fig. 3). Originally a 200 MHz digitizer was planned for use with the transformer signal but this could not be realized, due to delays in manufacture. The

Figure 3

digitizers are driven by a LeCroy 3500C microcomputer. This is a 8085 based, multibus-driven microcomputer with built in CAMAC minicrate and which can drive 7 external CAMAC crates via a parallel link. Another 8085 processor is used to drive the built in 256 × 512 pixel graphic display and a third built in processor, the 9511 Arithmetic Processing Unit, is used to speed up mathematical problems. The 3500C comes with a dual floppy disc unit and is using Digital Research CP/M 2.2 operating system.

The reconstructed analog signals are displayed on 10 Tektronix type SC502 dual-beam oscilloscopes. They are housed three in a crate, each one displaying two time-related signals (Fig. 4).

Timing is provided by the general timing system in the AA, i.e. a prepulse 300 ms before injection resets all digitizers and starts them digitizing. Another pulse, 1 µs before injection, stops them, keeping account of the selected number of pretrigger samples.

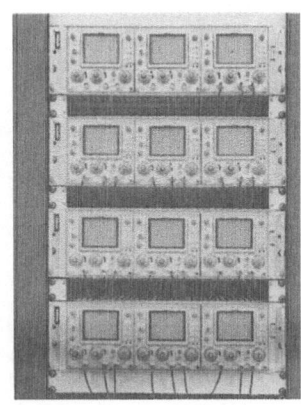

Figure 4

The display-DAC of all the digitizers are triggered externally in order to start the display memory at the same time.

Software

The software delivered with the LeCroy 3500C microcomputer is either BASIC or FORTRAN. Both are standard products of Microsoft Inc. and the BASIC is available as an interpreter and in a compiled version. CAMAC-functions, timing- and plotting-routines are available as relocatable files, which can be called from BASIC as well as from FORTRAN programs.

The program to drive the digitizers is kept very simple for the time being. It starts initializing all the digitizers, sets their clock frequencies, amount of memory and number of pretrigger samples. Then it waits for a LAM, generated by the injection prepulse, which starts the digitizers. Stopping of the digitizers is done hardware-wise as explained above and the program waits for the next prepulse to restart the cycle.

Results

Some of the results for injection can be seen in Figs. 5-8.

Figure 5 represents the transformer signal together with the integrator gate.

Figure 6 shows the kicksum together with the current of the magnetic horn. Note that only part of the horn current (only the top) is visible due to differences in length of both signals.

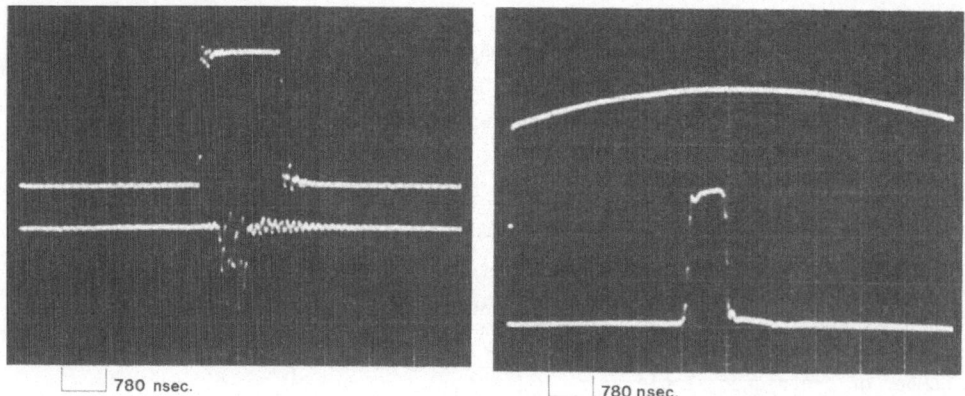

Figure 5   Figure 6

Figure 7 shows the same horn current, however now together with the transformer-signal.

Figure 8 finally shows the RF-voltage program with the shutter movement.

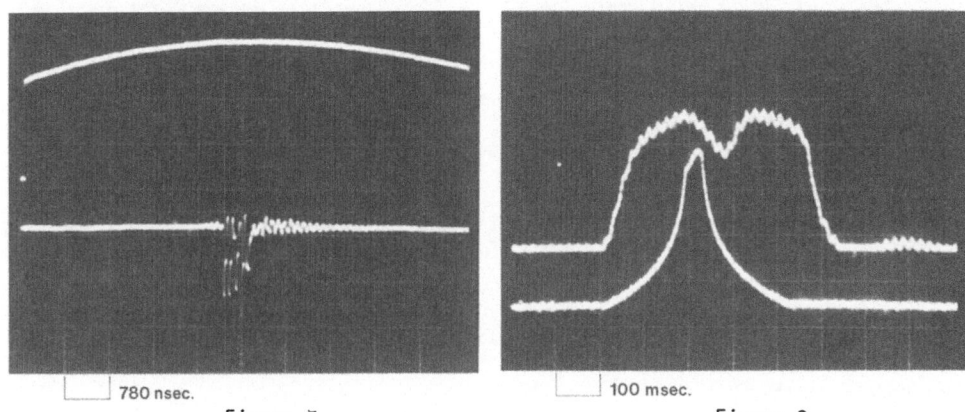

| 780 nsec. | 100 msec. |
| Figure 7 | Figure 8 |

Ejection results are shown in Figs. 9-12.

Figure 9 shows the signal of the ejection transformer with its integrator gate.

Figure 10 shows the ejection kicker signal and the signal of pick-up 22. The slight shift between the two signals is due to a difference in cable length.

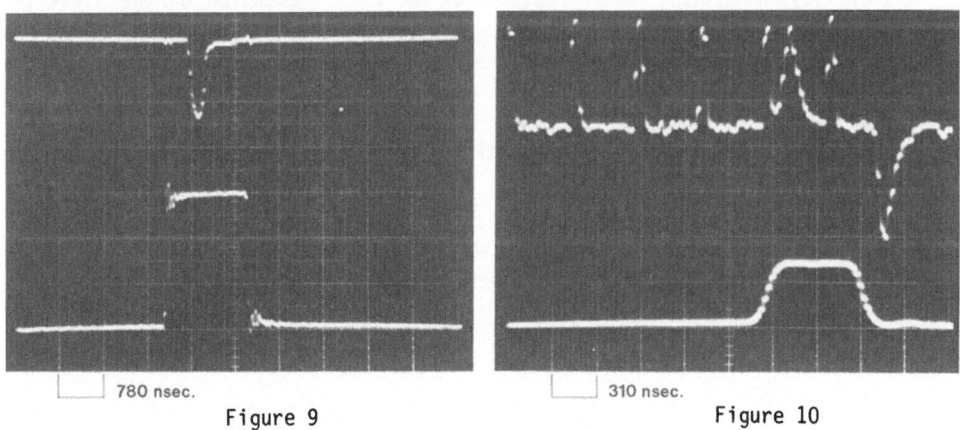

| 780 nsec. | 310 nsec. |
| Figure 9 | Figure 10 |

Figures 11 and 12 show two different signals from an electrostatic pick-up station.

Figure 11 shows the beam bunched into a 50 Hz RF bucket right in the middle of the ejection kicker signal. Figure 12 shows the same procedure, however with a 200 Hz bucket. As can clearly be seen it would have been advantageous if the kicker had fired 150-200 ns earlier.

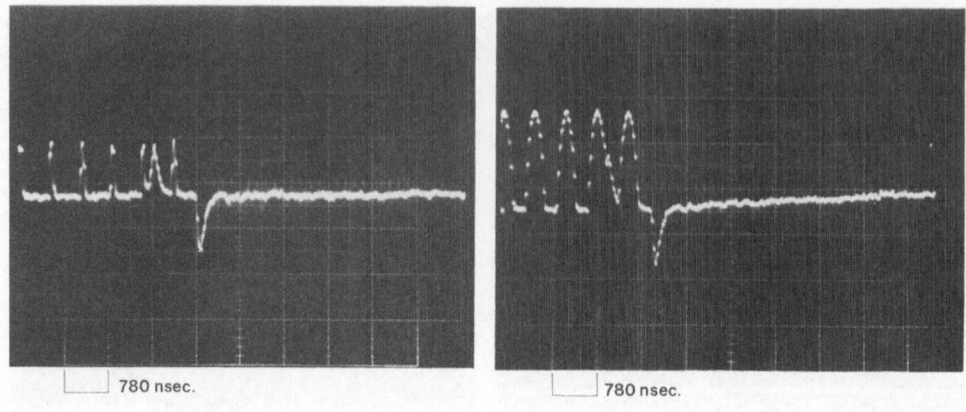

Figure 11 Figure 12

Future Developments

Although the system is now continuously used, it is not yet fully operational, in the sense that it needs no specialist intervention. One of the first improvements will consist of choosing a signal to be displayed on the video screen of the 3500 microcomputer; the selection will be done using the lightpen instead of the keyboard. It will ease the signal transmission to the PS main control room. This will include also the use of a special bit-slice processor in the 3500, which will speed up data treatment considerably. With this special processor an error detection on the digitized signal becomes also possible. Finally a communication link is foreseen with the PS controls network in order to control the 3500 microcomputer from the main control room.

FAST CAMAC-BASED SAMPLING DIGITIZERS AND DIGITAL FILTERS FOR BEAM
DIAGNOSTICS AND CONTROL IN THE CERN PS COMPLEX

V. Chohan, C. Johnson, J.P. Potier - CERN PS Division - 1211 Geneva 23/Switzerland
M. Miller - LeCroy Research System Corp. - N.Y./USA

Summary

The use of sampling techniques to reconstruct fast signals such as those generated by the fine bunch structures observed on beam position pick-ups in accelerators is well known. With sufficiently high sampling rate, the original signal is easily reconstituted without any loss of accuracy and subsequently analyzed. Indeed, at CERN such systems exist and are in use. However, all these existing systems rely on bulky instrumentation with its intermediate level slow interfaces for computer access.

In the PS Accelerator Complex with a new control system based on CAMAC serial highways, we are introducing fast (100 MHz) digitisers sitting directly in the serial highway CAMAC crates of the Process Control computers. The data access is therefore direct, easy and relatively fast even from remote computers and the sampled data may even be included in a closed loop system for control. The aim is to achieve some uniformity in hardware as well as in techniques to optimise the various injection processes in the Complex using these digitisers and find common solutions to similar problems.

Introduction

The chain of several interleaved accelerators which constitute the PS Complex are best illustrated in Fig. 1 which shows the nine machines operating, or under construction. The four particular injection processes that we deal with here are marked in terms of the PS straight section numbers SS42, SS58 and SS16 and, for the Antiproton Accumulator (AA) case as AA Extraction.

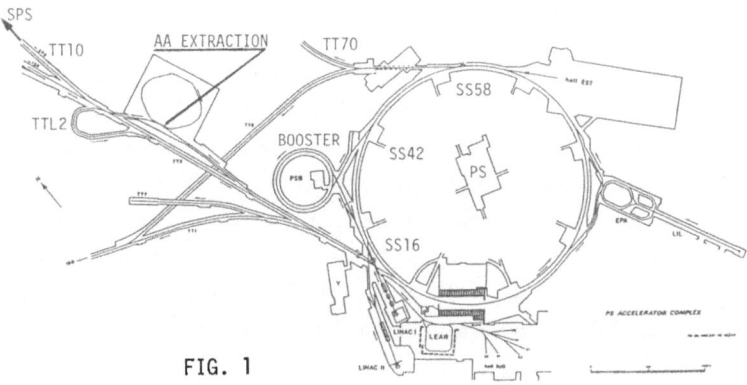

FIG. 1

Injection Processes

The currently operating interleaved circular machines in the Complex imply one machine feeding another essentially for two purposes namely (a) as a normal beam supply of protons or antiprotons ($\bar{p}$) for whatever ulterior usage and (b) test proton beams which test out the transport lines and injection/ejection in the opposite sense to the normal $\bar{p}$ usage. Four processes falling within the above two categories may then be identified :

i) The 4-ring, 20 bunch, 800 MeV/c Booster feeding the PS at SS42 (case (a))

ii) The single bunch $\bar{p}$ beam from the AA to the PS at 3.5 GeV/c at SS16 (case (a))

iii) The single bunch p test beam at 26 GeV/c from the SPS to the PS via the Transfer Tunnel TT70 at SS58 (case (b))

iv) The single bunch p test beam at 3.5 GeV/c from the PS to the AA via the Transfer Tunnel TTL2 (case (b)).

A badly injected beam has transverse oscillations and gives unwanted Emittance increase hence, for the cases i) and iii), it is the relevant injection process of the PS ring which needs to be adjusted and optimised while in the case iv) it is for the AA ring. Case ii) adjustment is not usually done because it is just the reverse of case iv) where adjustments are made. A fifth process of $\bar{p}$ beam from the production target to the AA is sufficiently different to be dealt separately (see below).

In all four cases it is the Injection process comprising of kicker/septum deflection magnets and vertical steering elements in the recepient machine and the transport line leading to it which needs to be adjusted for optimum beam trajectory. An added complexity for the PS ring is that each of the three injection processes are different in energy, and in physical location, implying different data reduction and analyses as well as adjustment of different kickers, septums and steering elements. In addition the beam intensity may vary from 10^9 to 2×10^{13} particles in 1 to 20 bunches on a pulse to pulse basis, necessitating the use of both difference (Δ) and sum (Σ) pick-ups for observation.

Present Adjustment Techniques

1) Recepient machine PS

i) The 4-ring Booster usually injects into the PS either 20 proton bunches of 50 ns each with all 4 rings or just 5 bunches with a single ring. The bunch separation is 125 ns. The injection process is adjusted by hand either (a) by looking at a position (Δ) pick-up on an oscilloscope over several turns and normalising for intensity variations by the Σ pick-up or (b) by looking at a fast orbit display over the first and second turns and correcting till the orbits overlap. In both methods, we are limited in diagnosis because several pulses are needed and we rely on the stability of the systems. In addition. the Booster is a stacked 4-ring machine and the 20 bunches are obtained by recombining 4 x 5 bunches and therefore may have different characteristics. Overall, upto an hour can be taken by such adjustments at machine start-up or set-up.

ii) For the p̄ physics in the SPS, ISR or LEAR, the AA sends a single p̄ bunch, 80 ns long at 3.5 GeV/c to the PS. The injection oscillations in the PS are not corrected at present because this is complementary to the case iv) below; however, we need to observe them for diagnostic purposes.

iii) To tune the p̄ trajectory, the SPS returns a single proton bunch less than 4 ns at 26 GeV/c to the PS via TT70. In fact the origin of this test bunch is the PS itself, which sent it via the TT10 tunnel to the SPS at 26 GeV/c on the previous cycle and is re-injected into the PS via TT70. The injection oscillations in the PS are observed on a high sensitivity pick-up and are manually corrected over several re-injections. Again, considerable time is involved in setting-up and adjustments.

2) Recepient machine AA

iv) The PS sends a bunch of 2 to 3 x 10^9 protons nominally 60 ns long at 3.5 GeV/c via the TTL2 loop. The injection oscillations in the AA are observed on a Δ pick-up over several turns (540 ns revolution period) and are digitised using a stand-alone instrument (200 MHz Tektronix digitiser) connected to the CAMAC Serial highway by a GPIB Interface. The corrections are applied after the acquisition and analyses of digitised data in the local AA computer[1]. However, for remote access from console computers in the Main Control Room, the existing closed loop correction is difficult to implement and is mainly governed by speed of data access over the CAMAC-GPIB interface.

Proposed Techniques

Hardware layout : The 100 MHz digitisers modules sit directly in a CAMAC crate, as shown in Fig. 2 which illustrates the PS Injection application. A pair of the digitiser modules is sufficient to handle all the 3 different PS Injections in conjunction with the routing modules to switch the appropriate high sensitivity pick-ups and timing stop triggers. Fig. 3 illustrates the diagramatic layout of the AA application which does not need the pick-up routing modules. In both cases, an attenuator module is desirable to bring down

FIG.2

the amplified pick-up signal levels within the dynamic range (0 to 510 mV) of the digitisers. Whilst in the AA the digitiser pair could be used for ΔH and ΔV signals for a fixed intensity single bunch test beam, it is more appropriate in the PS to use the Σ and Δ signals to allow individual bunch intensity normalisation and make separate acquisitions for each of the transverse planes, i.e., over two pulses. The 100 MHz standard configuration modules are capable of storing upto 32 K samples per module ; however for the AA application needing say the first ten turns, about 600 samples per module are sufficient while for the PS, about 2 K samples/module are needed. Fig. 4 gives an example of the typical ΔH and ΔV signals, reconstructed from the digitised data for the single bunch p test beam in the AA. For the SPS re-injection into the PS, the bunch is less than 4 ns long and therefore some intermediate level electronics would be necessary to expand the bunch prior to digitising.

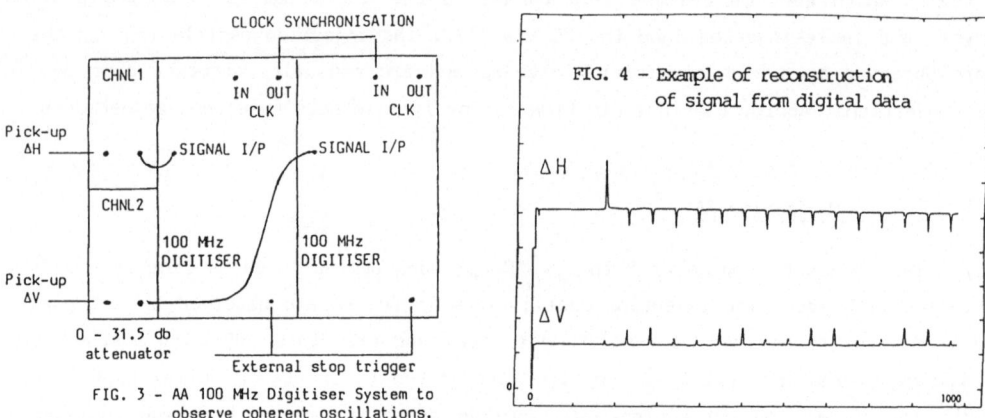

FIG. 3 - AA 100 MHz Digitiser System to observe coherent oscillations.

FIG. 4 - Example of reconstruction of signal from digital data

Data handling and correction : The digitised samples are accessible from the Interpreter language NODAL using standard intermediate level software to access CAMAC. Whilst for the single bunch applications the data is readily applicable for treatment, the multi-bunch application needs software filtering and reduction so that the coherent oscillations may be observed on one particular bunch over several turns.

Given the fractional part of the number of betatron oscillations per turn, q a general sinusoidal approximation may be applied to the amplitude modulated bunch peaks over the necessary number of turns; by minimising the summed least-squares error over this interval, values for the coefficients of the general sinusoid may be obtained. The changes in the kicker and Septum values or the vertical steering elements are then obtained from these coefficients using known and experimentally found conversion factors.

Fast Repetitive Analysis of Antiproton Injection

The beam of 6×10^6 antiprotons from the production target currently injected into the AA is too weak to be seen on even the most sensitive pick-ups. In any case, it is swamped at injection by the negative pions and electrons which are found in the spray of particles coming from the target. There are about 300 negative pions and roughly half that number of electrons accompanying each antiproton into the AA. The pions decay with a time constant of 680 ns, whereas the electrons lose energy by synchrotron radiation and spiral inwards to strike the inner aperture limits of the AA injection region. The radiation loss signal from 3 μs after injection, when most of the pions have decayed, until 22 μs, when all the electrons have been lost, is dominated by the electron showers.

This signal is naturally modulated at the AA bunch and revolution frequencies (9.5 and 1.85 MHz). Horizontal mis-steering at injection will cause a further modulation at the horizontal betatron frequency and higher harmonics given by : f x (n±q), where f is the revolution frequency, q is as defined earlier and n = 0, 1, etc. Injection timing errors causing all or part of an antiproton bunch to be missing from the AA result in an increase in the amplitude modulation at the revolution frequency.

To analyse these signals an electron shower detector, used in the DC mode and placed downstream from one of the electron loss points, is fed via a 1.5 MHz low-pass filter into a 100-MHz digitiser installed in an autonomous CAMAC crate; this is controlled by a LeCroy LRS 3500 microcomputer incorporating a bit-slice pre-processor that permits CAMAC transfers at approaching 1 MHz together with fast on-line signal treatment.

As the injection conditions must be monitored on each cycle, i.e. every 2.4 s, this microcomputer runs continuously, passing reduced data to the AA computer at the end of each cycle. By experiment we have found that the yield of electrons and the amplitude of their coherent oscillations are related to the yield and coherent oscillations of the antiprotons[2]. Consequently, the first application of this microcomputer has been to analyse the frequency spectrum of the electron loss signal, using FFT software written in microcode for the bit-slice processor. A 512-sample floating-point FFT takes 400 ms. The digitised waveform or the Discrete Fourier Transform (DFT) together with the magnitude and phase of four selected harmonics, the former multiplied by a selected scaling factor, plus digitizer frequency and offset information are displayed on the screen of the microcomputer using a routine written in FORTRAN.

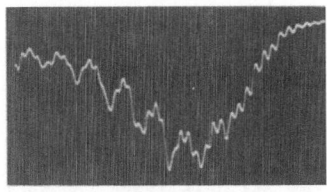

FIG. 5 - Electron loss signal

Examples are given in Figs 5, 6 and 7. Here the analogue form of the electron loss signal and the corresponding DFT are shown in two cases that illustrate i) large coherent betatron amplitude and ii) badly timed injection. The zero harmonic gives the electron yield, which is proportional to the antiproton yield. The 5th and 14th cofficients of the DFT correspond to the f × q and f × (1-q) frequencies, and the 19th coefficient to the revolution frequency, f. The sampling frequency and/or the number of samples may be adjusted so that the frequencies fall close to harmonic coefficients in the DFT. The overall shape of the signal (Fig. 5) provides a natural window function for the FFT and in practice we find that we do not have to sit exactly on the correct frequency in order to make use of a particular DFT coefficient.

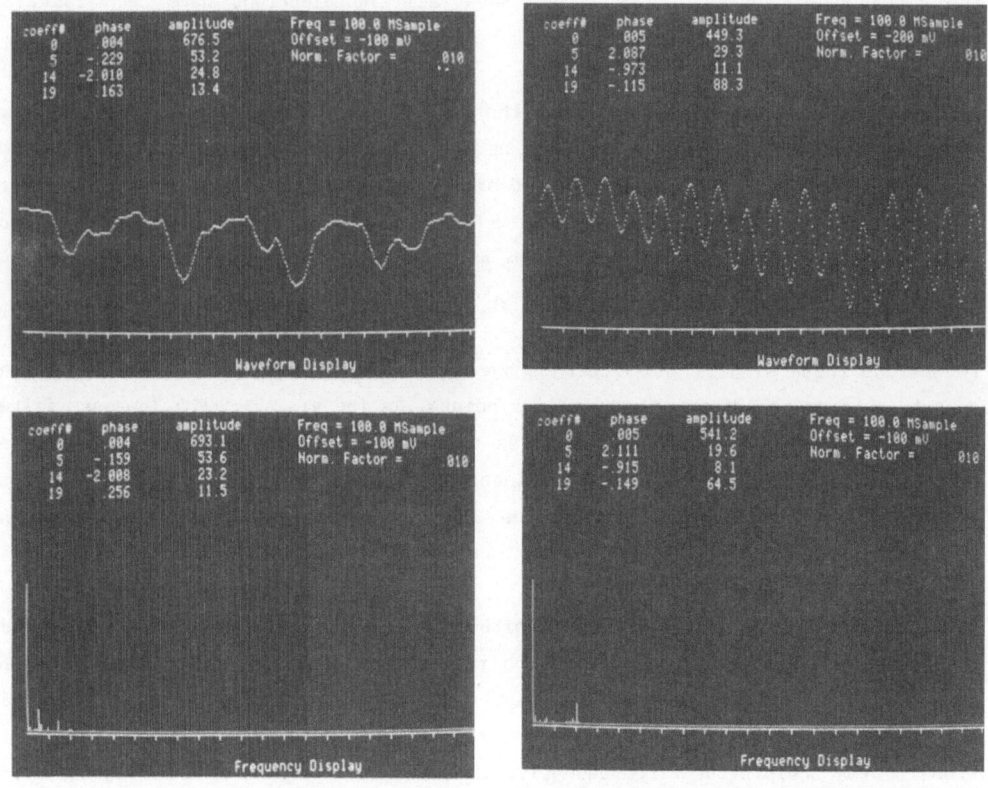

FIG. 6 - Part of the electron loss signal showing large coherent betatron oscillations and corresponding DFT

FIG. 7 - Part of the electron loss signal showing large revolution frequency modulation due to injection mistiming with corresponding DFT

Injection optimization is done manually, using this information passed to the AA computer. This is essential since many elements in the injection line can affect the steering into the machine and, without sophisticated software, a simple two element correction could lead to loss of aperture in the $\bar{p}$ injection line.

References

1. S. Van der Meer : private communication
2. C.D. Johnson : Antiproton yield optimization at the CERN AA; IEEE Trans. on Nucl. Science, Vol. NS-30, N°4, August 1983, pp. 2821.

AUTOMATED CYCLOTRON MAGNETIC FIELD MEASUREMENT AT THE UNIVERSITY OF MANITOBA

V. Derenchuk, J. Bruckshaw, I. Gusdal, J. Lancaster,
A. McIlwain, S. Oh, R. Pogson and J.S.C. McKee

Cyclotron Laboratory, Department of Physics,
University of Manitoba,
Winnipeg, Manitoba, R3T 2N2

ABSTRACT

The magnetic field of the University of Manitoba compact cyclotron has been measured in high vacuum by polar scanning with 52 flip coils. This was a unique in-vacuo operation was required because the Curie effect on invar material is used to trim the field. The data acquisition controller was a Digital Equipment Corporation LSI-11 with CAMAC and IEEE-488 interfaces. Filtering, display and conventional equilibrium orbit analysis were performed off-line by means of a VAX-11/750 computer. A description of the apparatus and software is given.

INTRODUCTION

The University of Manitoba Spiral Ridge Cyclotron is a compact 50 MeV cyclotron with a unique method of field trimming through the use of invar blocks, whose permeability changes rapidly with temperature. A detailed and accurate mapping of the cyclotron magnetic field was required due to recent problems with H^- depolarization and poor transmission of D^- beams. The mapping was carried out in vacuum in order to avoid disturbing the temperature distribution of the invar blocks, which are vacuum insulated, and to reduce heat transfer to the iron in the magnet. Except for the requirement of operation in vacuum with air cooling, the principle of operation of the mapping system is the same as described elsewhere[1,2,3].

The mapping apparatus (Fig. 1)[4] consists of 52 flip coils of the Chalk River design inserted into an acrylic rod at 12.7 mm intervals. This rod is housed within a quartz tube which provides support and an avenue for air cooling. The quartz tube, which also houses two temperature sensors, is inserted in a Delrin block that contains the flipping mechanism and provides the means for azimuthal rotation for the arm. Azimuth measurement was by means of a shaft encoder having a resolution of Pi/32768. The flip coil leads are brought out axially along the top central hole of the cyclotron and attached to their respective integrators. The field is measured by means of a precision voltmeter switched from one integrator to another.

The complexity of the mechanism and the number of measurements involved in a complete mapping make it necessary for a computer to manage all the repetitive operations in order to complete a field scan in approximately one hour. The information from the 52 flip coils, obtained at intervals of one degree azimuth, results in 18,720 field measurements per scan. To measure integrator voltage and to correct for integrator drift required 3 voltage measurements and 2 coil flips. Noise interference was avoided by the introduction of a time delay after each

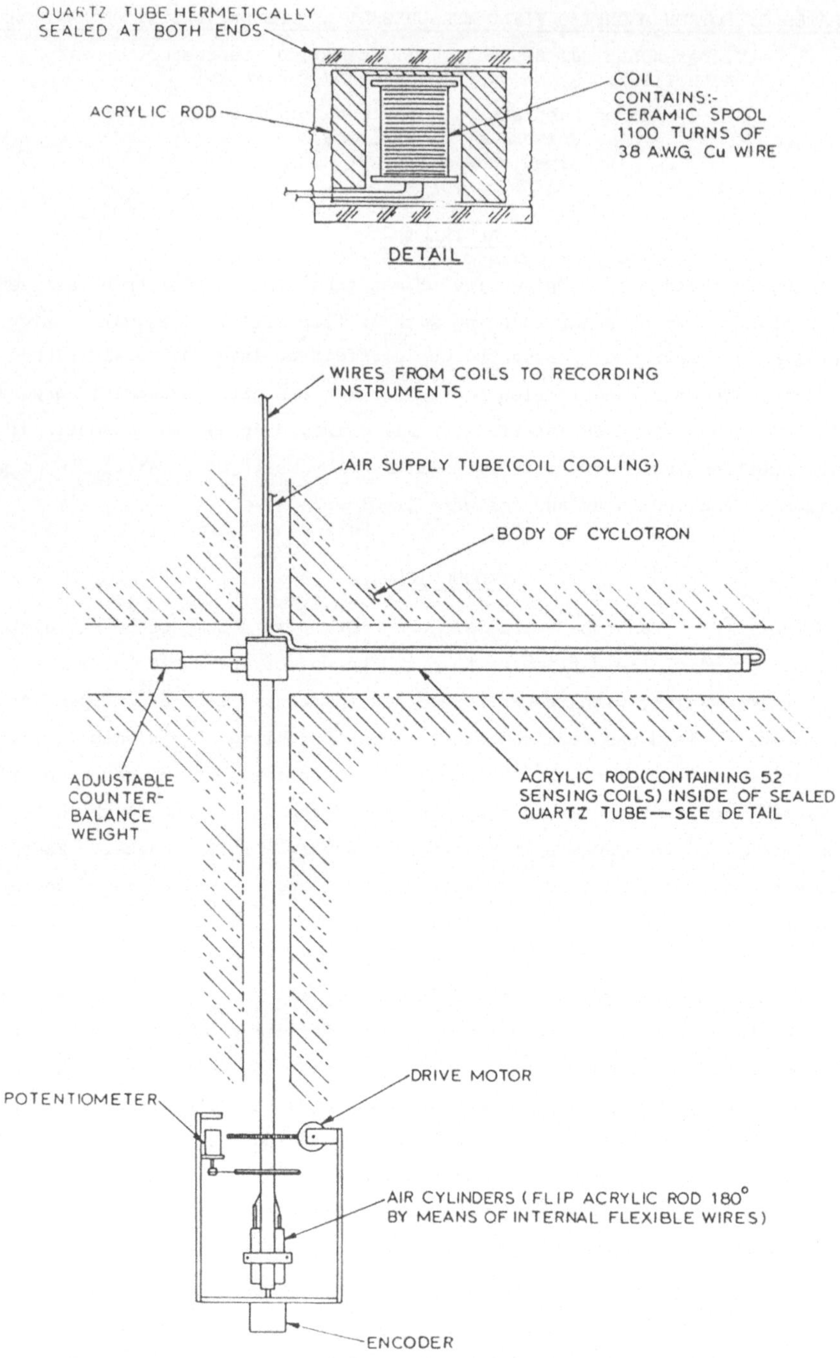

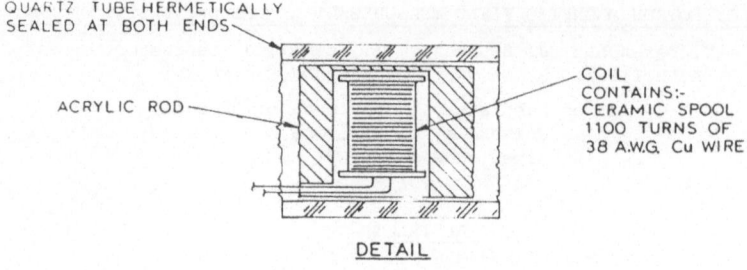

MAGNETIC FIELD MEASURING APPARATUS

fig. 1.

movement of the arm or change of voltmeter input by the multiplexer.

A CAMAC system was chosen as the interface standard for most of the electronics, with the voltmeter (an HP3456) using an IEEE 488 interface. All data and control signals were handled by a Digital Equipment Corporation LSI-11/23 microcomputer which was in turn linked to a VAX-11/750 computer. (Fig. 2).

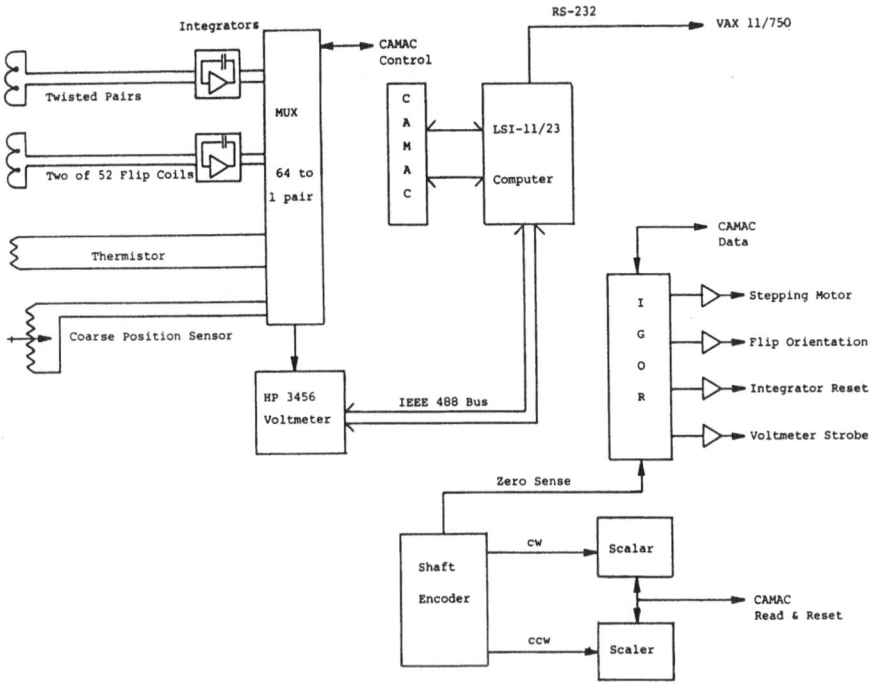

Fig. 2. Field Mapping Schematic

CONTROLLER

The controller computer programme was entirely written in FORTRAN except for some CAMAC library routines. The clock interrupt service and shaft encoder interrupt service routine were also written in FORTRAN. The LSI-11 architecture permits re-entrant routines to be written in a high level language but there are some restrictions on FORTRAN in the RT-11 system. Certain RT-11 system service routines are not re-entrant and calls to these had to be avoided.

There are four levels in the mapping programme. The highest level performs the initialization of the system including checks on power supplies, coarse positioning, and interfaces. After initialization the programme accepts a command from the user and scans a command table. If the command is present the appropriate command subroutine is executed. The third level contains sixteen such subroutines, each named after the command implemented, and which call combinations of very simple second level routines which manipulate a single element of the system. Elements are the voltmeter, the multiplexor, the integrator reset relays, pneumatic cylinders which

flip the arm, the azimuthal stepping motor, and the counting circuits of the shaft
encoder. Some of the second level routines merely set a flag so that the first level
of interrupt service routines will take the correct action when activated. The
second level routines often wait til the desired clock phase to take some action.

The synchronization of operations was accomplished using the 60Hz clock of the
LSI-11. On each clock interrupt an interrupt service routine was entered to advance
the state of a four state data structure. Higher level routines tested the system
state while waiting for settling times of multiplexor relays or stepping motors.
The interrupt service routine performed the triggering of the voltmeter, and the
switching of the multiplexors on different phases of the system clock when requested
by higher level routines which set global flags. (The mains period delay was
excessive in most cases but it is better to err on the safe side). Quoted settling
times for the multiplexor relays were about 2ms.

To begin operating the system, all power supplies are activated by the user and
the command given to start the programme. The programme checks the approximate
position of the flipping arm by reading the potentials on a potentiometer. If the
position is outside of a safe range, the operator must explicitly request a movement
in order to move the arm. If the arm is more than five degrees outside of the safe
circle, the computer will always refuse to move the arm further in the wrong
direction. This prevents the arm from damaging the cables leading to the rotating
apparatus. The user then has the responsibility to move the apparatus past the zero
reference of the shaft encoder to initialize the accurate position measurement.
Afterward, the programme will have three redundant position measurements, the potentiometer, the count of step commands given to the stepping motor and the count of the
number of angular steps measured by the shaft encoder. Each of these has successively
greater precision and all are reliable to prevent damage to the equipment. The last
two are sufficiently accurate to be used for measurement.

Once initialized, the programme can take charge of the major portion of the
mapping operation. By typing a single command and supplying an angular range and a
step size, the operator can scan any sector of the magnet. The programme makes the
decision about which end of the angular range is nearer the current position and
begins mapping from that point if the origin does not need to be crossed to reach the
starting point. Mapping can proceed in a clockwise or counter-clockwise sense. At
each point of azimuth, the programme generates a sequence of operations to reset the
integrators, read the zero voltage of the integrators, flip the coils, read the
integrator voltages, flip the coils and read the integrator voltages. The three
integrator measurements permit first order drift compensation. All voltage
measurements are stored in the Hewlett-Packard 3456 programmable voltmeter and are
transferred to the LSI-11 during the movement of the arm to the next point. The
LSI-11 transfers the measurements to a flexible disc storage device or the VAX
computer for further drift correction, conversion to Tesla and continuity checking.

Other useful features of the programme are the recording of the cyclotron magnet power supply shunt potential at each point and the automatic incrementing of a run number stored on a flexible disc. There is a calibration routine which generates a calibration table when the arm is in the gap of a standard magnet. The control sequences are identical to those carried out while mapping except that time delays replace movements in azimuth. An option to generate calibration pulses for the integrator inputs was conceived but never used because it was deemed too complicated to interchange coils in our system. Potentially, one could devise a system in which each coil and integrator would be calibrated and an interchange could be taken into account. This would permit replacement of a faulty integrator without the need to recalibrate the coil and integrator pair.

INTERFACES

The CAMAC crate held 6 single width CAMAC modules electrically connected to the measurement hardware. The crate controller (Bi-Ra Systems Inc. 1311-1) resided on the Q-bus of the LSI-11 and was accessed by means of standard calls to a software library provided by Effective Systems Inc. The CAMAC modules used in the measurements were one 16-bit Input Gate/Output Register (Kinetic Systems 3061T IGOR) which transmitted control signals to the flipping relay and stepping motor and which generated a LAM signal when the shaft encoder reached the zero reference position, one hex-scaler (Kinetic Systems 3610) which counted clockwise and counter-clockwise rotational pulses from the shaft encoder, and four analogue relay multiplexors of sixteen channels each, designed and fabricated in the Physics Department.

CONCLUSION

The computer control of the mapping system and data collection worked extremely well. Mechanical problems that were encountered in using the apparatus are being corrected prior to a further field mapping scheduled for November 1983. The original computer control system will remain intact.

REFERENCES

1. J. Ormrod, C.B. Bigham, K.C. Chan, E.A. Heifway, C.R. Hoffman, J.A. Hulbert, H.R. Schneider and Q.A. Walker, IEEE Trans. Nucl. Sci. NS-26 (1979) 2034.
2. P. Miller, H. Blosser, D. Gossman, B. Jeltema, D. Johnson and P. Marchand, IEEE Trans. Nucl. Sci. NS-26 (1979) 2111.
3. S.W. Mosko, E.D. Hudson, R.S. Lord, D.C. Hensley and J.A. Biggerstaff, IEEE Trans. Nucl. Sci. NS-24 (1977) 1269.
4. J. Bruckshaw, V. Derenchuk, I. Gusdal, J. Lancaster, A. McIlwain, S. Oh, R. Pogson and J.S.C. McKee, N.I.M. 207 (1983) 493-495.

On the problem of magnet ramping*

Eva S. Bozoki
National Synchrotron Light Source
Brookhaven National Laboratory, Upton, New York 11973

1. Introduction

In the accelerator rings, the momentum of the beam is being changed. It is necessary to ramp (up or down) all magnets in the ring in perfect synchronism and precisely according to a desired ramping function to prevent beam losses or unwanted changes in the beam characteristics (e.g. tune or chromaticity).

2. Function Generators

At the NSLS the power supplies are ramped by a Function Generator,[1] which is controlled and programmed by a μ-processor. When the FG is programmed by a so called vector information (see Fig. 1), it can change the current as specified by the vector, with a rate of change (slope) as also specified by the vector. Actually, the FG will change the current by 1 (or 16) count at every (1/S)-th "tick" of it's clock. This poses severe restrictions on the programming vector; both, the ΔI current change and 1/S = Δt/ΔI inverse slope has to be integers. In addition, the duration of the vector has to be longer than the time necessary to download the information into the FG.

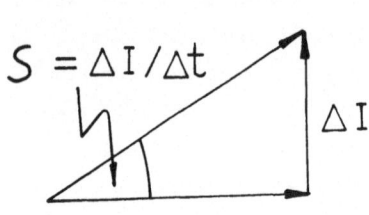

Fig. 1.

3. Ramping Functions

The basic requirement for ramping is given in terms of beam momentum. The P(t) ramping function is the specification of how one wants to change P in the accelerator ring as a function of time. This, in turn, specifies the necessary time dependence on the dipole's magnetic field since B_{dip} [KG] = 33.356 * P [GeV/c]/ρ[m]. On the other hand, the gradient field of the quadrupoles are determined at any time, by the dipole field, momentum and the tune. Thus, prescribing the time behavior of the tune (normally, one wants to keep it unchanged during ramp) in addition to the momentum, the time dependence of the quadrupole fields can be obtained. From magnetic field one can

---

* Work supported by the U.S. Department of Energy

easily calculate the exitation current in the magnets (using actual magnet measurements). Thus, the P(t) ramping function can be "translated" into a set of I(t) ramping functions, one for each magnet family* in the ring (for more detailes see [ref. 2]).

In the most general case the F(t) ramping function (P(t) or I(t)) can be described as a piecewise function, i.e. it consists of a number of regions in which F(t) is a different mathematical function (see Fig. 2).

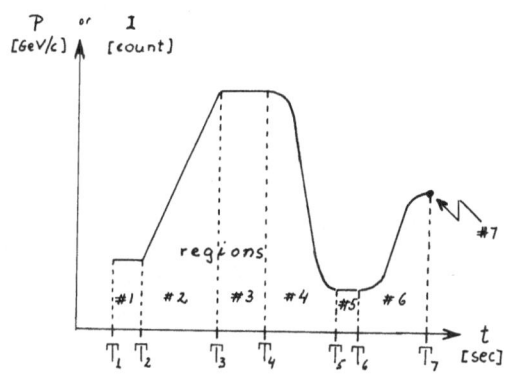

Fig. 2.

4. Segmentation

The problem is how to best approximate (in any accelerator/deceleration region) the I(t) ramping functions with a set of segments (vectors) while satisfying the hardware imposed restrictions. For detailed analysis see ref. [3].

The constraints   (a) ΔI = integer

(b) $\Delta t/\Delta I$ = integer and larger then 2

(c) $\Delta t > \Delta t_{min}$

suggest to solve the mathematical problem in the $x = I$, $y = \frac{dt}{dI} = t'$ coordinate system. The segmentation now is reduced to the following problem as shown on Fig. 3: Approximate the t'(I) function by a T'(I) step-function, which runs on the integer grid. Each step represents a segment. For the i-th step, the inverse slope is T'_i

---

*Identical (within manufacturing tolerances) magnets, powered by the same power supply and thus controlled together.

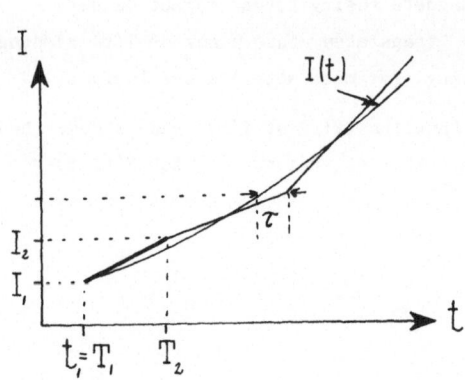 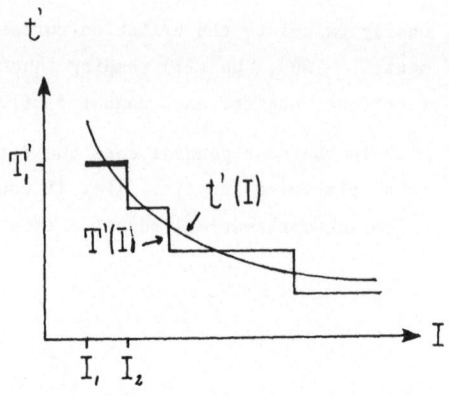

Fig. 3.

and the total change in the current is $\Delta I_i = I_{i+1} - I_i$. The goodness of the approximation for the i-th segment can be characterized by $\tau_i = t_{i+1} - T_{i+1}$, where $t_{i+1} = t(I_{i+1})$ is the time corresponding to the end of the i-th segment on the t(I) ramping function and T_{i+1} is the time corresponding to the end of the i-th segment. At each step, there are two possible values for the inverse slope:

$$T'_i = \begin{cases} \text{Integer}(t'_i) \\ \text{Integer}(t'_i) + 1. \end{cases}$$

One should always choose the value which makes $\varepsilon_i = \sum_{j=1}^{i} \tau_j$ smaller, thus forcing the consecutive segments to "overshoot" and then to "undershoot" the ramping function. The average goodness of the segmentation is defined as

$$\langle \varepsilon \rangle = \frac{1}{k} \sum_{j=1}^{k} \varepsilon_j .$$

$\langle \varepsilon \rangle$ is determined by the t(I) function and by the choice of a ΔI_{min} minimum stepsize*. Small ΔI_{min} yields small $\langle \varepsilon \rangle$, but the number of segments will be large, which might not be desirable. Thus for a given function one always has to choose a stepsize, which yields acceptable accuracy as well as acceptable number of segments.

---

*The resulting ΔI_i steps are a multiple of ΔI_{min}, since the adjoining segments with the same T'_i values are combined into one segment.

In Algorithm was developed [ref. 3], based on the above considerations.

5. Implementation

A program, RAMP, was written which can:
(a) "translate" a user defined $P(t)$ ramping functions into a set of corresponding $I(t)$ ramping functions for the dipole and all other ramping elements in the accelerator/storage rings and
(b) segment the ramping function while satisfying the imposed hardware critera.

The program has color display capabilities and offers a create/modify option for the input data file which contains the description of the ramping function(s) and hardware restrictions.

References

1. Om V. Singh, Introduction to NSLS Function Generator NSLS Tech Note #24, 1979.

2. Eva S. Bozoki, Automatic Generation of the Ramping Function for Dipoles and Quadru poles, BNL-26857, 1979.

3. Eva S. Bozoki: An Algorithm for Programming Function Generators Comp. & Math with Appls. 7, 1, 1981.

4. Eva S. Bozoki, A Program for segmenting Ramping Function NSLS Tech Not #52, 1980.
   Eva S. Bozoki, NSLS Tech Note #142, 1983.

High level control programs at NSLS*

Eva S. Bozoki
National Synchrotron Light Source
Brookhaven National Laboratory, Upton, New York 11973

Overview

The mathematical model of an accelerator can be used to control its operation and also to simulate its behavior. Through the use of such modeling programs it is possible to control a few important machine parameters rather than the many individual magnets. Such programs have been implemented for the current generation of accelerators.

At the NSLS, two control program systems have been developed; TRANCO[1] for the transport lines and RING[2] for the accelerator/storage rings. These systems are modular, can be used independently or in conjunction with each other, can communicate with each other via sequential and random access data files and has a common data base. There are modules for computation, control, display, measurements, data base manipulation, etc. [Fig. 1 shows the interaction of some of the main modules and the data flow.]

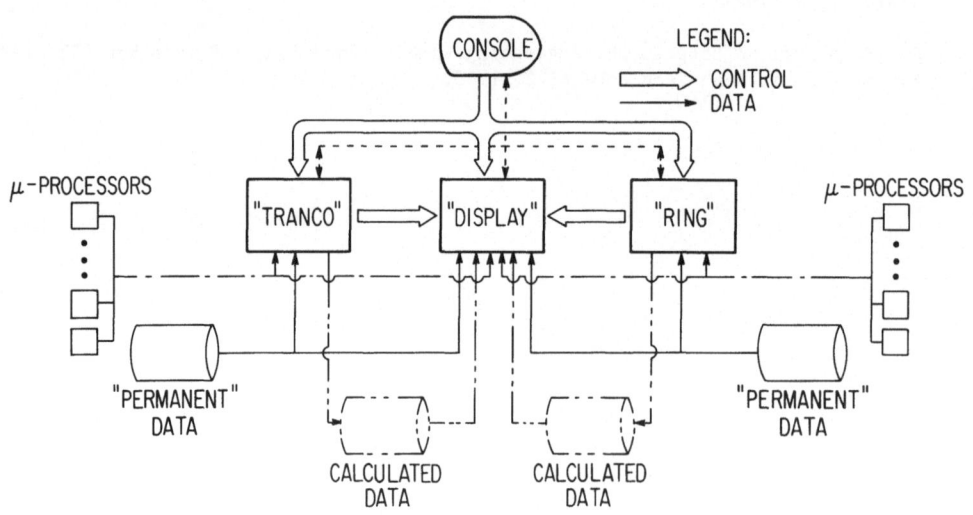

Fig. 1.

---

* Work supported by the U.S. Department of Energy

The programs are data-driven, so that they can be used for any tranport line or ring. The characteristics of the different lines and rings are in the system's data base and they "drive" the program to call on the appropriate models.

All programs can be used (a) <u>on-line</u> for measurements and control or (b) <u>off-line</u> as a tool for exploration and design. Color graphics are available in either mode.

In the <u>on-line</u> mode, they access the appropriate μ-processors to obtain the current settings of the magnet's power supplys and/or the current digitized monitor readings. They also accept operator input specifying desired conditions and options.

The programs then calculate a new group of settings and the corresponding derived variables which will result from these settings. Upon examination of the results (colored graphic and alphabetic displays are available) the new settings may be transferred to the power supplies.

Calculations & Operations

TRANCO uses first order 6 x 6 transport matrices (similarly to the TRANSPORT code) to calculate phase ellipses, beam matrices and emittances. The following operations can be performed. (1) phase ellipse matching at the end of the transport line by adjusting designated quadrupoles (2) waist can be made at the desired location by adjusting designated quadrupoles, (3) phase ellipses can be positioned at the end of the transport lines with respect to the acceptance elipse by using H & V correction dipoles (in some cases the septum magnet is used in lieu of one of the correctors) (4) beam can be steered through the desired trajectory by adjusting the strenghts of correction dipoles (5) emmittance can be calculated.

RING uses two (H & V) first order 3 x 3 transport matrices (similarly to SYNCH) to calculate beam and lattice parameters such as tune, Twiss parameters, dispersion, chromaticity, beam sigma, synchrotron integrals and the like. The following operations can be performed (1) tune (or other lattice parameters) can be changed by adjusting certin quadrupoles (2) the chromaticity of the ring can be changed by adjusting sextupoles (3) closed orbit can be measured (4) closed orbit can be globally corrected by finding the most effective correction dipoles and adjusting their strengths (5) closed orbit can be distorted locally to achieve a specified displacement or slope at a given point by adjusting the nearest correction dipoles, (6) orbit changes as a consequence of specified kicks around the lattice can be calculated.

Data Base

Conceptually, all data fall into one of four categories and are treated accordingly.[3]

1) Permanent data, characteristic of the transport line or ring. These data "drive" the program and are kept in disk data files which are read by the program.
2) Temporary data, including operation to be performed and options selected in addition to temporary values of accelerator parameters. These quantities are entered by the operator via the control desk keyboard or other terminal.
3) Variable data. These include magnet strengths and measured data from position monitors, profile measuring devices, etc. These parameters are read (a) from the microprocessors controlling the devices in the on-line mode or (b) from data files in the off-line mode.
4) Calculated data. Data calculated by RING or TRANCO are stored in random access disk data files. The data in these files can be printed or displayed and redundant calculation is avoided by sharing results between the program modules.

Model Calibration

The control and predictions from the above programs are as good as the models themselves which are used by the programs. All models represent a simplification, thus introducing deviation of the real machine from its mathematical model. The accuracy of the model is limited by the approximations made in describing the structure and components of the accelerator and the behavior of the beam, by the omission of some effects from the model and also by the errors and uncertainties of magnet calibrations which are incorporated into the model.

An alternative or complimentary method to the magnet calibrations, a method which empirically eliminates the effect of all inaccuracies is the calibration of the model[4] itself through calibration of "physics" quantities (e.g. tune, dispersion, chromaticity, etc.) from measurements on the operating machine (see Fig. 2).

This is possible, since all NSLS rings were designed to have modular control features; that is certain groups of magnets are used to control only specific "physics" quantities. There is one family of quadrupoles in each ring which is adjusted to control dispersion, there are 2 (or 3) families of quadrupoles to control the H&V tunes (and one other measurable machine parameter in the X-ray ring), there are 2 families of sextupoles to control the H&V chromaticities.

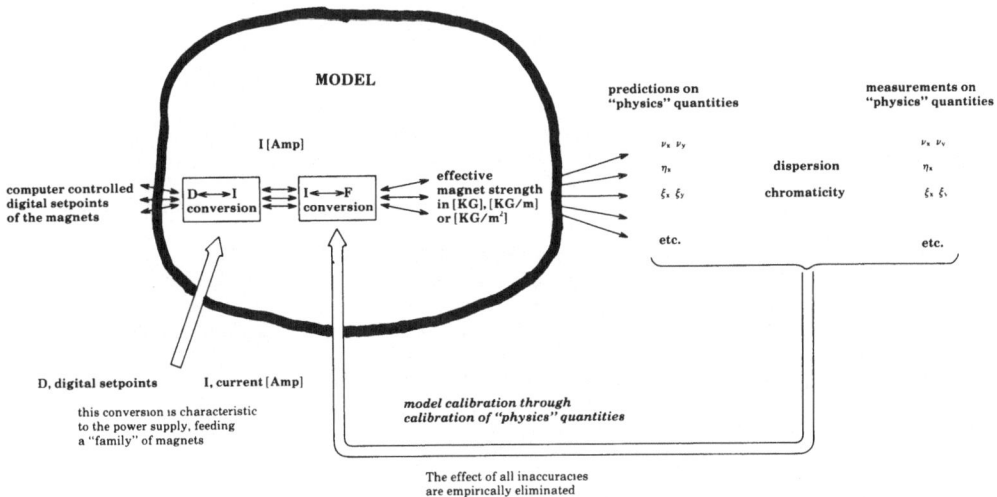

Fig. 2.

The advantage of this method is, that it compensates for any inaccuracy in the magnet measurement. Actually, there is no need for magnet measurements at all, which might be of great help for super large accelerators where accurate measurement of all magnets is very time consuming and costly.

There are effects, not directly included into the model, whose significance are different under different operational conditions. This introduces some inaccuracy into the model calibration. For example, as the beam intensity is increased, different beam instabilities will be important and they effect the measured tune.

References

1. Bozoki, E., Control Program for the Linac-to-Booster Transport Line (TRANCO), BNL-25473.
   Bozoki, E., Users's Guide of TRANCO Program, BNL-28752.
   Bozoki, E., The TRPLOT Display Program, BNL-29829.
   Bozoki, E., Emittance Calculation programs, Technote No. 118.
2. Bozoki, E., High Level Computer Control of the NSLS Accelerator and Storage Rings, BNL-31361.
   Bozoki, E., Color displays for the NSLS rings, BNL-31261, (1982).
   Bozoki, E., Incorporation of the coupled beam size and Touschek Lifetime into the RING Program, Technote No. 119.
   Bozoki, E., Orbit Measurments through the use of the PUEREAL Program, BNL-32899, (1983).
3. Bozoki, E., Some Thought on the Application control Program System, Technote No. 17.
4. Bozoki, E. Calibration of the RING Model for the NSLS Rings, IEEE Trans. of Nucl. Sci. 1983.

THE MINICOMPUTER NETWORK FOR CONTROL OF THE DEDICATED
SYNCHROTRON RADIATION STORAGE RING BESSY

G.v. Egan-Krieger, W.-D. Klotz and R. Maier

Berliner Elektronenspeicherring-Gesellschaft für
Synchrotronstrahlung mbH (BESSY)
1000 Berlin (West) 33, Lentzeallee 100

Summary

The architecture and the system software aspects of the BESSY-control system are presented. The network consists of four minicomputers (HP 1000F) which control the 800 MeV storage ring, the booster, a separated function alternating gradient synchrotron, and the preinjector, a 20 MeV microtron. All machines with their magnet-, vacuum-, rf- and interlock-, as well as their beam diagnostic components are connected via a simple bit-parallel byte-serial electronic interface system to the minicomputers. The software system handles about 1000 bytes of I/O-channels. With the exception of the producer's software the largest part of programming (about 90%) was done in FORTRAN. A description and summary of the organization and the features of the control system software is given.

I. Introduction

Equipments associated with the 800 MeV electron storage ring and the injector of BESSY[1][2] are controlled by a local area network of four minicomputers. The hardware (magnet power supplies, vacuum, RF-system etc.) is connected to this network by a commercially available electronic interface system[3][4]. The total amount of the I/O channels controlled by the software system is given in tab. 1.

II. The computer network

The process computer network consists of four identical Hewlett-Packard 1000F 16 bit processors. They are equipped with hardware floating point arithmetic, microcoded fast FORTRAN, scientific and vector instruction sets. The installed semiconductor memory ranges between 256 Kbyte and 1 Mbyte per processor. Two processors share a mass storage of 20 Mbyte. One of them has an additional disc space of 120 Mbyte. The minicomputers are linked by a local network called DS/1000. This network is a standard bit serial communication package for HP-1000 computers with a hardware transfer rate of 1 Mbyte/s. The two computers equipped with mass storage devices are running RTE 4B, a disc based real-time multiuser multiprogramming executive. A smaller memory resident version of this operating system, RTE MIII is running on the two satellite processors. The layout of the process computer network and the control system hardware is shown in fig. 1 and 2.

An additional link delivers the storage ring status online to the radiometry laboratory of the Physikalisch-Technische Bundesanstalt.

III. The software system

The software system is organized in a hierarchy of different layers. About 90% of the programming was done in FORTRAN. Mainly for reasons of time optimization the rest was written in ASSEMBLER.

III.1 Basic system software

Message transfer system

Despite the fact that the producer's software includes powerful tools for program-to-program communication in the network, a general message transfer system layer has been put onto the executives. This layer, which consists mainly of a message handling processor, a program called IPCCM (interprocessor communication & control monitor) is responsible for storing and retrieving messages through the whole network. The IPCCM was written in FORTRAN, uses 12 Kbyte for code and 4 Kbyte of dynamic buffer memory. It runs with high priority in real-time forground and is locked into semiconductor memory in order to prevent disc accesses for message transactions. Each node of the network contains one IPCCM program. All IPCCMs are permanently available to all programs, they are initialized on system start up. The main control flow in each IPCCM is shown in fig. 3. The IPCCM keeps track of two dynamic linked lists. One is the list of pending messages (LPM) the other the list of waiting receivers (LWR). For every request from a user's program an entry in one of these lists is created or deleted. If a user's program makes for instance a send request, to pass a message to another program, this request is either passed to the waiting receiver program or entered into the LPM, if no such receiver is waiting. On the other hand, if a program asks the IPCCM to receive a message an entry in the LWR is created, if no pending message is available at that moment. In this case the program, which has issued the receive request is put into a wait state by the IPCCM.

In order to avoid pile up of the entries in these two lists, a request time-out can be passed to the IPCCM. In case that a message is pending for a longer time than specified by the time-out without being requested by another program, it will be discarded by the IPCCM. A program which waits for a message longer than the specified time-out will continue to run on its point of suspension. For implementing this time-out feature the IPCCMs are repeatively scheduled by the operating systems in intervals of 0.5 seconds, which define the time-out unit.

The messages, which can be up to 256 words long at maximum, are actually stacked in the available memory of the operating systems, and only their message identifiers, called class numbers, are processed by the IPCCMs. Therefore the list entries which the IPCCMs are handling are rather short (10 words). Fig. 4 shows the contents of such an entry.

Message exchange between different nodes is automatically done by the IPCCMs based on a store-and-forward mechanism. Sending and receiving messages is completely transparent to the programmer. To achieve this, the IPCCMs are able to communicate directly with each other via remote EXECUTIVE calls (fig. 5). The path of the messages through the network, once the sender's and receiver's node are known, is found by a network description table, which is common to all IPCCMs.

Besides that the IPCCM can be scheduled to give status information of the message transfer system within its node on any terminal.

By means of this message exchange layer a transparent nucleus was established, which allows asynchronous program to program communication within the whole network. Program synchronisation can be simply achieved by performing message exchanges with acknowledge (handshake).

Test runs showed, that for message blocks of 64 words, the total turn around time for a message from a sender to a receiver and back to the sender is 20 msecs, if both programs are on the same node and 90 msecs, if they are on neighbouring nodes. Experience showed that this is more than sufficient for conventional control system applications, since hardware components like high current power supplies or vacuum valves are much slower.

Hardware access

I/O to the electronic interface

For I/O's without interrupt a subroutine, called DEV is loaded with the user program. A call to this routine switches off the executive's interrupt for the duration (typ. 10 microsecs) of the directly programmed I/O (fig. 6).

For interrupt driven I/O's a driver has to be generated into the operating system. This driver handles the interrupts and schedules interrupt service routines.

Depending on the different card types (tab. 2), especially for the IEEE-bus, a library of FORTRAN callable I/O-routines is used. For instance the call IA16 (<crate>, <slot>, <value>) outputs a 16 bit word, IE16 (<crate>,<slot>,<value>) inputs a 16 bit word, DEVIB (<crate>,<slot>) initializes the IEEE-bus, IBWRT (<bus logical unit>, <command string>, <length>) sends information to the bus listener.

Peripheral interface modules

A common problem in controlsystem design is the large number of different equipments and subsystems which have to be handled by the software system. All peripherals are grouped into classes of equal or very similar equipments. For instance: 4 groups of power supplies, each containing about 20 to 40 equipments, 2 groups of power supplies for ion getter pumps, 1 group of vacuum valves etc.

Access (I/O and control) to equipments of one group is performed by peripheral interface modules (PIM). In the system there is no other path to equipments than via PIMs and therefore only these programs contain device dependent code (fig. 7). The PIMs communicate with programs, performing control algorithms, through the message transfer system by exchange of messages in a standardized format. For each equipment a PIM owns a device description table (DDT). All DDTs of a group of equipments form the data table (DT). The DT can be visualized as a 2-dimensional sequential list.

The first index points to the equipment, the second to entries in the DDT. For instance:

    DT (1, 5) = I/O-address of power supply # 1
    DT (16,21) = status of power supply # 16
    DT (20,16) = actual current of power supply # 20 etc.

A user program can activate a PIM by sending a message to it, specifying the equipment # and an operation code (OP-code) in the message. The user specifies the OP-code by a 4 character string which is internally converted into a definite 16 bit integer:

    #CUR = read current
    #STA = read status
    #SET = set current
    #ON = switch on
    #OFF = switch off

When the PIM receives a message, it performs a binary search for the requested OP-code in an ordered index-table of all available OP-codes to get the appropriate JUMP-label for the operation to be done. Having performed the requested task, the PIM sends an acknowledgement to the caller and goes back into the receive state.

The data base

After initialization of all PIMs, all DTs in the different nodes represent a distributed data base. Access to this data base is exclusively done by the PIMs (fig. 8). Besides the actual data base of the running system represented by the DTs in the PIMs, there is need for a disc resident data base for the following reasons:

i) in order to initialize the system, the PIMs have to read their DTs from a back up file,
ii) in order to update permanent data in the data base, the DTs have to be resident on disc,
iii) There must exist a directory of equipments in the system, so that programs which want to perform a control operation on an equipment can get an identifier, specifying the equipment # and the PIM which is responsible.

Each equipment in the system is definitely defined by a 6 character string. A program which wants to perform an I/O operation has to do the following steps:

i) to get its own program identifier
ii) to ask the data-base-directory monitor (DBDM) to convert the equipment name into an identifier
iii) to send a message to the corresponding PIM
iv) to issue a receive request to get an acknowledgement from the PIM.

In the initializing phase the PIM asks the DBDM for the complete DT. The DBDM then reads the DTs from the data base file and puts all equipment names into its memory resident directory. After initialization the DBDM does not have to perform further disc accesses. All entries of the directory are accesses by an hash algorithm performed on the strings of the PIMs and the equipment names.

In addition there exists a memory resident string library of about 10 Kwords of buffer space. It contains ASCII strings for equipment descriptors, units, status-bit descriptors, formats, alarm-messages etc. ... In the data base only pointers to these strings are stored.

III.2 Operational software

As basic man-machine interface a monitor program exists which can be scheduled at any terminal on the site. This program allows the operator to step interactively through a tree structured menu field. Depending on the type of terminal this menu is either offered with programmable softkeys or by simple alphanumeric I/O. Through this dialog the operator finds all necessary tasks, like setting power supplies, reading status, switching polarities, opening and closing valves etc. ... He also has the opportunity to write simple procedure files in an unstructured command language, like switching on 20 power supplies one after the other. In order to perform control algorithms in a structured language, he has the possibility to schedule a real-time BASIC interpreter to which the library of system calls was added.

An example:

```
10 LET A$ = "DIPRR "
20 FOR C = 600 TO 700 STEP 0.5
30 CALL SET (A$,C): FAIL GOTO <Error>
40 WAIT (1000)
50 NEXT C
```

Within the loop the power supply named DIPPR will be ramped from 600 A to 700 A in steps of 0.5 A, one step per second.

To manage more advanced features the tree offers to the operator the possibility to schedule different utility programs. He can run an optics program which reads the actual settings of the storage ring and calculates online lattice functions and other physical ring parameters which are presented in graphical form on the operator console. Start up, stand by, ramping and protocol procedures are available as well as online hardware diagnostic tools. To produce closed orbit bumps or to vary independently the horizontal or vertical tunes at least two power supplies have to be ramped at the same time. This is an example of a generalized control parameter of vector type. In order to control vector type variables a multiparameter handler was installed. This feature consists of a configuration- and an online execution-processor. The configurator sets up the vector variable as a linked list of vector components. Each vector component holds complete information about one equipment like I/O-address, conversion and scaling factors. All these informations are read out from the distributed database in the configuration phase. The configurator sends this list to the execution-processor. Once the multiparameter processor is configured, on recept of the master parameter, it simply steps through this linked list. For each component of the vector it converts the master parameter to the scaled setting and performs the corresponding output operation. The master parameter can be entered to the execution processor either interactively, from computer controlled knobs or by other programs.

III.3 User libraries

To free the programmer from taking care about internal details, libraries of system calls have been established. All library calls are available in BASIC, FORTRAN and PASCAL.

Library call examples:

BASIC

```
10 LET A$ = "QUAP1R"
20 CALL CUR(A$,X): FAIL GOTO < Error>
```

FORTRAN

```
REAL X
INTEGER CUR

IERR = CUR(6HQUAP1R,X)
IF(IERR .LT. 0) GO TO <Error>
```

PASCAL

```
TYPE
STRING = PACKED ARRAY [1...6] OF CHAR;
VAR
ERROR: INTEGER; NAME: STRING; X: REAL;
FUNCTION CUR $ALIAS ˜CUR˜ $(NAME: STRING; X: REAL): INTEGER; EXTERNAL;

BEGIN
NAME: = ˜QUAP1R˜;
ERROR: = CUR (NAME, X);
IF ERROR   0 THEN .... ELSE ....

END.
```

All calls read the current of the quadrupole power supply named QUAP1R and return in X the actual current in amperes.

The system library is organized in three levels. On the lowest level there are the send and receive calls to the message tranfer system (fig. 9). On this level the user has to care about system details like message buffer format, getting program and equipment indentifiers and organizing message exchanges. On the next level a general purpose process I/O interface is available. Here the user has only to specify the ASCII strings of OP-codes and equipment names and the reply buffer. The message exchange and string conversion is done by this layer. On the highest level for each OP-code a seperate call is available. For example the FORTRAN statement
  CALL OFF (6HDIPPR)
switches off the dipole power supply named DIPPR.
  CALL SET (6HDIPPR , 900)
sets the current of the same power supply to 900 amperes.

References

1) G.v. Egan-Krieger, D. Einfeld, H.-G. Hoberg, W.-D. Klotz, H. Lehr, R. Maier, M. Martin, G. Mülhaupt, R. Richter, L. Schulz and E. Weihreter; IEEE NS-30 (1983) 3094

2) G.v. Egan-Krieger, D. Einfeld, W.-D. Klotz, H. Lehr, R. Maier, G. Mülhaupt, R. Richter and E. Weihreter; IEEE NS-30 (1983) 3103

3) FREY Analogtechnik, Marathonallee 33, D-1000 Berlin 19, FRG

4) G.v. Egan-Krieger, W.-D. Klotz and R. Maier; this conference

5) Donald E. Knuth, The art of computer programming, ADDISON-WESLEY, PUB.COMP., INC. (1973) 2nd Ed.

Tables

digital register outputs	173
digital register inputs	641
output pulsed (commands)	121

Tab. 1: Total number of I/O-channels in units of bytes.

typ	function	name
DOR	16 bit output register with handshake or interrupt	digital output register
IIR	16 bit input register with handshake or interrupt	isolated input register
SPO	16 bit input register with handshake or interrupt 8 bit pulse output	status input & pulse output
PSC	12 bit output register 8 bit input register 5 pulse output	power supply controller
MCC	17x16 bit counters with RAM-sequencer control and FIFO I/O-buffer	multichannel counter
HP-IB	16 bit open collector I/O	HP-IEEE bus interface

Tab. 2: standard I/O-cards

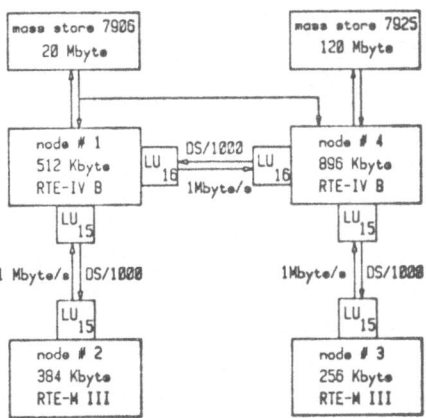

fig.1:Process computer network

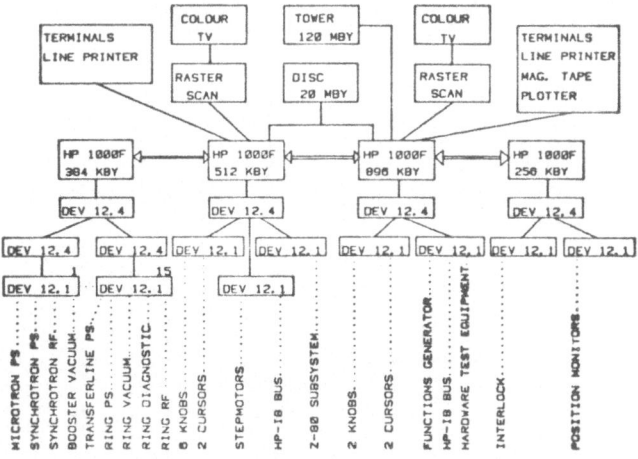

fig.2: Control system network

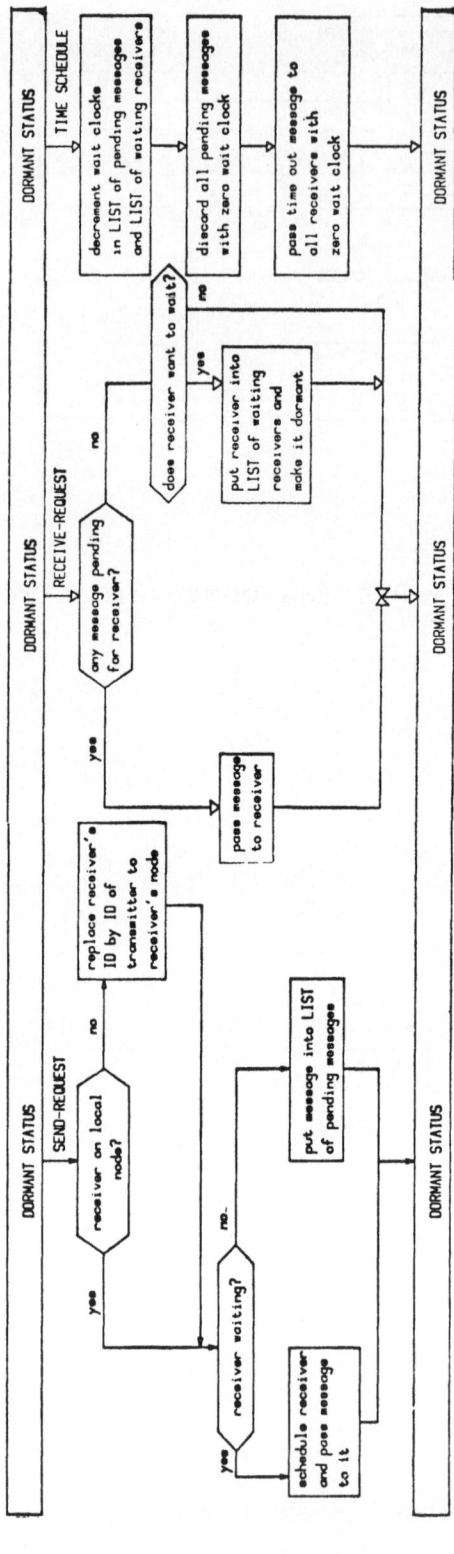

fig.3: Main control flow in the IPCCM

WORD	CONTENTS
1	link-word
2	message priority (0-32767)
3	local sender ID
4	local receiver ID
5	time out parameter
6	message sequence number
7	message origin (source ID)
8	wait clock
9	message class number
10	node of message origin
11	node of message destination

fig.4: IPCCM list entry contents

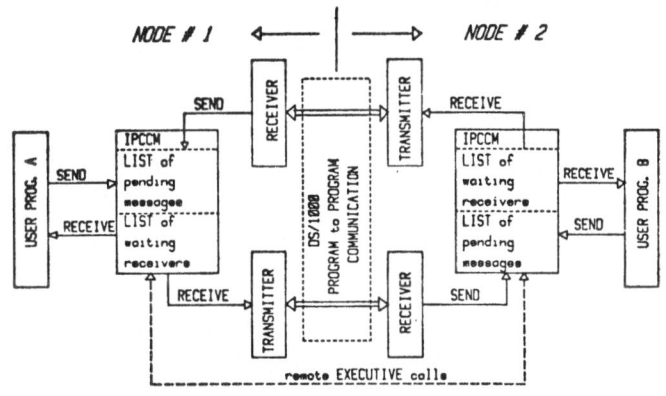

fig.5: Data paths in the MESSAGE TRANSFER SYSTEM

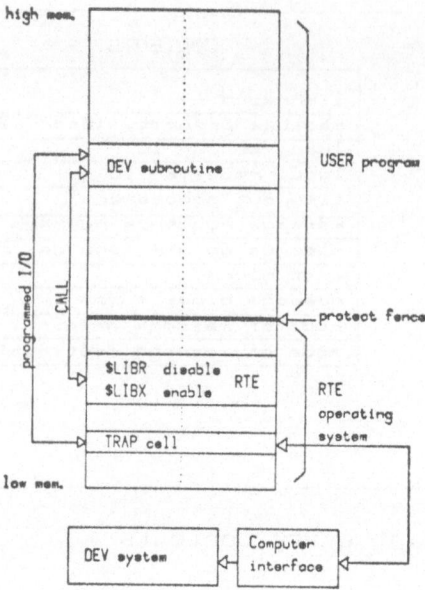

fig.6: Standard DEV subroutine
       INPUT / OUTPUT

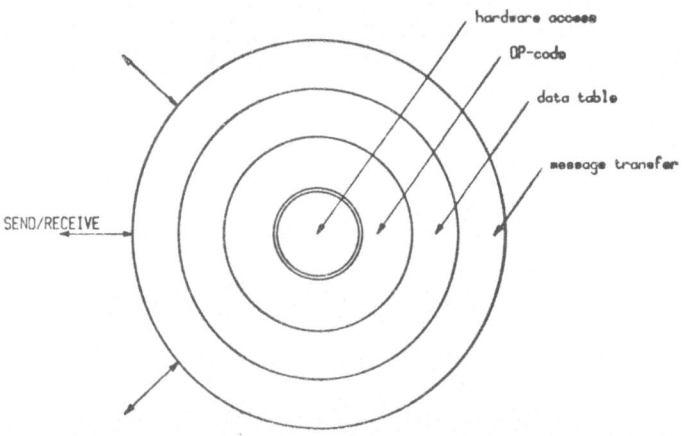

fig.7: Peripheral Interface Module (PIM)

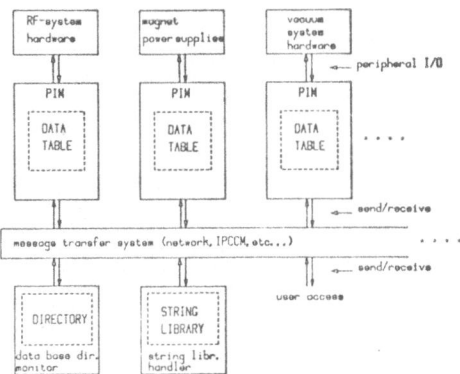

fig.8: The distributed DATA BASE

```
THE INTERFACE TO THE MESSAGE TRANSFER SYSTEM

A program, which wants to use the
message transfer system has to issue
either a SEND or a RECEIVE call.

CALL SEND (RN, SN, MSQN, PRIO, BUFR, LEN, TMO, ERR)

call parameters:

        RN   : receiver ID: = <identifier, node>
        SN   : sender ID: = <identifier, node>
        MSQN : message sequence number
        PRIO : message priority
        BUFR : message buffer
        LEN  : message length in words
        TMO  : time out parameter

return parameters:

        ERR  : error code

CALL RECEV (RN, SN, MSQN, BUFR, BUFL, LEN, TMO, ERR)

call parameters:

        RN   : receiver ID: = <identifier, node>
        SN   : sender ID: = <identifier, node>
               (zero if message from "any" sender expected)
        MSQN : message sequence number
        BUFL : receiver data buffer length
        TMO  : time out parameter
               -1  : no wait
                0  : eternal wait
               N>0: wait N units of time

return parameters:

        SN   : actual sender ID
        MSQN : actual message sequence number
        BUFR : message data buffer
        LEN  : actual message length in words
        ERR  : error code
```

fig.9: Send/Receive calls to the message transfer system

THE ELECTRONIC INTERFACE FOR CONTROL OF THE DEDICATED SYNCHROTRON
RADIATION STORAGE RING BESSY

G.v. Egan-Krieger, W.-D. Klotz and R. Maier

Berliner Elektronenspeicherring-Gesellschaft für
Synchrotronstrahlung mbH (BESSY)
1000 Berlin (West) 33, Lentzeallee 100

Summary

The commercially available electronic interface system, called DEV 12 and the developed BESSY-standard I/O-cards and hardware tools are presented. The DEV 12 system is a simple bit parallel byte-serial data multiplexing system, connectable in a tree or chain structure. Six different types of I/O-cards were developed. 3/4 of all equipments are interfaced only by a single card. The rest uses combinations of several cards. For reading back analog signals the voltage to frequency conversion is used. The computer and equipments are multiple optically insulated from each other. Computer controlled knobs were developed, becoming the most important hardware for machine adjustment. For units without digital remote control capability an interface module called 'General Purpose Interface GPI' was built to match any equipment to the standard I/O-cards. An intelligent 'GPI' is in construction to get rid of any specialized type of equipment and part of the so far used interface system. The general design philosophy, the hardware structure, and the features of the control system hardware is given.

I. Introduction

Equipments associated with the 800 MeV electron storage ring and the injector of BESSY [1)2)] are controlled by a local area network of four minicomputers[3)]. The hardware (magnet power supplies, vacuum, RF-system etc.) is connected to this network by a commercially available electronic interface system[4)], new developed I/O-cards, 'GPI' units, and computer controlled knobs. The total amount of the I/O channels controlled by the system is given in fig. 1.

II. General design features

The general philosophy for the interface system design was determined by boundary conditions like:
1) electrical insulation by opto couplers
2) current driven transmission lines up to 100 meters
3) symmetrical driven twisted pair interconnection
4) IEEE-bus facility which overcomes the range limitations imposed by the cabling rules of the standard IEEE-bus
5) interrupt capability
6) if possible, use only five different cards
7) locally present service work shop

III. The commercially available interface system

A 12 bit parallel data multiplexing system, called DEV 12, is used. The system is connected via full duplex interfaces to the computers. It uses 1 byte for data and addresses, 4 bits for control (fig. 2). Each I/O-card is addressed by a slot (0-15) and a crate (1-15) number. Up to 8 crates (DEV 12.1) can be connected to one distributor (DEV 12.4) The diagrams of the crate and the distributor are shown in fig. 3

and 4. The physical slot positions define the interrupt piority. All crates and distributors are optically isolated from each other. Cables of 25 symmetrically driven twisted pairs are used as interconnections. The system is completely program controlled by the computers.

IV. Standard I/O-cards

For data output registers differential line drivers are used. The input registers are optically coupled insulators with a reserve current photodiode to get a symmetrical circuit. That means, this circuit works as a differential receiver. All settings and status readings are done with this send/receive pairs.

The commands like ON or OFF switching are performed in pulse logic. The pulse outputs are driven by optically insulated darlington transistors with a maximum steady drive current of 0.5 A and a breakdown voltage of 70 V. That allows also direct coupling of relays.

The readout of analog signals was done in two different ways. Voltage to frequency conversion directly from the equipment is used up to an accuracy of 12 bits, with the advantage of one twisted pair line for every signal and galvanic decoupling. For signals with higher accuracy up to 16 bits, a digital voltmeter with analog scanner (IEEE-bus controlled) is in operation.

All peripherals are grouped into classes of equal or similar equipments. Five different cards were developed. The physical dimensions are approximately of eurocard format (100 x 225 mm).

1) The power supply controller 'PSC' (fig.5)
   controls about 85% of all power supplies. This card consists of
   12 bit data output register
   8 bit status input register
   5 pulse outputs for commands.
2) The status input and pulse output modul 'SPO' (fig.6)
   is the typical card for equipment without data input (like valves, pumps, etc.) it contains a
   16 bit status input register and
   8 bit pulse outputs for commands.
3) The digital output register 'DOR' (fig.7)
   handles 16 bit differential outputs with handshake or interrupt facility
4) The insulated input register 'IIR' (fig.8)
   is a board with 16 bit optocoupler inputs with handshake or interrupt
5) The multichannel counter card 'MCC' (fig. 9 and 10)
   contains 17 x 16 bit counters with FIFO buffer controlled by a RAM sequencer on an other board. The advantage of this modul is the parallel counting and the asynchroneous mode from the process computer

3/4 of all equipments are interfaced by a single card. The rest uses combinations of several cards. For the IEEE-bus a 16 bit open collector input/output card is in operation.

V. General purpose interface

To manage the interfacing of equipments without or with other electrical connections for remote control an adapter modul, called 'GPI', is used. This module consists of a 4/12 AEC NIM cassette and up to four eurocards, with digital-signal-conditioning, DAC and VFC functions (fig. 11).

VI. The operator console devices

Alphanumeric terminals

To enter commands interactively the operator simply has to push programmable keys in the keyboard to step down a tree-structured menu until he reaches a task like setting a power supply's current or reading an equipment status .

Colored TV-raster scan monitors interactive cursors

Independently from the terminals the operator has the possibility to get status information and to control the machines by stepping through a menu by means of interactive tracker-ball units connected to two colored raster scan monitors. The refresh memory of the raster scan monitor has a resolution of 256 x 512 pixels with 8 bit depth per pixel.

Computer controlled knobs

The most important devices for adjustment of machine parameters are six computer controlled knobs. The knobs are assignable to any controllable variable of the machines, so that the operator can for instance change power supply's current in a quasi analogeous manner. The actual and the demanded values of the controlled variable are displayed on-line on a small TV-monitor. The knob consists of an incremental angle encoder with 500 counts per revolution. The TV-monitor is connected to a character generator which produces an alphanumeric display of 8 lines with 16 characters per line. When the incremental encoder is turned, the first pulse of the pulse train triggers a gate circuit which opens an up-down counter input for 100 ms. When the gate circuit closes again, an interrupt to the computer is generated. After reading out the counter the computer enables the gating circuit again. With one knob in action a total turn around time, starting from the interrupt, sending a new value to the equipment through the message transfer system and receiving the actual value on the display, of about 100 ms is achieved. Additionally the device contains four programmable interrupt keys. The knobs have already successfully been used for multiparameter control tasks, like producing bumps on the closed orbit, where at least three power supplies have to be varied at the same time with definite proportions.

VII. Future hardware updates

At the time of construction no adequate ADCs for 16 bit were available. The market now offers 16 bit hybrid ADCs at reasonable prices. Therefore, now a solution of installing the high accuracy conversion inside of the equipments is preferred.

Under construction is an intelligent 'GPI' with the possibility of interfacing equipment without custom designed remote control facility. This 'IGPI's' may be connected directly to the process computers or to distributors. With the internal Z80 processor many tasks like magnet ramping, scaling, drift correction etc. can be done front end.

References

1) G.v. Egan-Krieger, D. Einfeld, H.-G. Hoberg, W.-D. Klotz, H. Lehr, R. Maier, M. Martin, G. Mülhaupt, R. Richter, L. Schulz and E. Weihreter; IEEE NS-30 (1983) 3094
2) G.v. Egan-Krieger, D. Einfeld, W.-D. Klotz, H. Lehr, R. Maier, G. Mülhaupt, R. Richter and E. Weihreter; IEEE NS-30 (1983) 3103
3) G.v. Egan-Krieger, W.-D. Klotz and R. Maier; this conference
4) FREY Analogtechnik, Marathonallee 33, D-1000 Berlin 19, FRG

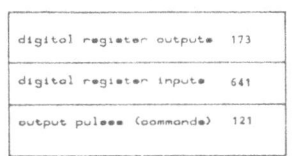

fig.1: Total number of I/O channels

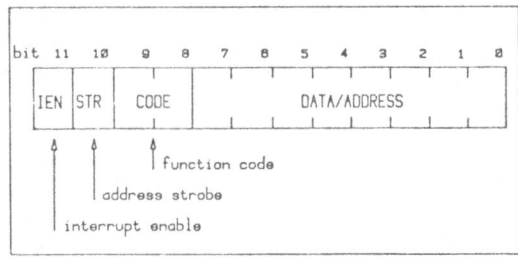

fig.2: DEV 12 transmission format

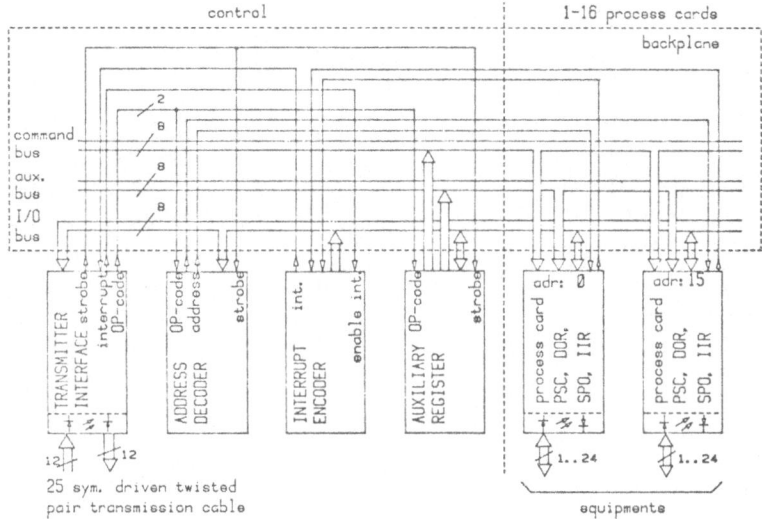

fig.3: Process crate DEV 12.1

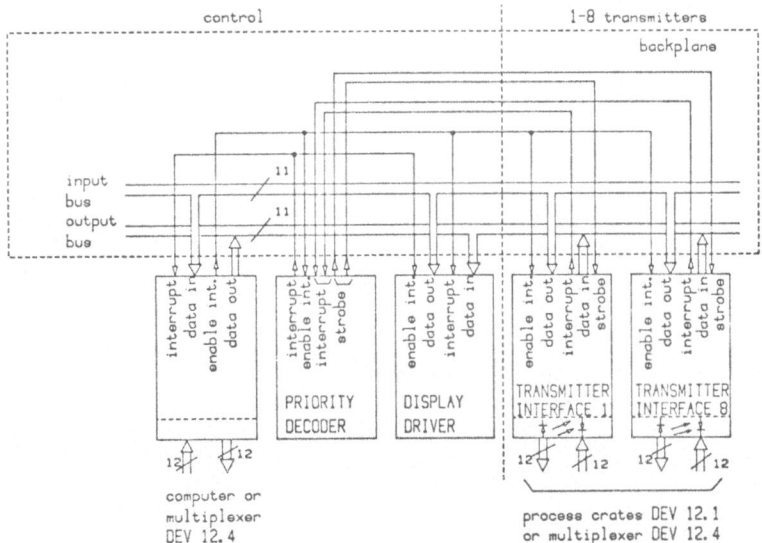

fig.4: Distributor DEV 12.4

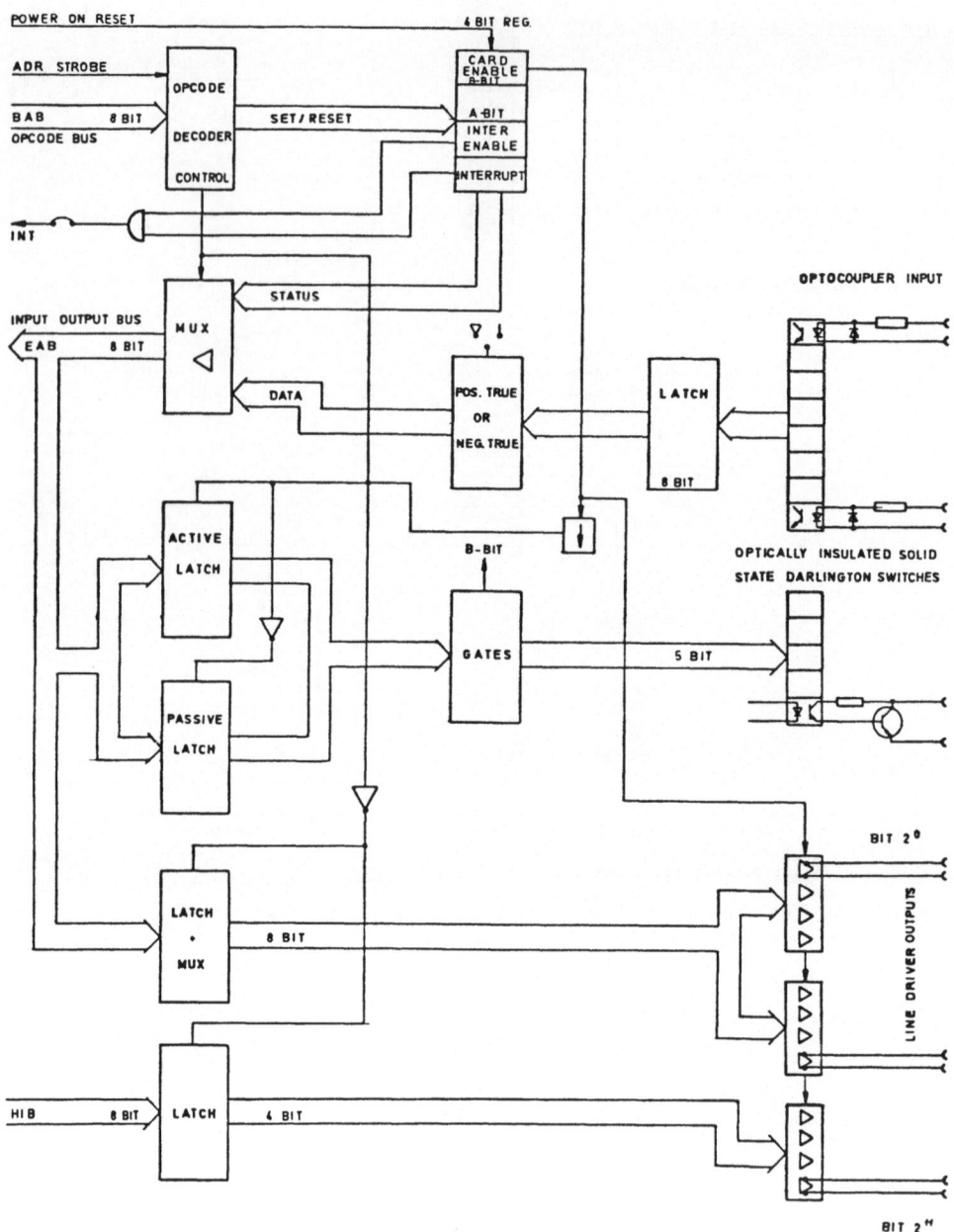

fig.5: Power Supply Controller PSC

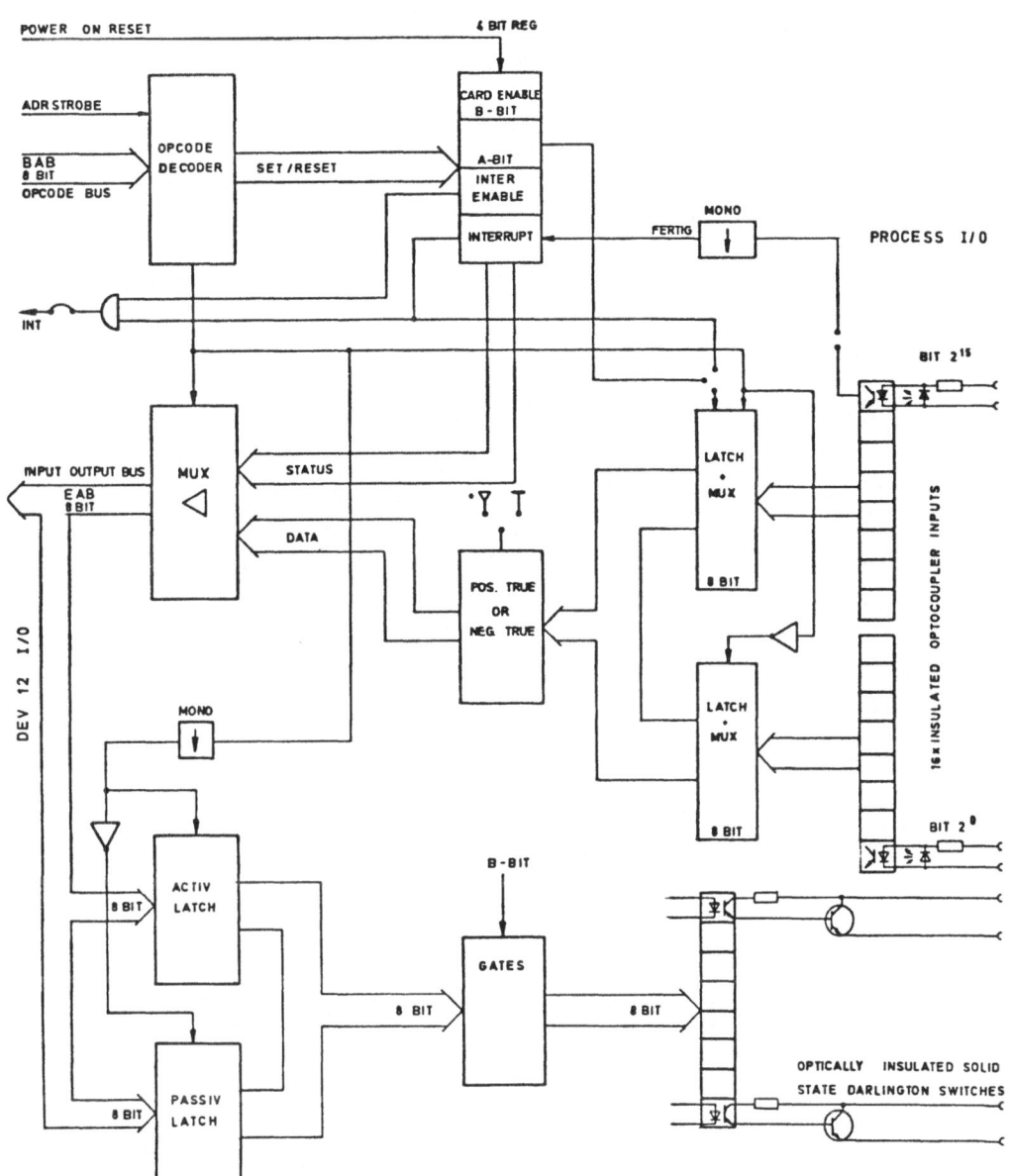

fig.6: Status input & Pulse Output SPO

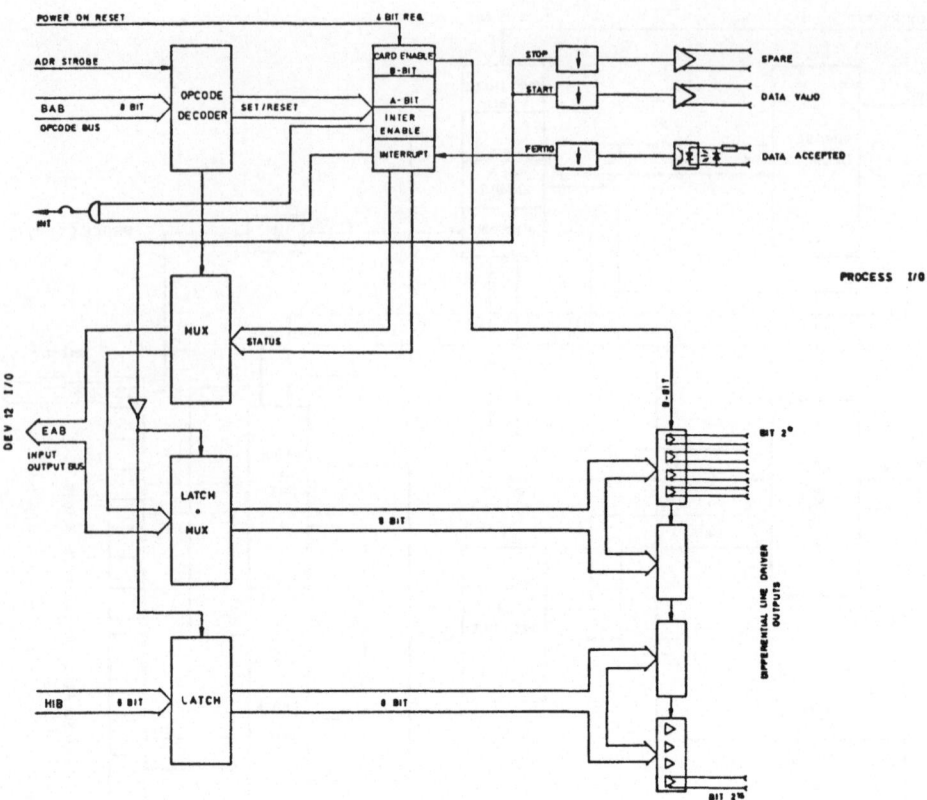
fig.7: Digital Output Register DOR

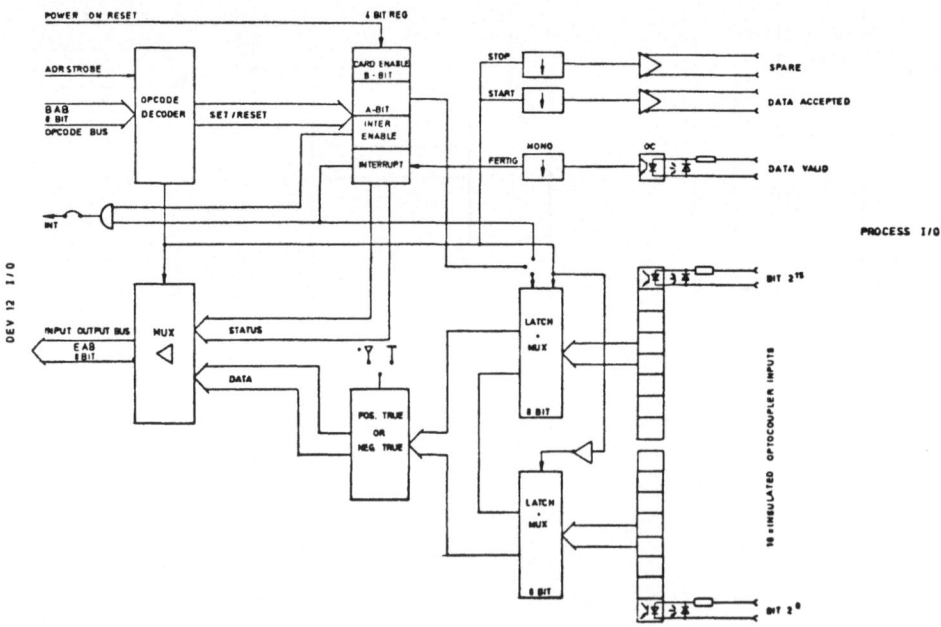
fig.8: Insulated Input Register I^{2}R

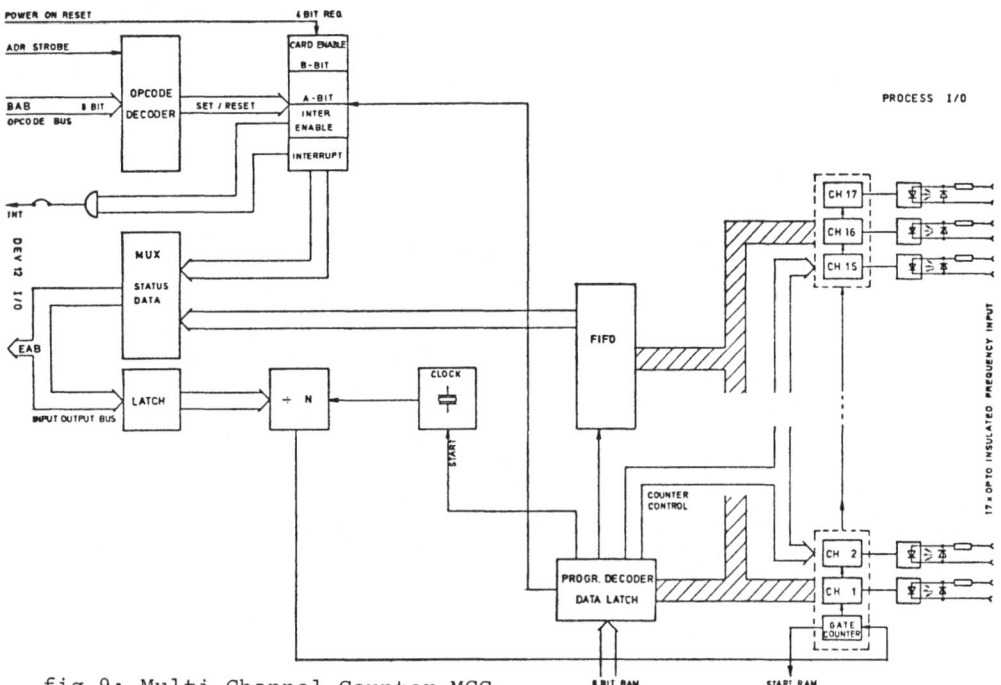

fig.9: Multi Channel Counter MCC

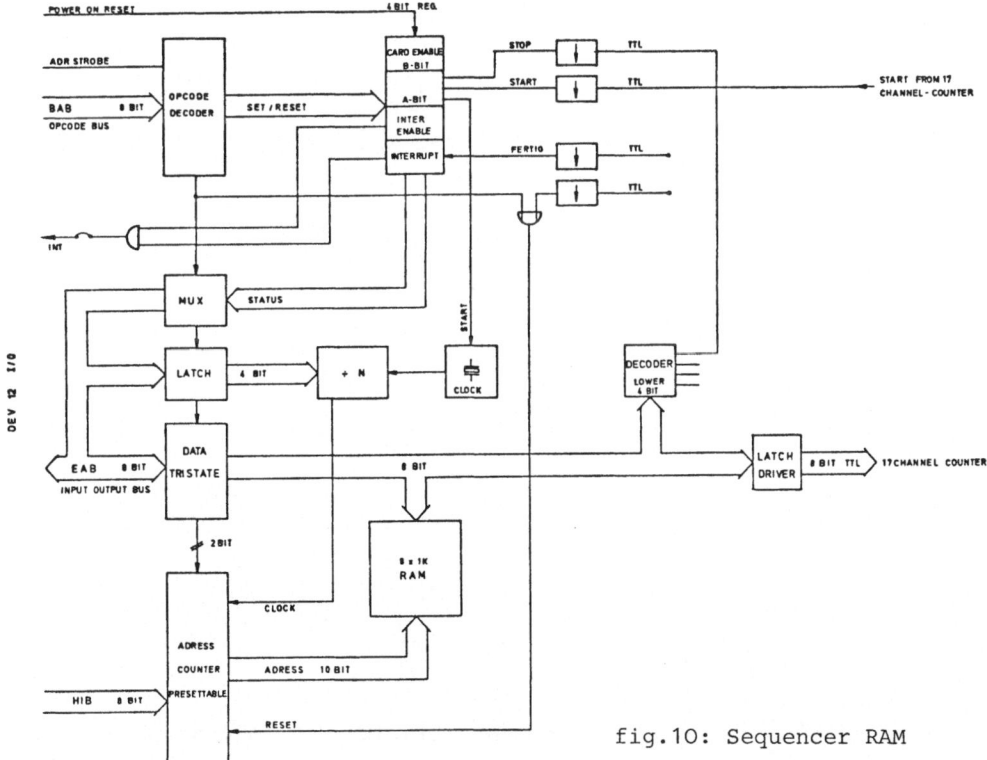

fig.10: Sequencer RAM

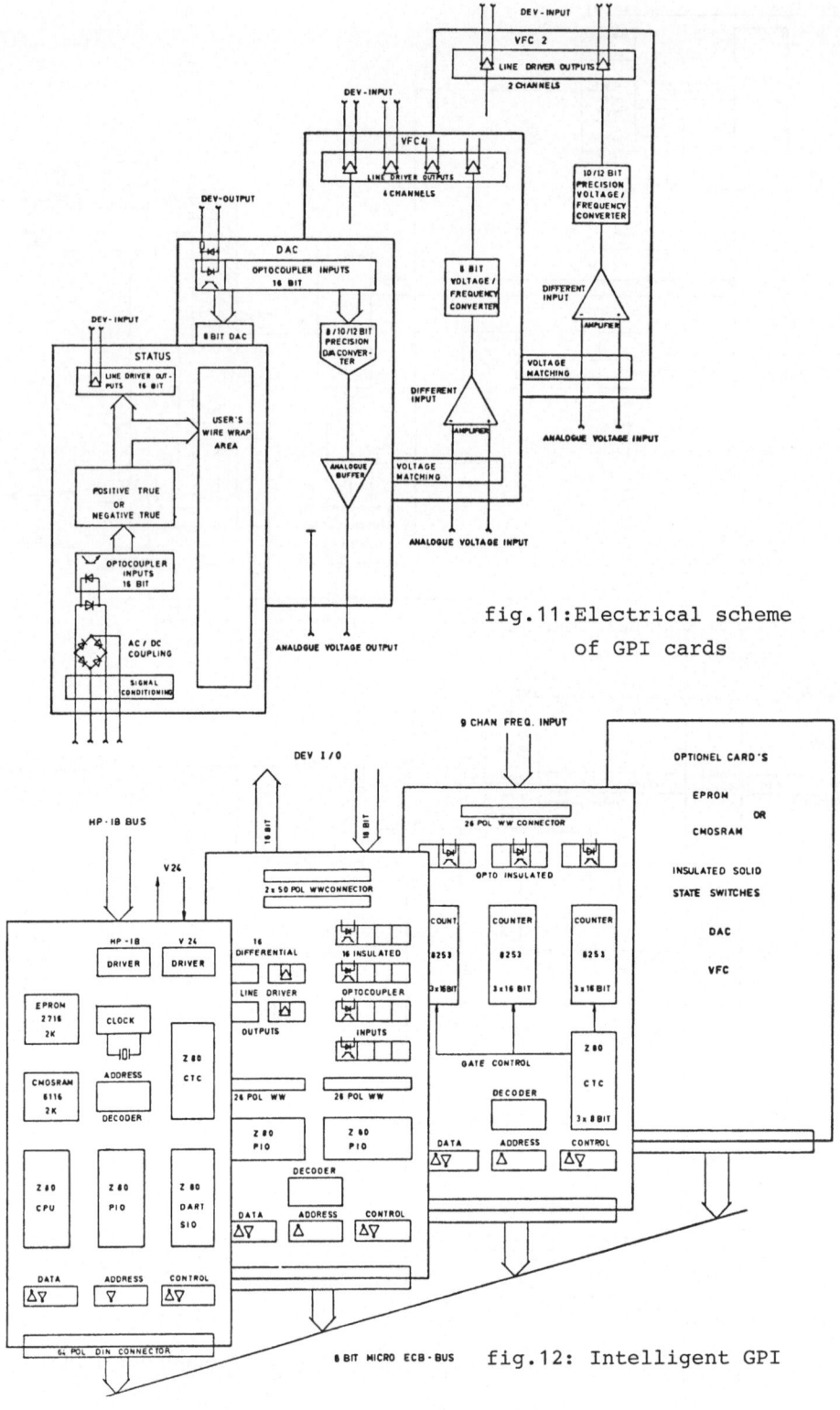

fig.11: Electrical scheme of GPI cards

fig.12: Intelligent GPI

WORKSHOP No.2: WHICH LAN TO USE FOR ACCELERATOR CONTROL

Governer: M, Crowley-Milling, CERN

The aim of the workshop was to find out whether accelerator control system designers could standardise on one type of Local Area Network (LAN) in future applications.

Unfortunately the discussion did not result in any agreement on this subject, mainly because of the diverging or missing experience in this field.

Although this result is not very encouraging it may still be worth while to mention that it took place.

W.Busse

INTRODUCTION TO COMPUTING FOR ACCELERATOR OPERATION

W. JOHO

SIN

CH 5234 VILLIGEN, SWITZERLAND

Today the key role of computers for the operation of an accelerator is fully recognized. The good old days of manual control are clearly over (compare fig.1 and 2) and the statement "no computer, no beam" (or at least "no good beam") has been finally accepted like similar statements as "no RF, no beam". As a beam physicist working on a cyclotron I am only a user of a control system, and therefore not qualified to speak about topics like distributed intelligence, file handlers or semiduplex modem couplers etc. The control of an accelerator is too important to be left to computer specialists only, hence I take the liberty in this introduction to look at the problem from my point of view as a machine physicist.

Dealing with the computer control of an accelerator we have to consider a wide spectrum of operating modes, users and computer tasks as shown below.

Operating modes

- setup of accelerator
- routine operation
- beam development
- trouble shooting
- maintenance

Users

- operators
- machine physicists
- hardware people
  (control+devices)
- software people
  (system+applications)

Computer control tasks

- device setting
- cook book procedures for setup
- logging
- status display
- alarm handling
- closed loop beam monitoring like:
  beam centering,
  correcting magnet drifts,
  ion source current stabilisation.
- beam development programs
  with measurements of:
  emittance, tune, chromaticity,
  acceptance diagrams,
  influence matrices etc.

With such a wide variety, it is clear, that the different users of a control system should enter the picture very early in the game. One can take as an example the organisation of the data base; this is at the heart of any control system [1] and requires careful planning and early exposure to all prospective users.

Figure 1: Control room for calutrons at the uranium isotope separation plant in Oak Ridge, 1944, showing an early and pretty example of distributed intelligence (courtesy U.S. Department of energy, ORO).

Figure 2: Control room for the SPS accelerator complex at CERN with five independent consoles (courtesy V.Hatton).

Optimizing the performance of an accelerator

The steps to bring an accelerator from the design stage to routine operation are shown in fig.3. For a discussion of computer models for the design of accelerators I refer the reader to the corresponding papers of this conference.

Due to inaccuracies in the accelerator models, in the calibration curves of components, in construction and alignment, there will be discrepancies between the ideal and actual beam parameters. To reach the ideal parameters one can use two main strategies which, in my opinion, complement each other.

The first strategy is to use an iterative cycle of beam development, fitting of beam measurements, modifications to models or calibrations, till agreement between theory and experiment is reached (see e.g. [2]). The most likely sources of discrepancy are the magnets (errors in effective length due to saturation or end effects, inaccuracies in gradient etc.)

The second approach is online optimisation, sometimes called "knob twiddling", and often disregarded by "purists". For this approach the operating crew should be provided with some means of evaluating quickly the status of their machine. It is thus essential to display continuously parameters like beam current, luminosity, beam losses, etc. At the SIN cyclotron, where we have a continuous beam, the introduction of a so called BONUS value (see figure 4) has been quite successful in stimulating the operators to minimize the activation induced by beam spill. For pulsed accelerators like synchrotrons the application of this bonus concept might be a bit trickier.

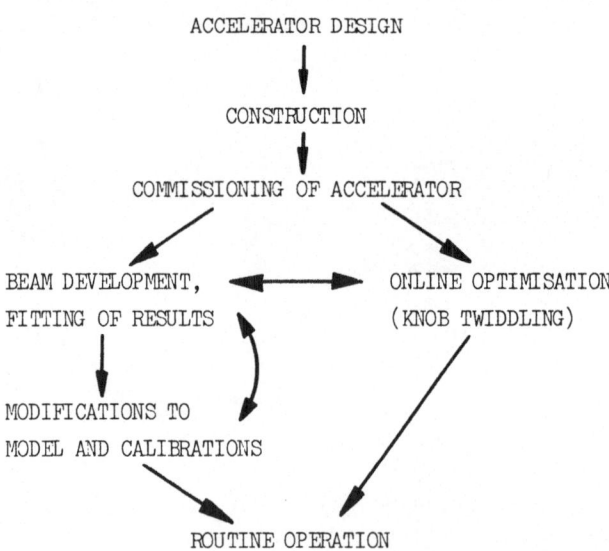

Figure 3: Flow chart for the steps to bring an accelerator from the design stage to routine operation.

Optimizing an accelerator is often done by using one knob at the time in random sequences. But close to an optimum the target functions are quadratic leading to elliptic contour lines for the case of two variables; using the ellipse axes as prefered directions for a two dimensional search, leads to a fast convergence towards the optimum (see fig.5). The correlations between parameters can be either calculated from models or measured directly. The desired parameter-couplings can be obtained by introducing pseudo devices which are then controlled by so called "superknobs". I could imagine that using the "mouse"-device, made popular by the personal computer APPLE LISA, would be quite useful for a two dimensional search. Pushing this idea of online optimisation even further I dream about a six dimensional search, where an operator controls three coordinates with each hand similar to a manipulator of a hot cell ! The danger is, of course, that too much "fiddling" may occur, which is unpopular with experimenters wanting stable beam conditions. Furthermore there is always a possibility that one gets trapped in a local optimum (mountain lake phenomena).

Computing for beam development

A generous allocation of beam time for accelerator development is generally a good investment. Progress in understanding an accelerator better usually pays off with faster set up times (time is money !) and higher intensities for production runs. A good example for this policy is the CERN antiproton accumulator (AA-ring), where only by careful tuning were the luminosities achieved that made the spectacular discoveries of the W and Z bosons possible! In order to fully benefit from beam development periods, one should have on the hardware side extensive beam diagnostic equipment, and on the software side online access to a large computer with the associated library of beam optics and accelerator codes like TRANSPORT [3], AGS [4] etc. As an example fig.6 shows the processing (in a matter of seconds) of a distorted beam profile at SIN with a fast fourier transform (FFT) [5]. The clean profiles can then be used immediately as input for beam fitting with the TRANSPORT code or for beam tomography as in fig.7 [6,7].
Let me make a side remark about fitting of beam results: We have to distinguish between calibration of magnets, which is hopefully done once and for all, and fitting of beam emmittances, which can vary from day to day. Calibration of magnets is best done by purposely displacing the beam, followed by fitting the measured displacements with a program like TRANSPORT. This procedure is quite accurate, since there is a linear relationship between displacements along the beam trajectory. Beam emittances on the other hand are obtained by fitting beam envelopes, which depend nonlinearly on other parameters. A relatively large number of measurements, especially in the neighbourhood of beam waists, are thus required to give meaningful results.
A special problem is encountered in low energy, high intensity beams, where spacecharge plays an important role. Neutralization by the residual gas is then an additional parameter, which has to be included in the fitting process.

Bonus as a function of beam loss L

1. encouragement part

   $B_1 = \frac{a}{L+c}$

2. penalty part:

   $B_2 = -d\, L^2$

   total Bonus: $B = B_1 + B_2$

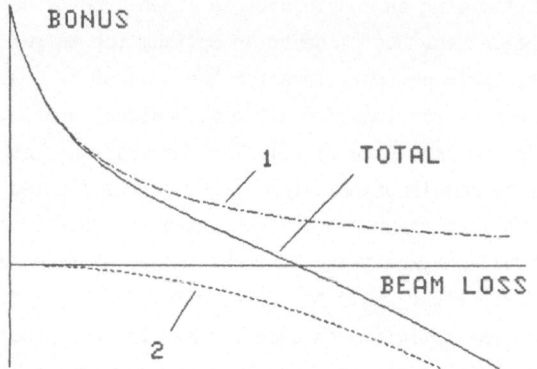

Figure 4: Bonus function as a measure of accelerator performance. It is desirable to give the operation crew a feeling for the overall performance of an accelerator complex. One possibility is to convert beam losses at crucial points, e.g. injection, extraction and beam transfer, into so called bonus values, which can be summed up to a single number. Proper Choice of the weights a, c and d of these individual bonus functions allows tuning of the accelerator towards an overall minimum of the induced activation level.

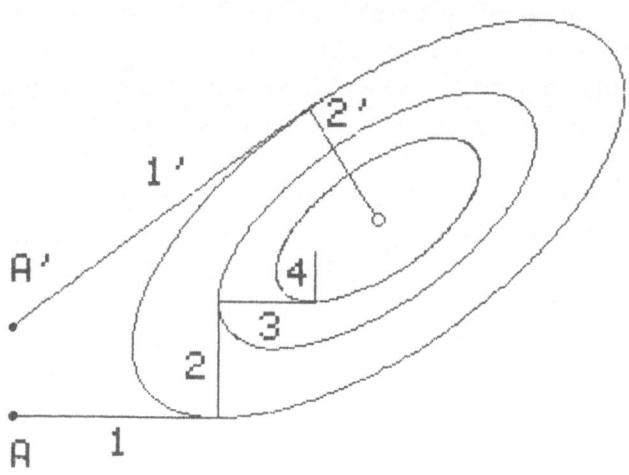

Figure 5:   Minimizing beam loss with two parameters x and y. Close to a minimum the beam loss is a quadratic function of x and y and contour lines of equal loss are given by ellipses. Searching a minimum alternatively in x and y from a starting point A   leads to an infinite path 1-2-3-4..towards the minimum. However if the search is conducted along the main axis of the ellipses , the minimum is reached in two steps from any arbitrary point   A'   over the path   1'-2' . The feature of a so called "superknob" can couple the parameters x and y in the desired directions of the main axis, enabling this fast search for a minimum.

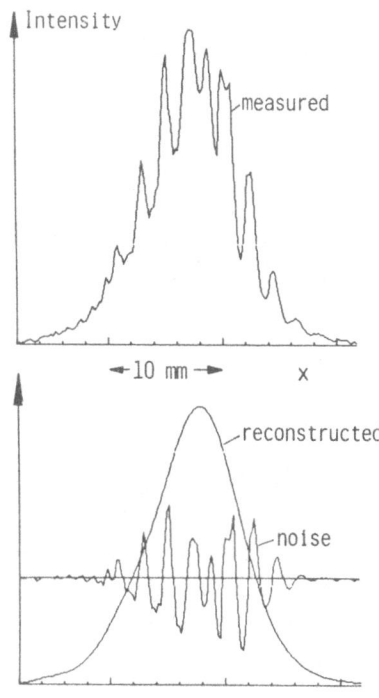
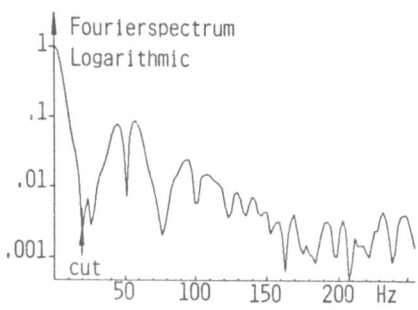

Figure 6: Online processing of beam profile. Top left shows measured profile of a 600 MeV proton beam with unusually bad 50 Hz ripple of .8 mm on horizontal beam position. Top right shows the corresponding Fourier spectrum and cut-off at 19 Hz. The picture on the left shows the decomposition into a reconstructed smooth profile and a noise contribution.

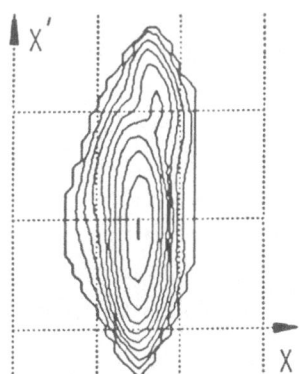
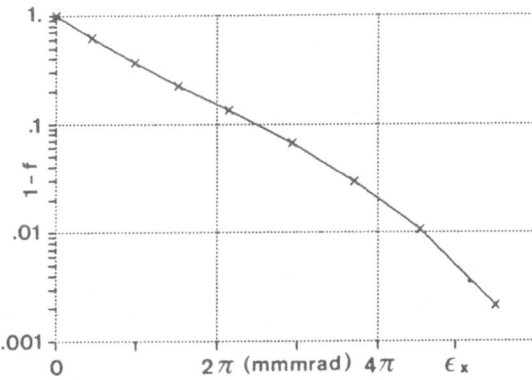

Figure 7: Emittance plots for a 600 MeV proton beam at SIN. The left part shows an example of a contour diagram of equal intensity in the horizontal phase space (x,x'). This plot was constructed from beam profile measurements using a beam tomography program developed at Los Alamos. The right part of the figure shows a plot of log(1-f) versus emittance, where f is the fraction of beam inside a given emittance. This representation has the advantage that Gaussian beam profiles lead to a straight line. Deviations from a Gaussian are thus easily spotted.

Some additional remarks on beam development:
- To use his allocated beam time efficiently, the beam developer should be able to assume that everything is working properly (including e.g. the hardcopy unit!), and he should not have to worry about data transfer, device access, interfaces etc.[8].
- The control system should provide working files besides reference and active files, and should have the flexibility to introduce pseudo devices quickly.
- Beam development time should not be misused for maintenance on the computer system; one should rather use the maintenance periods to test or upgrade programs.

Computing for routine operation

Computer codes for setup and routine operation should be very reliable and provide frequent feedback with intermediate results for the operating crew. Data should be presented in a simple form (as shown e.g. in figure 7 or 8) using calibrated values and engineering units, if possible. The programs should be selfprotected against "fatal operator errors" or at least terminate decently. Take as an example again the case of the AA ring at CERN : During the early stage of operation one could kill a priceless stack of antiprotons through the "itchy fingers" of an operator, but this deficiency has been corrected since.

The use of default values in regularly used set-up programs is both convenient for the operator (not so much typing) and reduces the possibility of wrong parameters being used. The growing confidence in online computing has resulted in the increased use of closed loop control functions, like keeping the beam centered in the accelerator or in the beam lines. A constant check on the quality of high intensity beams is so important, that a single computer may be dedicated for online beam tomography using the light from residual gas [9].

A controversial subject is the responsibility and standard for the different computer programs. One common problem is, that machine physicists tend to write programs for their personal use only and in an ad hoc (and hence a bit sloppy) manner. Very often these codes turn out later on to be useful for routine operations, but the original author is then quite reluctant to rewrite his piece of software with the standard necessary for regular use by the operating crew.

Interaction between a user and the control system occurs via input and output devices. Popular input devices are the touchscreen, knobs and alphanumeric keyboards. On the output side we have TV raster displays, high resolution graphic screens, oscilloscopes for analog signals etc. These gadgets are to be discussed in a special paper of this conference. Do we see soon some voice input/output in our control rooms?

Communication between the accelerator crew and the experimentalists is bidirectional: The experimentalist is interested in beam parameters like energy, intensity, cycle structure etc. The control room, on the other hand, can get valuable information from the users like interaction rates, beam jitter in time and position.

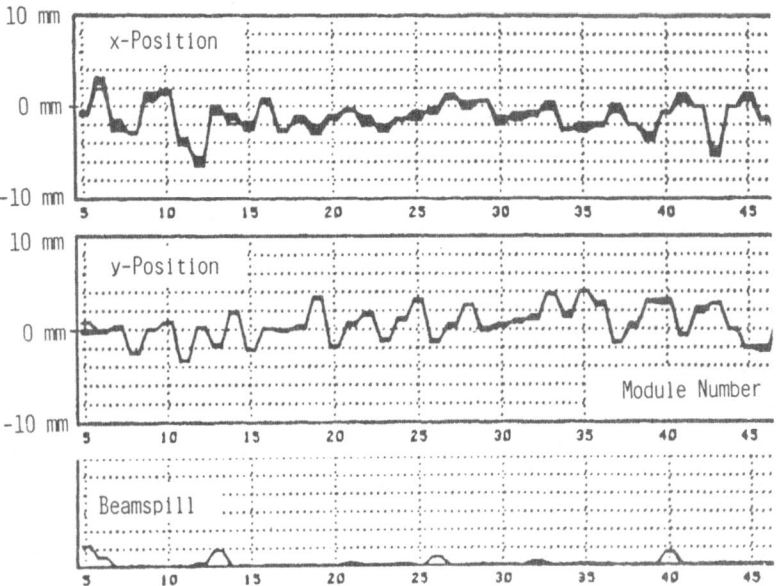

Figure 8: Example of a display of beam position and beam spill for the 800 MeV proton linac at Los Alamos (courtesy J.Bergstein). The center of the beam is measured with wire scanners after each of the 45 RF modules along the linac. The beam current is 1.2 mA. At this intensity it is essential to keep the beam well centered in order to reduce the beam spill to less than .01 %.

Miscellaneous

Here follows a somewhat arbitrary list of personal thoughts and remarks picked up elsewhere:

-Progress on the hardware side is so fast, that many programmers work on projects with planned obsolescence.
-Since software is expensive (manpower!), it should have priority over hardware. Shop for software on the market, then buy hardware that goes with it.
-Application programs should be as hardware independent as possible (e.g. no hardware adresses in programs) [10].
-It is easier to make a flexible system speedy, than to make a speedy system flexible [10].
-It is impossible to foresee everything, so build into the control system the potential to do new, unforeseen things [8].
-Switching from an existing control system to a new one is painful (parallel development necessary, hardly any time for online tests of new system)

-Do not try to invent the wheel again.

—Last but not least: BREAK THE JARGON BARRIER !
If you need help from software people they always work on things like :
remote debugging of relocatable data manager,
pipelined virtual memory task builder,
bootstrapping of partitioned foreground/background blocktransfer processor,
whereas the hardware people are fixing things like:
channel adapter for the bit sliced auxiliary crate controller,
pulse code multiplexed CMOS-logic circuit,
asynchronous handshaking with memory protected interrupt requests.
Beam dynamics people are no better of course, but here I can offer some help.
If next time somebody wants to impress you with his new computer code which
"corrects the chromaticity in the linear collider with interleaved mini beta
insertions" the proper answer is: "Did you include the beam beam tune suppressor
in the LANDAU-damped momentum compaction matching section ?"

References

[1]   M.C.Crowley Milling,   IEEE Trans. on Nuc. Sci. NS-30, 2142 (1983)
[2]   J.C.Sheppard et al. ,   ibid. p.2320
[3]   K.L.Brown et al. , CERN 80-04 (1980)
[4]   E.Keil et al. , CERN 75-13 (1975)
[5]   U.Rohrer, SIN , private communication
[6]   U.Rohrer,W.Joho, SIN Annual report 1982
[7]   O.R.Sander et al. , Proc. 1979 Lin. Acc. Conf. BNL-51134, 314
[8]   V.Hatton, CERN, private communication
[9]   D.D.Chamberlin et al. , loc. cit. ref.1   p.3247
[10] S.C.Schaller, P.A.Rose, loc. cit. ref.1   p.2308

EUROPEAN ORGANIZATION FOR NUCLEAR RESEARCH

CERN - SPS DIVISION

SPS/AOP/Note/83-9

MAN-MACHINE INTERFACE VERSUS FULL AUTOMATION

V. Hatton

ABSTRACT

As accelerators grow in size and complexity of operation there is an increasing economical as well as an operational incentive for the controls and operations teams to use computers to help the man-machine interface. At first the computer network replaced the traditional controls racks filled with knobs, buttons and digital displays of voltages and potentiometer readings. The computer system provided the operator with the extension of his hands and eyes.

It was quickly found that much more could be achieved. Where previously it was necessary for the human operator to decide the order of the actions to be executed by the computer as a result of a visual indication of malfunctioning of the accelerator, now the operation is becoming more and more under the direct control of the computer system. Expert knowledge is programmed into the system to help the non-specialist make decision and to safeguard the equipment.

Machine physics concepts have been incorporated and critical machine parameters can be optimised easily by the physicists or operators without any detailed knowledge of the intervening medium or of the equipment being controlled. As confidence grows and reliability improves, more and more automation can be added.

How far can this process of automation replace the skilled operator?. Can the accelerators of tomorrow be run like the ever increasing robotic assembly plants of today?. How is the role of the operator changing in this new environment?.

Paper presented at the
Computing in Accelerator Design and Operation Europhysics Conference
Berlin, September 20-23, 1983

1. INTRODUCTION

The operation of present day accelerators relies heavily on the use of computers. Their impact on accelerator control can be appreciated by noting how few operators are required on shift in spite of the considerable increase in amount and complexity of equipment to be controlled. The operators interface with the machine has been eased and made more sophisticated and some processes have been handed over to the computer for automatic control.

Using examples from the operation of the SPS this paper will show how the man-machine interface has been improved in all operations tasks, the directions in which automation have been made possible and those areas of operation for which there is still a long way to go.

2. WHAT IS ACCELERATOR OPERATION

The Cyclotron was one of the first accelerators and was manually operated before the introduction of computers. The philosophy of operation and the controls were much the same as was to be found in any production process plant in industry. The equipment was patrolled and surveilled by the operator and values recorded or adjustments made as he did his tour of the building. He decided what needed to be adjusted and interacted with the individual elements of the accelerator : a multifaceted role which required considerable and varied skills.

Today our accelerators have much more equipment than before, and is distributed over a greater area, sometimes in inaccessible and hostile places. The number of parameters in the process has increased considerably and the modes of operation and the performance requirements are invariably at the limit of what is technologically feasible. Even the cyclotron of today is heavily dependent upon the use of its computer [1] with its needs for different energy/particle modes of operation beam matching, etc.

The accelerator operation itself is multiphased, with separate yet overlapping control requirements. These phases include the initial and subsequent commissioning phases, the steady daily production and the study and improvement phase.

Used in its most general meaning, the term "Operator", means, a person interacting with a physical system. Todays accelerator operator is not only the classical machine minder, but also the equipment specialist and the machine physicist. The man-machine interface and automation must assist him in carrying out his separate operational tasks.

The question therefore for todays accelerator designers is not whether or not the computer should be used but rather in what way they should be used to control the accelerator complex. Among the many questions to be answered by the designer and user is to what extent the operator must be part of the control loop or how successfully can the computer replace him. The balance to be found is between the open and closed loop systems of control, the man-machine interaction versus full automation.

3. USE OF COMPUTERS, EXTENDED HAND AND EYE

The SPS consist of more than 10 km of tunnels on average of 40 meters below the ground containing thousands of individual elements. They are controlled by a network of computers to constrain, accelerate and deliver protons for the use of the European Particle Physics community. Protons at various energies up to 450 GeV are extracted for collision with stationary targets in external halls, caused to collide at 270 GeV with antiprotons inside the accelerator itself.

Individual computers are responsible for the control of the equipment in local zones or for particular parts of the process (acceleration, extraction, etc.) and communicate with the Main Control Room through the system of data links. The man-machine interface at the centre is a computer driven console whose programmes are in a tree structure and activated by means of a touch panel. The displays are all standard raster TV generated by microprocessor controlled CAMAC modules.

There are at present 35 computers in the network of NORD 10's and NORD 100's arranged in nodes which serve the Main Control Room of the SPS. A high level interpreted language called NODAL is used for the operations programmes which are written mostly by the equipment groups and the operations team.

How these many systems were brought together and used to control and operate the accelerator has been described in detail elsewhere.[2,3,4]

The SPS is an example of how computer control of accelerators is being approached today.

    a) The operator no longer has to patrol the building to check the state of the machine and change the equipment settings. An effective surveillance and alarm system checks the state of the equipment every few seconds and reports any variation from the desired state to the operator . The operator then calls up the control of that element on his console and makes the necessary adjustments. This has been referred to as "the extended hand and eye" function of the control system.

    b) Files of equipment settings, can be saved on the computers, loaded and compared with previous sets. Complicated equipment functions are easily generated and modified according to operational requirements.

    c) Operator training and skills are greatly improved by the use of console programmes written by system experts or by members of the operations group. Protection against faulty manipulation is assured by limitations built into the software at the computers-equipment interface.

    d) Machine physics concepts are included in the computer programmes in a natural way and the operator, is able to correct the orbit, adjust Q, chromaticity, etc without the need to know what current is being sent to which dipole, quadrupole or sextupole in the ring.

These are some of the more obvious ways in which the man machine interface has been improved by the introduction of computers and the increased complexity of operation compared to our early cyclotron model has been made possible.

4. DEVELOPMENT OF USE OF COMPUTERS TOWARDS AUTOMATION OF OPERATION

A. AUTOMATION IN THE SETTING OF THE MACHINE MODE OF OPERATION

The setting up of the accelerator from a previously optimised state by loading files of equipment settings is a familiar feature of computer control. In addition sequences of programmes can be run by the operator, or more likely automatically, to change from one machine state to another. This is used in cyclotron operation when a sequence of energy scans is needed and in synchrotrons (for example Petra) when it is necessary to move from one working point to another via intermediate points to avoid resonances.

It is clear that the accelerator must be operated as an integrated process. All equipment must function not only reliably but in the way requested by the central control room. The required mode of operation must be communicated to the equipment and the equipment must be able to report back any deviation from this state.

In the SPS the Engineer in Charge chooses the machine mode from a list of possible modes of operation based on precomputed magnetic cycle of the main power supplies. From his choice a Master File is generated which can be accessed by all the equipment systems software. The master file contains all the information concerning the machine physics parameters, times in the cycle of injection and extraction, which zones will receive beam and when, etc.[5] The applications programmes for the individual equipment systems have a button "Update to Master" which adapts all the settings for the equipment thus changing voltages, currents and timings accordingly. In addition the machine mode requirements are distributed to all the local computers so that the required states of the elements are defined, surveilled and communicated back to the control centre in case of any malfunction.

The changeover from one state to another within a given overall mode of operation was a problem highlighted when the SPS runs in collider mode. Up to 24 hours were needed to produce enough antiprotons in the Antiproton Accumulator (AA) for a reasonable fill of the SPS for Physics.

The processes for ensuring their reliable and efficient transfer had to be well defined and implemented with the minimum of equipment malfunction and human error.

Antiproton transfer to and acceleration in the SPS is the most critical period and so a series of preparatory states of operation before the final transfer were identified. The first was to extract protons from the SPS and transfer them to the CPS up the line that the antiprotons would use later. The second was to transfer pilot pulses of antiprotons and accelerate them. Finally the day's production of antiprotons would be transferred in 3 bunches at 26 GeV, accelerated and made to collide with 3 dense bunches of protons at 270 GeV within a time interval of about 25 seconds.

The change from one state to the next was carried out by a sequence of programmed tasks and checks in each local computer which was initiated from the centre. As operating experience grew so did the list of checks to be carried out and with it the efficiency and reliability of the transfer operation increased. A synchronous sequence of checks took care of the synchronisation of the 3 machines, AA, CPS and SPS.

This automatisation of the transfer process was very successful: there were no losses of antiproton bunches due to the malfunction of equipment covered by this procedure. More details of the "Sequencer" are reported elsewhere in this conference by C. Saltmarsh.[6]

However we are still a long way from the operations managers dream of a single button "SWITCH ON" which eliminates all the problems encountered after a long shutdown.

B. AUTOMATION OF THE STEADY STATE OPERATION

After a setting up or commisioning period the accelerator normally enters a phase of steady running. In this production phase the operation looks most like other industrial production processes.

Minor optimisations of the particle production occur during this time, the status of equipment and values are recorded, usually automatically by the computer system, and the alarm/surveillance system informs the control centre of any abnormalities.

The steady state is a desired state of operation but often it is not achieved due to equipment failure or operational faults.

Frequently the time taken to diagnose a fault is far greater than the time required to correct it. The fault diagnosis and remedy for the cyclotron operator was far simpler than in our present day complex accelerator. However today with a good, well maintained alarm and surveillance system the operator is kept well informed of the state of the equipment and from his experience - aided by the control system text facilities for example - he is able to decide on the correct cause of action to be taken. Most alarm systems have the facility to eliminate consequential alarms to some degree hence highlighting the root cause of the malfunction. Other more advanced systems based on previous known fault conditions offer the operator a recommended course of action; not many systems correct the fault themselves fully automatically. There are particularly interesting developments in the nuclear power industry, post Three Mile Island, on fault diagnosis and correction. In the field of the automation of fault diagnosis the computers can contribute in the future to improve efficiency of operation. With accelerator downtime at between 10 and 20% there is a strong incentive for further development of this application.

Production performance, equipment parameters recording, automatic log book entries are all aspects of the steady state operation that can and are being helped by the computers but are not yet fully automated. Control by voice commands seems a long way in the future. Perhaps the accelerator control would be a suitable test bed for the Fifth Generation Computers.

Some of the standard daily optimisation processes are being nearly fully automated and as an example of this evolution consider the steering of a beam down a beam line. In the SPS, this programme displays the position of the beam in mm in the horizontal and vertical planes each time it passes the monitors in the two independent planes. The operator chooses one position and corrects the displacement by changing the current through the corresponding steering magnet upstream of the point selected by the computer. The process of operation has been well modelled to make it simple for the operator to understand what is required from him.

However there are certain limitations in the model:
- a) there is no guide to the operator as to how much current should be put into the magnet corresponding to a given mm displacement at one point.
- b) his action at one place can cause displacements to be worse at others.
- c) throughout the time that he is steering with this programme the quality of the beam is reduced because of some blowing up of the beam by the monitors.

The skill and experience of the operator is relied upon to carry out the task quickly and efficiently. The task in fact has not been sufficiently well modelled and for this we have to return to the design of the beam line and include in the model more of the underlying machine physics.

An automatic programme has now been implemented which takes the measurement of the beam properties, emittance etc., and the displacement of the beam at all positions down the line. Knowing the lattice transfer matrix and its inverse, the required currents to be modified are computed. The whole line is then resteered. The model is sufficiently good that the steering process can be completed with two iterations taking only a few pulses of the machine.

At the SPS the operator is still offered the final decision to accept the new currents although with added confidence in the programme and equipment reliability this man machine interaction can be eliminated. The next step would be to let the control system fully optimise the line whenever it sees the need. The necessary algorithms for this step have not yet been worked out.

C. AUTOMATION, MACHINE PHYSICS AND MODELLING

There are immediate reactions from machine physicists and others when one talks about the automation of studies on the accelerator. After all, studies by definition, imply an uncertainty in direction and results. There is a logical sequence of studies first then automation. Nevertheless, machine physics investigations on the accelerator require man-machine interactions and with the power of the computers available, the potential for easing the burden on the machine physicists is as great as for helping the other accelerator operators. In the end we will not expect to see full automation in this field but will surely see much more and more powerful man machine interaction.

The machine physicist would like to communicate with the accelerator in his own particular way. His design model centres around the behaviour of the beam in terms of betatron Q values, bucket sizes, chromaticity, etc. He does simulations on larger computers which allow him to interact in terms of these beam parameters and produce "normalised" machine parameters which the equipment specialist translates into hardware. Once the hardware becomes a reality in the form of an accelerator, the machine physicists are again called upon to help commission and develop the machine. The machine physicists need to measure and modify the beam parameters using the control system to translate these requests into volts and amps in particular items of equipment. The modelling used for designing the accelerator can be then usefully transferred directly for its control and operation.

Take for example the correction of chromaticity in the SPS:

The machine physicists' model identify three major components in each plane contributing to the chromaticity of the machine, the natural chromaticity which is constant through the cycle, a term which reduces with increasing energy coming from remanent field effects and a term proportional to the rate of change of field, the eddy current contribution. Sextupoles in arranged family groups, are used to correct these and any other unknown effect which individually contribute to the horizontal and vertical chromaticity correction depending upon their physical location in the accelerator.

The computer, via the applications programme, automatically converts the desired correction into particular currents in the sextupoles at well defined times through the cycle. Once this relationship is established and programmed the machine physicist or operator can ignore the existance of sextupoles (the alarm surveillance system will warn him of any equipment faults) and can interact with the accelerator only in terms of his familiar machine parameter, chromaticity.

To set the chromaticity the physicist must first make measurements on the beam. These measurements require the concerted use of kickers and beam monitors, timing modules and computers, data links, CAMAC and multiplex. There are many possible sources of error which can lead to faulty measurements results. The principles of control used for setting the machine in a given mode need to be applied to these measurement subsystems. The computer system is better adapted to this complicated routine task than the human operator. The machine physicist should need only to specify his measurement and the system should guarantee the correct result. Until this becomes operative, no automatic correction will be possible.

CONCLUSION

The computer has made possible the efficient operation of todays accelerators and the operator role has considerably changed from the pre-computer days. The man-machine interaction has become more than just the ability to bring the control of equipment to the operator in the central control building. The tools for understanding the working of the accelerator have become more sophisticated and the accelerator equipment can be controlled as an integrated process. Many of the routine operator tasks are carried out by the computers and the operator relieved of the drudgery is thus freed to study those aspects of development of the machine performance not yet covered by the system with a view to including his findings in tomorrows automation.

REFERENCES

- The computer aided control system of the VICKSI Accelerator
  W. Busse, IEEE Trans Nuc Sci Vol NS-26 No.2 April 1979
- The Design of the Control System of the SPS
  M.C.Crowley-Milling, CERN 75-20
- Experience in the Control System of the SPS
  M.C.Crowley-Milling, CERN 78-09
- Controlling an Accelerator - The operation viewpoint
  V. Hatton & G. Shering, CERN SPS/80-12 (AOP)
- Master file 1983
  R. Lauckner, SPS/AOP/Note/82-9
- A Multi-Processor, Multi-Task Control Structure for the CERN SPS
  C. Saltamarsh, SPS/AOP/Note/83-8, This Conference

MODELS and SIMULATIONS*

M. J. LEE, J. C. SHEPPARD, M. SULLENBERGER, M. D. WOODLEY
Stanford Linear Accelerator Center
Stanford University, Stanford, California 94305

1. Introduction

On-line mathematical models have been used successfully for computer controlled operation of SPEAR and PEP. The same model control concept is being implemented for the operation of the LINAC and for the Damping Ring, which will be part of the Stanford Linear Collider (SLC).

Errors in construction and modification may cause an actual machine to be different from the ideal machine conceived in the design. When the machine parameter values calculated from the model are substantially different from the measured values, the model cannot be used for computer controlled machine operations such as changing operating configuration or correcting closed orbit errors or trajectory errors.

Simulations can be used to develop an empirical model based upon the measured values of beam parameters. The effects produced by suspected errors can be studied by simulating them with a model and comparing the results with measured values. In some cases, changes in the model can be found which minimize these differences. This is a possible method for finding an empirical model to represent an actual machine.

We have used this procedure to find an empirical model for the Damping Ring, which has been in operation for several months. Based upon the measured changes in closed orbits produced by known kicks from orbit correctors, an empirical model has been found which correctly predicts these measured orbit changes.

The purpose of this paper is to describe the general relationships between models, simulations and the control system for any machine at SLAC. The work we have done on the development of the empirical model for the Damping Ring will be presented as an example.

2. Modeling Programs and the Control System

For initial operation of a machine under computer control, it is very important that the on-line model of the machine be the same as the model used in the design calculations and error studies. Since the control model is the design model, the model works by definition albeit a system may not perform as predicted by the modeling. If such an event occurs, it would then be reasonable to investigate errors in other areas (design, fabrication, installation, or calibration), but not to be concerned with the accuracy of the modeling.

By the same reasoning, it would be desirable to incorporate the programs used in a machine's design into the control system in order to eliminate another possible source of error. A drawback to the use of the designer's programs for on-line machine control is that such programs are typically large, general purpose routines which require relatively large amounts of computer memory and are not as fast as one would like for an automated control system. Once a new system has been brought on-line and understood, faster and more compact programs should be developed and installed in place of the original design codes. To ease the process of code modification and replacement it is important that modeling programs be modular in nature. This is accomplished by requiring the specific modeling codes to accept input vectors and to return output vectors of information from or to appropriate driver programs. Modeling program replacement then becomes a minor localized perturbation on the control system.

---

*Work supported by the Department of Energy, contract DE-AC03-76SF00151.

Communication through the database is a tool by which modularity of modeling programs can be insured. Use of the database also allows simultaneous development of interacting models by several different people. The database provides a well structured method by which modeling programs can communicate with those portions of the control program associated with the actual adjustment of power supplies while also isolating the task of modeling from the remainder of the control system.

In general, each of the modeling programs can be considered to be a stand alone computer code which computes the value of an output vector corresponding to the value of a desired input vector. Input vectors to the models include the users's specifications, usually entered from options selected using a touch panel, as well as other necessary data which is stored in the computer database. The information in the output vector is in turn saved in the database or in library files. For example, in a storage ring lattice model an input vector may contain such values as tunes, β and η at some specific locations, beam energy, etc.; the output vector may contain the values of the strengths of the ring elements. Database entries have been reserved for information relating to the state of elements in the machine. This includes the locations, lengths, current settings, and integrated field strengths of magnets, the beam energy gain associated with each klystron and the locations of beam position monitors. Information resulting from a modeling calculation that is not involved with the setting of power supplies (such as the calculated machine functions or the results of calculations which will appear on a graphics display) is not included in the database but is stored in local data files.

From a modeling point of view, it is considered poor practice to operate a machine by "tweaking" power supplies. It is possible to adjust the value of any or all of the beam parameters which are elements of an input vector by using a model. In addition, given that the power supply setpoints have been changed manually, models enable a user to calculate the corresponding machine parameter values. For example, it is possible to find the values of tunes, β and η for a storage ring lattice from the known values of the setpoints of the ring elements. Furthermore, an on-line model can be used to study the effects of changes on the beam parameter values without actually changing the setpoints of the elements. Such an "ignore hardware" feature allows a user to read the extant system settings, to calculate a change, and to predict the results of such a change before it is implemented. In this "ignore hardware" mode, models enable the user to evaluate the effects of any desired change in the machine and the subsequent effectiveness of schemes designed to compensate for those effects; this can be applied, for instance, to the study of error effects and possible error correction schemes.

It is possible to summarize the relationships between the modeling programs and the control system in a block diagram as shown in Fig. 1. The portion of the control system relevant to modeling calculations is enclosed in the inner circle. The possible interactions between the users, machine physicists, modelers and operators are indicated by the arrows on the outer circle. This structure was developed for the model calculations of PEP and will be used for SPEAR and SLC.

3. Simulations

In the design stage, simulations can be used to study the effects of changes in the machine element strengths on the beam parameter values, which includes the study of error effects and their correction.[1] Once the machine is operating, the beam parameter values can be measured and the imperfections of the machine can be studied by simulating their effects using models. Errors can be introduced into the model and their effects calculated. Machine imperfections may be located by inserting errors into the model which yield the measured effects. This method has been used to locate imperfections in the Damping Ring. The results of this study will be described in the following sections in order to illustrate an application of model simulation and to demonstrate some of the interactions between the users and the modelers as described in Fig. 1.

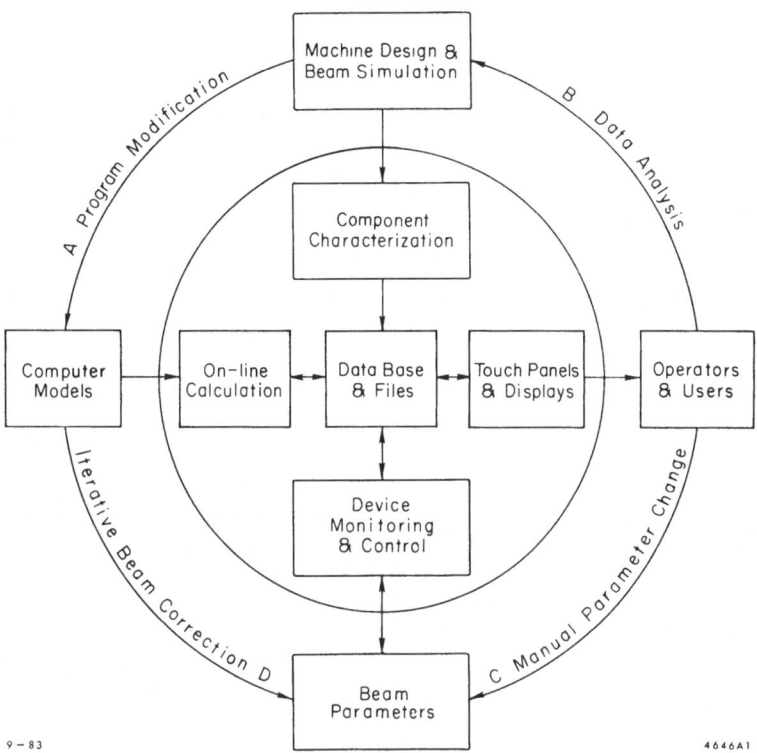

Fig. 1. A block diagram showing the relationships between model programs and the control system (inside the inner circle), and the possible interactions between users and modelers (on the outer circle).

3.1 LATTICE DESIGN

The ring lattice consists of two symmetric superperiods. Each superiod has five repeated FODO cells, two half insertions and two matching sections. The beam is injected and also extracted in the half insertions at the ends of one of the two superperiods. The circumference of the ring is 35.268 m. It has 40 bending magnets and six families of quadrupole magnets, which define the first order optics of the lattice.

The Damping Ring lattice was designed using the magnetic lattice design code MAGIC.[2] This code solves for the strengths of the six families of quadrupole magnets in the machine in order to obtain a desired set of six beam parameter values which define the operating configuration of the ring. Two of the configuration parameters are the horizontal and vertical tunes. Two other configuration parameters are the values of the hortizonal and vertical β function at the midpoint of the insertion region where the beam is injected into the ring. The fifth configuration parameter is the value of the energy dispersion function, η, at the same point. The sixth configuration parameter is α_x at the beginning of a FODO cell which is needed to impose the condition of periodicity on the horizontal β function in the periodic portions of the lattice.

Using MAGIC, the design of the lattice was studied and a suitable set of the tune, β and η values were chosen to be the design parameter values.[3] The object for the control system is to set the magnets to acheive these design parameter values in the actual machine.

3.2 ON-LINE LATTICE MODEL

We have developed a general purpose lattice computation program, COMFORT,[4] which is smaller and faster than MAGIC. COMFORT is intended to be the replacement of MAGIC for modeling and simulation of storage rings. In the control system, the Damping Ring model is a COMFORT dataset. The input vector to the model is the set of desired values of tune, β and η as described in the previous section. These values can be changed via touch panel commands as desired by the user. COMFORT calculates the the output vector, which contains the values of the magnet strengths, and sends it to the database. The control system converts these strength values to magnet current setpoint values and then adjusts the power supplies accordingly.

When the machine is operating under computer control, the values of the quadrupole strengths are converted into power supply setpoint values using magnetic measurement data which resides in the database. (See Fig. 1).

3.3 BEAM PARAMETER TEST

The Damping Ring has been operating for several months under computer control but not using the on-line model. When it was first turned on, the users tried to store beam using the design configuration but failed. As a last resort, they "tweaked" the power supplies manually via software knobs which can be assigned to each power supply with a touch panel command. They succeeded in storing beam in this experimentally obtained configuration. The beam parameters for this configuration have been measured and studied extensively. In addition, in order to study the variation of β functions in the lattice, changes in the closed orbit caused by known kicks at many of the orbit correctors have also been measured. It was found that the measured tune, β and η values were substantially different from the ideal values corresponding to the design configuration. Furthermore, when the strengths of the ring elements from the experimental configuration were used in the model, unstable tunes were predicted. It was impossible to change the machine tune, β and η values from the measured values to the desired design values or to correct the measured orbit errors conveniently without a working model.

3.4 MODEL MODIFICATIONS

In order to compensate the natural chromaticities of the ring, the pole faces of the bending magnets have been modified to produce sextupole fields. At one end, a "nose" piece was added, while at the other end a "hole" was cut from the pole face. It was discovered that the design lattice calculation had not been updated to include the effects of these modifications.

To correct this omission, the model was changed such that one thin-lens quadrupole, Q_n, was added to the end of the bending magnet with a nose and another thin-lens quadrupole, Q_h, was added to the other end. We assumed that the value of Q_n or Q_h would be the same for all of the bending magnets. Using COMFORT, we found the values of Q_n and Q_h needed to fit the measured horizontal and vertical tune values. The result was a configuration which was totally different from the design configuration, as can be seen in the differences between the machine functions of the design lattice and of this model lattice as shown in Figs. 2 and 3.

Using this model, we calculated the changes in closed orbit values at the beam position monitors for the corrector kicks which were actually used in perturbed orbit measurements. We found that the orbit changes produced by the horizontal corrector kicks agreed very well with the results of the model calculation, while the vertical orbit changes did not, as shown by two typical cases in Figs. 4 and 5. These results indicate that there are errors in the actual machine which effect the beam in the vertical plane and not in the horizontal plane.

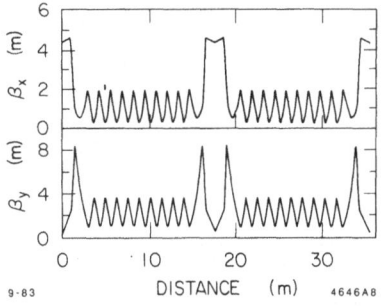

Fig. 2. A plot of the β functions along the Damping Ring for the design model.

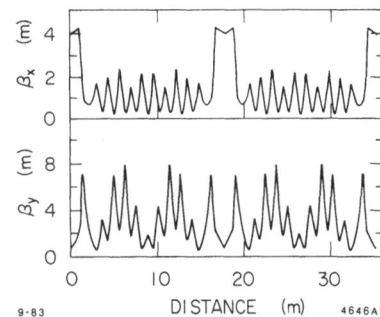

Fig. 3. A plot of the β functions along the Damping Ring after correcting the model to include the edge focusing effects from the bending magnets.

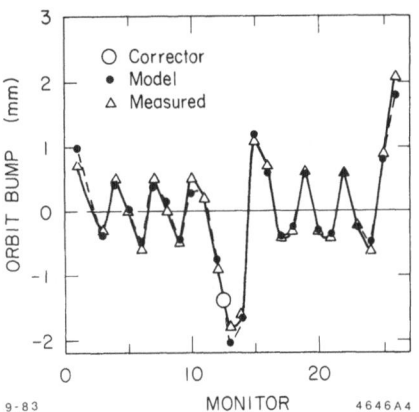

Fig. 4. A plot of the measured and predicted horizontal orbit changes caused by a given kick from a corrector near Monitor 12

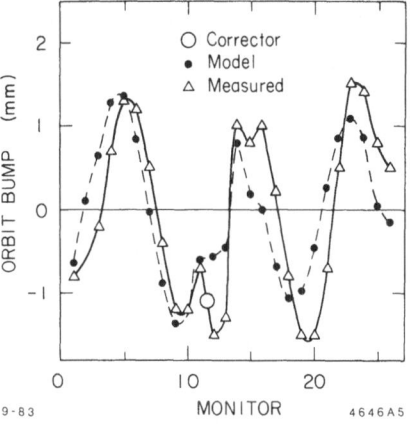

Fig. 5. A plot of the measured and predicted vertical orbit charges caused by a given kick from a corrector near Monitor 11.

3.5 Error Simulations

We assumed that the focusing error would be in one of the three families of horizontally defocusing quadrupoles where the vertical β function is large compared to the horizontal β function since the discrepancies were observed to be in the vertical plane.

A program, ORBFIT, was written to study the effects of this type of error on the closed orbit. The following tasks were performed by ORBFIT in this study:

1. For any given error to be studied the values of Q_n and Q_h are varied to fit the measured tune values using COMFORT.

2. COMFORT also computes the transfer matrix between a kick at any corrector and the change in orbit at any monitor for the lattice obtained in step 1.

3. Using this matrix, the value of corrector kick is adjusted to minimize the difference between the predicted orbit changes and the measured changes for a given corrector.

4. Step 3 is done for all of the correctors used in the measurements.

It was hoped that an error in one of the three families of defocusing quadrupoles could be found such that the changes in closed orbit predicted by the model would agree with the measured changes for known corrector kicks. The outputs from ORBFIT for cases with different errors in each of the three defocusing quadrupole magnet families can best be compared by considering the value of the rms ratio, which is defined as the rms of the difference between predicted and measured values divided by the rms of the measured values. The rms ratio would ideally be zero for a perfect fit between the prediction and the measurement.

The ORBFIT output for a reference case whithout any errors in the ring quadrupole magnet families is shown in Fig. 6. Since the rms values of the measured orbit changes are typically 0.5 mm, an rms ratio of 0.2 corresponds to about 0.1 mm, which is the order of magnitude of the errors in the measurements. It can be seen that the model predictions work well for the horizontal plane but not for the vertical plane.

Many different values of errors in each of the quadrupole families have been studied. Comparison of the results indicates that errors in the QDI magnet family give the smallest rms ratio values. QDI is the defocusing magnet nearest the insertion where the beam is injected or extracted. Since identical errors were introduced into each of the four QDIs, the resultant lattice has the same symmetry as the design lattice. The ORBFIT result for a -3% error in the QDI gradient is shown in Fig. 7.

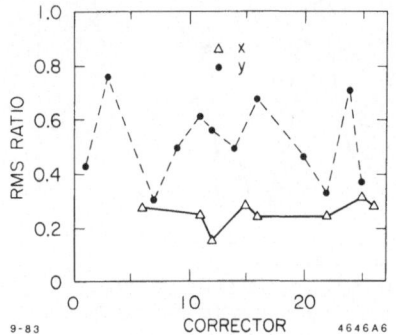
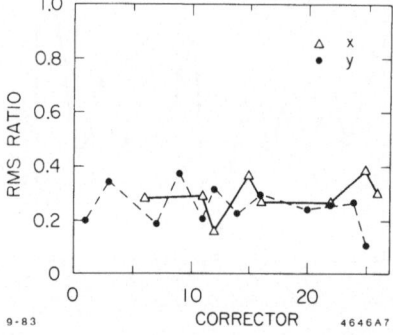

Fig. 6. A plot of the output from ORBFIT to be used as a reference case in the error simulation study.

Fig. 7. A plot of the output from ORBFIT for the case with an error of −3% in the strength of the insertion defocusing quadrupole magnets.

By comparing Figs. 6 and 7, it can be seen that the rms ratios for the vertical plane have been improved substantially in the empirically fitted model. For example, the orbit changes due to kicks by the vertical corrector in position 16 have an rms ratio of 0.608 using the reference lattice; the rms ratio has been reduced to 0.195 by using the empirical lattice. The orbit changes predicted for a kick by this corrector using the empirical lattice is shown in Fig. 8, which can be compared with the plot in Fig. 5 to see the improvement.

3.6 EMPIRICAL MODEL

We have seen that it is possible to modify the model empirically by studying the measured closed orbit changes due to known corrector kicks. In the empirical model, the QDI family is assumed to have an error in the calibration used to convert the quadrupole strength to magnet current. An error of -3% has been found to be optimum. The values of Q_h and Q_n needed to model the edge effects in the bending magnets in this model are equivalent to pole face rotations of -5.5 and -11.5 degrees, with the minus sign indicating horizontal focusing effects. These values are somewhat dependent upon the angular and positional displacements of the beam away from

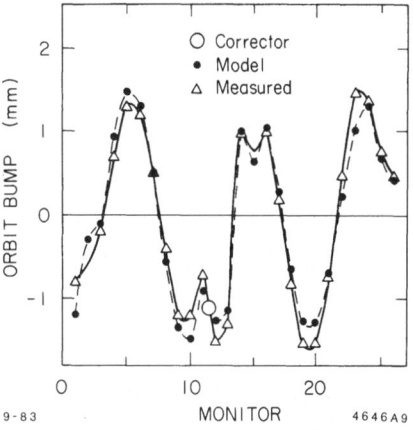

Fig. 8. A plot of the measured and predicted vertical orbit changes caused by a given kick from a corrector near Monitor 11. The predicted orbit change was calculated with the empirical model.

the design orbit at the entrance and exit of the bending magnets and hence could actually vary from magnet to magnet.

Changes in tune, β and η values can be predicted using this empirical model, as well as changes in the closed orbit produced by given corrector kicks. Orbit correction schemes based upon least square minimization of the closed orbit errors can also now be applied.

Thus far we have only included corrections in this empirical model which do not alter the symmetry and superperiodicity of the design lattice. Figure 9 shows a plot of the β functions for this empirical model. In order to improve the model further, it may be necessary to study the effects of errors which do not occur symmetrically.

Figure 10 shows a plot of the measured horizontal η function at the monitors. It can be seen that the η function does not have two-fold symmetric superperiods. An asymmetric model will be needed for our understanding of this anomalous η function. Development of a new code, ETAFIT, which minimizes the difference between measured η function values and values predicted by the model by introducing asymmetric errors into the model is in progress.

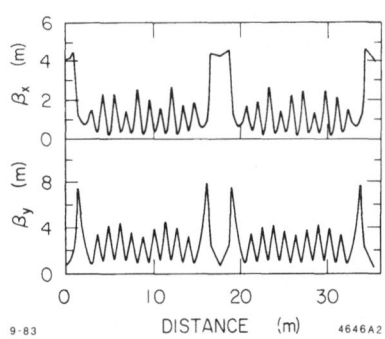

Fig. 9. A plot of the β functions along the Damping Ring for the empirical model.

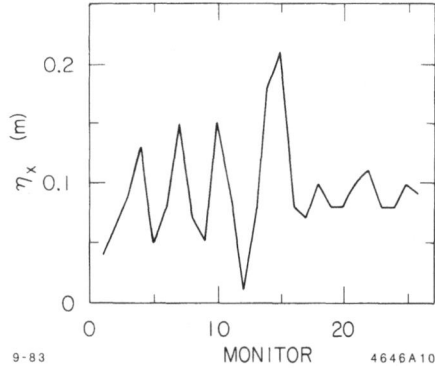

Fig. 10. A plot of the measured horizontal η function.

Acknowledgements

We would like to thank J. Jaeger for his help in analysis of the magnetic measurements and calibrations, G. Brown for writing the program ORBFIT and G. Hall for data analysis using ORBFIT.

The many hours of hard work on commissioning and operation of the Damping Ring, and on the measurement of its properties, by T. Fieguth, L. Rivkin, P. Morton and the rest of the Damping Ring crew are also greatly appreciated.

References

1. A. W. Chao, M. J. Lee, E. Linstadt and N. Spencer, IEEE Trans. Nucl. Sci. NS-28, March 1979.

2. A. S. King, M. J. Lee and W. W. Lee, SLAC-183, Aug. 1975.

3. Helmut Wiedemann, private communication.

4. M. D. Woodley, M. J. Lee, J. Jaeger and A. S. King, IEEE Trans. Nucl. Sci. NS-30, Aug. 1982.

OPERATIONS AND COMMUNICATIONS WITHIN THE DARESBURY NUCLEAR STRUCTURE FACILITY CONTROL SYSTEM

S.V. Davis, C.W. Horrabin, W.T. Johnstone, K. Spurling

Science Engineering research Council,
Daresbury Laboratory,
Daresbury Warrington WA4 4AD, UK.

INTRODUCTION

The Nuclear Structure Facility at Daresbury is a 20Mv Tandem Van de Graaff heavy ion accelerator. The design of the machine requires beam handling, vacuum, diagnostic and machine control equipment to be distributed within the accelerator in control sections maintained at different high potentials (fig 1).

Fig.1 The accelerator stack containing multiplex electronics

All equipment also has to withstand huge electrical discharges. It is not possible to communicate directly with equipment across wires and although in previous machines a limited range of control functions have been carried out using electro-mechanical means the number of control and monitoring channels required on the NSF, the limited space, and the hostile environment led to the decision to design a distributed analogue and digital multiplexing system housed in special double screened enclosures. The outstation crates of this system are interconnected by a serial digital communications ring using a mixture of free space or fibre optic infra-red light links, or transformer coupled coaxial cable transmission as appropriate. Each area of the accelerator is served by one A/D multiplex ring of this type which carries the entire data for control and monitoring within that area. The system makes extensive use of micro processors which organise communications over these rings. One crate on each ring provides a port for the connection of a control system computer. The NSF control system is thus based on a computer compatible connection with the accelerator over which <u>all</u> information passes, whatever its purpose within the overall control strategy.

OPERATIONS

The Analogue and Digital Multiplexing system and the special precautions applied to protect electronics in the hostile environment enable extensive monitoring and control features to be designed into the accelerator. For a machine of this complexity this is desirable not just to allow adequate control of the many machine parameters but to enable sufficient information to be presented about the operation of the accelerator to facilitate further understanding of the complex processes which determine overall machine behaviour. It is a prime function of the Control System to allow the collection of this information from the accelerator and provide for it to be processed and presented in the most easily understood and assimilated way. Add to this facilities to accept control requests in a manner convenient to the operating staff plus the ability to implement automatic control functions and sequences and the main requirements of a control system are satisfied. In this way a small operations team is able to oversee and direct the large complicated array of devices which together constitute the accelerator.

From the operations view point the system can be described by the way in which it tries to achieve these objectives. The NSF machine operations consoles in keeping with modern practice are designed to provide general control facilities so that together with application programs in the system computers any part of the accelerator may be selected and controlled.

Fig.2 One of the NSF Machine control consoles.

The optimum presentation of information depends on both the nature of the data and the use which is to be made of it. Reflecting this the NSF consoles cater for 2 basically different types of display. For normal running the operator must be informed of the condition of a wide range of parameters. This we achieve through colour displays showing schematic representations of parts of the accelerator indicating in real time the current values of selected parameters. Such displays require extensive processing of the raw information available but updating of this information need only be done at relatively long intervals. On the other hand investigation of a specific problem on the machine often requires detailed time varying information from one or a small number of parameters. In these cases it is essential to collect data at high sampling rates to preserve bandwidth. Presentation with a minimum of processing as a simple oscilloscope trace is then often sufficient. With such information at his disposal the researcher can proceed with his investigations, if necessary involving and directing further non real time processing of the signals as appropriate. The operators interact with the control system through the displays using light pen and touch screen and perform other control actions using control knobs and keyboard.

In addition to the real time communications with the operators through the control consoles the system allows the status of the system to be continually monitored, exception conditions being presented on an alarms screen. A continual log of machine operation is maintained, and various levels of automatic control and sequencing are performed.

THE NSF CONTROL SYSTEM

The control system used on the NSF to implement the operational facilities described above is best described in two sections, the distributed analogue and digital multiplexing system and the computer control network. Fig. 3 is a block diagram of the multiplex system. This is a network of distributed interconnected electronic crates each with a specific function.

The Outstation Crates (OC) are the interface with the accelerator signals. They are positioned around the accelerator each connecting the control and monitoring points within its locality. All conversions between analogue and digital form are done in these crates, the remainder of the network being all digital. Each Outstation Crate contains a micro-processor which controls the operation of the crate and allows the implementation of basic functions (ramping, local/remote control, interlocks, device control).

Each Plant Base Crate (PB) controls the serial communications loop connecting a number of Outstation Crates. Such a set manages the operation of each subsystem of the accelerator. The PB also contains a micro-processor. This provides some subsystem control functions (area interlocks, autonomous read/write channels) but its prime function is the organisation of communications within the subsystem. This includes the OC crate ring, a slave connection to a Main Ring interconnecting all base crates, and a parallel channel for a control system computer - the normal system access for accelerator information.

The function of the Main Ring is to provide fast channels to Console Base Crates (CB). These are located at machine control consoles and provide the simple high bandwidth displays of machine parameters on normal oscilloscopes. A parallel port is provided on these crates to allow an external control computer to control the selection of channels for display and take snapshot frames of display information for further processing or presentation. All base crates on the Main Ring are slave crates, communications on this ring being supervised by the Main Ring Controller (RC).

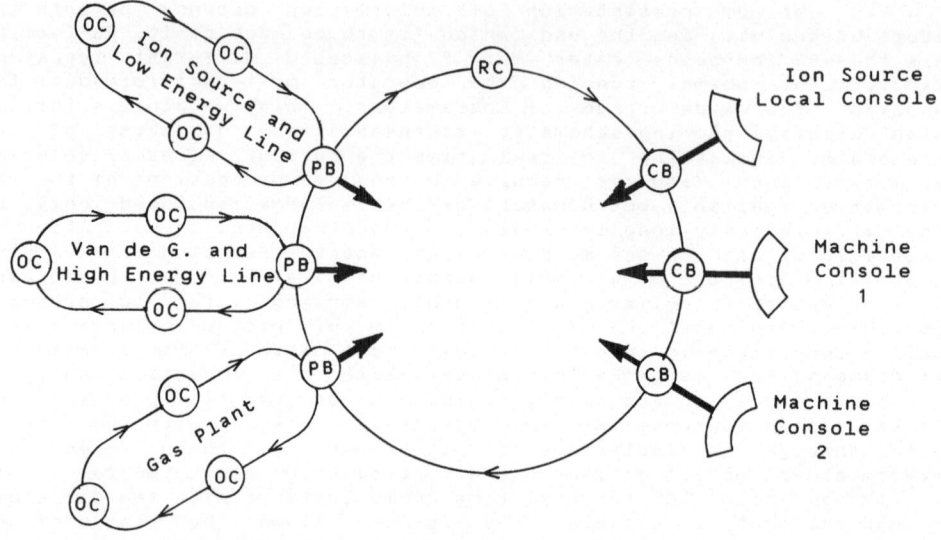

Fig.3 The NSF multiplex system network

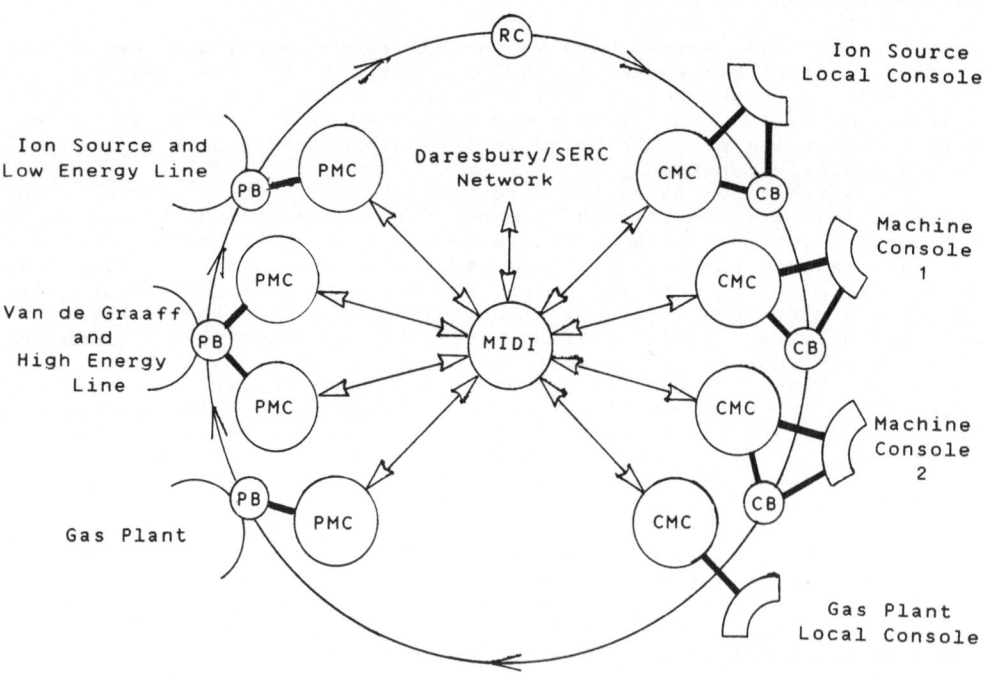

Fig.4 The complete NSF Computer Control System

The Main Ring Controller polls Console Base Crates (CB) for fast channel requests from which it sets up the appropriate Plant Base Crate(PB).It also generates the transfers for each active high bandwidth Plant to Console Base Crates channel over the Main Ring.

Fig.4 shows in block form the complete control system with the mini computer network included with the multiplex network described above. The computer network distributes the processing requirements over a number of 16 bit mini computers functioning as Plant Mini Computers (PMC), or Console Mini Computers (CMC) and a single larger 32 bit computer which we call the Midi computer (MIDI).

The Plant Mini Computers (PMC) provide front end processing for accelerator signals. Each contains a data base for its area of the machine so that requests for plant information from local control programs or from the Midi computer are made using a parameter naming system, and values are handled in appropriate engineering units, with protection limits applied. Other functions provided are local area surveillance, equipment test, and direct digital control.

The Midi computer is the main control computer in the system. This machine runs the main programs, for a wide range of applications including interactive operator control, alarm handling, logging, and accelerator control and control sequences. In addition it is the system communication node, the hub of the network. Through the Midi computer the NSF system has a connection to the Daresbury/SERC packet switched network. This gives access to other computer systems at Daresbury including the NAS 7000 mainframe and to other network resources for example interactive VDUs.

The Console Mini Computers (CMC) provides support for the machine consoles. Programs in these machines augment the basic console devices. This eases the burden on the Midi computer of manipulating the various operator devices and reduces the communications necessary for device control. They are local to the consoles and communicate with the Midi computer over network data links.

SYSTEM COMMUNICATIONS

As already noted our solution to the unusual control problem of a high voltage electrostatic accelerator have resulted in a system in which for much of the accelerator all information used in its operation is constrained to the data paths shown in the system block diagram, fig 4. As explained in the last section the actual path along which particular data flows depends not only on its source and destination but also the purpose for which it will be used. This section examines the transmission methods used over the various links and our operating experience with these.

Closest to the accelerator are the links forming a ring between a number of outstation crates and a Plant Base Crate. These links employ serial transmission at 5 Mbits/sec. using diphase encoding. Over these links all requests including those derived from blocks in the control computers become multiple independently addressed transfers. Autonomous transfers for local control functions within the loop are also single in nature. Thus the links are required to read or write single values to a single accelerator monitor or control point. Consequently a protocol is chosen which optimises a single read or write operation on a single address in a single crate. The format of the 32 bit data packet used is shown in fig. 5. These packets are transmitted round the ring by the Plant Base Crate. Each packet contains a 4 bit crate address (C0 - C4) which selects the target Outstation Crate, from a maximum of 15 on a ring. Within each crate a single addressing scheme is used for both local memory and hardware. A packet is able to address a 1024 word window (A0 - A9) of this address range which covers all the functional hardware input/output addresses (512 words) and a 512 word area of memory. With the mode bit (M) and 16 data bits (D0 - D15) the packet can thus access a single address in a selected crate as required.

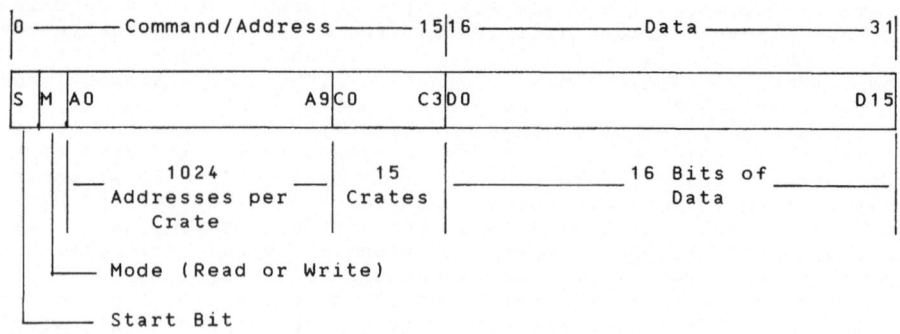

Fig.5 The Multiplex system transmission packet format.

The generation, reception and decoding of these packets is performed by hardware in each crate. All communication originates in the Plant Base Crate, Outstation Crates are not able to initiate transfers.

The Main Ring uses the same hardware, transmission technique, and protocol as above. This again works well as by far the greatest volume of traffic over these links is the repetitive sampling data for each active channel. To give the maximum bandwidth in real time, data for each channel is passed one value at a time, the Main Link control micro-processor uses time division multiplexing to provide up to 18 channels. Only when the parameters of a channel are redefined does the need arise to pass small blocks of information between the participating base crates and the Main Ring Controller. These are easily handled by the Main Ring Controller working into a small predefined fixed memory block in each base crate.

Closely coupled to the Plant Base Crates by a conventional parallel Camac interface are the Plant Mini Computers. Each has a link to the Midi computer. Their prime function is the collection and processing of data from their local area of the accelerator in response to Midi requests. Most of these requests are for display update or logging data of many machine parameters and are thus ideally suited for organisation into blocks for transmission between processors. Other logically independent processes carried out by the Plant Mini Computers require block communications with the Midi (e.g. Plant Surveillance). These links use a protocol able to handle block organised data and carry a number of independent calls. The standard packet switched protocol in use at Daresbury at the time these links were implemented satisfied the requirements and was used. The links are implemented with pairs of commercial Camac modules (Sension 1121 fast serial data link). These provide full duplex operation using a four twisted pair cable - one data and one response pair for each direction. Sixteen bit data words are transmitted as a twenty bit serial data stream. These bits are a start bit, 16 data bits, 2 parity bits, and a stop bit. A handshake response from the receiving end is used to control word transmission timing. Data rates up to 5Mbits/sec. are possible with these modules if suitable cable is used, but in our system we operate at 625 Kbits/sec.

The network protocol is implemented in software as special purpose drivers and operating system modules in the mini and Midi computers. In practice the links are a weak point in the system. The communications are always between predefined tasks, and their special nature has so far precluded the use of a standard higher level protocol. The data links used are of high integrity and the transfers short, so sophisticated recovery techniques are unnecessary. Many transfers including virtually all used to set values require very short blocks. The result is that the communications over these links incur a high penalty in both machine resources and performance for very little advantage.

Communication between the control system and the operations staff passes between the Midi and Console Mini Computers. The latter augment the basic hardware of the console input/output devices to generate intelligent text oriented devices. This has the effect of reducing the communications required, and is highly suitable for block data transfers to the advantage of both the processors and communications system. In our implementation the console devices look logically independant and require logically independent communications. The standard Daresbury packet switched protocol is again used. The results for these links are a much better justification for the choice. Console devices organised to operate in an interactive text oriented mode can use standard higher level protocols designed for this purpose. Efficient use of data transmitted in large blocks reduces the overheads of using the protocol. The ability to vary the communicating parties is also beneficial. An operator's screen can be coupled to an accelerator control program in the Midi computer, or just as easily access facilities in the Daresbury main frame (for NSF documentation for example). Conversely programs in the Midi computer can, if designed to operate from a basic interactive terminal, be controlled from any terminal on the Daresbury Network. In particular this makes possible access to the control system from staff offices or over a dial up line by staff at home. Access to these latter resources and to other computing resources outside the control system itself come through the Midi connection to the Daresbury/SERC network. This packet switched network serves the whole SERC computing community using a range of standard protocols. Our connection uses the same link hardware and Daresbury standard protocol as for our other links with the Midi computer.

FUTURE DEVELOPMENTS

The NSF control system uses different communications paths and protocols to carry control system information to meet the overall function requirements of the different aspects of accelerator control. Other communications are necessary to successfully set to work, operate, and maintain a geographically distributed accelerator. Voice and video channels are necessary for commisioning, fault finding, calibration, and safety often between the same points as control information. Today high bandwidth multi-channel systems are becoming available for use in local area networks in a wide variety of applications. On the NSF we believe that such equipment will be eminently suitable for accelerator control systems. Together the development of suitable protocols and communications hardware should make it possible to implement for the next generation of accelerator control systems a totally integrated, standardised communications network.

FURTHER REFERENCES

1/ Operational Experience of the Computer Control of an Electrostatic Generator. - T. W. Aitken, C. W. Horrabin, W. T. Johnstone, K. Spurling. - Control Systems and Methods.

2/ Stabilisation of an Electrostatic Generator using a Digital Computer - T. W. Aitken, I. Goodall, K. Spurling. - Nuclear Instruments Methods No. 153.

3/ The Organisation and Support of Colour Displays for a Nuclear Research Accelerator. - K. Spurling. - IEE Publication No. 172, Trends in On-line Computer Control Systems - 1979.

4/ Single Line Serial Transmission at 5 Mbits in Di-phase Code using Phase Locked Clock Recovery for Minimum Clock Degradation in Transmission. - C. W. Horrabin. - Daresbury Laboratory Technical Memorandum DL/NSF/TM 23.

5/ Remote Real Time Oscilloscope Displays for Accelerator Diagnostics on the Tandem Van de Graaff at the Daresbury Laboratory. - J. C. Beech, S. V. Davis, C. W. Horrabin, W. T. Johnstone, W. Siversides, K. Spurling. - Proceedings of the 6th Tandem Conference, Chester U.K. 1983 - Nuclear Instruments Methods.

CONSOLES AND DISPLAYS FOR ACCELERATOR OPERATION

G. Shering

CERN, 1211 Geneva 23, Switzerland.

1. <u>The operator's control desk</u>

Before the introduction of computers, most accelerator control rooms had rows of racks with separate controls and indications for each piece of equipment. Some of the most important controls and indications were often grouped together with the telephones and intercom equipment on a central control desk, but the role of this control desk in the operation of the accelerator was very limited. Computers were initially used in applications where the human operator was not very effective, such as orbit correction, or the logging and comparing of many accelerator settings. A simple set of hardware buttons to select the programs, and a printer or storage scope to display the results, sufficed as operator communication.

An important step forward came from the first accelerator designed from the start for complete computer control. This was the 800 MeV proton linac at Los Alamos[1]. The accelerator equipment was connected to the computer by a multiplex system and the operator had access, via the computer, from a control desk containing displays and other devices designed to interface the operator to the computer. This was the first of the "consoles" as we know them today, and sparked off a lot of interest in console design. The challenge was to replace the myriads of conventional hardwired controls and indications with "remote hand and eye" operation from a console, through the computer.

Important advances were the touch panel from the switchyard control at SLAC[2], the use of an interpretive language from beam line control at the Rutherford Laboratory[3], and the tracker ball and cursor from the bubble chamber film scanning experience. Many of these ideas, together with experiments with displays, were tried as additions to existing accelerator control systems. A completely new accelerator, however, provides the best opportunity for a fully thought out synthesis, and this took place in the early 1970's for the SPS control system[4,5]. Figure 1 shows an SPS console as used during the commissioning of the SPS. This was a milestone in console development and provided inspiration for subsequent machines such as VICKSI, PETRA, and JET.

2. The Mobile or Mini-Console

Mobile or mini-consoles have played an important part in the commissioning of the CERN Antiproton Accumulator (AA), JET, and the CERN Low Energy Antiproton Ring (LEAR). This is a more important development than it might seem at first sight. In the early days of computer control it was difficult to persuade equipment builders to economise and leave local controls and indications off their equipment. With the improvement in computer literacy and the microprocessor revolution, people now expect to interact with their equipment through some form of computer device. The problem is to maintain some form of coherence despite the large numbers and variety of the controls required. This coherence should be maintained between local and central controls, both from the point of view of the user and of the implementor/programmer.

One approach is to make the mini-console a subset of a full console, as shown in figure 2. Three of these mini-consoles were used for the commissioning and initial operation of the CERN AA. These consoles are quite similar to those used at PEP in that they consist simply of a touch screen and a display screen. This might suggest that a simple console is adequate when the underlying applications software is powerful and comprehensive. Another three of these mini-consoles, further reduced by using the touch screen also for the graphic display, are in use in the SPS control room for subsidiary duties. The main criticism of this approach is that the software is not compatible between the mini-console and a main console. A solution to this

Fig. 1 SPS Console during Initial Operation 1976-77.

problem is to make the mobile console an exact miniaturised version of a full console as shown in figure 3. This ensures complete software compatibility but is expensive and so is limited in its range of applications. Figure 4 shows the mobile console used in the commissioning of LEAR. It is perhaps intermediate between the two and nearest the optimum.

LEP will require many local and mini-consoles. Separate consoles for machine components and supporting services may be required at each access point "village". Also some of the machine subsystems, such as the RF plants, will require dedicated consoles for local maintenance and repair. Central operational control will be from the SPS control building, but as the office complexes are spread over several buildings in both the Swiss and French sites, additional consoles will be required.

The major challenge in the design for LEP is to achieve the triple objectives of low cost for the many mini-consoles, adequate power for central operational control, and compatibility between the central and local consoles. Subset compatibility would be a minimum requirement, but there are strong feelings that complete compatibility is required, so that any program can run on any console. The rest of this paper develops ideas on how this can be made possible.

3. The Personal Work Station

The recent development of personal work stations such as APOLLO, PERQ and SUN, is of great significance for the design of accelerator consoles. Up till recently most work on console design has been in specialised areas such as accelerator control, aircraft cockpits, nuclear power station control, computer aided design, where the numbers involved are quite small. The cost per unit has not been an important factor, but the total manpower devoted to their development has been limited. The personal work station, however, is finding mass applications in office automation, education, research, and business. A large amount of effort, both from industry and research establishments, is being devoted to their development, and this will have a big impact on accelerator control.

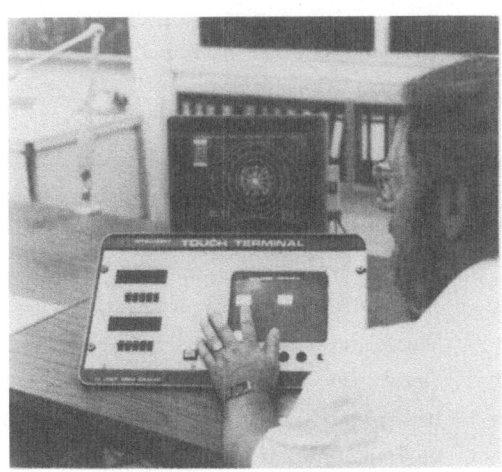

Fig. 2 Touch Terminal in Control of Antiproton Accumulator.

The personal work station consists of three main elements: a high performance CPU; a display with high (1000 point) resolution and cursor capability; and a network connection. Figure 5 shows an APOLLO, one of four different makes of personal work stations currently installed at CERN for evaluation. More important even than the hardware are the software concepts being developed for, or applied to, the personal workstations. Multi-processing, display windowing and network wide random file access, are the most relevant to accelerator consoles.

How this impact will be realised is not yet clear. Will consoles be built on the core of a personal workstation? Will some of their modules, hardware and/or software, be incorporated in a console built in a standard bus system? Or will they simply provide new directions and set new standards? Whichever way it goes, console design is poised for a big leap forward.

4. Networking

Early systems had the operator interface and the control of the accelerator equipment integrated into a single computer. This approach can be quite successful as it is easy to obtain a high bandwidth between the equipment and the operator. As the system becomes larger, however, this approach breaks down partly due to the complexity involved, partly due to lack of CPU power. Where the equipment to be controlled can be broken down into subsystems, or separate machines, the integrated approach can be given new life by having separate semi-autonomous computer systems loosely linked by a network for transmitting shared parameters.

Fig. 3 Mobile console in Control of PS Booster.

A radically different technique is being developed in other fields of computing centred round the emerging technology of local area networks. Here each "node" or computer is function oriented, i.e. dedicated to, and optimised for, a particular job. Examples are a network with a file server, a print server, a gateway to other networks, and several personal work stations.

In accelerator terms this translates into separation of the equipment management functions and the console functions into separate machines connected by the network. This approach has been used successfully by the SPS control system since the beginning. For economy, however, the same type of computer and operating system were used for both the consoles and the process computers. For LEP a modular system based on a standard (VME) bus will be used[6].

This will provide more flexibility in configuring the hardware and software for the varied tasks of the console computer assembly, the process computer assembly, and other assemblies required in the control system.

5. Multi-Tasking and Windows

The screen in figure 5 is divided into three "windows", two being used for graphics and one for text. The text window is temporarily "overlapping" the graphics windows so that all the text can be read. Each of these windows is the output of a separate process running in the work station. The user can interact with the process of his choice simply by moving the cursor into a window controlled by that process. The other processes can continue in the background, however, unless they are waiting for user input.

The touch terminal shown in figure 2 is at the other extreme, and interacts with only a single process in the AA control computer. This does not mean a single program, as programs can be run one after the other or in "tree" fashion. If several actions or displays are required concurrently this must be built into the single program, or another touch terminal must be used.

The SPS consoles have used a multi-processing system with some good results, but have been limited by two main problems. The first is how to handle the interaction. The solution was to divide the processes into one interactive process which alone could use the touch panel, ball and knob, and several (up to six) so-called "real-time" processes which could only access the accelerator and create displays. As the interactive process is the more powerful most jobs were programmed as interactive processes so if two were required at once, two consoles had to be used. The APOLLO feature of using "Pads" for the displays so that they can be overwritten and re-made visible at any time could be applied to the touch screen, knob, and ball, so that all processes could be "interactive processes", with the interaction directed by the user rather than the programmer.

The other main problem with multi-processing in the SPS consoles is display allocation. The philosophy was to give each programmer the full console facilities, with the not surprising result that most processes use all the console facilities, particularly the two main displays, leaving nothing for any other process. Other systems, e.g. JET, pre-allocate the screen into different areas for different processes, but

Fig. 4 Mobile Console in Control of CERN LEAR.

this of course reduces the effective resources available to any single task. This problem is solved by the "window" technique. Each process can be given a display window, the equivalent of a whole screen, and can even "borrow" additional windows if necessary. Thus the programmer has no a priori restrictions. Whether these windows use the full physical screen(s), and whether they are overlapped or at the top of the pile, is not specified to the programmer, but determined by the user.

Another problem with multi-processing is the sharing and communication of data between processes. In the SPS this is mainly done using files on the library computer. With the increasing use of multi-processing in the consoles there has been a demand for more rapid and random access to common data. Experiments are under way to provide this for cooperating processes within one console. This could lead to two separate techniques for shared data, one for processes in the same console, another for processes in different consoles. The personal work-stations have attacked this problem by supporting "paging over the network". This is made possible by the high speeds of modern local area networks, and of course sophisticated software. This could form a mechanism for any process to have rapid random access to common accelerator data.

6. The modular control desk

Three consoles of the type shown in figure 1 were originally installed in the SPS control room, so that one could be used for primary operation, another for auxiliary control duties, and one say as a spare or for program development. After a few years' operational experience a fourth console was added and the control room was re-arranged to give two operating areas, each with two consoles. One of these operating areas is shown in figure 6. This was done partly because the console limitations discussed previously often required the use of two consoles together, but also because many aspects of accelerator operation, particularly machine development and setting-up, are team efforts. Space and facilities for up to half a dozen people are often required at such times. Similar concepts of a large control desk with duplicated or repeated control facilities can be found elsewhere, for instance VICKSI[7].

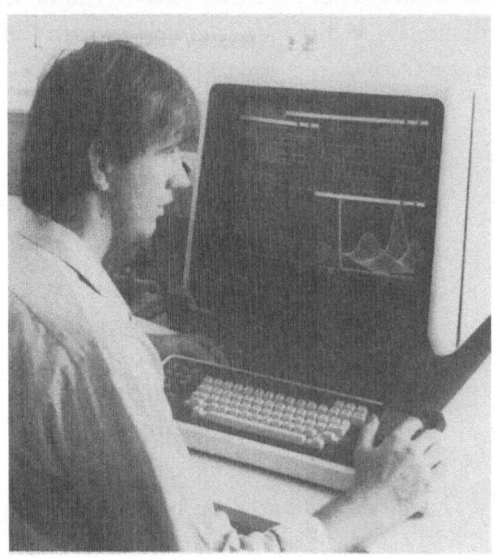

Fig. 5 APOLLO Personal Work Station at CERN.

The control area shown in figure 6 provides a number of separately identifiable facilities. These are :

1) Analogue Signal display and processing
2) Alarm display and manipulation
3) Interactive accelerator control
4) Permanent (wall mounted) displays for "at a glance" status
5) Closed Circuit TV screens.

Figure 7 is a schematic diagram of a console module which, it will be argued, can provide all the above functions. The main display is, of course, a high resolution colour screen. Below this is the touch screen (probably also colour), and an auxiliary screen for programming and computer messages. As the technology improves this latter screen might be generalised into a full function auxiliary screen. Then there is a separate unit incorporating the keyboard, knob and ball, although the ergonomic placing of these might vary between a local control console and a central console.

The ability of such a console module to perform all the duties required for central control can be examined as follows :

Analogue signals. As accelerators have become larger the traditional technique of attaching a scope or spectrum analyser to the output connections of a sensor has become more and more difficult. This problem was tackled for the fast kicker signals on the SPS by using remote digitisers. This proved very satisfactory, and with increasing software support gave

Fig. 6 Control Area in Revised SPS Control Room.

additional advantages of automatic setting of gains, offsets, and associated timings, and at the display end the addition of scales, explanatory notes, calibrations, and zoom capability. Even for small machines such as the AA, transient digitisers in the economic CAMAC module form are finding extensive use. In the multiprocessing console of figure 7, one process (or more) might be devoted to reading an analog signal and displaying it on a window in some corner of the screen.

Alarms. The SPS alarm computer provides two functions. It accepts and manages alarm information from the accelerator; and it displays this information to the operator with the possibility of interaction at two separate points in the control room. For LEP two alarm computers may be used, one for services and one for machine components. It will be advantageous to separate out the operator interface part of the alarm system into a standard console module. A program in the console module will communicate over the network with the appropriate alarm computer. This will simplify the alarm computer software and will make alarm information available on any console, although in the control room some console modules will be dedicated to this function.

Interactive accelerator control. This is the main role of the console, and in the SPS the now NORD-100 based consoles have been steadily improved over the years to fill this role as well as possible. The new console would have all the same facilities as the old one, such as knob, ball, touch screen, program screen, and some special purpose buttons. There is a question over the number of main display screens, however. The existing console has three main display bays containing four small monochrome screens, a low resolution colour graphics central screen, and a high resolution monochrome screen. Theoretically all of these screens could be represented as windows on a single high performance display, but physically this may not provide adequate display area for typical operations scenarios. The solution is to generalise the window capability from N windows on one screen to N windows on M screens. N would be some suitably high number such as 15, whereas M could be chosen according to the physical requirements. Even in a single bay local control console the auxiliary small program screen, and even the touch screen, could be extended to hold one or more of the logical windows.

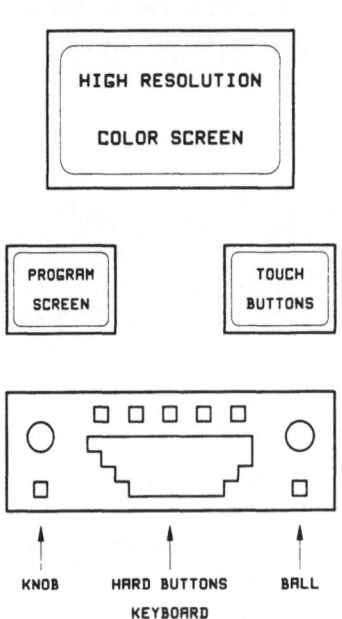

Fig. 7 Proposed Console Module Layout.

Accelerator computations. In the SPS the large computations required for modelling, data reduction, and other calculation oriented control functions, are done in the console, usually as FORTRAN sub-routines called by the NODAL interpreter. The proposed console module must also have this capability, even though an array processor may be attached to the LEP network for the heaviest particle tracking computations. This should not be a problem as the console module will be based on a powerful 16/32 bit micro-processor. The MC68000 is the current CERN 16/32 bit standard and is also used in the APOLLO and SUN personal work stations. At CERN trials have shown that it is possible to take large programs from the central computers and run them with little modification on APOLLO. From the speed point of view, there should also be no problem. The current prototype of the console module uses a standard 8 MHz 68000 and can interpret NODAL faster than the NORD. Substantial improvements in the 68000 line of microprocessors are expected to increase the performance to levels approaching super-mini CPU's. Also the console will use multiple processors[6], one processor per process. Local control consoles may support only a few processes for cost reasons, whereas the main central consoles could have more.

Closed circuit TV. Ideally we would like to transmit TV pictures over the standard console digital connection, then display them on any window on a screen. The bandwidth and processing power required for this are at the limits of current technology. Two areas of development are promising, however. The first is TV "CODEC" techniques of data compression already in use for video-conferencing. The second is the use of cameras with low resolution and low scan rate based on new semiconductor digital video sensors with resolution in the 300 x 200 region.

7. Conclusion

Given that all the functions described can be supported over the network by the single type of console module shown in figure 7, then complete compatibility between local and central control can be achieved as :

a) The central control area can be made up of a number, say 6, of the modules.

b) Local control with only one module can have complete access to all data, restricted only by the number of windows reasonable to use at one time.

References

1) H.S. Butler, B.L. Hartway, D.R. Machen, and T.M. Putnam, An Operator's Console for the LAMPF Accelerator, IEEE Trans. on Nuclear Science, NS-18(1971), p. 419.

2) D. Fryberger and R. Johnson, An Innovation in Control Panels for Large Computer Control Systems, IEEE Trans. on Nuclear Science, NS-18(1971), p. 414.

3) Peter Adams, Beamline Computer Control by Interpreter, IEEE Trans. on Nuclear Science, NS-18(1971), p. 361.

4) M.C. Crowley-Milling, The Design of the Control System for the SPS, CERN 75-20.

5) F. Beck, The Design and Construction of a Control Centre for the CERN SPS Accelerator, CERN SPS-CO/76-1.

6) M.C. Crowley-Milling, The Control System for LEP, IEEE Trans. on Nuclear Science, Accelerator Conference 1983.

7) W. Busse, The Computer Aided Control System of the VICKSI Accelerator, IEEE Trans. on Nuclear Science, NS-26, 1979.

OPERATOR INTERFACE TO THE ORIC CONTROL SYSTEM

C. A. Ludemann and B. J. Casstevens
Oak Ridge National Laboratory*
Oak Ridge, Tennessee, USA

Introduction

The Oak Ridge Isochronous Cyclotron (ORIC) was built in the early 1960s with a hard-wired manual control system. Presently, it serves as a variable-energy heavy-ion cyclotron with an internal ion source, or as an energy booster for the new 25 MV tandem electrostatic accelerator of the Holifield Heavy Ion Facility. One factor which has kept the cyclotron the productive research tool it is today is the gradual transfer of its control functions to a computer-based system beginning in the 1970s.

This particular placement of a computer between an accelerator and its operators afforded some unique challenges and opportunities that would not be encountered today. Historically, the transformation began at a time when computers were just beginning to gain acceptance as reliable operational tools. Veteran operators with tens of years of accelerator experience justifiably expressed skepticism that this "improvement" would aid them, particularly if they had to re-learn how to operate the machine. The confidence of the operators was gained when they realized that one of the primary principles of ergonomics was being upheld. The computer software and hardware was being designed to serve them and not the computer.

The undertaking, in fact, was aided by information not usually available when one designs a new accelerator and control system simultaneously. The idiosyncracies of the accelerator were well-known and the areas in which the computer could provide assistance were readily identifiable. Furthermore, the operators had developed "natural" tuning techniques that a system designer would not necessarily have thought of. Typically, in a new installation these aspects are discovered only after a few years of operating experience. It is usually within the first revision of the software (but hopefully not the hardware) that trouble spots are eliminated and unanticipated features added.

The ORIC computer control system manages the operation of a moderate number of elements (~200). They were transferred to the control of the computer system gradually while working within the constraints of an active research program. In fact, there are many elements that remain hard-wired today because it would not be cost effective to transfer them to the system unless they are modified or expanded upon.

The Operator Console

The cyclotron operators communicate with the computer by means of an alphanumeric keyboard and a "manual" control panel. CRT displays and a voice synthesizer provide status information from the computer. While this approach to operator-machine interface is relatively standard, the details of the control panel are not.

---

*Research sponsored by the Basic Energy Sciences Division, U.S. Department of Energy under contract W-7405-eng-26 with the Union Carbide Corporation.

Physical Layout. The elements of the operator interface were installed in one 100 cm "wing" of the existing U-shaped hard-wired console (Fig. 1). The components replaced approximately 40 potentiometers, 40 panel meters, and 80 switches. The hard-wired console had been designed to minimize operator fatigue by making all controls readily visible and accessible. The computer components were organized to conform to this example. Not shown in the foreground of the figure is a 30-cm horizontal shelf that permits the operator to rest log books or arms while tuning the machine. This shelf is 77 cm above the floor with adequate clear space for legs and feet. The 30 cm by 48 cm panels that hold the alphanumeric keyboard and one portion of the control panel are inclined approximately 10 degrees above horizontal. The CRT faces and current metering panel are almost vertical. The 11 cm transition piece that supports hard-wired controls above the keyboard on the left, and the second portion of the control panel on the right, are inclined approximately 35 degrees. Since the console height is only 112 cm above the floor, the operator has an almost unobstructed view of the remainder of the control room. The shield that was placed above the CRTs to reduce their illumination by overhead room lighting, also serves as a convenient shelf for resting materials during conversations between the operators and other staff members.

CRTs. The three 23 cm (diagonal measure) monochromatic CRTs are refreshed by CAMAC modules. Each screen is capable of presenting 24 lines of 64 alphanumeric characters. This information is easily readable by most operators at the console. Some difficulty in reading has been experienced, however, by a staff member who uses trifocal spectacles. Intensity variation and blinking characters are used rather than color to inform operators of equipment status and error conditions. This is because the fringe magnetic field of the cyclotron is of the order of 1–2 mTesla in this area of the control room. The left-hand CRT records major equipment malfunctions detected by the system, echoes characters from the keyboard, and displays "prompt questions"

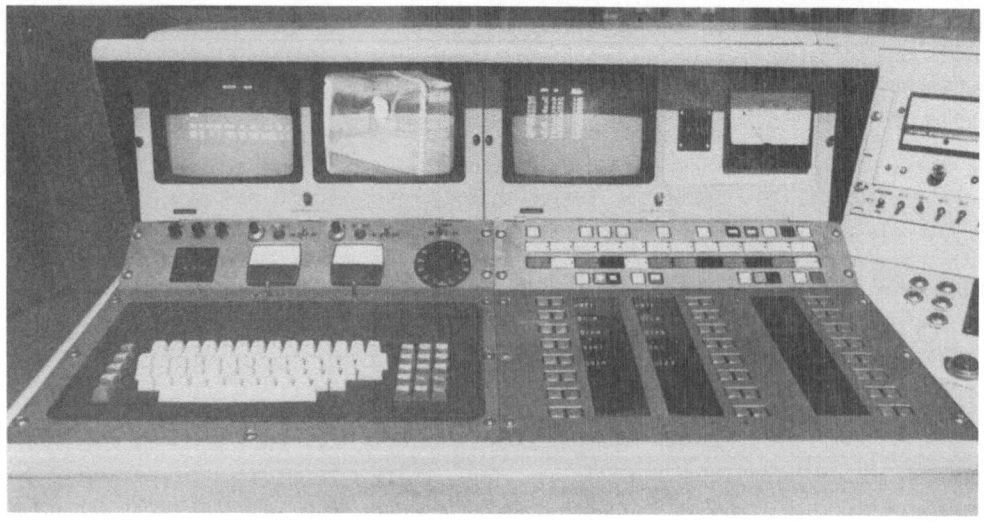

Figure 1. The components of the ORIC operator's computer interface mounted in the existing control console.

and reply messages from programs that the operator has entered by means of the keyboard. The center CRT displays equipment status or can be switched to monitor television images of equipment in the experimental caves. The third CRT presents the status of equipment that the operator is tuning with the control panel.

Control Panel. In order to tune the accelerator, operators need to have a large number of controls available to them rapidly and in logical order. Menu-picking and cursor manipulation to assign controls were considered to be undesirable distractions to the tuning process and were to be avoided. These considerations led to the design of a panel comprised entirely of LED-lighted and LED-labeled pushbutton switches.

The upper portion of the control panel has four rows of LED-lighted switches. On the top row are the "command" switches, the next row is for "page selection", the third row performs functions that modify the format of the CRT pages or the control page that has been selected, and the final row determines the rate and size of changes that will be performed by the controls on the lower portion of the panel. This last row also furnishes "save" and "restore" capabilities.

Eleven of the twelve page selection switches permit the operator to rapidly select functional groups of controls. The left to right order of these switches corresponds to sub-systems that accelerate or transport the ion beam from the center of the ORIC to the final target station. The controls selected by two of the switches are determined by target station information entered by the operator via the keyboard. The beam from the ORIC can be directed to 14 different areas, and the tandem beam to 8 stations under control of this system.

The depression of a page switch selects up to 16 operating parameters (coil currents, coefficients of functional relationships, etc.) for operator manipulation. The name of each control (up to eight characters) appears in LEDs next to a pair of pushbutton switches in the two leftmost columns of switches on the lower portion of the panel. Each pair of switches becomes the "manual" control for a parameter. These controls appear on the panel in the order of beam transmission, alternating from left to right columns. The switches are recessed below the surface of the panel so they will not be depressed inadvertently by placement of books or arms on the console. The LED-labels are not coplanar with the red plastic covers but are mounted at an angle to permit easier viewing by the operator seated at the console.

The labeled switches permit the operator to raise or lower the currents in magnets, move motor-driven elements in or out, etc. The sense, or direction, of control is that the left (red) switch generally increases the value of the operating parameter, and the right (blue) switch decreases it. The rate and increment of change is determined by the states of the LED-lighted switches on the fourth row of the upper portion of the control panel. Changes occur once or sixteen times per second and can have three magnitudes: one bit, approximately one quarter percent full scale, and one percent full scale. The actual magnitude of change depends on the element being controlled. The rate and increment size switches make it possible for the operator to change the current from a power supply from zero to full output in as little as six seconds, or as long as 4.3 minutes if the supply is driven by a 12-bit DAC.

The settings of all parameters on a control page can be "saved" by depressing a LED-lighted switch next to the increment rate and size switches. The settings for each page are saved independently and remembered as the operator moves from page to page. If a parameter is altered from its saved setting the difference is recorded on the CRT next to its present setting. An operator may return a single element to its saved value by adjusting its controls to make the difference zero, or may restore all elements on a page to their saved values by depressing the switch next to the save switch — the "restore" switch.

The column of eight pairs of LED labeled switches on the right side of the control panel allows the operator to control any 8 functions regardless of the page that is selected on the left portion of the control panel. These controls are assigned by use of the keyboard and have an independent set of rate, increment size, save, and restore switches.

The top row of LED-lighted switches represents commands that can be performed on most of the other switches on the panel. Once depressed these commands remain active for up to three seconds as indicated by their flashing LEDs. For example, by touching the "POF" command key and then, within three seconds, a page selection key, all power supplies associated with that page are slewed to zero current and turned off. Similarly, the operator can touch POF and a single switch on the control panel and that individual supply will be turned off. Other command keys that affect power supplies include:

ZAP - set the reference voltage to zero and turn the supply off immediately,
PON - turn the power supply on,
RSET - slew the supply's current to its value when last POFed or ZAPed.

Other command keys are used to open and close vacuum valves, etc.

The twelfth page selection key is labeled "ALL". It is used in conjunction with the command keys to affect all elements connected to the system. As an example, depressing ZAP and then ALL effectively SCRAMs the accelerator.

Keyboard. The alphanumeric keyboard is used to communicate with tasks that are entered by the use of keywords. These tasks permit the operator to ramp any power supply to a prescribed current, calculate the beam energy, file all the operating parameters on disk storage, search for and set the machine up from recorded settings, etc. All information entered through the keyboard is examined for commands (i.e., ZAP, POF, etc.) before it is passed to the task. This preprocessing permits the operator to perform critical operations (e.g., ZAP ALL) at the keyboard regardless of the task being executed. To limit the keystrokes needed to perform these command sequences, the function keys on the left side of the unit have the same labels as the command switches on the control panel and the keypad on the right side includes the names of all the page selection switches.

Use of the Control Panel

The design goal of any control system, hard-wired or computer-assisted, is to have the resulting system support the operating process and not confuse it. In describing the operator's console in the previous section it would appear that the principle just mentioned was being violated. The transformation of single-turn, ten-turn, and three-decade potentiometers into pushbutton switches was the initial concern of the operators and system designers as well. The concept was proposed because it provided the high density of well labeled controls the operators desired. Furthermore, it would eliminate the awkward and tiring arm and wrist positions necessary to tune with an individual potentiometer or shaft encoder without inadvertently disturbing others. In order to test the approach, a single trimming coil power supply was operated using a panel with a pair of control switches. When it was found that the operators had no difficulty adjusting the supply, a panel with rate, increment size, and twelve pairs of control switches was constructed. The ten trimming coil power supplies were controlled with this panel. The question as to whether the new panel led to confusion was in reality the question whether the machine could be tuned as well, or better, than with the hard-wired controls. The answer was "as well". The operators wished to know why so much effort was being expended to do something they could do previously. Their question was answered when the panel was used to control the harmonic coil power supplies.

The ORIC has three sets of three pairs of windings that are used to cancel imperfections in the magnetic field caused by asymmetric cyclotron elements such as the extraction system. Proper tuning of each set requires adjustment of their currents in such a manner that their contribution to the azimuthally averaged field is a constant while their resultant cancellation field can be altered in strength and azimuth. This was accomplished in the hard-wired system by modifying the voltage references to the electrically floating power supplies. Batteries, linear potentiometers, and potentiometers whose resistance was proportional to the SINE of the angle of their shaft rotation were used. Reproducible operation was possible providing the coupling between potentiometer shafts remained fixed and the batteries were new. Under the computer system, the same strength, azimuth, and average field level controls were provided. However, the data values sent to the DACs providing the reference voltages to the supplies were generated by the appropriate trigonometric expressions modeling the process. The results were a significant increase in reliability, reproducibility, and resolution in control of the harmonic coils. It was at this point that the final version of the control panel could be completed and installed in the control console. It was the computer's computational capability that provided the endorsement from the operators.

It is interesting to observe the operators fine tune the ORIC today. Their concentration is directed primarily to the beam current metering panel and the video display of diagnostic devices, as it should be. They can move rapidly from one control subsystem to another as they work the beam through the machine to the experimental station. It often appears that a Braille system is being used by the operator as his hands move down the lower panel balancing one control element against another. The fear of "getting lost" has almost disappeared because of the save and restore features. This lack of intimidation is unfortunate in some instances. If these features are used indiscriminately, one can rapidly detune the machine from a pre-calculated configuration with little hope of return.

Functional Control - Additional Remarks

Since the original implementation of the harmonic coil controls, other elements of the ORIC have been linked by functional relationships. There are two comments to be made which might be helpful to others who are about to attempt such a control strategy. First, the use of pushbutton switches to increment and decrement functional parameters provides a software advantage over an absolute control (e.g., absolute shaft encoder). If the operator attempts to exceed the range of a parameter, further requests for change can be ignored easily. With absolute devices it is necessary to maintain the new "dial reading" in RAM or in the external hardware. Second, regardless of the hardware method, when functional parameters are used as controls, the designer should always provide means for the operator to break the relationship. As with the harmonic control outlined above, if a power supply fails, the operator must be able to control that supply for testing and repair purposes. In our system, the supply can be assigned to one of the right hand pairs of control switches. When this is done provision must be made in the software to calculate the new functional parameters using inverse relationships (even if they are outside what would be considered normal bounds). Another equally practical reason for permitting the relaxation of functional control, especially during its original implementation, is to provide the accelerator physicists the ability to test the validity of the function itself.

Concluding Remarks

The considerations that went into the design of the human interface for one of the first applications of a computer to the control of a cyclotron have been described. Space does not permit an outline of the error detection and reporting schemes, the file management approach taken for the recording of operating data and automatic setup of the machine[1], or the extensive software for the operator interaction with tasks entered by means of the keyboard.

Perhaps the most important practice exercised in this undertaking is one that is valid for the design of any new system: It is imperative that the system designer work in close communication with the operating staff through all stages of the project. After all, they are the ones that must use the system day after day. The success of the entire operation rests upon their ability to operate the machine.

Acknowledgements

The authors wish to express their appreciation for the early contributions by S. W. Mosko and J. M. Domaschko of the Physics Division, and E. Madden and the late E. McDaniel of the Instrumentation and Controls Division to this endeavor. We wish to thank also C. L. Viar and the entire ORIC operations staff, past and present, for their patience and constructive criticism and suggestions.

Reference

1. B. J. Casstevens and C. A. Ludemann, Database Automation of Accelerator Operation, ORNL/CSD/TM-191.

COMPUTER AIDED SETTING UP OF VICKSI

W. Busse, B. Martin, R. Michaelsen, W. Pelzer, B. Spellmeyer, K. Ziegler
Hahn-Meitner Institute, Postfach 39 01 28, D-1000 Berlin 39

Introduction

The VICKSI accelerator facility at the Hahn-Meitner Institute in Berlin has been in routine operation since 1979. It is a combination of a single ended 6 MV Van de Graaff accelerator and a four-fold symmetry separated sector isochronous cyclotron. The beam line between the injector and the cyclotron contains two bunchers, a stripper and all the other necessary elements to optimally adapt the Van de Graaff beam to the requirements of the cyclotron. After extraction from the cyclotron the beam may be transported to different target locations. A plan view of the installation is given in fig. 1. It also shows the Tandem injector which is presently being added to the system as part of an improvement program.

Detailed descriptions of the specific design considerations and features of the VICKSI installation have been given in earlier reports (1-4). Therefore it is only recalled that positive ions are extracted from an axial Penning ion source in the high voltage terminal of the Van de Graaff accelerator. After charge state selection, prebunching and optical adaptation to the accelerator tube the beam is accelerated with termimal voltages of up to 6 MV. In a gas or foil stripper the ions are stripped to higher charge states and then accelerated by the cyclotron to about 17 times the injection energy. The prebuncher in the terminal and the two bunchers in the beam line between the two machines allow to compress about 60 % of the ion-source dc-output into a 6^0 phase interval of the cyclotron RF.

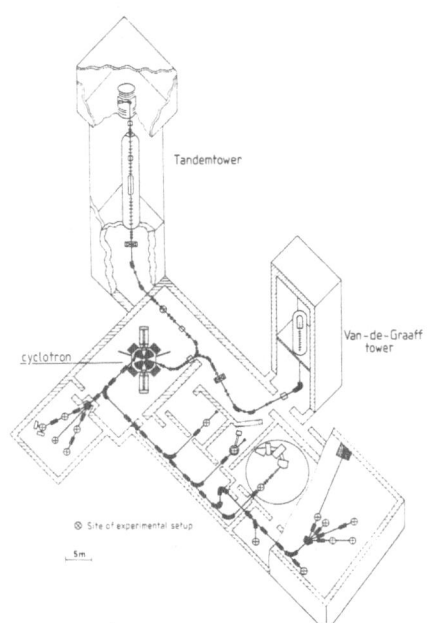

All the existing equipment is interfaced to a computer assisted control system which provides access to any parameter for both, the operator and the automatic setting up or tuning procedures.

Fig. 1:
Layout of the VICKSI facility with 6 MV type CN Van-de-Graaff and 8 MV Pelletron Tandem injectors

VICKSI Control System

1. The VICKSI Control System Features

The control system is based on a PDP-11/44 (Digital Equipment Corp.) as control computer and the parallel and serial CAMAC system standards as control interface. The operator consoles and several serial CAMAC loops are connected to the control computer via a parallel CAMAC highway. The serial loops run each along part of the accelerator system and the beam lines where the machine components are almost evenly distributed. Fig. 2 gives a schematic view of the control system hardware.

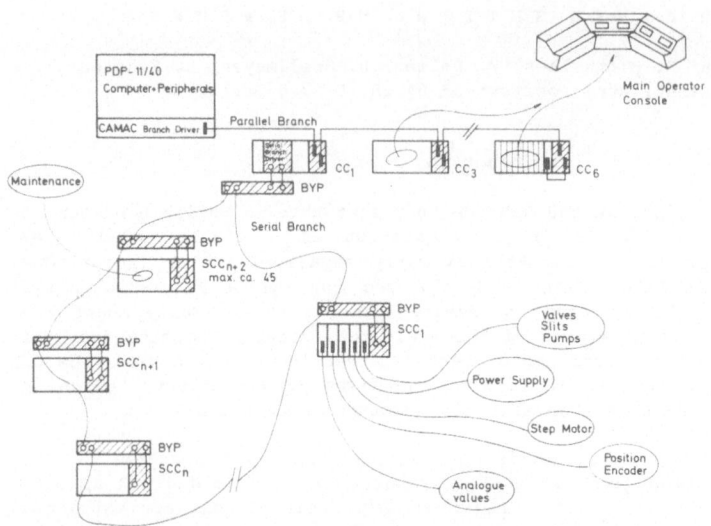

Fig. 2: Schematic view of the VICKSI control system

As detailed descriptions of the design concept and control philosophy of the VICKSI control system have been given in earlier publications (5-8) only the main features will be repeated here.

The interface hardware and software are strictly standardized. The hardware is commercially available or has been made commercially available by tendering with exact specifications of the requirements. The specifications are based on interfacing standards taking into account hardware and software design aspects. A comprehensive operator interface is provided which is easy to use and which reflects site and operating aspects. Touch panels lead the operator to the various subsystems, to beam line sections and to the final set of distinct device or parameter names, for which the specific control variety is displayed on the final 'service page'. This also applies to control procedures or to groups of parameters which may be controlled simultaneously to facilitate the overall operation.

To be able to standardize the software all system information is held in a central data base with entries for each accelerator parameter. All accelerator parameters are identified by their physical names in engineering or physical units to simplify the study and the discussion of beam properties and behaviour. These names may be directly used when accessing the system or programming procedures with the help of an interpreter (9). The interpreter handles these names as predeclared variables (System Variables) like any other variable but implying automatic process access in acquisition and control depending on the syntactical context.

2. VICKSI Control Philosophy

The accelerator and beam line components are manufactured and aligned to such a precision that initial setting values for a given beam (particle, final energy, beam current and quality) may be precalculated by beam transport programs to allow for almost optimum transmission.

The initial setting values are stored in a file for access by various setting programs which run on operator request for adequate subsystems or beam line sections. Among these are complex startup procedures performing all the necessary operations in cases where systems (e.g. RF-systems) require special attention.
The beam is then optimized either by operator interaction on a single or grouped parameter basis or by automatic procedures which measure the beam properties, compute and execute the necessary corrections to the initial setting. If the latter case requires an iterative process each step of the iteration is initiated by the

operator taking his decisions according to the acquired beam quality.

Operating experience shows that this procedure is at present the quickest and most effective for beam development. Storing and recalling the setting data of a given beam once it has been developed does not deliver results of comparable quality. This is mainly due to the insensitivity of the beam diagnostic system in the case of low beam currents and the large number of parameters influencing the final beam properties. Quite some development work remains to be done in this area.

Setting and Tuning Procedures

In the following the setting procedures are described in more detail according to the general and most common way of operation:
- beam parameter and initial cyclotron setting values   - PARSET
- injection beam line setting   - DIP, OPTIN
- extraction beam line setting   - OPTEX
- setting of RF-systems   - HF

Fig. 3 may be used for orientation.

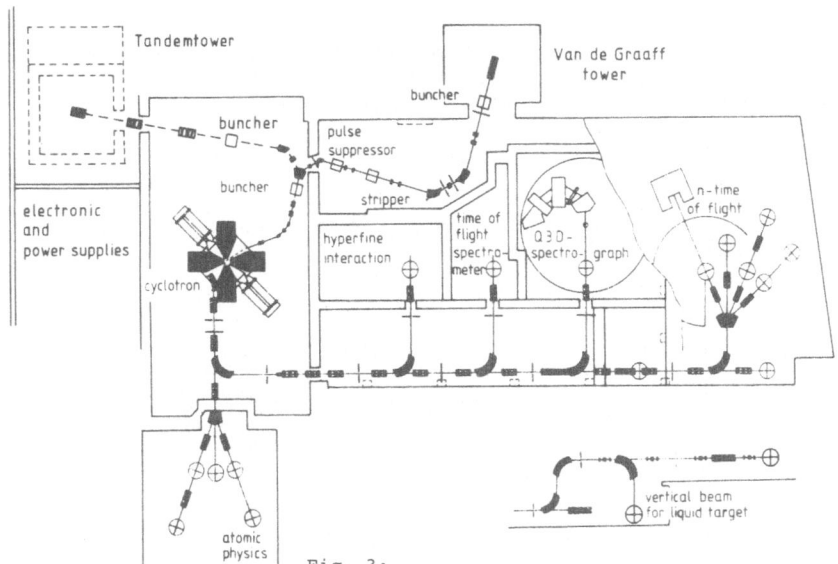

Fig. 3:
Layout of the VICKSI injection and extraction beam lines

1. Ion Beam Parameters, Cyclotron Settings

Each time a new accelerator set up is required for a beam requested by an experiment the Fortran code PARSET is run to determine the ion beam parameters and the initial setting of the whole transport system including the cyclotron. The output values are stored in a file on disk to be available for the various setting procedures.

In general the operator may transfer these values to the 'theoretical setting buffer' in the above mentioned data base of accelerator parameters. The initial setting of transport elements may then be done section by section, again on operator request. At present, mainly for historical reasons, this is only done for the cyclotron setting. This part of the code is as developed by SCANDITRONIX S. A. and is based on the precision field map measurements done in the cyclotron before it was delivered. The beam lines are still handled by the routines described later.

The basic input parameters to PARSET are the ion species and desired extraction energy. It calculates the necessary ion charge states (before and after the stripper) and the various magnet and RF parameters, based on the field maps of the cyclotron, as well as the parameters for the bunchers. The parameters do also include the output of the beam rigidity at injection and extraction and the Van de Graaff terminal voltage.

2. Setting Up and Tuning the Van de Graaff

The Van de Graaff accelerator still undergoes "manual" beam development, i.e. all beam optical and acceleration parameters are slowly (stepwise) tuned up to their (calculated) setting values one by one always waiting for the beam to stabilize. This is due to the fact that most of the parameters involved are driven by unstabilized power supplies which are mounted in the high voltage terminal of the Van de Graaff.

In addition tuning up the Van de Graaff acceleration voltage to its required value may also include conditioning of the tube if the setting value is far above previous running conditions.

3. Injection Beam Line Settings

The interpreter codes DIP and OPTIN calculate the settings of all optical elements in the injection beam line using the beam parameters given by PARSET and a setting of a well optimized beam by scaling this setting to the required beam rigidity. The setting of the elements is done on request by tuning the dipoles to the exact value of the magnetic field with NMR-probes which are located in the different magnets to acquire the magnetic field values.

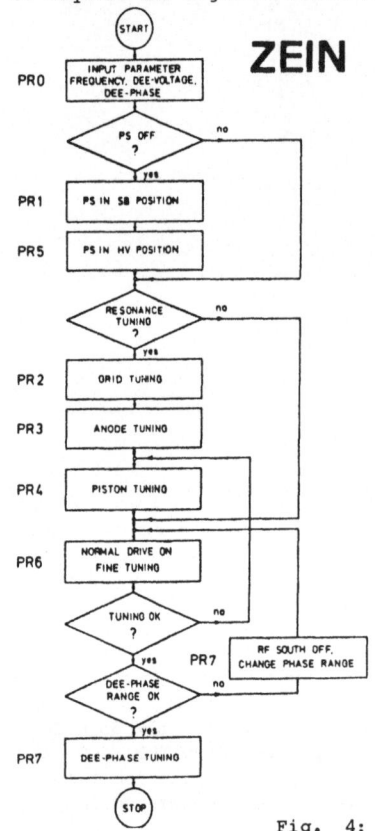

Fig. 4: Flow chart of RF acceleration voltage set up

4. Selection of the Target Area, Extraction Beam Line Setting

The desired target area is selected via the operator's console (touch panel) causing a microprocessor to initiate the switching of the different magnet power supplies and to start the degaussing procedure for inactive dipoles to be passed by the beam. The interpreter code OPTEX calculates the setting of the required elements in the extraction beam line system and sets the elements on request.

5. RF Settings

All necessary operations such as calculation of settings, the switching of power supplies, the tuning of the resonant circuit, the control of amplitude and phase values, the monitoring and logging can be performed by a set of interpreter codes available under the header HF. This code declares the common parameters and lists all available commands. As example fig. 4 shows a flow chart of eight different overlays to set up the two cyclotron RF acceleration voltages using the command ZEIN.

6. Beam Line Tuning

In general the calculated setting of the injection and extraction beam lines is sufficient to set them properly for optimum transmission of the beam. If tuning is necessary, this is done "manually" by the operator either by tuning single parameters or by tuning a set of parameters with

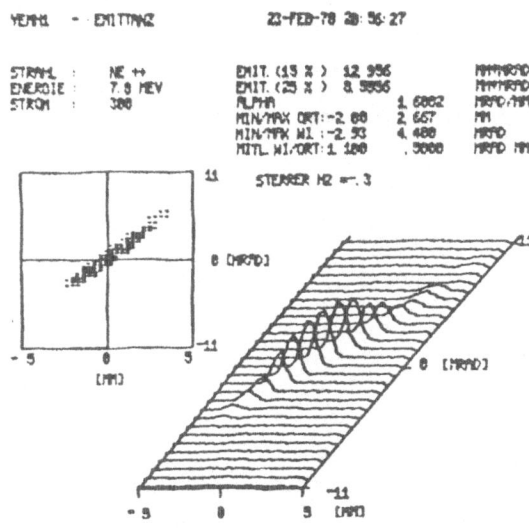

the help of several programs which alleviate the fine tuning. Among these are for example:
- the coupling of injection beam line elements between the stripper and the cyclotron to correct for energy loss of the beam in different stripper foils
- the coupling of the last three quadrupole triplet lenses in front of the cyclotron to optimize the emittance and dispersion matching to the cyclotron
- the coupling of quadrupole triplet lenses to the operator's choice

Emittance measurements can be done after the Van de Graaff and before and after the cyclotron. The emittance measurement devices are driven by a special interpreter program which also evaluates the data and displays the result on a graphic storage scope (see fig. 5).

Fig. 5: Example of an emittance display

7. The Cyclotron Tuning

The injection of the beam into the cyclotron onto the first turn is done by the operator without any special computer assistance. However, to extract the beam with high energy resolution and good stability, isochronisation of the particle revolution frequency with the frequency of the acceleration voltages and good centering of the internal beam must be achieved. Various interpreter programs as part of the main code EVA (10) are available for this purpose.

Isochronisation:
The code dedicated to this task is called PHOP. Using a correction matrix from PARSET and a phase measurement of the beam versus the RF reference at ten different radii the necessary trim coil current changes can be calculated to optimize the magnetic field shape for constant revolution frequency. The correction setting is done automatically if the values appear to be acceptable. In general 2 to 5 iterations are sufficient to minimze the phase deviation.

Centering:
After injection the beam is usually not on the equilibrium orbit. Therefore centering has to be achieved by creating a radially localized first harmonic field distortion ("field bump") by means of current changes in two harmonic trim coils. Several programs are involved in this procedure. The following list illustrates the sequence of subsequent steps:
- Run the radial differential probe to acquire the present turn pattern (code IRPSAV initiating a microprocessor to acquire the data and transferring them to a disk file for evaluation)
- Analyse the pattern and calculate the turn separation (codes AUTOTURN, DRPL, SAVENULL)
- Change the field bump and remeasure the turn separation (codes SAVESIN, SAVECOS and others)
- Calculate, from the resulting turn separations of the different field bumps, the trim coil settings which center the particle turns (code AUTOCENT).

The criterion for a well centered beam is a smooth curve without oscillations in the graph of the turn separation versus radius. An example is given in fig. 6.

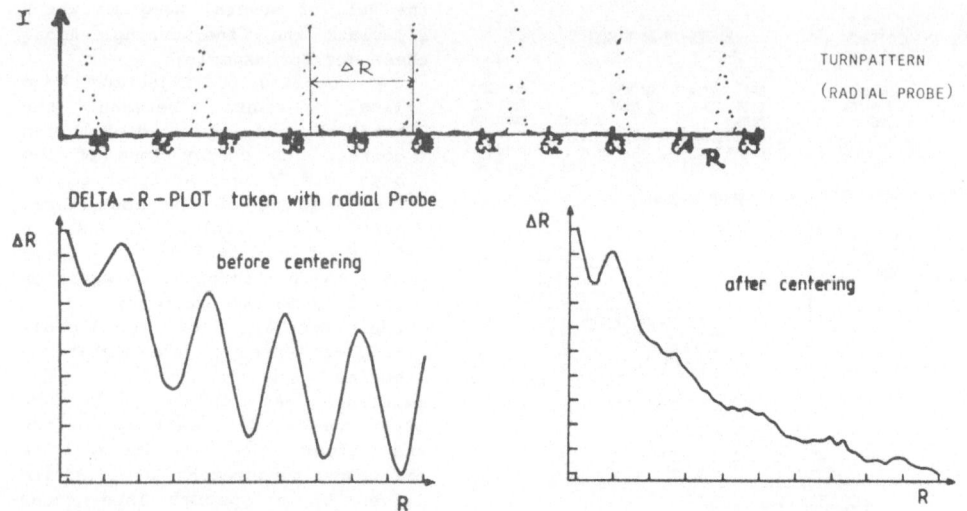

Fig. 6: Turn pattern delta-R-plot taken with the radial differential probe

Extraction:
The optimisation of the extraction parameters is done by varying the first harmonic component of the magnetic field in the extraction region by adjusting the current in the outermost harmonic trim coil. Phase and amplitude of this first harmonic field distortion are available as control system variables and can as such be assigned to knobs on the control desk. Changing the setting values of these variables automatically implies the setting change of the corresponding trim coil power supplies.

Conclusion

In general beam development following the described setting and tuning procedures takes about 8 hours including hardware changes which must be done locally by the operator (as e.g. coil changes in the bunchers). It may be possible to cut down this time by still improving the computer assistance, by installing a still more sensitive beam diagnostic system which might allow for automatic beam line tuning and last not least by replacing some of the hardware.

With respect to computer assistance, the introduction of further system variables for the coupled operation of beam line elements can replace the more tedious running of corresponding computer programs. Further automatic starting up and tuning procedures may also help in gaining time.

Experience has shown, however, that always a certain amount of time is used to keep the accelerator hardware in shape to cope with the requirements of automatic control.

References

(1) K. H. Maier, Proc. 7th Int. Conf. on Cyclotrons and Their Applications (Birkhäuser, Basel, 1975) p. 68
(2) VICKSI Collaboration, IEEE Trans. on Nucl. Sci., Vol. NS-24, No.3 (1977) 1159
(3) K. Ziegler, IEEE Trans. on Nucl. Sci., Vol. NS-26, No.2 (1979) 1872
(4) VICKSI Group, IEEE Trans. on Nucl. Sci., Vol. NS-26, No.3 (1979) 3671
(5) W. Busse and H. Kluge, IEEE Trans. on Nucl. Sci., Vol. NS-22, No.3 (1975) 1109
(6) W. Busse and H. Kluge, Proc. 7th Int. Conf. on Cyclotron and Their Applications (Birkhäuser, Basel, 1975) 557
(7) W. Busse, IEEE Trans. on Nucl. Sci., Vol. NS-26., No.2 (1979) 2300
(8) W. Busse and H. Kluge, IEEE Trans. on Nucl. Sci., Vol. NS-26, No.3 (1979) 3401
(9) W. Busse and K. H. Degenhardt, HMI Report B-251 (1978)
(10) G. Hinderer, IEEE Trans. on Nucl. Sci., Vol. NS-26, No.2 (1979) 2355

GANIL BEAM SETTING METHODS USING ON-LINE
COMPUTER CODES

OPERATION GROUP AND COMPUTER CONTROL GROUP

GANIL. BP 5027. 14021 CAEN-CEDEX. FRANCE

ABSTRACT : Two families of routines have been developed to help operation group.

The first family using off-line pre-calculated parameters (**PARAM Code**) sets all the current values in the Separated Sector Cyclotron (SSC) magnets (main-coils and trim-coils) and in the transfer lines (bending-magnets and quadrupole magnets). In this family are included the automatic RF voltage and RF phase control processors as well as the magnet sectors balancing routine.

The second routine family which requires the beam acceleration makes easy the fine beam parameters adjustments specially into two SSC. The essential automatic on-line computered parameters are : injection beam phase, RF resonators and RF buncher phase, isochronization with use of beam central phase probes, ejection turn centering.

It is explained how independent processors are used to perform local tasks and how various independent beam, electrical and mechanical sensors are associated to the main computer for on-line operation.

I . THE GANIL CONTROL SYSTEM (1,2)

It consists of three types of processors, linked one to the other via CAMAC. First, there are two MITRA125 minicomputers, one being used on line and the other for back up and developments. Second, there are 15 JCAM10 microprocessors, 6 of them taking care of the consoles, the 9 others being devoted to special subsystems like RF cavities, phasing mechanisms, and beam data collection. Third come 15 programmable controllers mainly in charge of the vacuum system ; every request from a console is sent to this computer ; in turn, it sends the relevant CAMAC order to perform what has been asked for either directly to the equipments, or to a JCAM10 which will take over.

At the present time 1200 equipments (motors, power supplies...) can be controlled by the computer system. Each equipment is known by its "operational name" which is chosen so as to be easily memorised by an operator. Moreover, the operator has never to type such a name (except when writing a program); taking benefit of a large number or touch panels, the console proposes sets of names between which the operator chooses with a single finger ; the console has no alphanumerical keyboard.

Consoles can be used at two levels. At the elementary level one deals with equipments on the basis of one hand - one equipment. At the more sophisticated level one runs programs (called via touch-panels), selecting options in the program by running a cursor on a TV screen. All the programs are written in LTR, a real time structured language. It proved to be very convenient not only for application programs, but also for the layer of software which had to be added to the monitor to take care of the CAMAC : Look-At-Me treatment, equipment handlers and data base, data links management.

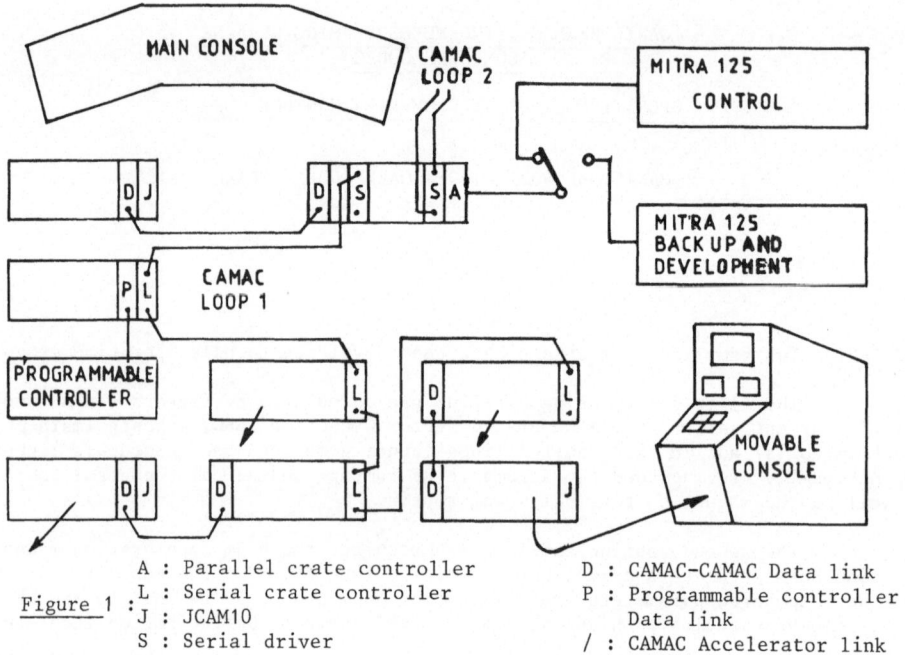

```
              A : Parallel crate controller    D : CAMAC-CAMAC Data link
Figure 1 :    L : Serial crate controller     P : Programmable controller
              J : JCAM10                          Data link
              S : Serial driver              / : CAMAC Accelerator link
```

II . INITIAL SETTINGS

II.1. Calculation of initial values of the parameters (PARAM Code) :

A fortran code (PARAM) computes all the parameters necessary to start the machine. This code is at the present time operational for 3 cyclotrons (the injector I_1 and the two Separated Sector Cyclotrons). Calculations for the beam lines will be included later. Results are stored on a disk file for use by the on-line tasks.

For a given particle (charges Q1 at the source, Q2 after stripping, atomic mass M, RF frequency, harmonic numbers) the program computes a large number of parameters :

a. general parameters such as energy, Bρ at the extraction of each cyclotron, stripping efficiency etc.

b. for the injector I_1
- RF and magnetic parameters by interpolation between magnetic field maps (27 magnetic field levels obtained with DC currents ranging from 500A to 1900A).

c. RF voltage for the buncher R1

d. for each SSC
- RF parameters : voltage and phase deviation between the two cavities (zero degree on even harmonic number, 180 degrees on odd harmonic number).
- magnetic parameters :
  . the required field level on the sector axis at the reference radius R_{axe} = 2.330m and the associated currents for the main and auxilliary coils.
  . the remaining injection perturbation δB_{cor} (R) on each sector at the required level by interpolating between the six reference levels data, and then using a least mean square method and the appropriate matrix TC (N), the currents in the "nose" coils for each sector.
  . the currents for isochronism coils using a very efficient method (3,4). This method requires very few memory for its data, the code being itself very fast and small.
- injection and ejection parameters (not yet included).

c. beams characteristics such as energy spread, emittances, theorical magnetic field along the sector axis etc...

The small size of the data table (only about 15000 reals for the three cyclotrons) and the short running time are very useful and allow this code,at present running on UNIVAC computer, to be executed on the GANIL computer.

II.2.Methods of setting :

All the individual equipments ($\cong$ 1200) are designated by an operational name and can be controlled one by one with a shaft encoder. In order to reduce the work of the operators, specific user tasks have been written to control a group of equipments of the same family (for instance power supplies).

Sophisticated systems which are controlled by microprocessor are managed only by user tasks (for instance RF system).

a.power supplies : about 160 regulated current power supplies for magnets and quadrupoles (not including those of the experimental area) are automatically set up by different tasks. Interactive facilities allow the control to be exercised section by section and relevant messages appear on the TV screen for operation guidance.

The magnetic field in the SSC main magnet is first cycled under the control of a microprocessor. The other power supplies are set up only when the final main coil current is reached.

b.RF system : five RF systems are controlled by three local microprocessors (one by cyclotron). The RF phases adjustment system centralized for all resonators is controlled by a dedicated microprocessor. All RF parmeters (Dee voltage, RF phase value), RF starting and stopping processing and status are controlled by user main computer tasks. These systems are similar to the beam phase measurement system described in another paper presented in this conference /5/.

c.magnet sectors balancing routine : four Hall probes movable with a tight arm through the yokes are used to measure the field level on the four sector axis. Routine EQUSEC controls the probes. It reads the four values of magnetic field at the reference radius 2.330m, computes and sets the current in each auxiliary coils to balance or to reach the field level introduced in data for each sector.

In SSC2, the field along the sector axis can also be measured using integrated signal of a moving coil. The two gaussmeters are on line controlled, the accuracy of the balancing or desired field being ± 1 gauss.

The Hall gaussmeter uses a 6809 microprocessor to interpolate in calibration table, and adequate field value is directly sent to the main computer.

III . SSCs ON-LINE TUNING

Major procedures using beam diagnostics have been carried out for automatic tuning of the SSCs.

III.1.Determination of the cavity RF phase (PHAREG Code)

Using the method described in /5/ the code PHAREG computes for each SSC the cavity RF phase according to the buncher phase and the beam absolute phases measured in different points along the beam line.

The ± 10 degrees accuracy is good enough to tune the injection system and to accelerate the beam until the extraction.

III.2.Isochronization (ISOGRO Code)

Starting from precalculated values the magnetic field is then automatically isochronized via the beam phase history measured by 15 capacitive probes located in the SSC valley. An on-line code (ISOGRO) developped for this purpose is described in details in /5/.

Figure 2 gives the results obtained for the first beam accelerated on SSC1 (Ar $^{+4}$ at 3.4 MeV/A). Remarks about this figure are :

- the beam was accelerated to extraction radius after one adjustment of the main magnetic field (about 5.10^{-4}).

- a total deviation of $\cong$ 10 degrees of the beam central phase was measured as expected. After two iterations tuning the trim-coil currents with the help of ISOGRO, the magnetic field has been isochronized and the desired beam central phase law fitted within ± 1 degree.

Figure : 2
Beam central phase curves measured after each step of the isochronization procedure.

a) with precalculated trim-coil values (PARAM Code).
b) after one iteration of trim-coil currents with ISOGRO.
c) after a second iteration of the trim-coil currents.

The isochronization process as described above takes about 15 minutes including central phase measurments, calculations and power supply current settings. It has to be done only once for each type of particle and energy, since the final trim-coil current ensure a good reproductibility of the magnetic field.
Similar results have been obtained on SSC2 for Ar^{+16} accelerated at 44MeV/A.

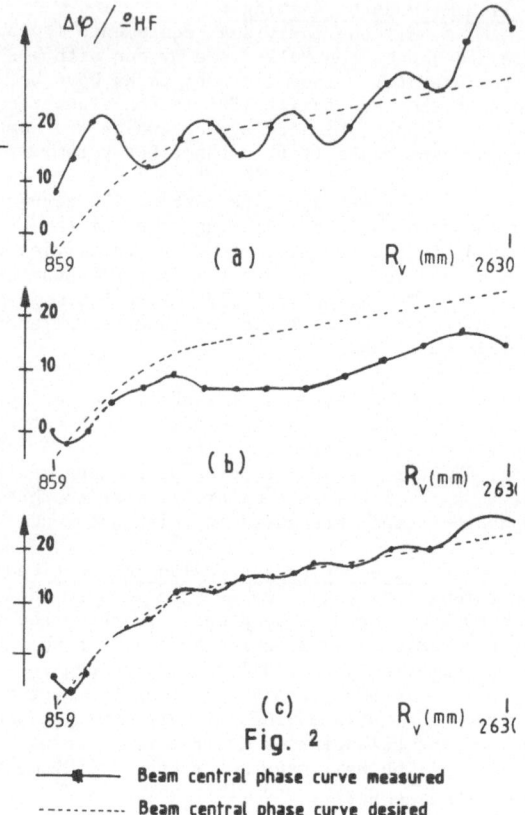

Fig. 2

— Beam central phase curve measured
---------- Beam central phase curve desired

At the beginning of beam experiments on SSC1, acceleration tests have been done without using bunch length compression. This process gives expected results and is now used to operate SSC1. The program can adjust the required phase law for bunch compression as shown on figure 3.

```
*ISOG2  33    MESURE PHASES CENTRALES VALLEE CSS      C1.DIA.PC
31 AOU 1983 **  17HR  49MN              AUTRE ECHELLE?    NON

  - 32.8           *   *   *   *   *   *   *   *   *
  - 36.3.              *   .       .       .       .
  - 39.0       .   .   .           .                       +  273
         .   .   .       .   .           .             .   
                                                      .    +  270
  - 47.0   *                                               
                                                           +  266


  - 60.9   *        Figure 3 : Bunch length compression    +  258
                               phase law

                                                           +  248

                                                           +  245
                                                           +  243
                                                           +  242

                                                           +  236
  - 86.3*                                                  +  234
       --1.0---------1.5---------2.0---------2.5-------RAYON(M)
       1-2--3-----4-------5------6-------7----8---9--10--11-12
```

III.3. Beam centering. (TROPIC + CENTRE Codes)

Field being established and then isochronized following the procedures described in § II.2.c. and III.2, a slight unbalancing between sectors can remain which gives a corresponding beam off-centering. An interactive on line task (TROPIC) allows to move successively the radial differential probes along the axis of the 4 sectors and to obtain the corresponding I(r) values which after treatment lead to the turn characteristics and in particular give the center of gravity of the successive turns along each sector axis. These values are stored and a second task (CENTER) performs the calculations giving the orbit perturbations and the corresponding corrections to be applied on the magnetic field /6/.

Results obtained in the case of a beam of Ar^{+4} accelerated in SSC1 are shown on figure 4.

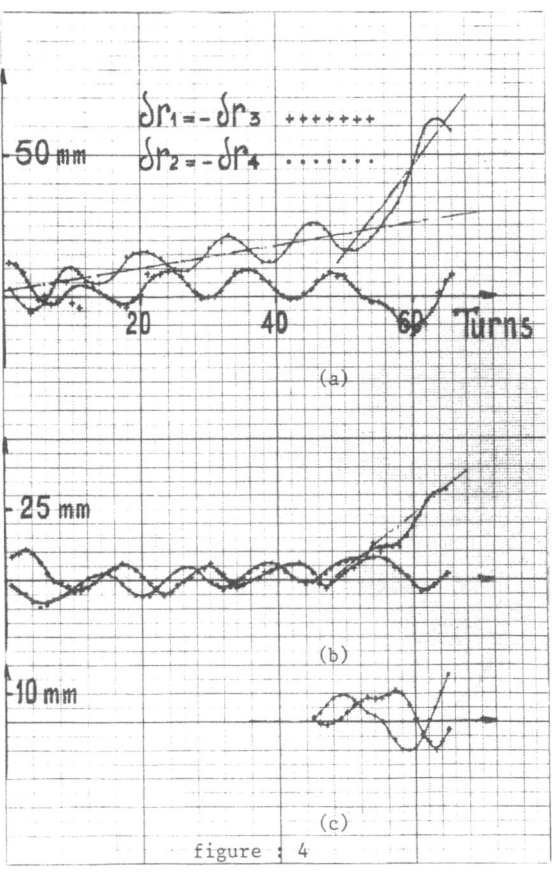

Figure 4 : orbit perturbations calculated by TROPIC and CENTRE codes:

a) without any correction on the initial auxiliary coil currents.
b) after balancing of sectors 2 an 4 (B and D) with auxiliary coil ($\Delta B/B = \pm 1.1 \, 10^{-3}$)
c) after correction of perturbation near ejection by rear coil currents in sectors 2 and 4 (B and D).

The linear increase of the off-centering in the sectors B and D during the 50 first turns corresponds to an unbalancing $\Delta B/B = \pm 1.1 \, 10^{-3}$ between these sectors which is cured by means of the auxiliary coils. A perturbation still remains near extraction. In this case there is no particular perturbation near injection.

Results given by our method are quite satisfatory. The treatment has to be performed for each new case of acceleration.

III.4. Optimization of the RF phase and extraction turn centering (PICNEJ Code)

The RF phase determined by PHAREG code for the two SSCs as described in III.1. must be refined to obtain the best turn separation and the largest possible gain per turn. A code (PICNEJ) moves a probe on 60mm just in front of the ejection electrostatic deflector, measures the current I(R) with a path of 1mm and displays the peaks on the TV screen as shown on figure (5). An interactive dialogue submits different options which allows a slight variation of the RF phase under control of a microprocessor and the code makes a new record. So we find the RF phase which gives the best turn separation and pushes the last turn towards the larger radius. The phase is optimized with an accuracy of 1 or 2 degree by this method.

Then with the same task RF voltage variations are made to center the last turn just on the deflector axis.

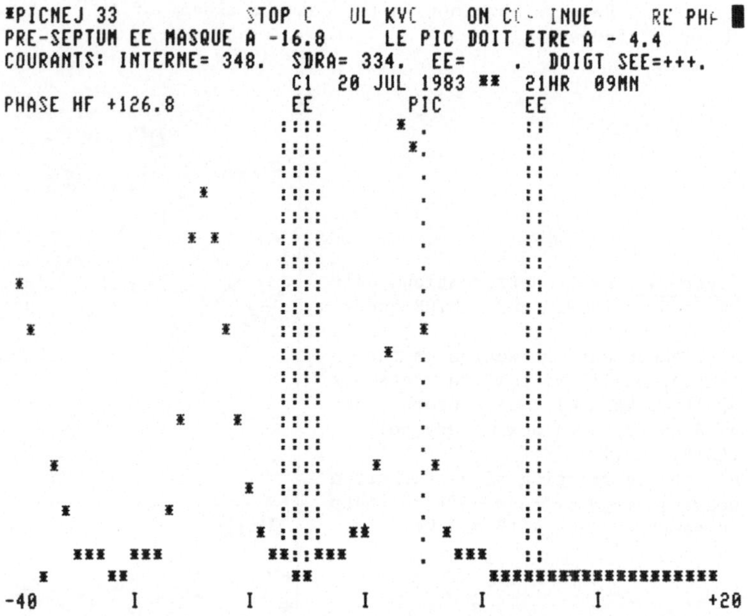

```
*PICNEJ 33            STOP C   UL KYC    ON C( - INUE    RE PHF ■
PRE-SEPTUM EE MASQUE A -16.8      LE PIC DOIT ETRE A - 4.4
COURANTS: INTERNE= 348.  SDRA= 334.  EE=    .  DOIGT SEE=+++.
                         C1  20 JUL 1983 **  21HR  09MN
PHASE HF +126.8          EE         PIC         EE
```

Figure 5 : shows the 3 last turns in SSC1 with PICNEJ code. Every millimeter test point correspond to 1 column on the screen.

(vertical double dash-line draws the deflector window and vertical single dash-line represents the deflector axis).

CONCLUSION

The first beam (Ar^{+4}) from SSC1 was obtained in June 82 and accelerated in SSC2 (Ar^{+16} at 44MeV/A) in November 82. Since these dates user tasks have been performed on the GANIL computer system to optimize the main parameters and to help the operator's job. Now we intend to develop facilities and beam control procedures to increase the reliability of the machine during the operation time. Later all these independent tasks could be gathered to make beam tracking.

REFERENCES

(1) M. PROME " The GANIL control system" IEEE. Transaction on Nuclear Sciences vol NS-28 M 3 June 1981
(2) L.DAVID and al. Proceedings of the Conference "Real Time DATA 82" Versailles - November 1982
(3) M. BARRE and al. "Main results on the SSC's magnetic Field mapping at GANIL" 9th Int. Conf. on Cyclotrons and their Appl. - CAEN - FRANCE -September 1981
(4) A. CHABERT "Isochronization du champ dans les SSC" GANIL Internal Report 81R/071/TP04.
(5) J.M. LOYANT and al. "The computerized beam phase measurement system at GANIL and its Applications" This Conference September 1983.
(6) "Status Report on GANIL" Particle Accelerator Conference .IEEE Transactions on Nuclear Sciences. August 83 . Vol NS 30 Number 4 page 2102.

EUROPEAN ORGANIZATION FOR NUCLEAR RESEARCH

CERN - SPS DIVISION

SPS/AOP/Note/83-8

A MULTI-PROCESSOR, MULTI-TASK CONTROL STRUCTURE
FOR THE CERN SPS

C.Saltmarsh

ABSTRACT

In response to particular operational needs of the part-time use of the CERN SPS as a proton-antiproton collider, a multi-processor, multi-task control structure has been developed.

Known as the "SEQUENCER", it is used to prepare the SPS configuration just before injection and to ensure data collection in cases where single programs prove too cumbersome or too difficult to operate. It controls the execution of existing application programs which are still usable as "stand alone" code.

The preliminary version was coded almost entirely in the SPS high-level language, NODAL, and required very little disturbance to either the single processor systems or to the network.

The structure itself and the user service routines are described, with some comments on lessons learnt and refinements foreseen. A properly configured system will require more low-level restructuring.

GENEVA, August 1983

INTRODUCTION

The increasing demands on the running time of the CERN Super Proton Synchrotron – as a fixed target physics machine, a pp collider, soon as an electron-positron injector for LEP and always as a test bed for the machine physicists – require that the operation group be able to put the machine into different well understood states as quickly and reliably as possible. A few examples illustrate this point:

1. ### PROTON/ANTIPROTON INJECTION

    In collider operation, energy matching of the CPS and SPS and setting up of the antiproton transfer is done by extracting protons back to the CPS via the antiproton transfer tunnel. Moving from one mode to the other involves changing the electrostatic septum polarity and adjusting timings for the RF, the inflector and position monitors. These operations should be done in a given order to facilitate consistent software images of the machine and thus avoid confusing alarm reports.

2. ### ANTIPROTON INJECTION

    Before antiproton injection, a large amount of hardware must be checked for correct operation. Data acquisition hardware must be set up and armed and acquisition software used in normal operation must be stopped. Immediately after successful injection, the main magnet power supplies must be told to stay at coasting energy, various hardware must be switched to a safe state and data acquired at injection must be taken from local temporary storage and saved for future reference. Software for the acquisition of coasting beam data must be launched. Once the machine is in coast, final tuning and checks must be made before data taking by the physics experiments can start.

3. ### UPDATING OF THE "MASTER" DATA BASE

    A primary operational date base, called the MASTER FILE [1], exists to define the machine state for the current operational requirements. (The actual state may not be the same, although it should be a subset of that defined by the MASTER file). Editing facilities are used to generate future master files as operational requirements change.

At the time of such a change, the operation team will update the master file from the previously prepared "NEXT" file. For small changes, this is all that is required, as the majority of software refers to the master file for operational parameters. Thus alarm messages are generated with reference to the master data-base - irrelevant alarms do not appear whereas emergent faults will be reported according to the change of master file.

However, larger changes in the machine definition - changes of cycle length or flat top duration, for instance - imply re-configuration of both hardware and software; some of these changes are by no means obvious and many are interconnected.

The expertise necessary to carry out operations or "sequences" such as described above is distributed among the SPS groups. In particular, software control of the equipment of many hardware groups - RF, Beam monitoring and so on - must be executed under the control of a structure designed by the operations group.

The environment under which these operations are to be carried out is the SPS control network plus some specialised central computers in the control room, all connected through a few message handling computers to a star-configuration switching network.[2] Previous operations of this sort have been done by large programs with full control localised in the central console computer, and using the remote execution possibilities built into the SPS NODAL control language. Problems can arise from this approach: typically, one programmer must cope with a wide range of hardware; the program is vulnerable to small faults which should not stop the whole of the operation; and the program structure is confused by the necessity for complex book-keeping.

THE SEQUENCER

It was decided to construct a control structure to allow these sequences to be defined and implemented by the appropriate experts whilst freeing them from as much of the overall control of software as possible - a sort of multi-processor Job Control Language. Certain guidelines were recognized:

1. The separate "tasks" going to make up a sequence should have little or no reference to the formal sequence structure. These tasks are more or less complex, and may be used, perhaps with different input parameters, in more than one sequence. Where relevant they should give sensible results if run outside the sequence structure.

2. The overall sequence structure - the logical connection between tasks - should be definable in a simple fashion and this structure should be available for easy inspection and modification.

3. It should always be possible for operators to drive a sequence by hand - to bypass faults for instance, or inhibit task execution until faults outside the sequence structure are cleared.

4. Real-time reporting of the status of a sequence should be made by the control software and should be available to individual task programmers.

In addition, two implementation details were deemed important.

5. Parallel processing should be used as much as possible, and traffic over the links minimised. This meant that polling was not used to drive the sequence - for instance, the message "RUN TASK A" is generated when the conditions for running are satisfied, by the CPU that completes these conditions, and is sent to the CPU which has responsibility for TASK A. In turn, this mean that there is no single "MASTER" CPU, although at a given time the whole sequence may be dependent upon a single computer's actions.

6. The project was experimental but could not be satisfactorily tested except under real operating conditions. Thus, for speed of implementation and to avoid expenditure of effort on a possible white elephant, NODAL, the SPS interpretor, was used for almost all coding. Execution speed requirements and space limitations meant that some of the coding is far from elegant, but NODAL coped well enough to encourage further work.

Rather than attempt a formal description of the sequencer, I shall describe its use in the implementation and running of the (hypothetical) sequence illustrated in fig.1. Here, in each of four computers a checking task is run to verify that software and hardware are correctly configured for this sequence. On a satisfactory report from this program in computers 1 and 2, a setting routine loads hardware controlled by 2. A hardware fault in this task will provoke a full reset of that hardware, otherwise the reset task is skipped. In computer 3, a successful check routine will trigger the running of data gathering software, but only when an operator gives authorisation. Finally, hardware on computer 4 is activated when all preparation reports success.

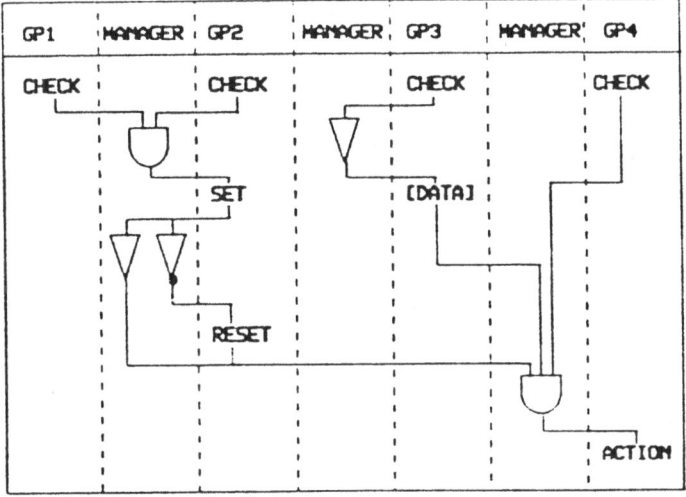

"A hypothetical sequence using four computers to control
equipment after preparatory hardware checks and data gathering."

Having decided that this is the sequence of operations needed, the user responsible for the sequence definition uses an editor to create a description file for each task:

```
TASK     : (20) <123> RESET
DESCRIPTOR: FULL RESET GP2 HARDWARE
RUN IN GP2
EXEC OF TASK 5 FAULT
IGNORE IF TASK 5 OK
START BLOCK 20
PRIORITY 0
OPERATOR AUTORISATION: NO
SET ARGS = 1,3
OWNER    : D.SPART
```

This example details:

- The computer number, disk index and filename of the program.

- A description string

- The computer in which the task is to be run

- The condition for running the task (if task 5 - SET - reports a fault)

- The condition for ignoring the task. That is, task 5 giving the status "OK" will cause a status "OK" to be given for RESET without reference to the program RESET itself.

- Program RESET will be started from block 20 - i.e coding in lines 1.01 to 19.99 will be skipped -

- A priority level is set low.

- The task can run without waiting for operator intervention.

- The first two standard input arguments are set to 1 and 3 (there are 16 floating-point arguments).

- The program RESET is the responsibility of D. SPART.

These descriptions files are then treated by two processes somewhat similar to compilation and link-editing. The result is a series of disk files, one for each computer used in the sequence, containing the description file information in a compressed and reorganised format. The most important reorganisation translates statements such as "EXEC IF TASK 5 FAULT" into an instruction to computer 2 to set a given trigger bit associated with RESET when task 5 reports a fault. Comparison of the resulting trigger bit pattern with a locally stored reference pattern shows if the conditions for running RESET are fulfilled or not.

The message is translated from the input format, implying that the target task asks if the conditions are right for starting, into the systems format. The system stores each component of a trigger condition until it is fulfilled, and only then is the target task contracted.

The coding of individual tasks is now completed and tested by those responsible.

Various facilities are offered:

- By allowing different starting points, a given file can contain code for different parts of the same task, or similar tasks for different computers, with entry points defined in the sequence description.

- Access is given to the standard arguments mentioned above; either private arguments for a given sequence reference or global arguments defined at sequence run-time

Each task reports to the local sequencer management by simple statements like:

SET SEQ(#ST1) = #OK (summary status is "OK")

SET SEQ(#ST1) = #FLT (summary status is "FAULT")

SET SEQ(#ST1) = #RER (the task may be restarted at a new entry point,
               after a demanded elapsed time during which the CPU can be used for
               other tasks).

Task reporting to the operator can be made by ASCII strings sent to a real-time status screen, or to a disk file, or to both.

Test software exists for running single tasks, for de-linking the output files of the "edit-compile-link" processes, and so on.

Finally, the sequence can be run. A short initialisation program is run in one of the central SPS console computers. This sets up initial arguments and loads the "edit-compile-link" output files for the required sequence into data areas in the relevent computers. A management program is then started in each of these computers and the sequence is launched, with the console now free of sequence control and available for monitoring the operation.

The real time report screen shows, on one line for each active computer, the actions of the management program as it starts tasks and reports their finishing status. User-generated messages may also be shown here, and as each computer runs out of work the management program reports END on the appropriate line. Additionally, the operator may acquire the status of all programs at any time, and may act on the sequence by re.running tasks, ignoring faults, inhibiting or authorising tasks and so on.

All control and reporting of the sequence tasks is done by a single management program, a copy of which is kept in each active computer's memory. This program runs at the same priority as the tasks themselves and only when necessary. The exact CPU in which the decision to trigger a given task is completed is not in general known when the sequence is constructed. It is dependent upon run-time execution details. This is symbolised in fig. 1 by putting the logical connection symbols outside the task lists.

PRESENT USE AND FUTURE DEVELOPMENTS

At the present time, there exist five major sequences of 20 to 70 tasks which were used during the two major collider operational periods.These sequencers carried out all the main operational changes such as example [1] above. It controlled the preparation and final countdown to antiproton injection. Although a large minority of countdowns contained one or more individual task faults or crashes, no antiprotons were lost due to sequence failure; showing that the integrity of the software structure is not threatened by the majority of hardware faults or task module bugs. The sequences were continually modified as new requirements were recognized and logical faults in the definition of the sequence showed themselves.

The application of the sequencer to fixed target operation is now being considered – in particular example 3) above. This is in fact a more complex problem – the fixed target machine is more complicated, and there is a huge amount of existing software to be brought under sequencer control.

For the future, there are a number of facilities which could be added to the existing SPS computer systems to make the sequence easier to use and more comprehensive in the structural definitions possible. Finally, however, more control is needed at the system level to configure and control suites of task modules driven by program-program data communication rather than operator intervention – echoes of the "processes" and "pipes" of UNIX.

ACKNOWLEDGEMENTS

Thanks are due to many colleagues in the SPS Division for help and arguement, and in particular A. Ferrari, R. Hopkins, R. Lauckner and D. Thomas for more concrete aid in coding and implementing the sequence.

REFERENCES:

1) R. Lauckner, SPS/AOP/Note/82-9
2) M.C. Crowley-Milling, The design of the control system for the SPS
   CERN 75-20

COMPUTER CODES FOR AUTOMATIC TUNING OF THE BEAM TRANSPORT AT THE UNILAC

L. Dahl and A. Ehrich
GSI, Gesellschaft für Schwerionenforschung mbh
D-6100 Darmstadt / Fed. Rep. of Germany

Abstract

For application in routine operation fully automatic computer controlled algorithms are developed for tuning of beam transport elements at the Unilac.
Computations, based on emittance measurements, simulate the beam behaviour and evaluate quadrupole settings, in order to produce defined beam properties at specified positions along the accelerator. The interactive program is controlled using a graphic display on which the beam emittances and envelopes are plotted.
To align the beam onto the ion-optical axis of the accelerator two automatic computer controlled procedures have been developed. The misalignment of the beam is determined by variation of quadrupole or steering magnet settings with simultaneous measurement of the beam distribution on profile grids. According to the result a pair of steering magnet settings are adjusted to bend the beam on the axis.
The effects of computer controlled tuning on beam quality and operation are reported.

Introduction

Numerous changes of ions accelerated by the Unilac to energies between 3 and 20 MeV/u as well as changes of the experiment stations demand computer-aided operating procedures to improve the efficiency of the Unilac. In addition these algorithms preserve the transverse beam properties and transmission by reproducible tuning of the beam emittance, e.g. in case of changing the ion source within an experiment run, variations of ion source parameters, time depending influence of the stripper foil or instability of magnet currents. The automatic fitting codes also enable to adjust dispersive, achromatic or isochronous beam transport, corresponding to the experimenters requirements on beam quality.

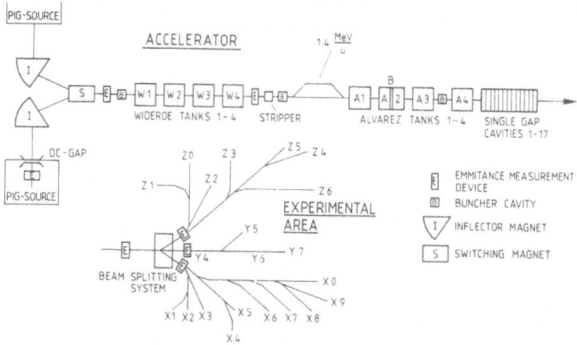

Fig. 1: Schematic diagram of the Unilac.

Fig. 1 gives a view of the Unilac in order to explain the operating strategy. Emittance measurement devices mark the beginnings of the tuning sections. The idea is to fit the beam parameters as closely as possible onto the acceptance of the successive beam transport line or accelerator tanks. Then nominal currents in the quadrupoles of these subsystems should lead to maximum transmission and to the desired beam quality. The successful application of the emittance matching code is based on the assumption that the beam is aligned onto the ion optical axis of the Unilac. For this purpose two complex algorithms for steering corrections have been developed. Alignment sections are related with every emittance measurement device. Additional alignment sections are adapted in the post stripper and experimental area.

In the following the beam emittance matching code and the beam alignment codes for automatic tuning of about 300 beam transport elements at the Unilac are described.[1][2]

Beam Envelope and Emittance Matching

The on-line computer code TSO (Transversale Strahl-Optimierung) evaluates magnet currents from a simulated reference beam transport system, which is build up by a read out of the data base representing the geometrical and physical structure of the selected Unilac section. The input parameters are measured by a computer controlled emittance measurement device. Output values for the beam transport elements are directly transferred to the power supplies on command.

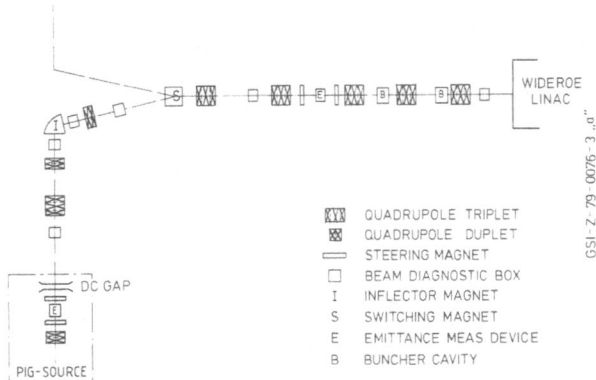

Fig. 2: Layout of the Unilac low energy transport system.

Fig. 2 shows the low energy beam transport system of the Unilac in detail. The first emittance measurement device gives the beam data for fitting the envelope to the mass separator consisting of the inflecting and switching magnet. Another emittance measurement by the second device delivers beam parameters for further matching to the Wideröe tanks acceptance using three quadrupole triplets in between. Adaption of emittance and acceptance leads to maximum transmission through the nominal adjusted preaccelerator.

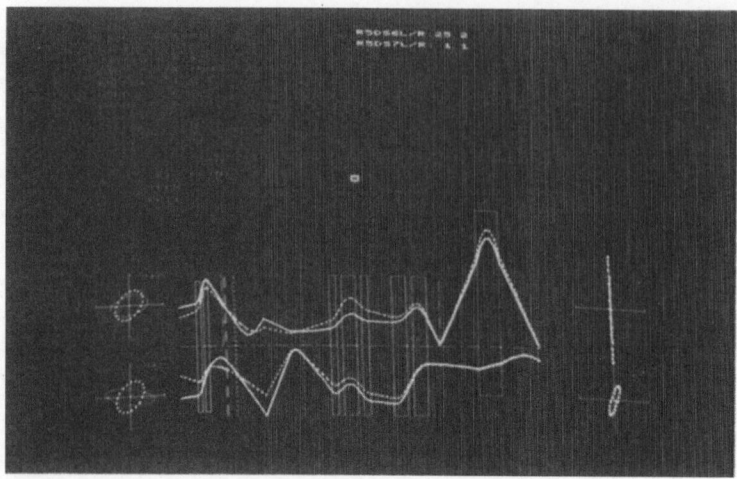

Fig. 3: Computer simulated envelope matching for mass separation.

Fig. 3 is a copy of the screen displayed representation of the fully automatic program. Reference input-, output-emittance and envelope are indicated by dotted lines. Measured input emittance and optimized envelope are drawn in solid lines. Cursor interrupt by the operator executes the setting of the monitored magnet currents as well as the nominal magnet currents of the connected subsystem.

The mathematical problem is to minimize the error function of real and reference parameters of the ion beam:

$$F(p) = \sum_{i=1}^{m} f_i^2(p) = \sum_{i=1}^{m} (r_i(p) - r_i^s(p))^2 = \min$$

$r_i(p), r_i^s(p)$ measured, reference beam properties
$p = (p_1, ..., p_n)$ beam line parameters.

In general the system of equations

$$(p_j - p_j^s) = (A^T \cdot A)^{-1} \cdot A^T \cdot (r_i - r_i^s)$$

has to be solved to match the beam to design values.
A is a functional matrix with partial derivatives $\delta f_i / \delta p_k$
$i = 1,...,m$ and $k = 1,...,n$.
Because $f_i(p)$ are complicated transcendent functions, the functional matrix A is approximated by difference quotients $\Delta f_i / \Delta p_k$.

The computer code TSO uses the "conjugate gradient method" to solve the linearized system of equations in an iterative procedure. A solution is found, if one exists, within N (order of the system) iterations. The reference quadrupole currents are taken as initial guess. The coefficient matrix A may be arbitrary. Because the areas of emittances are constant over the defined matching sections, the number of transverse beam parameters is four. The number of quadrupoles is minimumly four, which means that the system of equations is exact or underdetermined. In the latter one of several solutions is calculated. All computations are carried out with double precision (64 bit).

Different errors affect the success of the tuning procedure, e.g. limited precision of the emittance measurement, deviations in magnetic field calibrations, tolerances in physical and geometrical data. The extensive beam diagnostic system of the Unilac proved to be extremely valuable for a postcorrection of the transport elements.
The computed envelope will be checked by numerous profile grids.[3] While the emittance orientations at the position of the profile grids considered correct, unpermissibly deviating beam widths are transformed back to the entrance of the defined focusing channel. For subsequent treatment they are used as input and form a consistent dataset together with the optimized quadrupole gradients.
The residual error

$$F^*(p) = \sum_{i=1}^{m} (r_i^*(p) - r_i^s(p))^2 = \min$$

is minimized as before. This second matching of the beam transport line usually results in quite rapid convergence of computer simulated and real beam behaviour.[4]

Beam Alignment Algorithms

To prevent restrictions in the computer evaluated adjustment of quadrupole lenses two procedures for automatic beam alignment have been developed.

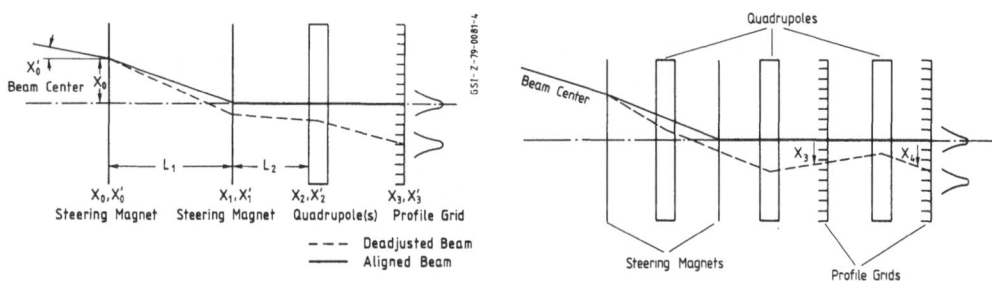

Fig. 4: Alignment section with one profile grid

Fig. 5: Alignment section with two profile grids

The first one is illustrated in Fig. 4. The goal is to determine displacement x_0 and slope x_0' of the beam center at the beginning of the section by measuring the beam center on a profile grid at different quadrupole settings. According to the result a pair of steering magnets are adjusted so that the beam crosses the center of the second steering magnet which bends it onto the accelerator axis.
The beam center is defined as the center of gravity under the current distribution of the profile grid. To calculate the beam center deviation x_0 and slope x_0' at the entrance of the system two different quadrupole settings and profile measurements (n = 1,2) are sufficient by theory.
From the single particle transformation

$$\begin{pmatrix} x_3^{(n)} \\ x_3'^{(n)} \end{pmatrix} = \begin{pmatrix} a_{11}^{(n)} & a_{12}^{(n)} \\ a_{21}^{(n)} & a_{22}^{(n)} \end{pmatrix} \begin{pmatrix} x_2 \\ x_2' \end{pmatrix}$$

which describes the transport of the beam center from the first quadrupole to the profile grid, a relation follows for the beam center parameters at the entrance of the first quadrupole:

$$x_2 = (x_3^{(1)} a_{12}^{(2)} - x_3^{(2)} a_{12}^{(1)})/Z$$
$$x_2' = (x_3^{(2)} a_{11}^{(1)} - x_3^{(1)} a_{11}^{(2)})/Z \quad \text{where} \quad Z = a_{11}^{(2)} a_{12}^{(1)} - a_{11}^{(1)} a_{12}^{(2)} .$$

The changes in transverse divergence induced by the deflecting magnets is described by the known product $k_i I_i$. This enables to transform x_2, x_2' back to the entrance of the section:

$$x_0 = x_2 - x_2'(L_1 + L_2) + k_2 I_2 L_1$$
$$x_0' = x_2' - k_1 I_1 - k_2 I_2 .$$

The condition $x_1 = 0$ delivers the current of the first steering magnet:

$$x_1 = x_0 + L_1 x_0' + k_1 I_1 L_1 \stackrel{!}{=} 0 . \qquad \text{Hence} \qquad \boxed{I_1 = -\frac{x_0 + L_1 x_0'}{k_1 L_1}} .$$

The remaining transverse angle $x_0' = -x_0/L_1$ is compensated by the second steering magnet:

$$\boxed{I_2 = x_0/k_2 I_1} .$$

The second beam alignment option is based on a more empirical method.[5] Displayed in Fig. 5 the alignment arrangement consists of a pair of steering magnets, followed by two profile grids and arbitrary located quadrupoles.

The following linear system of equations describes the steering magnet effects on the beam center displacement at the two profile grids considering also the influence of the quadrupoles:

$$\begin{pmatrix} \frac{\delta x_3^{(1)}}{\delta I_1} & & & 0 \\ & \frac{\delta x_4^{(1)}}{\delta I_1} & & \\ & & \frac{\delta x_3^{(2)}}{\delta I_2} & \\ 0 & & & \frac{\delta x_4^{(2)}}{\delta I_2} \end{pmatrix} \begin{pmatrix} dI_1 \\ dI_1 \\ dI_2 \\ dI_2 \end{pmatrix} = \begin{pmatrix} dx_3^{(1)} \\ dx_4^{(1)} \\ dx_3^{(2)} \\ dx_4^{(2)} \end{pmatrix}$$

dI_n (n = 1,2) variations of the steering magnet currents

$dX_m^{(n)}$ beam center displacement at the profile grid m caused by steering magnet n.

The partial derivatives are approximated by difference quotients which are measured. In order to compensate the deviations x_3 and x_4 of the beam center from the accelerator axis at the two grids, the composed effects of both steering magnets have to be considered:

$$x_3 = x_3^{(1)} + x_3^{(2)} = \Delta I_1^* \frac{\Delta x_3^{(1)}}{\Delta I_1} + \Delta I_2^* \frac{\Delta x_3^{(2)}}{\Delta I_2} , \qquad x_4 = x_4^{(1)} + x_4^{(2)} = \Delta I_1^* \frac{\Delta x_4^{(1)}}{\Delta I_1} + \Delta I_2^* \frac{\Delta x_4^{(2)}}{\Delta I_2} .$$

The difference quotients constitute matrix elements a_{ij} (i=j=1,2). From the two equations the modifications ΔI_1 and ΔI_2 of the actual steering magnet settings necessary to produce an aligned beam, can be found:

$$\Delta I_1^* = (x_3 - x_4 \frac{a_{12}}{a_{22}})/Z_1 \quad \text{where} \quad Z_1 = (a_{11} - \frac{a_{12} a_{21}}{a_{22}}) \neq 0 , \qquad \Delta I_2^* = (x_4 - x_3 \frac{a_{21}}{a_{11}})/Z_2 \quad \text{where} \quad Z_2 = (a_{22} - \frac{a_{12} a_{21}}{a_{11}}) \neq 0 .$$

The slopes x_3' and x_4' of the beam center at the profile grids automatically are zero unless the transfer matrix between the two grids satisfies a special symmetry condition.

Experiences at the Unilac turned out that accumulation and propagation of errors affect the reliability of both alignment procedures in case of only two variations of magnets. Therefore the computer codes execute up to ten variations of quadrupole or steering magnet settings and analyze the measurements by the least-squares-method. Both alignment algorithms are designed as fully automatic computer codes using the same data base as the emittance matching code. Horizontal and vertical plane are optimized simultaneously. Magnet current variations with corresponding observation of the profile grids and setting of the evaluated steering magnet currents are carried out just by pressing a button related to the section to be treated. In case of failure another button permits to reproduce the original state of the transport line adjustment.

Operational Experiences and Planned Improvements

The beam emittance matching code TSO is applied all over the Unilac. In addition to the above mentioned matching channels in the pre-stripper part, another emittance measuring device in front of the stripper permits automatic transverse beam matching onto the gas stripper tube and subsequent adaption for the charge separator and post-accelerator acceptance. The beam splitting system at the high energy end demands different options for emittance matching, depending on the number and positions of experiments running simultaneously. Finally emittance measurement devices in each of the three primary experiment beam lines give the possibility to match the beam achromatically, dispersively or isochronously onto aperture configurations of the experiment.
Beam alignment with one profile grid is mainly used for transport lines, whereas the second algorithm described above is well suited for steering corrections through accelerator tanks because of the arbitrary arrangement of quadrupoles.
All computer codes are simply applicable and reliable. The operator decides on a further treatment or setting the computed values. The procedure takes about three minutes. Also of short duration are the activities of the beam alignment codes, which can be observed by a monitored profile grid measurement.
The programs improved the efficiency of the Unilac tuning procedure as well as the beam quality especially in the experimental area and unburden operators from manual knob rotating and try and error procedure. Because of the reproducibility of the methods quadrupole control knobs of the Wideröe accelerator have been taken out from the operators console. It is planned also to eliminate the Alvarez accelerator quadrupole control knobs. The latter would be a first step to the future time share operation for different energies with pulsed magnets only in the intertank sections.

References

[1] L. Dahl, Computer Aided Tuning Procedures at the Unilac, Ninth International Conference on Cyclotrons and their Applications, Caen 1981.
[2] L. Dahl et al., Longitudinal and Transverse Beam Optimization at the Unilac, Proceedings of the 1979 Linear Accelerator Conference, Montauk, BNL 51134.
[3] V. Schaa et al., Interactive Testprogram for Ion Optics, these proceedings.
[4] W. Kneis, Rechnergesteuerte Strahldiagnostik und Strahloptimierung am Karlsruher Isochronzyklotron, KfK 2835 (1979).
[5] B. Franczak, Einstellung von Steerern, GSI Arbeitsnotiz 220776 (1976).

INTERACTIVE TESTPROGRAM FOR ION OPTICS

V. Schaa, G. Fliss, P. Strehl, J. Struckmeier

GSI, Gesellschaft für Schwerionenforschung mbH D-6100 Darmstadt, Fed. Rep. of Germany

Abstract

For testing of beam transport systems, elaboration of standard data sets, and tuning procedures during accelerator experiments an interactive program has been developed.

The structure of beam transport sections, which have to be studied, are built up automatically by a read-out of data bases representing the accelerator structure.

In on-line mode the calculation of beam envelopes is always based on measured emittances and quadrupole currents. Therefore, consequences of manipulations on beam transport elements or even the usefulness of various models for the mathematical description of quadrupoles are discernible immediately for the accelerator physicist. In order to check the envelope calculation, on-line measured beam profiles, slit positions and beam currents can be shown simultaneously on the same display.

In off-line mode the program offers various features and options for envelope calculation and optimization, which will be described in detail.

Introduction

There are numerous excellent programs like **TRANSPORT, TURTLE** or **MIRKO** for the design and optimization of beam transport systems. However, for the accelerator physicist, confronted with the hardware including all the tolerances, inaccuracies of measured data and imperfections the use of such off-line programs will be a troublesome, time consuming procedure. Taking the complexity of a multiparticle, variable energy machine like the UNILAC into account an on-line program including options as well-known from the programs mentioned above can improve the efficiency of machine studies and accelerator experiments considerably. Therefore at GSI an on-line program (**ENV**elope **TR**ansformation) has been written which offers some interesting features for on-line manipulations on the beam including simultaneous observation of envelopes and other beam parameters.

Structure of program

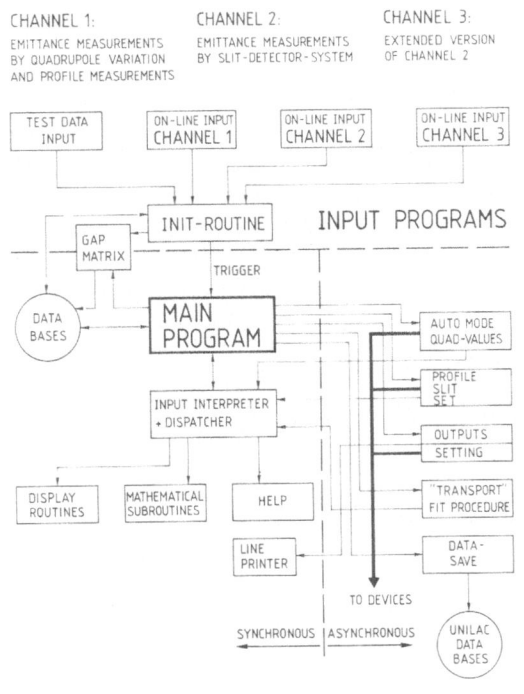

Figure 1

The block diagram in fig.1 shows the organization of the program. The main program may be triggered by the **INIT** routine when emittance data are transmitted from one of the 4 input channels. Channels 1-3 are coupled directly to the on-line emittance measurement systems positioned along the UNILAC. After measurement has been finished a picture as seen in fig.2 will be built up automatically on a large graphic display. The section of the beam transport system displayed will be determined by the position of the emittance measurement system. Type of elements as well as their arrangement within the selected section with respect to drift spaces, apertures and beam diagnostic elements like harps, Faraday cups and slits are determined by read-out of the corresponding UNILAC data bases. In on-line mode the displayed envelopes are based on actual quadrupole settings received from the computer controlled data processing system. As can be seen in fig.2 the complete menu list of all interactive options is also displayed (fig.2 left hand side, top), offering a multitude of possibilities for manipulations on the beam or envelopes, respectively.

Menu

All options of the menu field may be classified into 3 groups:

1. on-line :   (OFFL)   1X L   AUTO   STOP

2. on-line or off-line: (ON L)   (OFFL)   SAVE   PRINT
                          ELLIP   BETA   SET   PROF
                          SLIT   ENABLE

3. off-line :   (ON L)   QUVA   CFIT   SFIT
                     INTG   RESET   ENVD   UPDT
                     KURZ   ORIG   ENABLE

1. ON L-MODE

In the on-line mode all quadrupole settings will be updated every minute or on request (**1X L**) and the new calculated envelopes will be displayed immediately. The refresh time can be shortened to about 1-3 s during the **AUTO**-mode. Therefore the consequences of variations on the quadrupole settings may be observed nearly simultaneously on the screen. There is another more general on-line function **IST/SOLL** which may be helpful to control the data processing system as well as the closed loop control of the power supplies. In **IST**-mode all calculations are based on the actual digitized quadrupole currents measured by a shunt impedance and normalized to 10 volts. In **SOLL**-mode calculations are based on data transmitted to the devices.

2. ON L or OFFL-mode:

The options **SAVE**, **PRINT** and **SET** are self explanatory. Selecting the functions **ELLIP** and moving the cursor to a defined position at the displayed beam center line will result in an emittance plot as shown in fig.2 (insert right hand side, top). For orientation two arrows are shown on the center line:
- ↑ = position for which the emittance plot holds
- ↓ = position where the emittance has been measured.

Normally the β-value {β=v/c; c = velocity of light} is calculated by a read-out of the extraction voltage of the ion source as well as the terminal voltage. Selecting the field **BETA** gives an update by a new read-out. In addition various β-values may be entered from the keyboard. This additional possibility may be very helpful studying effects of particle momentum or even the effect of the accelerating gap in dependence on the terminal voltage.

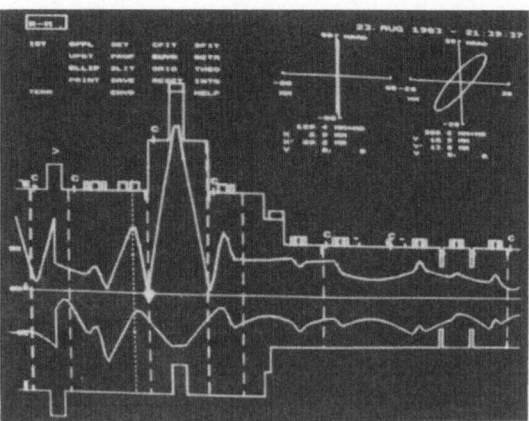

Figure 2

Initializing the functions **PROF** or **SLIT** by cursor results in a measurement of all profiles (**PROF**) and a read-out of slit (**SLIT**) positions in the displayed section. Measured data will be shown on the screen (blinking arrows '←' for all slit-positions and bright stars for profile FWHM-points).

3. OFFL-mode

The function **QUVA** offers two possibilities for envelope manipulations:

a) After selection of a quadrupole by cursor and positioning the cursor within the limits of the transfer lines (comp. fig.3) the quadrupole gradients can be changed with a percentage proportional to the distance of the cursor from the center line. Cursor positions above the center line correspond to changes within +0.25% and +5%, where-as positions below result in decreasing gradients (-0.25% to -5%).
b) The quadrupole gradients can be changed also directly by entering values between 0.005 and 9.998, which correspond to the range of normalized voltages (0-10V).

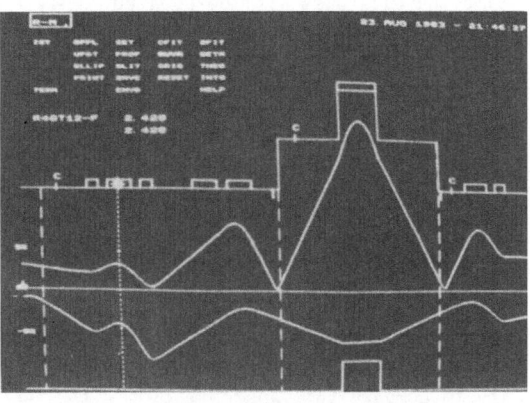

Figure 3: HELP-Display

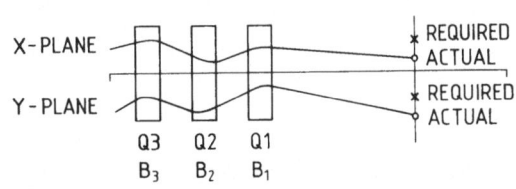

C-FIT

MATHEMATICAL PROCEDURE [EXAMPLE FOR A DUPLET] :

1. STEP: CALCULATION OF

$$R_0 = \begin{pmatrix} REQUIRED_x - f_1(B_1^0, B_2^0) \\ REQUIRED_y - f_2(B_1^0, B_2^0) \end{pmatrix} \quad f_1, f_2 \text{ TRANSFER-FUNCTIONS}$$

2. STEP: CALCULATION OF

$$A = \begin{pmatrix} a_{11} & a_{12} \\ a_{21} & a_{22} \end{pmatrix} \text{ with } \begin{array}{l} a_{11} = [f_1(B_1+\Delta B_1, B_2) - f_1(B_1, B_2)] / \Delta B_1 \\ a_{12} = [f_1(B_1, B_2+\Delta B_2) - f_1(B_1, B_2)] / \Delta B_2 \\ a_{21} = [f_2(B_1+\Delta B_1, B_2) - f_2(B_1, B_2)] / \Delta B_1 \\ a_{22} = [f_2(B_1, B_2+\Delta B_2) - f_2(B_1, B_2)] / \Delta B_2 \end{array}$$

3. STEP: SOLUTION OF

AX = R with X = $(X_1, X_2)^T$ = VECTOR OF CORRECTION

4. STEP: REPLACING B^0 BY

$B^1 = B^0 + \varepsilon X$ with $\varepsilon = /B/ \cdot 0.001 / |X|$

5. STEP: CALCULATION OF

$\chi^2 = |R|^2$

6. STEP: $\varepsilon \rightarrow 2\varepsilon$ IF $\chi_1^2 < \chi_0^2$ ELSE NEXT ITERATION (IT)

7. STEP: END OF ITERATION (IT) IF IT>100 OR $\chi^2<0.01$ OR $\chi_{1T}^2 > \chi_{1T-1}^2$

With the option **CFIT** a mathematical fit procedure will be started, which is explained schematically in fig.4. The position where the envelopes have to be fitted is selected by cursor on the beam center line. After this selection has been done the required values have to be specified by cursor, too. A maximum of 3 lenses will be varied automatically by the program using **CFIT**-option. For optimization of larger sections the option **SFIT** is recommended, using the well-known **TRANSPORT**-algorithm.

Figure 4: CFIT-Algorithm

By using fit procedures (**QUVA, CFIT, SFIT**) the 3 options **RESET, UPDT, ENVD** may be very helpful to return to defined conditions. **RESET** displays the last envelope before a fit procedure has been started. **UPDT** results in a new read-out of the quadrupole gradients and displays the envelopes belonging to the actual values. At the same time the data base will be updated. **ENVD** can be used to return to the last **SAVE** or **SET**-values stored in the data base.

As well-known, programs for design and optimization of beam transport systems mostly use the hard edge model in which the action of a quadrupole lens is described by the product $B' \cdot l_{eff}$ (B' = quadrupole gradient, l_{eff} = effective length of the quadrupole). Selecting the function **INTG** offers the possibility to compare this approximation with numerical integration of measured $B'(s)$-values, which are stored in a special file. Depending on the aspect ratio of the quadrupoles considerable differences can arise. A typical example is shown in fig.5 and 6.

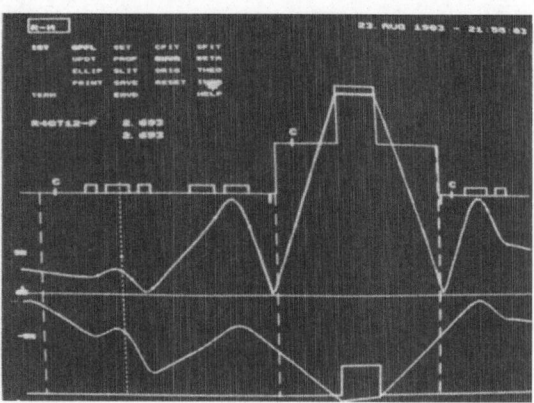

Figure 5: Hard Edge Model

In fig.5 and 6 the option **KURZ** has been used, to demonstrate the ZOOM-function within the program. Only start point and final point have to be selected by cursor to shorten the displayed beam transport section to the specified length. The option **ORIG** (**+ENABLE**) is provided to return to the original display with full length.

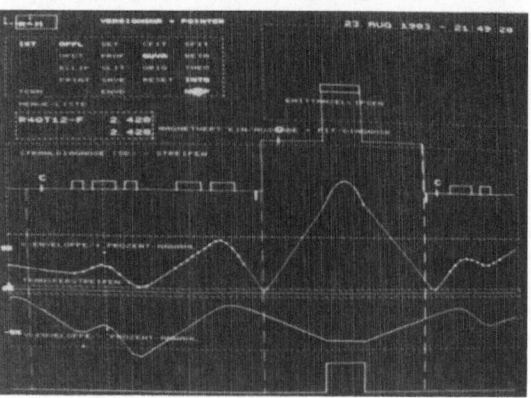

Figure 6: Numerical Integration of $B'(s)$-values

Example of application

For the implementation of computer controlled optimization procedures the correct mathematical description of transfer matrices becomes essential. At the UNILAC the accelerating gap behind the ion source acts like a very strong electrostatic lens. A gap matrix was derived on the base of constructional details using well-known programs for numerical integration of Laplace- and Poisson-equation, respectively. The experimental test of the solution was performed with the program **ENVTR**.

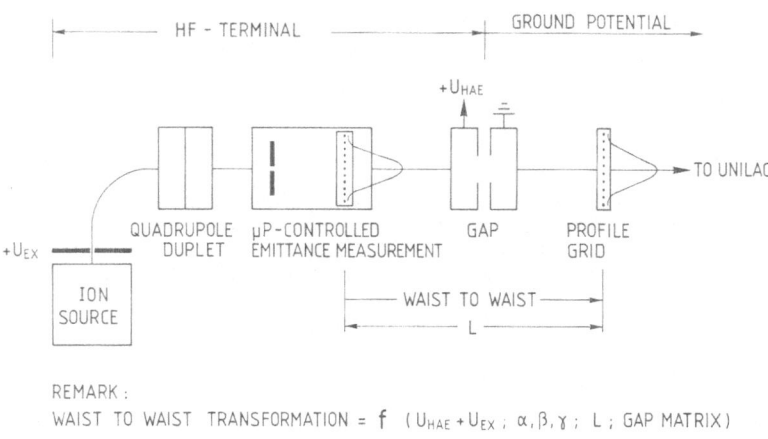

REMARK:
WAIST TO WAIST TRANSFORMATION = f ($U_{HAE} + U_{EX}$; α, β, γ ; L ; GAP MATRIX)

Figure 7 shows a layout of the section under discussion. For comparison between theory and experiment the following procedure was applied: A waist was produced at the position of the first profile grid (belonging to the emittance measurement system) by means of the quadrupole duplet in front of the grid. As may be seen in **Fig. 7** the transfer matrix to the next profile grid is determined only by drift spaces and the action of the gap. By varying the gap voltage a waist was also produced on the second profile grid. Detection of a waist by means of a grid is very sensitive. In comparison to this experimental procedure the envelopes displayed by **ENVTR** correspond to the theoretical action of the gap, including also the change of β-value within the whole range. Excellent congruence was achieved. In fact a waist was calculated by program at the exact position of the second grid, which confirms the correct calculation of the gap matrix.

NUMERICAL ORBIT CALCULATION FOR A LINAC AND
IMPROVEMENT OF ITS TRANSMISSION EFFICIENCY OF A BEAM

A. Goto, M. Kase, Y. Yano, Y. Miyazawa and M. Odera

The Institute of Physical and Chemical Research (RIKEN)
Wakoshi, Saitama 351, Japan

Abstract

A computer program named LINOR has been developed that can deal with orbit calculations for a linear accelerator of Wideröe type. This program is applied to the RILAC. Description on LINOR is given as well as comparison between the results of the measurement and the calculations obtained using it.

1. Introduction

A variable frequency heavy ion linac (Riken Linear Accelerator, RILAC)[1] is under operation at RIKEN. Its acceleration frequency can be set between 17 and 45 MHz. The operation is going on without any serious troubles. The transmission efficiency of an accelerated beam, however, has not yet been so high compared with the design goal, especially at the low energy stage.

In general, an impulse approximation becomes less reliable in the region of low beam velocity. Therefore, we have developed a computer program named LINOR with which beam trajectories are traced by numerically integrating the equation of particle motion in the realistic electro-magnetic fields. LINOR will be very useful to improve transmission efficiency and beam quality in the RILAC.

2. The computer program LINOR

2.1. CONTENTS OF PROGRAM

The RILAC consists of an array of six acceleration tanks (cavities), in each of which 10 to 18 drift tubes are incorporated[1]. Focusing quadrupole magnets are arranged in the tubes at the earth level. The first tank operates in π-3π mode and the others in π-π mode. The length of each tank is 3 m and the tanks are located 1.5~2.0 m apart from each other. In each of the sections between the tanks are set quadrupole singlets as well as such beam diagnostic devices as a beam profile monitor, a bunch pick-up monitor and a Faraday cup.

Using the program LINOR the following calculations are made for the above six tanks individually: 1) the rf phase excursion of an on-axis particle, 2) the acceptance of a tank and 3) a beam behavior when injected with certain emittance. In 3) a transmission efficiency through a tank, the energy distribution and the beam profile at the exit of the tank are obtained with the Monte Carlo simulations.

In order to get proper results with this kind of program, it is essential to use an electro-magnetic field distribution as close as possible to the real one. We used the measured magnetic field distribution for a quadrupole magnet. On the other hand, as for an rf electric field, such a distribution is difficult to measure, especially for its radial component. Therefore, an effort was devoted to deducing the electric field distribution which should be used in the calculation. We determined independently its axial and radial components (E_z and E_r, respectively) in the following way;

E_z-distribution

As for this distribution, we have the data measured using a model cavity. We also made calculations for it by a finite element method (FEM)[2]. It was found that the calculated field distribution reproduces well the measured one. Therefore, we decided to use in LINOR the distribution approximated by such a simple function as to fit the above distributions. An example of these distributions is shown in Fig.1.

In order to check the justification of the above E_z-distribution used in LINOR,

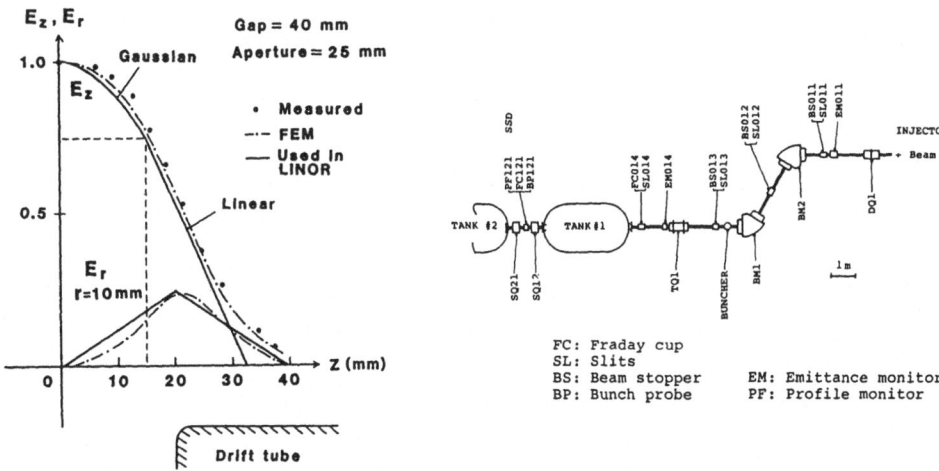

Fig.1 Example of the distributions of E_z and E_r used in LINOR.

Fig.2 Schematic layout of injection line and the first tank.

we made comparison between the results of the calculation using it and those of the measurement with respect to the behavior of an on-axis particle. A schematic layout of the injection line, the first tank and the diagnostic devices used for the measurement is given in Fig. 2. The on-axis beam was produced using the double slit system consisting of the SL013(1 mm x 1 mm) and the SL014(1 mm x 1 mm). We measured the dependence of energy spectrum and bunch shape at the exit of the first tank with respect to the rf voltage. The advantage is that these quantities measured using this double slit system hardly depend on the distribution of E_r. The energy spectrum was measured with a solid state detector (SSD) using a faint beam obtained in two sequential charge exchange reactions. The bunch shape was measured with a bunch

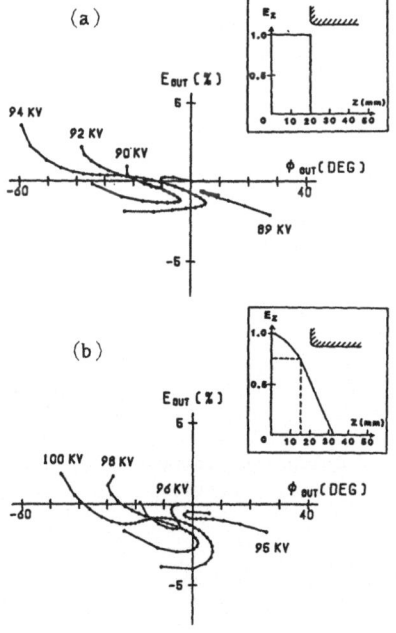

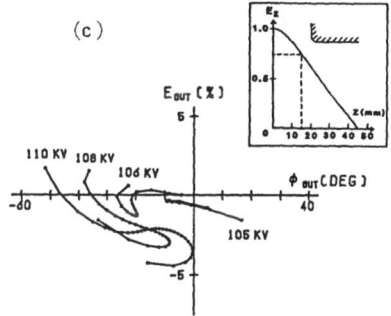

Fig.3 Output energy versus output phase calculated for three cases of distributions of E_z. The distribution of E_z used is shown in each inset.

pick-up device of beam-destructive type collecting the signals of secondary electrons. Both the rf voltage at which the bunch shape is sharpest and the threshold rf voltage at which the peak corresponding to the properly accelerated beam appears in the energy spectrum were found to be 96 kV. For comparison with the measurement we made calculations of output energy versus output phase. The calculations were made for three cases of E_z-distributions: a) a uniform distribution, b) the distribution shown in Fig. 1 and c) a distribution whose width is wider than that of case b). These calculations are shown in Fig. 3 together with the supposed E_z-distribution exhibited in each inset. It can be seen that the rf voltages mentioned above are 90, 96 and 106 kV for the cases a), b) and c), respectively. The calculation in case b) gives the result that is closest to the measurement. Consequently, we can conclude that the use of the distribution of E_z shown in Fig. 1 is reasonable.

E_r-distribution

It is quite difficult to measure the radial component of an electric field. To determine this distribution, we studied the dependence of the transmission efficiency of a beam through the first tank with respect to the rf voltage. The double slit system was also used and the beam current was measured with Faraday cups inserted at the entrance and exit of the tank. The reason why we used this narrow double slit system is that the uniformity of beam intensity in the emittance defined with it is very good. Figure 4 shows the measurement together with the Monte Carlo simulations in two cases that the quadrupole magnets are excited and not excited. In the calculations, we used the distribution above for E_z, and as for E_r, we supposed the following distribution : the distribution of E_r has a triangular shape with its top located at the edge of a drift tube. The half width of the triangle was fixed to the value of 0.78 times the aperture of the drift tube, which was obtained with the calculations by FEM. Then the height of the triangle E_r^{max} was given as follows,

$$E_r^{max} = \alpha \cdot (\frac{1}{2}\frac{\partial E_z}{\partial z} r), \tag{1}$$

where $\partial E_z/\partial z$ is the gradient of the linear part of E_z-distribution, r the radial coordinate of the particle and α the proportional coefficient. In Fig. 4 are shown

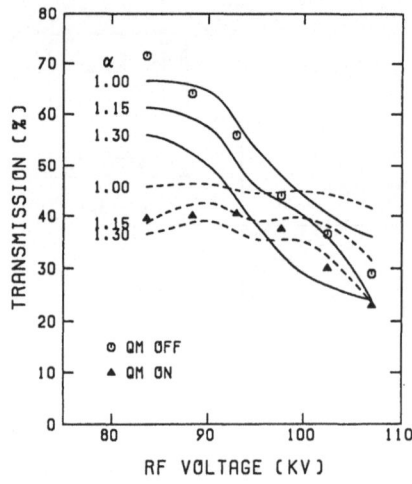
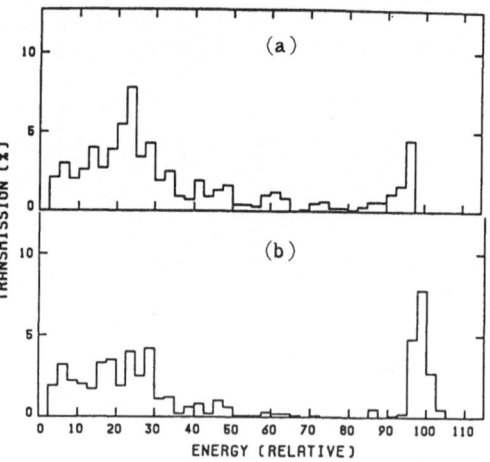

Fig.4 Transmission efficiencies of an on-axis beam through the first tank in two cases that the quadrupole magnets are excited and not excited. It is noted that the output beam consists of all the particles having different energies.

Fig.5 Comparison between the calculated energy spectra for two rf voltages: a) 85 kV and b) 95 kV.

the calculations for three values of α's. From this figure we decided to adopt the value of 1.15 for α. Here the reason why the transmission efficiency for the case of quadrupole magnets not excited increases with the decrease of the rf voltage is that the portion of the particles getting insufficient acceleration increases in the region of lower rf voltages. The calculated energy spectra in the cases of 85 and 95 kV are compared in Fig. 5.

An example of the distributions of E_r thus determined is shown in Fig. 2.

2.2. EVALUATION OF AN IMPULSE APPROXIMATION

It is interesting to compare the results of LINOR and those of the calculation using an impulse approximation. So we calculated with LINOR the phase excursion of an on-axis particle and the acceptance of a tank using the acceleration parameters obtained by an impulse approximation. The calculations were done for two tanks, the first (low energy stage) and fifth (high energy stage) ones. Results of the calculations are shown in Fig. 6. A star indicates the phase at the center of each acceleration gap for a particle with an injection phase equal to -25°, which is the designed synchronous phase of the RILAC. It is seen that the phase of -25° is indeed synchronous in the fifth tank but there is no complete synchronous phase in the first tank. The beam acceptances of the first and fifth tanks are presented in Fig. 7. As can be seen in the figure, the results of the two kinds of calculations agree quite well with each other for the fifth tank but disagree a little for the first tank. The implication of these results is that the calculation using an impulse approximation gives proper results in the high energy region but becomes less reliable at the low energy stage.

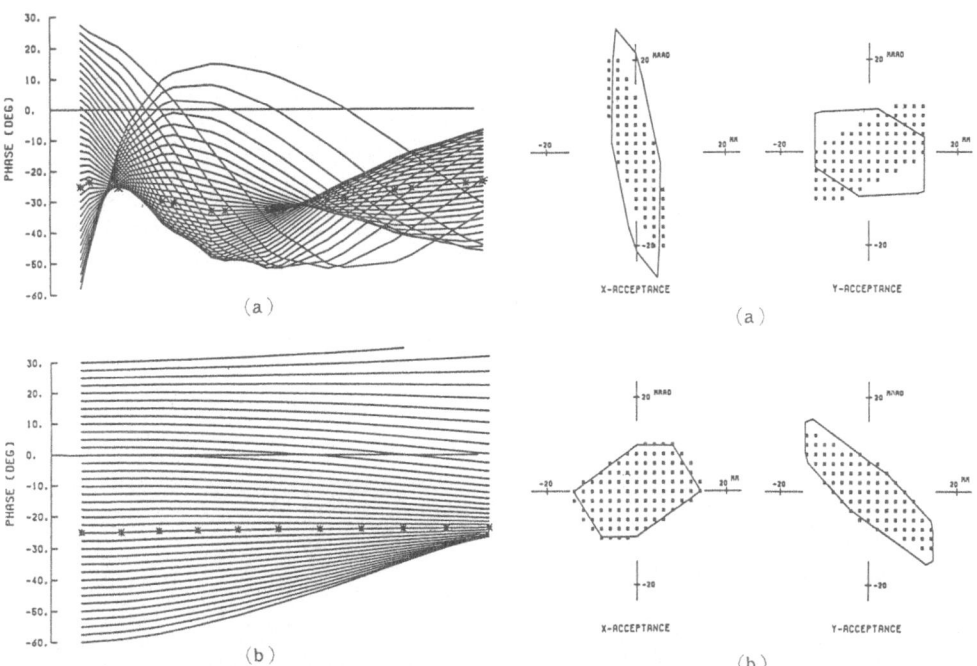

Fig. 6 Phase excursions of an on-axis particle calculated using LINOR; a) for the first tank and b) for the fifth tank. Acceleration parameters are those obtained by an impulse approximation.

Fig. 7 Comparison between the beam acceptances calculated using LINOR (dots) and those calculated by an impulse approximation (solid line); a) for the first tank and b) for the fifth tank. Injection phase is taken to be -25°.

3. Applications and future problems

We studied using LINOR on the matching of beam emittance and acceptance. For this purpose the transmission efficiency and beam profile were measured at the exit of the first tank when injecting two kinds of beams with different shapes of emittance. Figures 8 a) and b) show the emittances then measured at the EM014 together with the calculated acceptance for a beam with an injection phase of -25°. The measured transmission efficiencies for the cases of a) and b) are 20 and 29 %, respectively. The Monte Carlo simulations with LINOR give a value of 18.7 % for the case of a) and a value of 23.5 % for the case of b). The calculations reproduce well the measurements. Figures 9 and 10 show the measured and calculated beam profiles for the case of b), respectively. In Fig 11 and 12 are shown examples of the injection emittance used in the simulations and the calculted energy spectrum for the case of b), respectively. Returning to Fig. 8, it is interesting to see that the transmission efficiencies in two cases are different from each other in spite of the fact that both emittances are matched to the acceptance. This difference arises from the fact that the radial acceptance changes according to the initial phase of a beam. As shown in Fig. 13, the calculated acceptances for beams with different initial phases are changed. From this figure it can be seen that the emittance of case b) is matched better to the acceptance in a wider range of initial phases than that of case a).

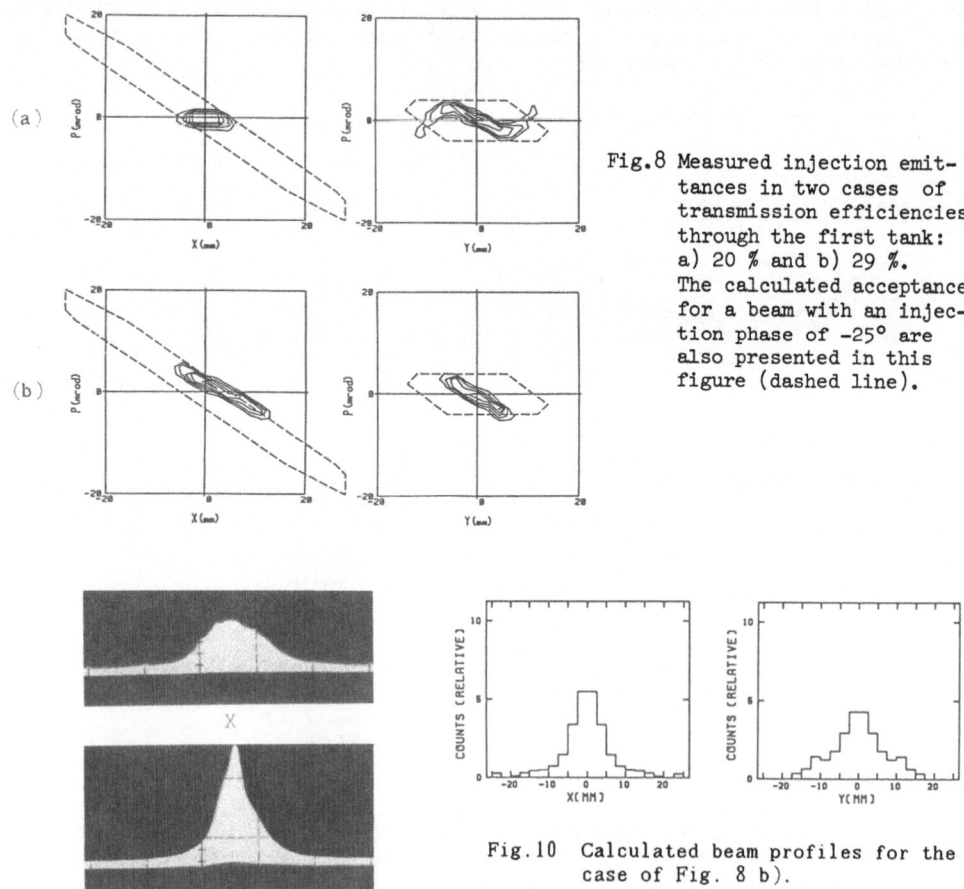

Fig.8 Measured injection emittances in two cases of transmission efficiencies through the first tank: a) 20 % and b) 29 %. The calculated acceptances for a beam with an injection phase of -25° are also presented in this figure (dashed line).

Fig.10 Calculated beam profiles for the case of Fig. 8 b).

7 mm/div

Fig.9 Measured beam profiles for the case of Fig. 8 b).

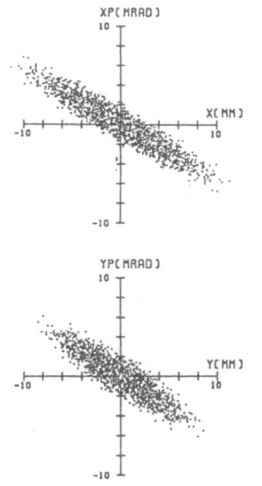

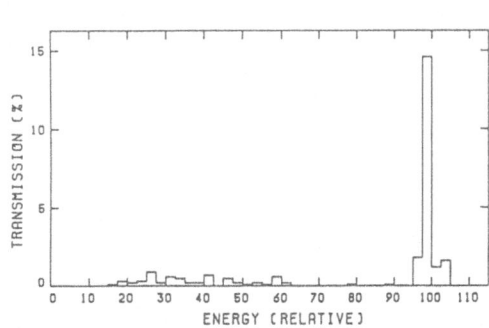

Fig.12 Calculated energy spectrum for the case of Fig. 8 b).

Fig.11 Injection emittances used in the Monte Carlo simulations for the case of Fig. 8 b).

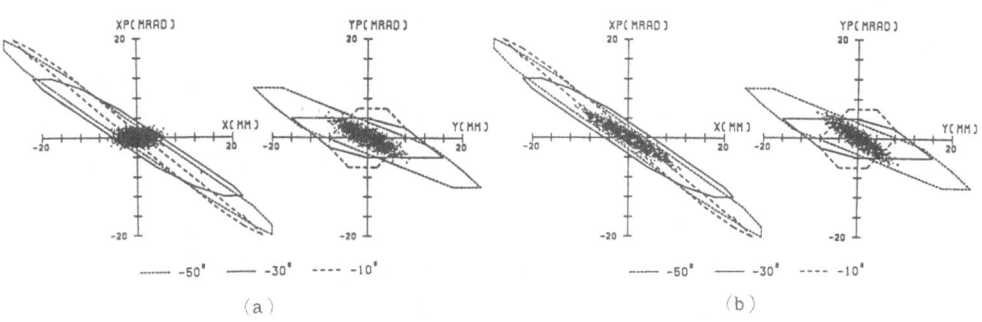

Fig.13 Calculated beam acceptances of the first tank for beams with different initial phases. The injection emittances used in the Monte Carlo simulations for the cases of Fig. 8 a) and b) are also presented.

We have not yet obtained the compleltely optimal operational parameters of the RILAC. LINOR, however, has led to an improvement in the transmission efficiency of a beam. In order to achieve higher transmission efficiency and get better beam quality, we are now developing LINOR to have the function to optimize the currents of quadrupole magnets and the RF voltages.

References
1) M. Odera, Y. Chiba and T. Kambara, Proc. of 1979 Lin. Accel. Conf., p128, 1979, "Status of the Variable Frequency Heavy Ion Linac, RILAC".
2) M. Hara, T. Wada, A. Toyama and F. Kikuchi, Sci. Papers I.P.C.R., 75, No.4 (1981) 143.

THE COMPUTERIZED BEAM PHASE MEASUREMENT SYSTEM at GANIL
ITS APPLICATIONS TO THE AUTOMATIC ISOCHRONIZATION IN THE SEPARATED
SECTOR CYCLOTRONS (SSC) AND OTHER MAIN TUNING PROCEDURES

J.M.LOYANT-F.LOYER-J.SAURET and the Control and Theory GANIL Groups

GANIL. BP 5027.14021 CAEN-CEDEX. FRANCE

I. THE BEAM CENTRAL PHASE MEASUREMENT PROCESS

I.1 Where and why ?

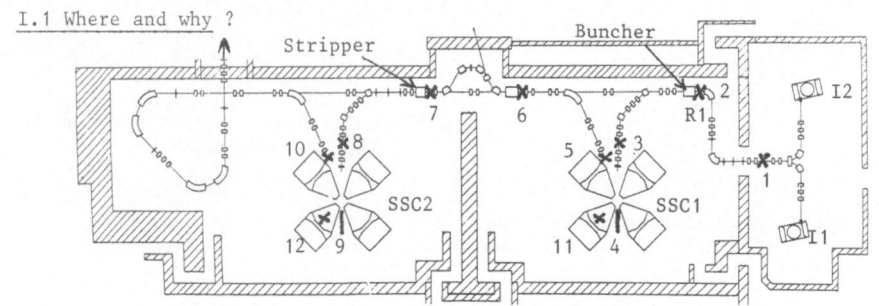

Figure 1 : Location of Beam Central Phase Probes

The beam phase measurements (figure 1) are performed /1/,/2/

- inside each SSC, using 15 pairs of capacitive probes (4 and 9) to adjust the isochronism (see application II.3.). These probes, loaded by 50 ohms, are located in a valley. An UHF target, using secondary electron emission, fixed on a yoke probe, allows the same phase measurement but turn by turn (11 and 12).

- along the beam line using 8 capacitive high impedance probes :
  * at the input (3 and 8) and the output (5 and 10) of each SSC to adjust and control the SSC parameters (RF phase, magnetic field,...) (application II.2 and II.4).
  * at the input of SSC1 (3) to adjust the buncher RF phase (application II.1)
  * at the input of the buncher (2) to control the buncher RF phase or the injector beam phase (application II.4).
  * generally speaking, to measure the beam characteristics such as the central energy and its variation, and to estimate the beam intensity since the measurement system also gives a rough value of this intensity.

I.2 How ?
The phase measurement is made between the second harmonic of the pick-up signal and the injector double RF frequency.
Two types of measurements are used :

- absolute measurement which gives the components of the pick-up signal with beam (x,y) and without beam (x_0,y_0). Therefore the beam phase is $\phi = \tan^{-1}\frac{y-y_0}{x-x_0} + \phi_0$. Constant ϕ_0 is the sum of the 2 phase shifts : ϕ_1 from the electronic devices and ϕ_2 from the transmission cables. ϕ_1 is given by measuring the components of a reference signal (x_R, y_R) and of the ground signal (x_G, y_G), assuming $\phi_1 = \tan^{-1}\frac{y_R - y_G}{x_R - x_G}$.
ϕ_2 is obtained by measuring the beam phase at the buncher input where it is well known (-7.9° with respect to the buncher RF). Note that all probes are connected with cables of the same length.

- relative measurement which gives, as a function of time, the phase variation y/x when y ≅ 0 and the beam intensity x.

The results consist of data digitalized by CAMAC interface and analog data (maximal bandwidth 3 kHz).

I.3 Hardware

Figure 2 shows the analog electronic device and the associated digital hardware for one probe in beam line or for 15 pairs of probes plus the UHF target in each SSC./1/

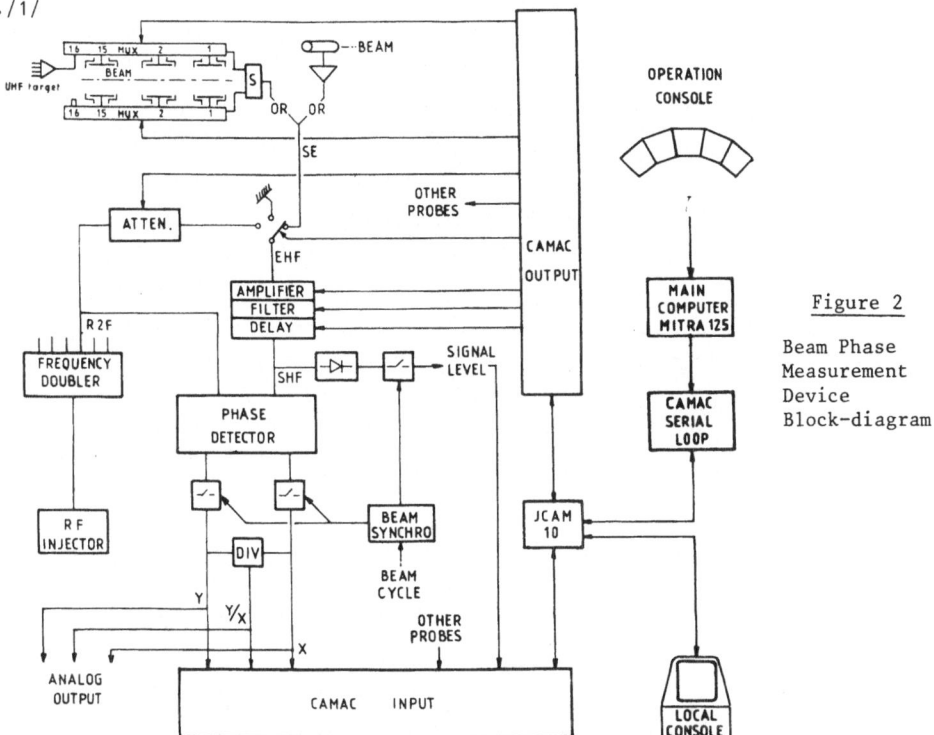

Figure 2

Beam Phase Measurement Device Block-diagram

Filters and delays are made with cable delay lines. The gain of amplifiers is programmable by 8 dB steps within a 40 dB range. The phase detection is performed by analog multiplying of the pick-up signal(SHF) by the reference signal (R2F) to give x and by the 90° out of phase reference signal to give y. Its input level ranges from -30dBm to -10dBm. y/x is obtained by an analog divider when $y \cong 0$ by shifting the phase with the delay pine. The analog outputs (x,y,y/x and signal level) are synchronized with the beam cycle. The sensitivity and accuracy of the electronic device allow accurate phase measurements (±0,1°) with a beam intensity of a few electrical nanoamperes.

This whole device is connected to the CAMAC interface which is controlled by an autonomous crate controller JCAM10 equipped with an Intel 8080 microprocessor. This controller is linked to a local console for use in stand alone mode or to the main CAMAC loop for use on line with the main computer MITRA125 and the main console.

I.4 Software

I.4.1 JCAM10 software /2/

The software has been built around a minimonitor which manages the system tasks and the process tasks.

4 system tasks manage the messages from MITRA to JCAM and their associated reply messages. They are started as soon as the corresponding message arrives. Those tasks are : starting a process task, killing a process task, sending back the status and present data, initialization of the system.

1 system task, trigged by any process tasks, sends the results to MITRA (data and status) and receives the reply message from MITRA.

The process tasks perform the specific actions and run when the minimonitor asks them to do so. These tasks are :
* initialization of the process : adjusting the filters and checking the electronic devices.
* absolute measurement : adjusting the gain, picking up the components $(x,y), (x_o,y_o), (x_R,y_R)$ and (x_G,y_G). Options are : without new gain adjustement, averaging several measurements and multiplexing several probes.
* relative measurement : adjusting the gain and shifting the delay line to get $y/x = 0$. The probe electronics keeps on giving y/x until the task is killed by the operator or by an interrupt signal from other systems.

I.4.2 MITRA and console software /3/,/4/

The main computer software, named GANICIEL, is written in LTR, a high level Real Time Language similar to Pascal. In particular, it takes care of the dialogue between the JCAM and the MITRA user tasks which operate the phase measurement system and which use for that purpose a specific routine (MESPH).

The operator can choose the tasks with one of the touch panel selectors and interact with them via a TV screen.

I.4.3 Check task

A check task (TESTPH) allows to perform all the actions with the phase measurement system and to display all the results (data and status). It also gives informations about the messages between MITRA125 and JCAM10.

I.4.4 Example

Figure 3 shows in a simplified way, the different actions accomplished when an user task (here ISOGRO) wants to make an absolute measurement.

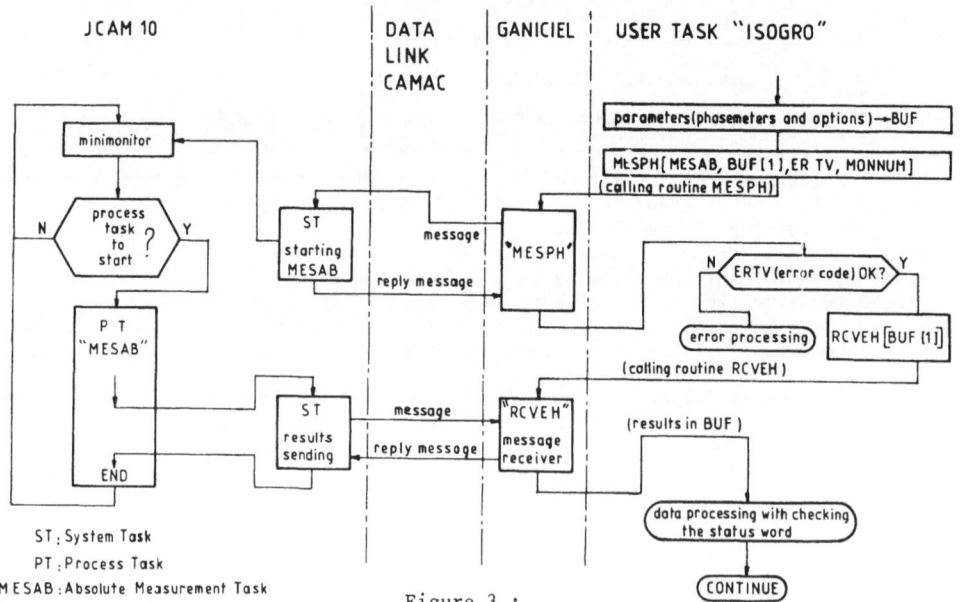

Figure 3 :
Example of an user task using the beam phase measurement system.

II. APPLICATIONS

II.1 Application 1 : buncher RF phase adjustment

Input energy of the SSC1 has to be the same with and without the buncher R1. Beam phase is characteristic of the beam energy. So the beam line L1 is first tuned

with R1 off and relative beam phase is measured on probe 3 (fig. 1) using TESTPH
(§ I.4.3). Then R1 is turned on and its RF phase is tuned until the beam phase on probe 3 becomes the same as previously. Doing so it is sure that the beam crosses R1 at 90° out of RF phase.

II.2 Application 2 : cavity RF phase determination for the two SSCs

The method consists in reproducing the same phase ϕ between the beam and the SSC RF as this which has been obtained for a previous good tuning.

A task (PHAREG) allows to set up this phase ϕ for a new tuning. It performs the following actions :

* determining the constant ϕ_2 by measuring the beam phase relative to the buncher RF at the buncher input with the probe 2 according to the method described §1.2
* measuring the beam phase relative to the injector RF at the SSC1 (or SSC2) input with the probe 3 (or 8).
* calculating the SSC1 (or SSC2) RF phase relative to the injector RF to reproduce the phase ϕ.

Previously the phase ϕ has been determined by a similar processing when the tuning seemed to be the best one.

II.3 Application 3 : automatic isochronization in GANIL SSCs

An user task (ISOGRO) fits automatically a given beam central phase law into the SSC and specially achieves the isochronism. Starting from the beam phase measured along the valley, the code computes the required corrections on trim-coils and main coil current to reach the theoretical phase law introduced in data.

Figure 4 presents the block diagram of the code.

The routine ISOGRO reserves the full screen (28 lines and 64 cocolumns) to write specific messages and display data.

It begins by the choice of the SSC, a few seconds later results are sent back to the console (figure 3), the task can run on using these data. The radius dependence of the relative phase between the beam and the RF pilot is then displayed on the TV screen.

Watching the shape of the resulting curve ϕ_c (R) and taking into account some simple criteria, the operator can select what parameters have to be optimized to reach the desired curve if necessary.

The following suggestions and an interactive dialogue lead the operator's job. If the curve is not good enough, two processing methods are proposed to the operator.

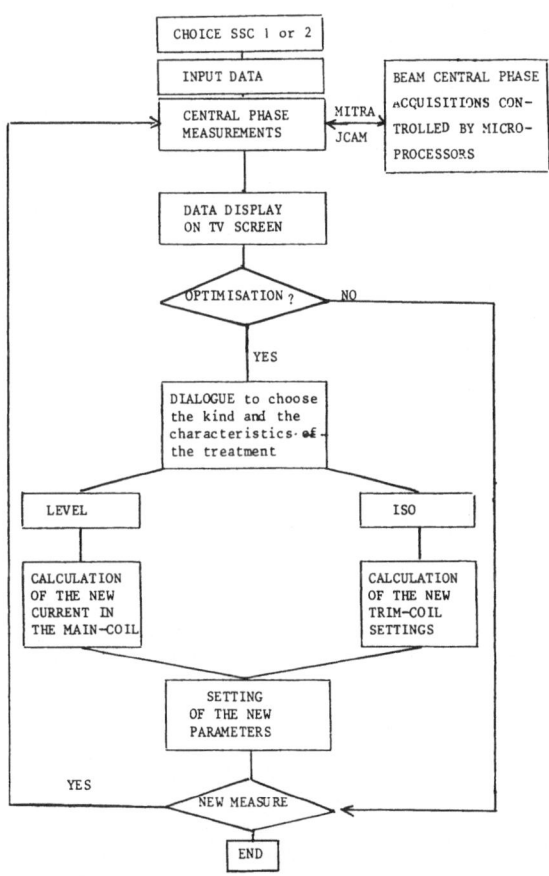

Figure 4

Block-Diagram of the on-line routine ISOGRO

a) a linear radial dependence of the central phase can be corrected only by adjusting the current in the main coil. In this case the operator points the cursor on the option "NIV" and valids the question. Then he has to choose the probe numbers to delimit the radius range for calculation. The program computes the needed change on the main coil current and writes the value in mA on the screen. An example is given on figure 5.

```
*ISOG2  33    MESURE PHASES CENTRALES VALLEE CSS        C2.DIA.PC
31 AOU 1983 **  17HR  48MN                  AUTRE MESURE? OUI      N
                   PREMIERE SONDE+    2        DERNIERE SONDE+   15

              .        .   .   PENTE DE LA DROITE C(1)*1000 +   13
                .  .   .       COEFFICIENT DB/DI     (G/10*A) +   66
                               CORR. NIVEAU DE CHAMP     (G) +    0
                               COR COURANT C2.M.BP(MILLI-A) +    30

-109.6                       *
-111.0                                *    *    *    *    *  + 323
-112.5           *    *                               *    * + 318
-115.1    *  *
-116.1 *     *    *  *

-119.9*
       --1.0----------1.5---------2.0----------2.5-------RAYON(M)
        1-2--3-----4-------5-------6--------7----8----9---10--11-12
```

Figure 5

Example of processing "NIV" to refine the current in the main coil

b) a more complicated variation of the measured curve from an ideal curve must be corrected by adjusting the isochronism currents (see example figure 6).

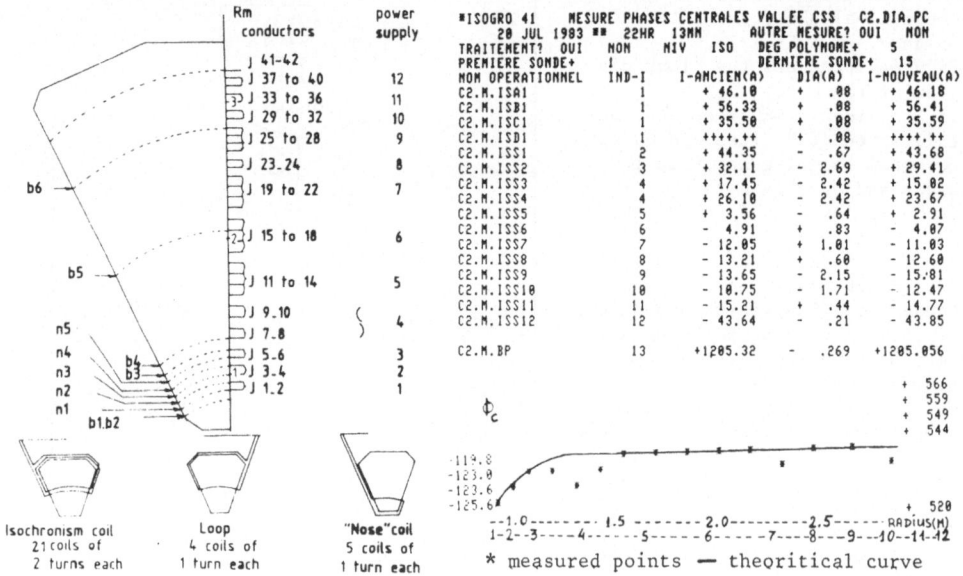

Figure 6 : Lay out of trimming coils and example of treatment "ISO" to optimize the isochronism currents with ISOGRO code in SSC2 (Ar^{+4}/+^{16} 44MeV/A)

The operator chooses the treatment (fig 4) option "ISO" and the degree of the polynominal used to fit the experimental curve. The task computes the new trim-coils currents in the following way :

The effect of a magnetic field change upon the beam central phase is given by the relation $\delta B/B \sim - \delta\phi / (360.h.N)$ (1) with h = RF harmonic number, N = turn number, so that difference ϕ_2 between two measurements made at the radius R_2 and R_1 can be reduced to theoritical value by adjusting the magnetic field of a quantity δB.

From relation (1) and orbit data, the magnetic field defect upon radius $\delta B(R)$ is deduced from the measured central phase deviation. The orbit data used have been obtained by computer simulation of acceleration in measured isochronous field map /5/. At each probe is affected a reference orbit with all the characterics : its turn number its radius in the valley, its radius in the sector axis, the average isochronous field and the ideal central phase at the azimut where the probe is located. Taking into account the actual trim-coil pattern, their characteristics and the number of power supplies, the control matrix T giving the magnetic effect along the axis for a given current has been calculated by the program BOBO/6/7. As we have more measuring points (15 beam phase probes) than parameters (12 power supplies) a least square method using the control matrix T as input data gives the new currents to minimize the defect δB (R). After roughly 30 seconds the new trim-coil settings appear on the TV screen and the operator can valid the automatic setting of the new parameters. As the beam central phase device is non-destructive, a new measurement is proposed to control the effect on the new beam and to go on in tuning the acceleration if necessary.

II.4 Application 4 : Beam Control Loop

Small drift of some parameters as Dee voltage, magnetic field and input beam phase of the SSCs results in intolerable alterations of the tuning, up to the point of losing the beam.

Beam phase is very sensitive to these alterations. So the beam phase shifts can be used to control these parameters in order to increase the tuning stability.

The following control loop, presently used for SSC_1 and planned for SSC2 cancels the magnetic field shift with an accuracy of some 10^{-6} :

$\Delta B = K_B (\Delta \phi_O - \Delta \phi_I)$ where ΔB is the main magnetic field correction
$\Delta \phi_I$ (or $\Delta \phi_O$) is the input (or output) SSC1 beam phase variation
K_B is the gain loop.

Trials performed up to now showed that this control is disturbed by the input beam energy variation. It is therefore necessary to stabilize this energy and, in particular, the beam energy at the injector output.

For this purpose, we intend to add another control loop on the injector dee voltage :

$\Delta V_D = K_V (\Delta \phi_{R1})$ where ΔV_D is the injector dee voltage correction.
$\Delta \phi_{R1}$ is the beam phase variation at the buncher input
K_V is the loop gain.

We also intend to add a third control loop which relates the RF phase of each SSC to the beam phase at their input.

IN CONCLUSION : the beam control phase measurement system has proved to be an efficient tool to tune and to control the whole accelerator. Main procedures at present used have been described in this paper. Other applications already exist such as beam line or SSC efficiency measurements, or will be developed such as the different control facilities which will be very helpful to keep a good beam tuning. This system is also very useful to watch the beam during the operation time.

REFERENCES

/1/ F. LOYER and al. Main beam diagnostics at GANIL . 9th Int. Conf. on Cycl. and Appl. Caen (France) Sept. 81

/2/ F.LOYER-J.M.LOYANT-B.PIQUET-M.PROME-M.ULRICH. Mesure de la phase centrale Int. Report . GANIL 83R/010/CC/01 . Feb. 83

/3/ M.PROME . The GANIL Control System . IEEE Trans. on Nuclear Science Vol NS-28, N°3, June 81

/4/ L.DAVID-E.LECORCHE-LUONG T.T.-B.PIQUET-M.PROME-M.ULRICH . Le Systeme de Controle du GANIL . Real Time Data 82. Paris (France) Sept.82

/5/ A.CHABERT and al. "Multiparticle Codes developped at GANIL" This Conference September 1983

/6/ M.BARRE and al. "Main Results on the SSC magnetic field mapping at GANIL" 9th Int. Conf. on Cycl. and Appl. Caen (France) Sept. 81

/7/ D.BIBET- A.DAEL."BOBO : programme pour le calcul des Trim-Coils" GANIL. Internal Report 77N/078/AI/19

ON-LINE OPTIMIZATION CODE USED AT SATURNE
JM. LAGNIEL, JL. LEMAIRE
Laboratoire National Saturne, SACLAY.

ABSTRACT

A computer code has been developped in order to make the tuning of the injection process easier in the Saturne synchrotron accelerator and search for sets of new values of parameters leading to the optimum of any criterion. The usual criterion being mainly the beam intensity given by current transformers or any non-destructive measurement device. Acquisition of the criterion is made at each cycle of the acceleration. The technique used has many advantages :
- It is well suited to solve problems dealing with several parameters. 10 parameters are usually used but the code has provision for 50.
- There is no need for prior knowledge of the analytic relationship between the parameters and the criterion.
- No differential calculus and manipulation of matrices or determinants are required. The algorithm used, given by R. Hooke and TA. Jeeves fits with many optimization problems encountered during accelerators operation.

INTRODUCTION

Particle accelerator facilities include ion-sources, beam transport lines, accelerating systems (linacs, RFQ's, synchotrons, cyclotrons...), storage rings and experimental areas whose usual tuning methods are made stage by stage corresponding to artificial boundaries between these different parts. This seems to be fixed once forever. Indeed, it is hard to worry about the direct dependance or relationship between a given criterion and a set of parameters located very upstream of the beam when one is performing the setting up of an accelerator machine. It is also difficult to quickly demonstrate this dependance otherwise, by correlation records because these manipulations take a lot of time during normal operation of a machine. It can be more commonly done during machine studies.

Since particle accelerator machines have to produce a maximum of particles with given beam qualities one has to reduce as much as possible time spent to search for these requirements. The overall disponibility of the machine will be increased.

The problem is to find an optimum as fast as possible keeping in mind that the optimum must be as close as possible to the optimax (best solution). The theoretical operating optimax is determined either by calculations needing sometimes approximations or just theoretical considerations. Consequently in the real world the process is going to be

different to the expected theoretical optimax. That is to say, it is impossible to find
a maximax when all the parameters can fluctuate, when the measure of the optimum lies
within a given accuracy and when there are still parameters remaining uncontrolled.
At this step the operator procedure becomes ineffective. His method of iteration is
done step by step in agreement with a model that he keeps in mind and the tuning
procedure that he is undertaking can only be done with a limited set of parameters
(usually 2 and sometimes more if he is wise).
Thus human tuning methods suffer from
- unacceptable time spent for tuning,
- uncertainty of reaching the optimum.

We have to solve the problem of running a process having many parameters and to search
for the optimum control of this process. To find the answer, we have developped a
system which led to.
- the acquisition of the main parameters of the process in order to have a fast control
on them and informations if some of them present defaults.
- the command of the parameters, one by one or by groups.
- the automatic search for the optimum command.

All of those requirements were made possible in using the computer control system
developped around the SATURNE accelerator facility and we describe the third point in
this paper (ref. 1).

SEARCH FOR AN OPTIMUM SOLUTION

1) DEFINITIONS

The strategy is to minimize (or maximize) the fonction J (Y) that we call the
criterion. Y is defined by its components $y_1, \ldots y_n$ which are the controlled parameters
belonging to the domaine $\mathcal{D} \subset \mathcal{R}^n$

Searching for the OPTIMUM SOLUTION (best solution) means that we have to find a new set
Y^* resulting from the move so that

$$J(Y^*) = r^* < J(Y), \forall Y \in \mathcal{D}$$

r^* is called the "best revenu" : this is the value of the criterion when $Y \equiv Y^*$
It may arise that relative minima exist so that

$$J(Y^*) < J(\tilde{Y})$$ and the search stops.

For most of the problems that we deal with, the criterion will represent a beam
property at a given location of the accelerator (position of the beam, beam size, beam
intensity, emittance). The strategy is to look for a set of parameters upstream of this
location which leads to the "best revenu".

2) REVIEW OF DIFFERENT STRATEGIES

We distinguish :

- a) the methods which require evaluation of partial derivatives of the criterion function at a given point of the search in order to find the direction of the travel toward the appropriate optimum (gradient search). They are very improved when the analytical form of the criterion is known (ref. 2).

In the other hand they suffer from inaccuracy for evaluation of the derivatives when the numerical acquisition of the criterion is unstable, or because they require to progress at a very small increment when a very narrow ridge is detected.

Acceleration improvements of these methods exist and results are quite satisfactory (Partan's method).

- b) direct search methods which do not require any analytical calculations. It is a succession of trials where each solution is compared to the solution obtained at the previous step. The simplest of these methods is the one used by an operator. Parameters are incremented one at the time keeping the rest of them constant. Comparisons are sequential. With the use of modern computers it becomes possible to use better procedures (Rosenbrock method, Hooke and Jeeves method) which have the property of continuously progressing along ridges whatever they are straight or curved.

3) PATTERN SEARCH

We abandonned the usual derivative method because the problems that we are dealing with do not exhibit any analytical formulation of the criterion. Another advantage of direct method is to write on a computer only one routine for the optimization : this routine being the same for all the parts of the accelerator.

Among the direct methods, we have choosen the pattern search of Hooke and Jeeves (ref 3) which does not require many complicated calculations and whose the travel of the search is easy to follow.

The Rosenbrock method which is also a very good one was given up since it needs a very precise search of the minimum of the criterion in a given direction (ref. 4).

We have compared different methods on a test proposed by Rosenbrock. The criterion function correspond to a very narrow and curved ridge represented by the analytical form :

$$J(y_1, y_2) = 100 (y_2 - y_1^2)^2 + (1 - y_1)^2$$

Results are given on table 1 and show that both Rosenbrock and Hooke-Jeeves methods are very improved ; on the other hand gradient method or sequential method led by an operator show poor results.

TABLE I

base point $Y_1 = -1.2$ $Y_2 = -1$
optimax $Y_1 = 1$ $Y_2 = 1$ $J(Y^*) = 0$

method	Y_1	Y_2	J (Y)	total number of iterations
ROSENBORK	0,995	0,991	0,000022	200
HOOKE AND JEEVES	1,006	1,012	0,000096	134
	1,004	1,008	0,000024	200
SEQUENTIAL (operator)	- 0,970	0,945	3,882	200
GRADIENT				
variant 1	- 0,605	0,371	2,578	200
variant 2	- 0,235	0,068	1,542	340
variant 3	0,219	0,046	0,611	338

As an example fig. 1 and 2 represent a more realistic 2 dimension problem. The travels along the detected ridges for the operator sequential method and for the Pattern Search method and drawn. Base points are the same and both gradient search or grid search (operator method) fail while pattern search find the optimax (ref. 5). Nevertheless, depending upon the starting point as shown, the optimax may not be reached. Explanations are given in appendix 1.

OPTIMIZATION CODE - OPTINJ

1) FLOW DIAGRAM

Flow diagrams are shown for pattern search in appendix 2 and 3. Sequences are given and correspond to different tests and subroutines.

2) HOW TO USE THE COMPUTER CODE

This program is implemented on a mini computer MITRA 125 (SEMS Manufacturer) comparable to a PDP 11. The code is accessible on a display terminal which is connected to the remote control system of the machine. Once the program is loaded, the operator has to answer to a set of questions (identify the criterion, the parameters, ...) and send the order to start the iteration. He continously sees which parameter is changing and the amount can be simultaneously observed on a graphic color display. By just watching, one is possible to remember which parameters are more efficient and at what time the acceleration process occurs. New set of values are displayed after each acceleration step.

The operator can stop whenever he wants, definitively if something wrong is happening or continue the action .

Reduction of the magnitude of the increments is asked if no improvements are made.

At the very end, the operator can either send the best solution obtained or the initial set of parameters.

3) PRESENT LIMITATIONS

For on-line use of optimization process, one has to take into account perturbations which means that at a given command does not always correspond exactly the same revenu.

Those perturbations which have a stochastic behaviour may be random variations of the beam or uncertainties on the criterion.

Let us note P the random vector representing the perturbation. The new criterion is defined to be J (Y,P) and is now a statiscal function. The revenu is not going to be optimized anymore since it is a statistical value.

We have to set a new definition of the revenu based on the mean value or a modified criterion which takes into account the fluctuations.

This is partly done and work is underway to make more improvements.

CONCLUSION

The beam parameters optimization code is currently used at Saturne , it gives better and quicker results than manual tuning. If the program does not find a better solution, the operator does not either. This code is very usefull when the problem to solve is function of a large number of parameters (currently ten parameters are used). For example, we run the code to tune the matching between the LEBT line and the LINAC, to perform the multiturn injection process or to find the best circulating accelerated beam. Instead of progressing stage by stage as usually done we can optimize for instance the accelerated beam intensity by tuning the LEBT which is rather unusual on other accelerator machines. Results are interesting. The code is also usefull for solving theoretical problems where an analytical function exists.

Some improvements of this code are underway, we presently try to make a faster convergence ; it seems also interesting to take into account in the calculus of the criterion the stability of the criterion parameter which is to be optimized.

In its present form, the optimization code gives good results at Saturne where we run different formats (tuning of protons, deuterons, alpha, He_3, polarized p and d) ; it is also used whith success at the ALS (electron linear accelerator of Orme les Merisiers) for the tuning a positron beam line. Direct search techniques constitute an approach to a variety of numerical problems for which classical methods of optimization are unusable.

As a general concluding remark, people who want to turn to this new way of tuning have to put a real effort on the reliability of the acquisition system of the criterion which is choosen. If not, they really will be in trouble and it might be the reason why these direct methods have not been used earlier for one line optimization.

REFERENCES

1. Conduite informatisée du Synchrotron SATURNE II et de son environnement,
   internal report LNS-SSG 80-51 INEL-93, Juillet 1980
   J.L. HAMEL

2. HOPI : On line injection optimization program
   BNL 50741 UC-28, 1977
   JL. LEMAIRE

3. "Direct Search" Solution of numerical and statistical problems
   Journal of the Association for Computing Machinery, vol 8 n° 2, 1961
   R. HOOKE, TA. JEEVES

4. An automatic method for finding the greatest or least value of a function
   Computer Journal, vol 3, 1960
   H.H. ROSENBROCK

5. Informatisation de la commande et du contrôle des paramètres de l'injecteur du synchrotron SATURNE II : OPTIMISATION
   thèse JM. LAGNIEL, juin 1982.

APPENDIX 1

Let Y_0 be the starting point. Y_1 is the best exploratory result around Y_0. Acceleration takes place in the Y_0Y_1 direction leading to the point T_1, and search continues.

INITIAL PHASE : SEARCHING FOR A VALLEY - FIRST ACCELERATIONS

Y_0 starting point

Y_1 is the exploratory result around Y_0 giving the acceleration in the Y_0Y_1 direction whose result is T_1

Y_2 is the exploratory result around T_1 giving the acceleration in the Y_1Y_2 direction whose result is T_2

PHASE TWO : THE VALLEY IS DETECTED AND SUCCESSIVE ACCELERATIONS OCCUR

Y_3 is the exploratory result around T_2 giving the acceleration in the Y_2Y_3 direction whose result is T_3

Y_4 is the exploratory result around T_3 giving the acceleration in the Y_3Y_4 direction whose result is T_4

Y_5 is the exploratory result around T_4 giving the acceleration in the Y_4Y_5 direction whose result is T_5

PHASE THREE : IF NO IMPROVEMENTS ARE MADE - NEW STARTING POINT IS Y_5

Y_6 is the exploratory result around Y_5 giving the acceleration in the Y_5Y_6 direction whose result is T_6

Y_7 is the exploratory result around T_6 giving the acceleration in the Y_6Y_7 direction whose result is T_7

FINAL PHASE : REDUCTION OF THE INCREMENTS ARE NEEDED - OPTIMUM REACHED

Y_8 is the exploratory result around Y_7 giving the acceleration in the Y_7Y_8 direction whose result is T_8

Y_9 is the exploratory result around Y_8 giving the acceleration in the Y_8Y_9 direction whose result is T_9

Y_{10} is the exploratory result around T_9 giving the acceleration in the Y_9Y_{10} direction whose result is T_{10}.

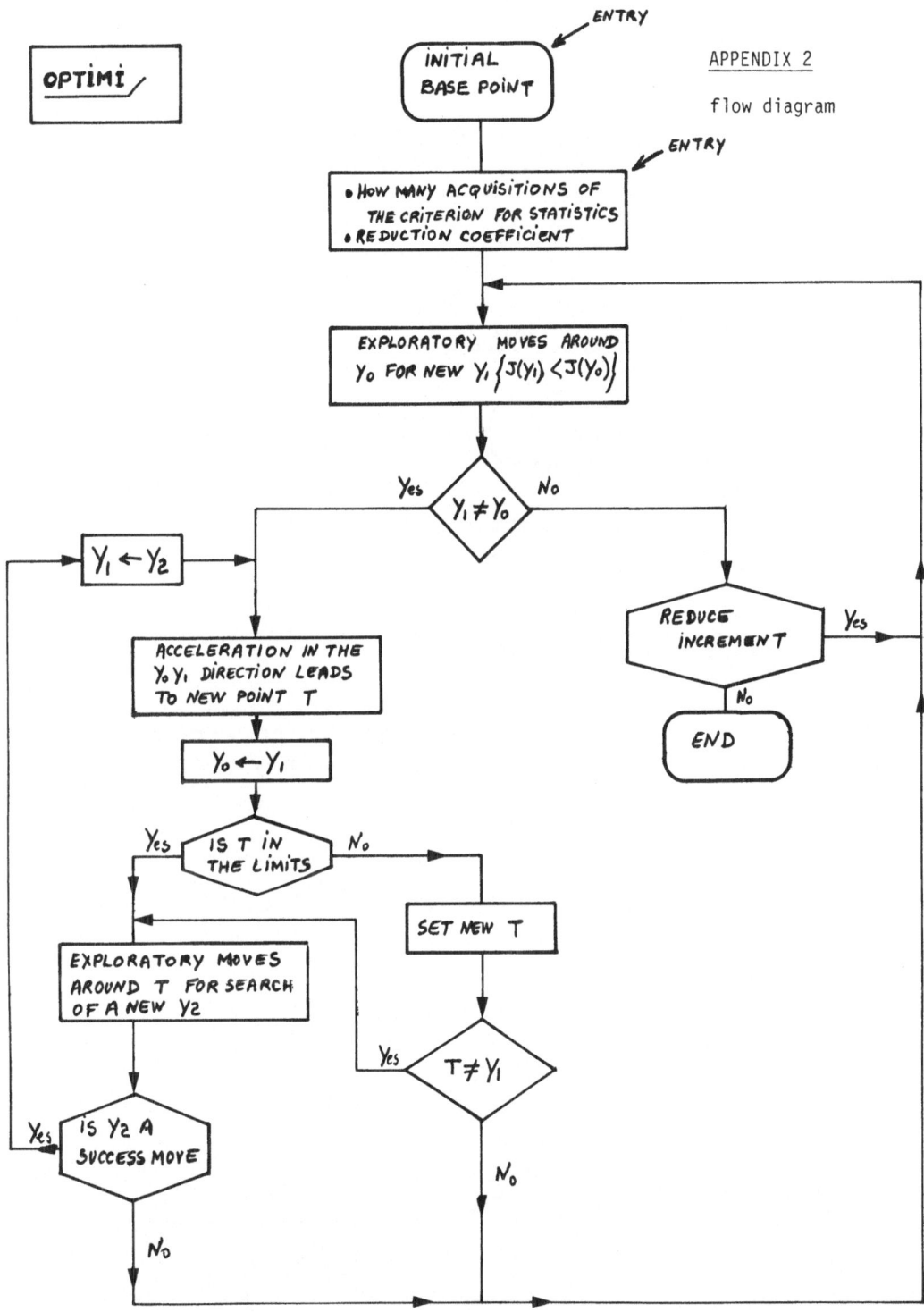

APPENDIX 2

flow diagram

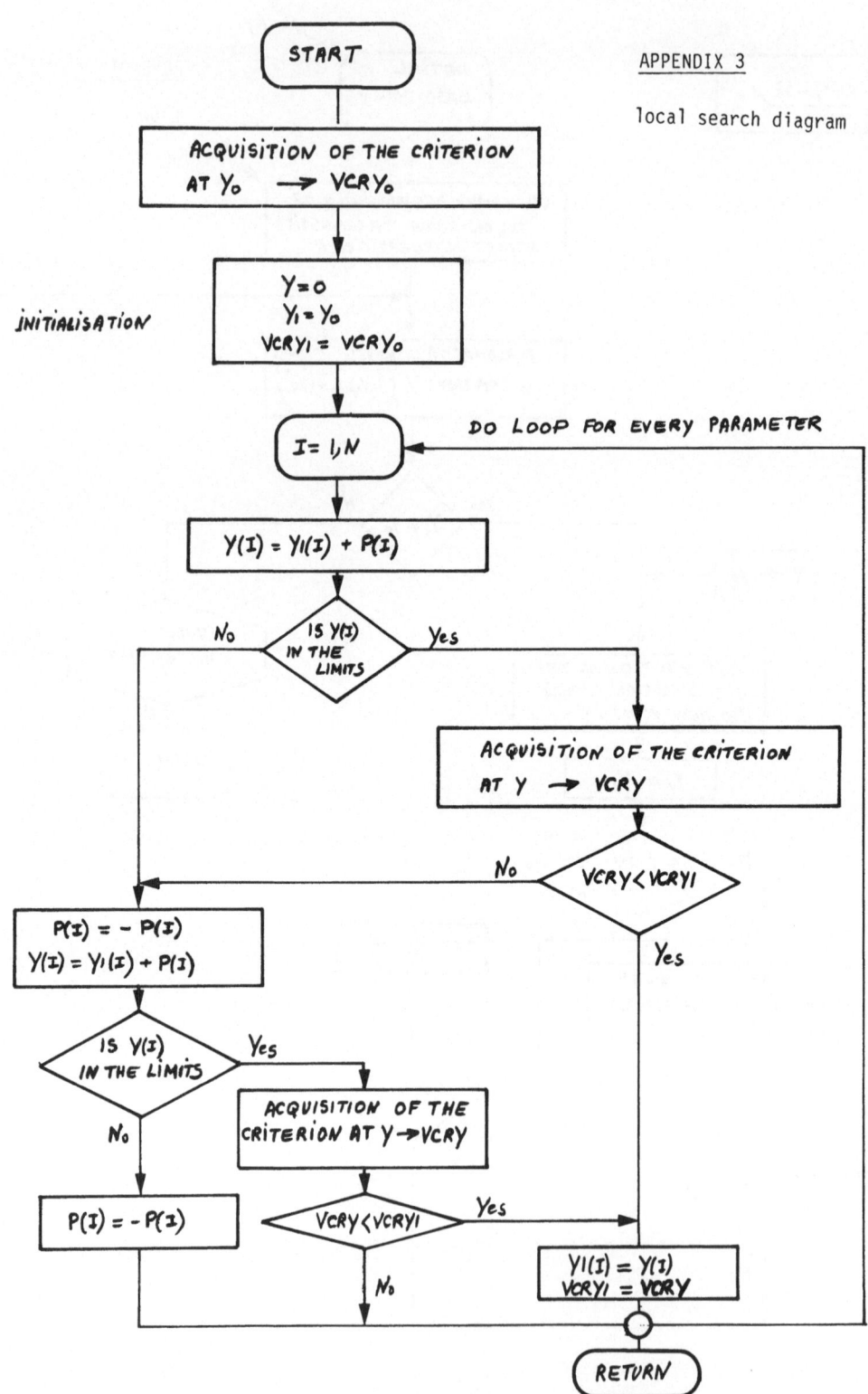

APPENDIX 3

local search diagram

Figure n° 1

gradient method

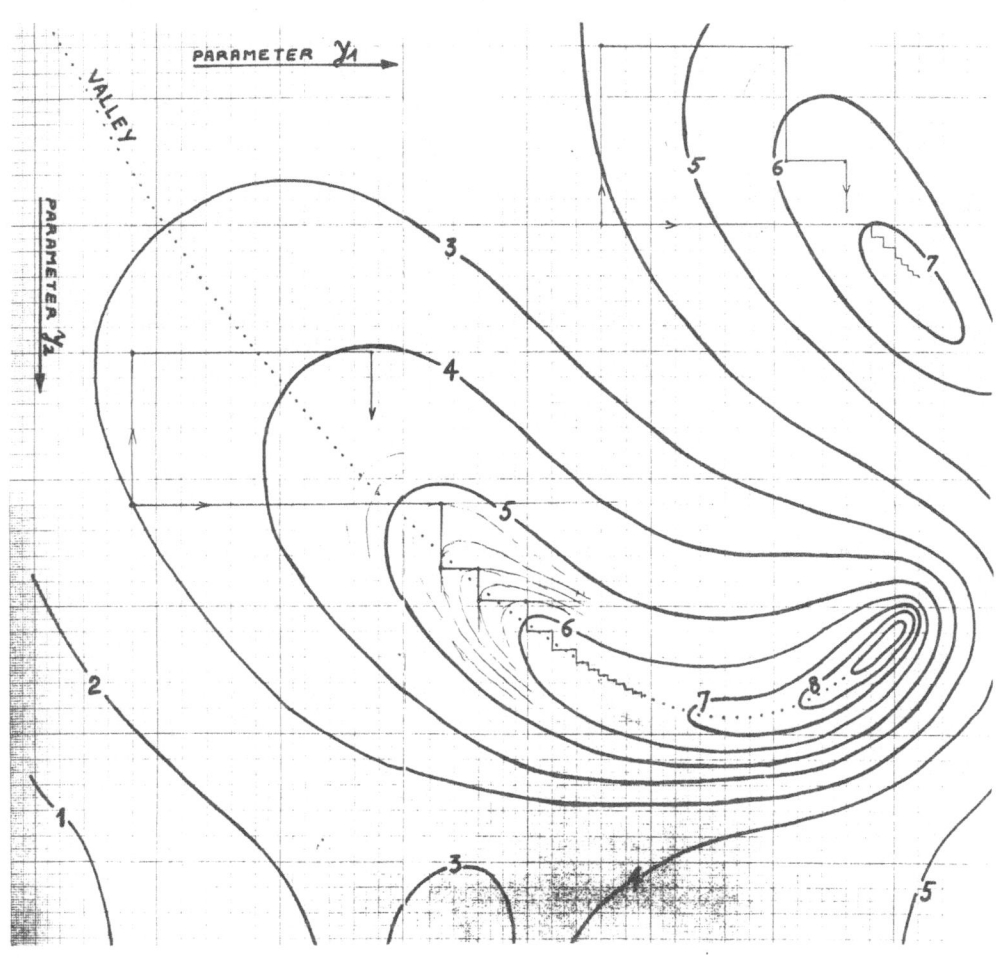

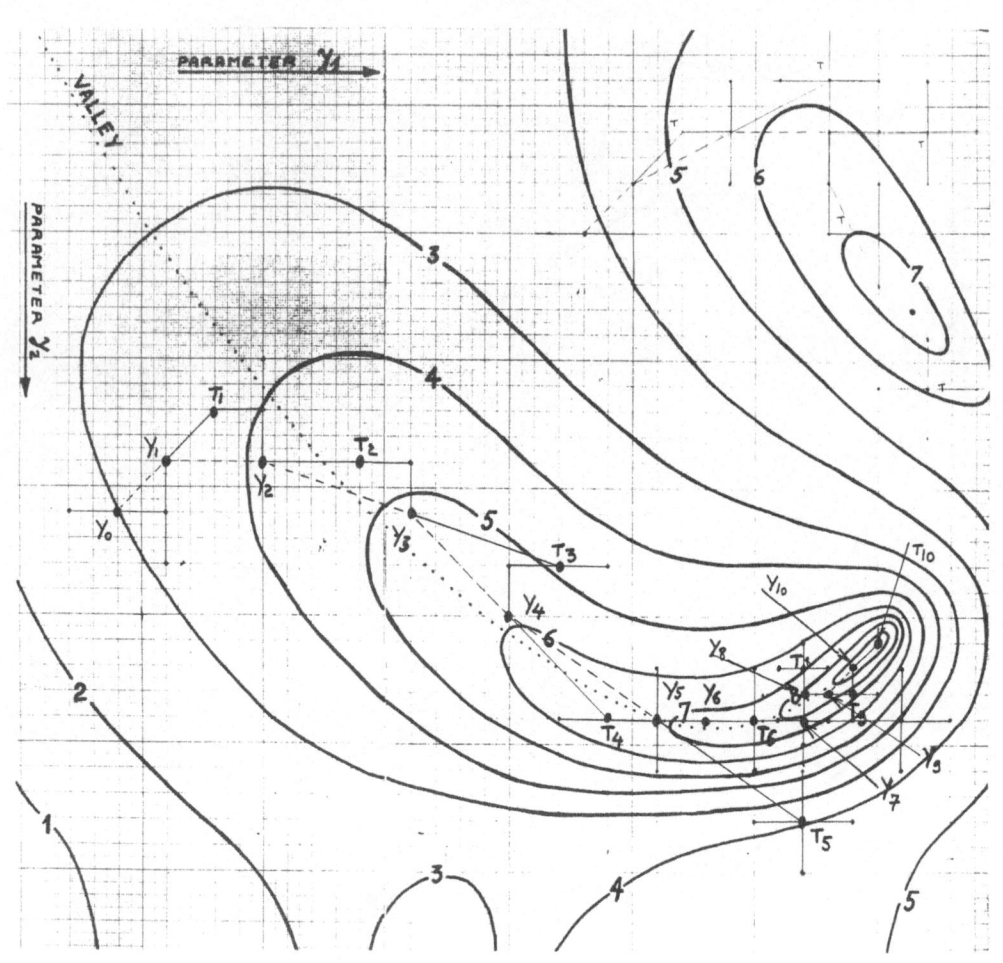

Figure n° 2

pattern search method

AUTOMATIC SUPERVISION FOR SATURNE

C.Fougeron, J.Gontier, J.M.Lagniel, P.Mattéi,

LNS, Saturne, CEN-Saclay, France

The Synchrotron SATURNE II can accelerate actualy several kinds of particules (α, He^3, p, d, polarized d and p) and, in a short time, a great number of heavy ions. It is able to extract these ions in two different lignes, with all the energies between 150 MeV and 3000 MeV.

It furnish these beams with a repetition rate around 1 second to experiments.

The large number of operation types asked many automatizations, for the starting, the tuning and so for the operation watch.

The automatization has been organized around and with the existing computer control system, and the Camac standard working jet in the accelerator aera.

SATURNE COMPUTER SYSTEM

It was built around a 3 Mitra 125 network in accordance with the shematic in annex.
One of the Mitra 125 is particularly assigned to the machine control.
All the accelerator parameters are known by identification sheet in a file into the computer memory. Many program use these parameters for :
    GENERAL CONTROL
    STARTING AND INITIALISATION
    WATCH
    TUNING
    MACHINE STUDIES
All the terminals can ask all the programs in the accelerator aera.

PRINCIP OF SUPERVISION

The supervision of the operations has for object to decrease the time of the starting, and to diagnostic as quikly as possible the detuning and the hard-ware fault and brackdown.

It decreases in the same time the difficulty of the operation and so discharges the operation personal.

The system was built with three differents ideas :
- 1 - oversee all the hardware,
- 2 - watch the stability of ions beams characteristics,
- 3 - log and dispatch the information.

HARWARE SUPERVISION (AJ)

All the parts of the accelerator, excepted ions sources and injector, are controled by this system.
Each part, each material has a status circuit, sometime very detailed, connected at the Camac system by many standard interface "MAB".
So, the status of a R.F. amplifier for exemple, is known by the central computer. Seven hundred status are connected.
All the MAB plug-in are hardware inter-connected and able to call a program in the computer, by a Camac "Look at me". So, the computer is used only when it needs, when a fault appears, and stay almost ever free for all the others uses.
The states are grouped in a software hierarchic organization and in several families according with the main parts of the accelerator (f.e. MAGNETIC FIELDS, VACUUM, ..).
The parts in failure of the accelerator are displayed automaticly on a color TV monitor, with the main informations concerning the material out of order. All the details can be called on the other computer terminals.
Nine parts of the machine are separatly watched :
    ION-SOURCE, LINAC, VACUUM, COOLING, ACCELERATION,
    MAGNETIC FIELD, EXTRACTION 1, EXTRACTION II and SECURITY.
This system is started by an operator after estimation of the accelerator good working.

WATCH OF THE BEAM CHARACTERISTICS (SURV)

The precedent system is no suffisant for insure a total watch of the good work of the machine. Some detuning cannot be taken in account by a binary pick up.
Therefore are so watched different signals given by beam measurements along the particles journey, between the ion source and the extraction lines.
- exit of ion source
- exit of linear accelerator
- input in the synchrotron
- radiofrequency beam
- high energy internal beam
- extracted beam
- spilling time

The start of this watch program is given by an order of the operation personal.

The computer keeps in memory the values of these 7 characteristics at the start time. If one of them changes more than 20 %, an alarm appaers, and a detailed exploration is made around all the analog parameters witch can be the origin of the breakdown, and only these.

Each parameter having changed more than 2 % is logged.

LOGGING AND INFORMATIONS DISPATCHING

By the Camac system, the informations concerning the accelerator state are available in the accelerator area.

They are logged on a tape writter in the main control room on a logbook watevever origin they have (SURV, AJ or other).

An analog representation can be made on a graphic display, day by day.

In the other hand, a clear information concerning the actual state of the machine is continiously sent to the experiments by a harware slifting display system, by AJ program and the central computer.

We looked for automatize the Saturne supervision, and let at the operators to take the initiative of the choise to start them or not.

The system uses more than 1000 pick up, and only one of them can distroy its good utilization without an interpretation and adaptation.

All the programs were written by the users, the operation group. So they can adapt, if it is needed, the programs at the actual situation.

We have been already systematically using this system for two years. Its structure permits the necessary evolution due to the new materials building.

That is a permanent state around our accelerator.

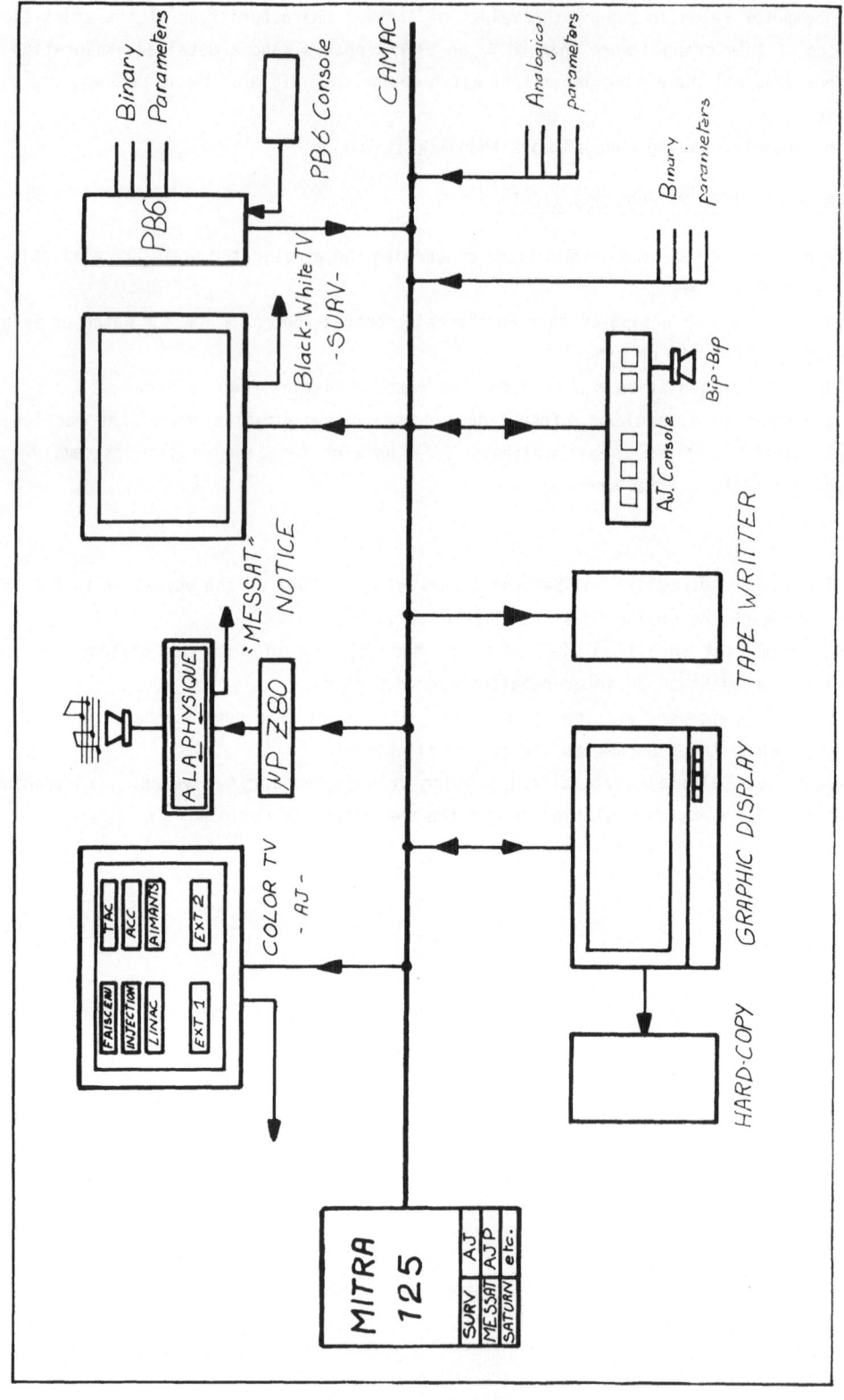

A LOCAL COMPUTER NETWORK FOR THE EXPERIMENTAL
DATA ACQUISITION AT BESSY

W. Buchholz

Berliner Elektronenspeicherring-Gesellschaft für
Synchrotronstrahlung mbH (BESSY)
1000 Berlin (West) 33, Lentzeallee 100

Summary

For the users of the Berlin dedicated electron storage ring for synchrotron radiation (BESSY) a local computer network has been installed: The system is designed primarily for data acquisition and offers the users a generous hardware provision combined with maximum software flexibility.

I. Introduction

For experimental data acquisition and data processing BESSY offers its users a micro-computer-network which consists of 15 PDP-11/23 or /03 computers. One PDP-11/23 is used as a host-computer with peripherals: hard and floppy-disks, plotter, printers and magnetic-tape. The other 14 computers are connected directly to this central computer via a star-shaped-network. Twelve of these satellite-computers (PDP-11/03) are designed to stand adjacent to the experimental equipment. They are free standing and include various experimental input/output interfaces as well as a graphic terminal and a dual-drive floppy-disk. Two other satellite-computers (PDP-11/23) are used for three terminal working places (program-development, data-analysis) and for dedicated system-jobs (print and plot-spooling, monitor-system etc.). The network is based on the STAR-eleven software, a multi computer implementation of the RT-11 operating-system. At the experimental computers a single-job working place with full real-time capabilities is provided.

Fig. 1 shows the block diagram of the system which was designed according to user demands as expressed at the first BESSY-Workshop in April 1980. The system supplies the user's with:

   i) A quasi-independent micro computer system at the experimental working place
  ii) Printer, plotter, mag-tape and hard-disk access via a transparent network
 iii) Terminals for program development, data analysis and user information.

In the following three sections these features are described in more detail.

II. Experimental Computer

II.1 Hardware

The experimental satellite computers are installed in portable cabinets and include a PDP 11/03 CPU with 64 Kbyte memory, dual floppy-disk, graphic terminal and the following experimental I/O modules with BNC or 25 PIN Cannon-Connectors at the front panel:

8-Channel ADC: This is a 12-Bit analog to digital converter with 8-channel differential multiplexer. The input voltage range is from 0 to 10 Volt at a maximum sample rate of about 35 KHz.

4 Channel DAC: This module supplies 4 independent analog output signals from 0 to 10 Volt using a 12-Bit digital analog converter. The outputs can drive a maximum current of 20 mA.

Real-Time-Clock: This module can generate pulse outputs with programmable rates between 1 micro-second and 320 seconds. Two Schmitt-Trigger inputs are supplied allowing external signals to start the clock.

RS-232: Three full duplex channels with EIA-Level are available.

16-Bit Parallel I/O: Two 16-Bit Input and Output Registers are available. TTL-signals are used. Handshake-lines are supported.

IEEE-Bus: IEEE Standard 488-1975 is available with a 3 meter cable to the first device.

Counter: A counter unit with one 32-Bit counter and 4 16-Bit counters is available. All counters are gated synchronously from the real-time-clock.

II.2 Experimental Satellite Software

The operating system is the STAR-eleven network implementation of the RT-11 monitor. The resident part of the satellite monitor occupies about 3 Kwords of memory. If support routines and the handler for a local mass-storage device like the floppy-disk are loaded there are about 24 Kwords of memory free for user programs. The full RT-11 SJ command language and all program requests are supported so that every standard RT-11 SJ program can run without modification. All system utilities (PIP, DIR, etc) as well as editors, FORTRAN-compiler, system-libraries and BESSY-programs can be loaded from the host computer's "system-disk" and executed.

Because of the single-job environment satellite programs can directly access the I/O registers of the computer which is ideal for real-time and measurement applications. There is no CPU-time sharing in the satellites so that interrupt driven data rates up to 10 KHz are possible. A few programs for collecting pulse-data from the counters, sampling analog signals with the ADC's and controlling BESSY monochromators are now available. Raw data output on the graphic terminal and the plotters is always part of these programs. For user-programming MACRO-11, FORTRAN IV, BASIC-11 and an assembly language written library of FORTRAN callable subroutines to control the various I/O modules (Counters, ADC's etc.) are available and increasing in number.

II.3 Disk Access

All satellites have access to one "system-disk", namely the one at the host. This makes the software easy to maintain and supplies the users always with the latest version of any given program. For program development or temporary data storage each satellite has access to an extra disk area at the host computer which is inaccessible to all other satellites. If files on common devices are generated users must include their I.D. code in the file name to avoid conflict with others.

III. Host Computer and Network

III.1 Host

The host computer is a PDP 11/23 with 64 KByte of memory and the following peripherals: dual floppy disk, dual 5 Mbyte-harddisk (Cartridge), 10 Mbyte-harddisk (Cartridge), 70 Mbyte-harddisk (Winchester), Printer/Plotter and a 9-track mag-tape (800/1600 Bpi).

The operating system is the RT-11 V04 foreground/background monitor which is easy to use and fast in I/O processing. The foreground job is used to run the network supervisor program "SUPER". SUPER performs all satellite specified directory and I/O requests and returns data and status to the satellites. Virtual disks, line-printer spoolers and record locking are also supported by the foreground job. The background job can be used as operator console but is mostly used for a program called "CACHE" which includes various network control functions as well as a technique to use all available memory as "cache-buffers" to increase satellite transfer speeds. The operating system bootstraps on "power-up" automatically and executes a "command-file" to load all handlers and network utilities including date and time from a battery driven calendar-clock.

III.2 Network

The network connection is as modular as the rest of the system. For each satellite connection there is one interface module in the host and one in the satellite computer. Both are connected with a 10*2 shielded twisted pair cable. Differential drivers and opto-coupler receivers supply a error-free transmission over distances of up to 200 meters. 16-bit serial data is transferred at 1.2 Mbaud. The transmission rate is regulated with interlock signals. With this technique the sender always waits until a receiver has accepted the last word sent before sending more data.

The connection software is transparent to users and programs, there are no special transfer commands. Messages are transferred between satellite and host using interrupt service routines. Each message consists of a 14 word header and up to 256 words of data. In the satellite data are transferred directly to or from the user-program buffers while the host end uses one 256 word data buffer for each satellite. If the program "CACHE" is running in the host background additional buffers are used by the satellite connection software.

Each working place at an experimental satellite (single-job) or at an analysis satellite (four-jobs) is provided with a software-tool to send and receive messages or commands to or from other working places in the system, even the host terminal. Send and receive functions are implemented in the keyboard command language and in a system library to call them from user written programs. At the keyboard the commands are "SEND" to send a message and "TELL" to send a command to be executed by an other computer. This technique can be used for both independent jobs controlled by one user or communicating jobs executed in different machines.

III.3 System Extensions

For further user support with computing power for memory or time-consuming software the concept of the BESSY data acquisition system includes a connection to a larger computer system as well as a connection to a regional network. In the realisation of this complete concept the wishes of the users, questions of capacity and financial considerations will have to be weighed against each other.

IV. Data Analysis and User Information

Satellite No. 1 is used to run the SHARE-eleven implementation of the RT-11 monitor which allows four user programs to execute in one PDP 11/23 with 256 Kbye of memory in a time-sharing environment. This software is designed for fast terminal I/O and each job gets a static memory region of 48 Kbyte so that no swapping occurs. All 4 jobs can have access to the local and to the host peripherals simultaneously. Jobs can execute all standard RT-11 SJ programs, but access to the I/O registers is restricted to special library calls.

Satellite 2 runs the same operating system as satellite 1 but all four jobs are used for system work. Job 1 has a terminal attached and is used mainly as operator console because the working space of 48 Kbyte of memory is much greater than the host computer's background. Job 2 has no terminal attached but can be controlled from any other terminal in the system. It executes a program which searches every 3 minutes a particular disk area for files which are sent from other satellites in the system to the LA-120 printer. After the output to the printer is done the "spooled" file is deleted from the disk. Job 3 performs the same function than job 2 but the output device is the HP 7220 plotter. Job 4 has a graphic terminal attached and runs a program for the BESSY "Monitor-System" which displays machine-parameters, information and the actual beam current on the terminal. The main task of the program is to sample beam-current data via an ADC-channel and update the display every 30 seconds. The video-output of the terminal is used to transmit the information using RG58 coax-cables to about 15 monitors (modified TV sets) in the experimental hall as well as in the office areas.

Acknowledgements

I would like to thank Prof.W. Peatman and G. Skerra and other members of our department for their help and co-operation.

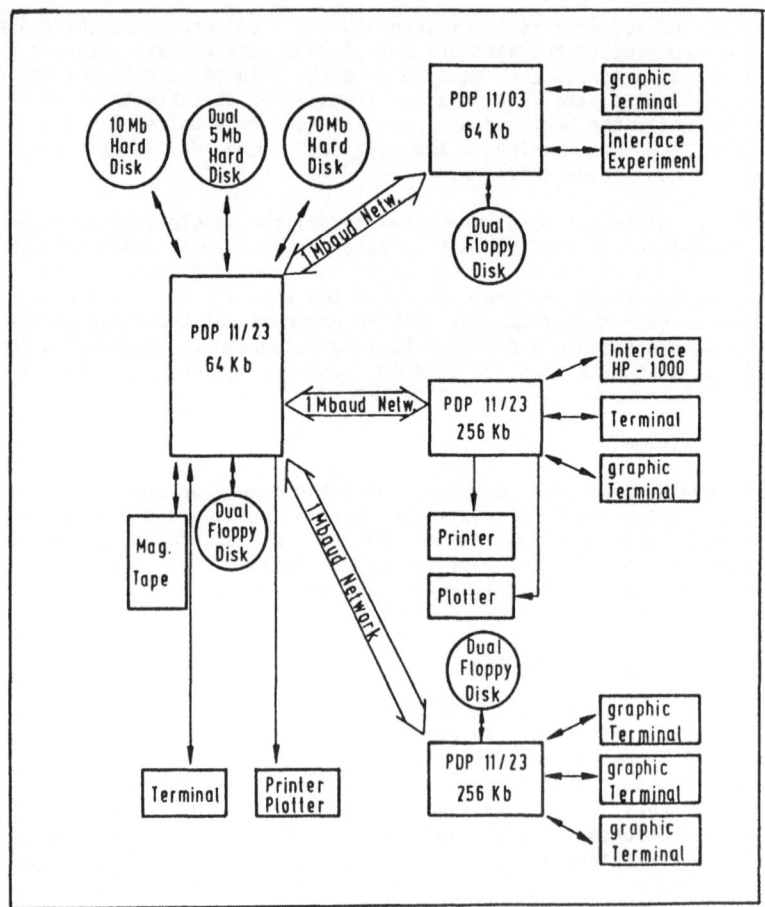

Fig. 1: System Layout

CLOSING REMARKS

M.C. Crowley-Milling, CERN, CH-1211 Geneva 23

I am glad to see that, according to the program, I am down to give the concluding remarks, which leaves me free to decide what these should be. Often in conferences, the final speaker is expected to give a resumé of all the papers that have been presented at the conference, weighing his remarks according to the standing of the speaker or the organization to which he is affiliated. Thankfully, I am spared that commitment, which I would not take on willingly. Incidentally, I attended one physics conference where the opening speaker gave a review with the highlights of the papers to be presented, so that those attending could go an enjoy themselves without having to attend the sessions!

Shed of this duty, I can take a broader view of this conference, and we were certainly urged to take a broad view, from Sir John Adams' trackless desert in the Physics field, to Dr. Joho's Treasure Island in the southern seas. This has been the first attempt to bring together all the aspects of the use of computers in the design, control and operation of accelerators. This encompasses a wide spectrum of people and activities, who may not have been as successful in communication between themselves as they have been in designing systems for communications between computers. This conference has succeeded in bringing together these different aspects of the use of computers. We have some general reviews of the subjects and their problems, which I hope have enabled specialists in one field to gain an insight into the other fields, and we have had specialised papers that have taken us up to the frontiers of present day knowledge. I hope you will agree with me that this has produced a reasonably well balanced mixture.

I also think that the decision of the Scientific Advisory Committee that we would only have invited talks, where the speaker was given a reasonable time to explore his subject, and poster sessions, was a good one. One of the curses of conferences is the ten minute contributed paper. It needs a very skilful presentation to give a significant amount of information, and at many conferences I have attended, the speaker has only got through his introduction by the time the yellow light comes on! With the poster sessions, those who are interested can discuss the subject of the paper to any depth desired with the author(s), while those with only a passing interest in the subject can get the gist of the paper in just passing by.

Where do we go from here?

I am tempted to speculate that, as the design and simulation programs become more and more powerful, we will be able to predict exactly the performance of an accelerator before it is built.

The next stage is to get our colleagues on the experimental side to predict the results of their experiments to the same accuracy, and then it will be no longer necessary to build accelerators at all!

However, I must have second thoughts about such a prediction as long as we have the situation, pointed out in the introductory speech, that only a tiny minority of high energy accelerators are famous for the discoveries that were predicted when they were constructed. So I think we will still have to build the accelerators in the future, and we will not all be out of work!

Rather will I recall an advertisement for Fiat cars. This said that the cars were designed by computers and built by robots. To this the graphitisti had added "and driven by morons". As a result of this conference I could change "Fiat cars" to "accelerators", but with a very significant difference. We have seen how much computers are used in the design of accelerators. Sir John has suggested that accelerators should be built by robots in the future, but the graphiti will have to change. The contributions to this conference have shown that accelerators are driven by very intelligent people with very intelligent programs.

LIST of CONFERENCE ATTENDEES

Adams, J. B.
CERN

CH-1211 Geneva 23
Switzerland

Aghion, F.
University of Milano
Cyclotron Laboratory
Via Celoria 16
I-20133 Milano
Italy

Albrand, S.
Institut des Sciences Nucléaires
de Grenoble
53 Ave. des Martyrs
F-38026 Grenoble Cedex
France

Anderson, J.
University of Manitoba
Cyclotron Laboratory
Department of Physics
Winnipeg, Manitoba R3T2N2
Canada

Arhippainen, J.
Âbo Akademi
Accelerator Laboratoriet
Porthans Gatan 3-5
SF-20500 Âbo 50
Finland

Âsberg, A.
Gustav-Werner-Institute

P.O. Box 531
S-751 21 Uppsala
Sweden

Astarlioglu, T.
Chalk River Nuclear Laboratories
Atomic Energy of Canada Ltd.

Chalk River, Ont. KOJ 1JO
Canada

Barber, D. P.
Deutsches Elektronen Synchrotron
DESY
Notkestrasse 85
D-2000 Hamburg 52

Blechschmidt, D.
CERN

CH-1211 Geneva 23
Switzerland

Blumer, Th.
Schweizer Institut für
Nuklearforschung - SIN

CH-5234 Villigen
Switzerland

Bogert, D.
Fermi National Accel. Laboratory

P.O. Box 500
Batavia, IL 60510
U.S.A.

Bombi, F.
Joint European Undertaking (JET)

Abingdon, OX14 3EA
United Kingdom

Botman, J.
TRIUMF - University of B.C.

4004 Wesbrook Mall
Vancouver, B.C. V6T 2A3
Canada

Bozoki, E. S.
Brookhaven National Laboratory
National Synchrotron Light Source

Upton, NY 11793
U.S.A.

Brand, K.
Ruhr-Universität Bochum
Dynamitron-Tandem-Laboratorium
Universitätsstr. 100
D-4630 Bochum

Brandis, H.
Gesellschaft für
Schwerionenforschung mbH - GSI
Postfach 11 05 41
D-6100 Darmstadt 11

Brefeld, W.
Universität Bonn
Physikalisches Institut
Nussallee 12
D-5300 Bonn 1

Buchholz, W.
Berl. Elektronen-Speicherringges.
für Synchrotronstr. mbH - BESSY
Lentzeallee 100
D-1000 Berlin 33

Busse, W.
Hahn-Meitner-Institut Bln GmbH
Bereich P   VICKSI
Glienicker Str. 100
D-1000 Berlin 39

Chabert, A.
GANIL

B.P. 5027
F-14021 Caen Cedex
France

Chao, A. W.
Stanford Linear Accelerator Center
Stanford University
P.O. Box 4349
Stanford, CA 94305
U.S.A.

Chohan, V.
CERN
PS Division

CH-1211 Geneva 23
Switzerland

Ciapala, E.
CERN

CH-1211 Geneva 23
Switzerland

Crowley-Milling, M.
CERN

CH-1211 Geneva 23
Switzerland

Crowley-Milling, M.
CERN

CH-1211 Geneva 23
Switzerland

Czosnyka, T.
Institute for Nuclear Science and
Technology

PL-05 400 Otwock-Swierk
Poland

Dahl, L.
Gesellschaft für
Schwerionenforschung mbH - GSI
Postfach 11 05 41
D-6100 Darmstadt 11

Degenhardt, K. H.
Hahn-Meitner-Institut Bln GmbH
Bereich D/M
Glienicker Str. 100
D-1000 Berlin 39

Delgado, R. M.
CERN

CH-1211 Geneva 23
Switzerland

Dell, Jr, G. F.
Brookhaven National Laboratory
Building 902 A

Upton, NY 11973
U.S.A.

Dohan, D. A.
TRIUMF - University of B.C.

4004 Wesbrook Mall
Vancouver, B.C.   V6T 2A3
Canada

Donald, M. H. R.
Stanford Linear Accelerator Center
Stanford University, Bin 26
P.O. Box 4349
Stanford, CA 94305
U.S.A.

Dorenbos, T.
CERN
PS-Division

CH-1211 Geneva 23
Switzerland

Dragt, A. J.
University of Maryland
Dept. of Physics and Astronomy
Center for Theoretical Physics
College Park, MD 20742
U.S.A.

Edwards, T. R. M.
Rutherford and Appleton Laboratory

Chilton
Didcot, Berkshire OX11 0QX
United Kingdom

Egan-Krieger, G. von
Berl. Elektronen-Speicherringges.
für Synchrotronstr. mbH - BESSY
Lentzeallee 100
D-1000 Berlin 33

Egan-Krieger, G. von
Berl. Elektronen-Speicherringges.
für Synchrotronstr. mbH - BESSY
Lentzeallee 100
D-1000 Berlin 33

Eversheim, P. D.
Universität Bonn
Institut für Strahlen und Kernphys.
Nußallee 14-16
D-5300 Bonn

Farrell, J. A.
Los Alamos National Laboratory
AT-2, MS H 818
P.O. Box 1663
Los Alamos, NM 87545
U.S.A.

Fougeron, C.
Laboratoire National SATURNE
CEN-Saclay
B.P. 2
F-91191 Gif-sur-Yvette
France

Franczak, B.
Gesellschaft für
Schwerionenforschung mbH - GSI
Postfach   11 05 41
D-6100 Darmstadt 11

Frese, H.
Deutsches Elektronen Synchrotron
DESY
Notkestrasse 85
D-2000 Hamburg 52

Glatz, J.
Gesellschaft für
Schwerionenforschung mbH - GSI
Postfach   11 05 41
D-6100 Darmstadt 11

Gournay, J. F.
Centre d'Etudes Nucléaire de Saclay
Service de l'Accélérateur Linéaire
B.P. 2
F-91191 Gif-sur-Yvette
France

Gurd, D. P.
TRIUMF - University of B.C.

4004 Wesbrook Mall
Vancouver, B.C.   V6T 2A3
Canada

Gusdal, M. I.
University of Manitoba
Cyclotron Laboratory
Department of Physics
Winnipeg, Manitoba R3T2N2
Canada

Halling, H.
Kernforschungsanlage Jülich GmbH
Zentrallabor für Elektronik
Postfach 19 13
D-5170 Jülich 1

Hartwig, D.
Gesellschaft für
Schwerionenforschung mbH - GSI
Postfach 11 05 41
D-6100 Darmstadt 11

Hatton, V.
CERN

CH-1211 Geneva 23
Switzerland

Hellborg, R.
Department of Nuclear Physics
University of Lund
Sölvegatan 14
S-223 62 Lund
Sweden

Heymans, P.
CERN
PS-Division

CH-1211 Geneva 23
Switzerland

Hoffmann, J.
Gesellschaft für
Schwerionenforschung mbH - GSI
Postfach 11 05 41
D-6100 Darmstadt 11

Hofmann, I.
Gesellschaft für
Schwerionenforschung mbH - GSI
Postfach 11 05 41
D-6100 Darmstadt 11

Iselin, F. Ch.
CERN

CH-1211 Geneva 23
Switzerland

Jacoby, W.
Gesellschaft für
Schwerionenforschung mbH - GSI
Postfach 11 05 41
D-6100 Darmstadt 11

Jaeschke, E.
Max-Planck-Institut
für Kernphysik
Postfach 10 39 80
D-6900 Heidelberg 1

Jahnke, A.
Max-Planck-Institut
für Plasmaphysik
Boltzmannstr. 2
D-8046 Garching

Joho, W.
Schweizer Institut für
Nuklearforschung - SIN

CH-5234 Villigen
Switzerland

Jowett, J. M.
CERN
LEP-Division

CH-1211 Geneva 23
Switzerland

Junior, P.
Universität Frankfurt am Main
Institut für Angew. Physik
Robert-Maier-Str. 2-4
D-6000 Frankfurt a./M.

Kaspar, K.
Gesellschaft für
Schwerionenforschung mbH - GSI
Postfach 11 05 41
D-6100 Darmstadt 11

Keil, E.
CERN

CH-1211 Geneva 23
Switzerland

Kewisch, J.
Deutsches Elektronen Synchrotron
DESY
Notkestrasse 85
D-2000 Hamburg 52

Kiehne, T.
Hahn-Meitner-Institut Bln GmbH
Bereich P
Glienicker Str. 100
D-1000 Berlin 39

Klotz, W.-D.
Berl. Elektronen-Speicherringges.
für Synchrotronstr. mbH - BESSY
Lentzeallee 100
D-1000 Berlin 33

Kluge, H.
Hahn-Meitner-Institut Bln GmbH
Bereich P   VICKSI
Glienicker Str. 100
D-1000 Berlin 39

Kölbig, K. S.
CERN

CH-1211 Geneva 23
Switzerland

Kost, C.
TRIUMF - University of B.C.

4004 Wesbrook Mall
Vancouver, B.C.   V6T 2A3
Canada

Kraus-Vogt, W.
Kernforschungsanlage Jülich GmbH

Postfach 19 13
D-5170 Jülich 1

Krusche, A.
CERN

CH-1211 Geneva 23
Switzerland

Kugler, H.
CERN
PS-Division

CH-1211 Geneva 23
Switzerland

Kulinski, S.
Institute for Nuclear Science and
Technology

PL-05 400 Otwock-Swierk
Poland

Langenbeck, B.
Gesellschaft für
Schwerionenforschung mbH - GSI
Postfach 11 05 41
D-6100 Darmstadt 11

Larsson, J. E.
Instrument AB Scanditronix

Husbyborg
S-75590 Uppsala
Sweden

Leaux, P.
Laboratoire National SATURNE
CEN-Saclay
B.P. 2
F-91191 Gif-sur-Yvette
France

Lee, M. J.
Stanford Linear Accelerator Center
Stanford University
P.O. Box 4349
Stanford, CA 94305
U.S.A.

Lemaire, J.-L.
Laboratoire National SATURNE
CEN-Saclay
B.P. 2
F-91191 Gif-sur-Yvette
France

Lindenberger, K. H.
Hahn-Meitner-Institut Bln GmbH

Glienicker Str. 100
D-1000 Berlin 39

Lobb, D. E.
TRIUMF - University of Victoria
Physics Dept.

Victoria, B.C. V8W 2Y2
Canada

Ludemann, C. A.
Oak Ridge National Laboratory
Physics Division, Bldg. 6000y
P.O. Box X
Oak Ridge, TN 37830
U.S.A.

Lustig, H. D.
CERN
PS-Division

CH 1211 Geneva 23
Switzerland

Mackenzie, G.
TRIUMF - University of B.C.

4004 Wesbrook Mall
Vancouver, B.C. V6T 2A3
Canada

Maier, K. H.
Hahn-Meitner-Institut Bln GmbH
Bereich P   VICKSI
Glienicker Str. 100
D-1000 Berlin 39

Maier, R.
Berl. Elektronen-Speicherringges.
für Synchrotronstr. mbH - BESSY
Lentzeallee 100
D-1000 Berlin 33

Mais, H.
Deutsches Elektronen Synchrotron
DESY
Notkestrasse 85
D-2000 Hamburg 52

Marti, Y.
CERN

CH-1211 Geneva 23
Switzerland

Martin, S. A.
Kernforschungsanlage Jülich GmbH

Postfach 19 13
D-5170 Jülich 1

Martin, S. A.
Kernforschungsanlage Jülich GmbH

Postfach 19 13
D-5170 Jülich 1

Martin, B.
Hahn-Meitner-Institut Bln GmbH
Bereich P   VICKSI
Glienicker Str. 100
D-1000 Berlin 39

McMichael, G. E.
Chalk River Nuclear Laboratories
Atomic Energy of Canada Ltd.

Chalk River, Ont. K0J 1J0
Canada

Meads, P. F.

7053 Shirley Drive
Oakland, CA 94611
U.S.A.

Melen, R. E.
Stanford Linear Accelerator Center
Stanford University
P.O. Box 4349
Stanford, CA 94305
U.S.A.

Michaelsen, R.
Hahn-Meitner-Institut Bln GmbH
Bereich P   VICKSI
Glienicker Str. 100
D-1000 Berlin 39

Nettesheim, M.
Hahn-Meitner-Institut Bln GmbH

Glienicker Str. 100
D-1000 Berlin 39

Niederer, J.
Brookhaven National Laboratory
Building 515

Upton, NY 11973
U.S.A.

Nielsen, B.
Danfysik A/S

DK-4040 Jyllinge
Denmark

Nietzel, Ch.
Universität Bonn
Physikalisches Institut
Nussallee 12
D-5300 Bonn 1

Nuhn, H.-D.
Universität Bonn
Physikalisches Institut
Nussallee 12
D-5300 Bonn 1

Panzeri, E.
University of Milano
Cyclotron Laboratory
Via Celoria 16
I-20133 Milano
Italy

Pelzer, W.
Hahn-Meitner-Institut Bln GmbH
Bereich P   VICKSI
Glienicker Str. 100
D-1000 Berlin 39

Peters, F.
Deutsches Elektronen Synchrotron
DESY
Notkestrasse 85
D-2000 Hamburg 52

Piwinski, A.
Deutsches Elektronen Synchrotron
DESY
Notkestrasse 85
D-2000 Hamburg 52

Poole, D. E.
Daresbury Laboratory
S.E.R.C.

Daresbury, WA4 4AD
United Kingdom

Poppensieker, K.
Gesellschaft für
Schwerionenforschung mbH - GSI
Postfach 11 05 41
D-6100 Darmstadt 11

Potier, J. P.
CERN
PS Division

CH-1211 Geneva 23
Switzerland

Prasuhn, D.
Kernforschungsanlage Jülich GmbH

Postfach 19 13
D-5170 Jülich 1

Priesmeyer, H. G.
Inst. f. Reine u. Ang. Kernphysik
Universität Kiel
c/o GKSS-Forschungszentrum
D-2054 Geesthacht

Promé, M.
GANIL

B.P. 5027
F-14021 Caen Cedex
France

Reich, K. H.
Universität Dortmund

Postfach 500 500
D-4600 Dortmund 50

Remmer, W.
CERN
PS-Division

CH-1211 Geneva 23
Switzerland

Repnow, R.
Max-Planck-Institut
für Kernphysik
Postfach 10 39 80
D-6900 Heidelberg 1

Reuber, C.
VDI nachrichten
VDI-Verlag GmbH
Postfach 1139
D-4000 Düsseldorf 1

Riedel, C.
Gesellschaft für
Schwerionenforschung mbH - GSI
Postfach 11 05 41
D-6100 Darmstadt 11

Ripken, G.
Deutsches Elektronen Synchrotron
DESY
Notkestrasse 85
D-2000 Hamburg 52

Rohrer, L.
Beschleunigerlabor der
Universitäten

D-8046 Garching

Roßbach, J.
Deutsches Elektronen Synchrotron
DESY
Notkestrasse 85
D-2000 Hamburg 52

Roy-Poulsen, K.
Niels Bohr Institutet

Blegdamsvej 17
DK-2100 Kopenhagen
Denmark

Saltmarsh, Ch. G.
CERN
SPS Division

CH-1211 Geneva 23
Switzerland

Sauret, J.
GANIL

B.P. 5027
F-14021 Caen Cedex
France

Scandale,
CERN

CH-1211 Geneva 23
Switzerland

Schaffner, E.
Gesellschaft für
Schwerionenforschung mbH - GSI
Postfach 11 05 41
D-6100 Darmstadt 11

Schreuder, H. W.
Kernfysisch Versneller Institut
der Rijksuniversiteit
Zernikelaan 25
NL-9747 AA Groningen
Netherlands

Segler, S. L.
Fermi National Accel. Laboratory
MS 307
P.O. Box 500
Batavia, IL 60510
U.S.A.

Sethi, R. C.
Hahn-Meitner-Institut Bln GmbH
Bereich P   VICKSI
Glienicker Str. 100
D-1000 Berlin 39

Shering, G. C.
CERN

CH-1211 Geneva 23
Switzerland

Sherman, J. D.
Los Alamos National Laboratory
MS H 818
P.O. Box 1663
Los Alamos, NM 87545
U.S.A.

Sherman, H. J.
Daresbury Laboratory
S.E.R.C.

Daresbury, WA4 4AD
United Kingdom

Simrock, S.
Technische Hochschule Darmstadt

Schloßgartenstr. 9
D-6100 Darmstadt

Spellmeyer, B.
Hahn-Meitner-Institut Bln GmbH
Bereich P   VICKSI
Glienicker Str. 100
D-1000 Berlin 39

Spurling, K.
Daresbury Laboratory
S.E.R.C.

Daresbury, WA4 4AD
United Kingdom

Stoff, H.
Kernforschungsanlage Jülich GmbH
Zentrallabor für Elektronik
Postfach 19 13
D-5170 Jülich 1

Strehl, P.
Gesellschaft für
Schwerionenforschung mbH - GSI
Postfach 11 05 41
D-6100 Darmstadt 11

Struckmeier, J.
Gesellschaft für
Schwerionenforschung mbH - GSI
Postfach 11 05 41
D-6100 Darmstadt 11

Swanson, E.
University of Washington
Nuclear Physics Laboratory GL-10

Seattle, WA 98195
U.S.A.

Tenten, W.
Kernforschungsanlage Jülich GmbH
Zentrallabor für Elektronik
Postfach 19 13
D-5170 Jülich 1

Thouw, H.
Kernforschungszentrum Karlsruhe
Zyklotronlab., Inst.f.Kernphys. III
Postfach 36 40
D-7500 Karlsruhe 1

Tran, D. T.
THOMSON - CSF
Division Tubes Electroniques
B.P. 305
F-92102 Boulogne Cedex
France

Tronc, D.
CGR MeV

B.P. 34
F-78530 BUC
France

Trowbridge, C. W.
Rutherford and Appleton Laboratory

Chilton
Didcot, Berkshire OX110QX
United Kingdom

Valero, S.
Centre d'Etudes Nucléaire de Saclay
DPhN-BE
B.P. 2
F-91191 Gif-sur-Yvette
France

Volmer, P.
Centre de Recherches Nucléaires
de Strasbourg
B.P. 20
F-67037 Strasbourg Cedex
France

Vos, L.
CERN

CH-1211 Geneva 23
Switzerland

Vukanovic, R.
Department of Physics
Boris Kidric Inst.of Nucl. Sciences
P.O. Box 552
YU-11001 Beograd
Yugoslavia

Warren, J. L.
Los Alamos National Laboratory
MS H 847
P.O. Box 1663
Los Alamos, NM 87545
U.S.A.

Weiland, T.
Deutsches Elektronen Synchrotron
DESY
Notkestrasse 85
D-2000 Hamburg 52

Wermelskirchen, C.
Universität Bonn
Physikalisches Institut
Nussallee 12
D-5300 Bonn 1

Werner, K.
Max-Planck-Institut
für Kernphysik
Postfach 10 39 80
D-6900 Heidelberg 1

Wilhelm, M.
Hahn-Meitner-Institut Bln GmbH

Glienicker Str. 100
D-1000 Berlin 39

Wilhelm, W.
Technische Universität München
Physik Department, Teilinstitut E12
James-Frank-Str.
D-8046 Garching

Wilson, E. J. N.
CERN

CH-1211 Geneva 23
Switzerland

Wrulich, A.
Deutsches Elektronen Synchrotron
DESY
Notkestrasse 85
D-2000 Hamburg 52

Zech, E.
Technische Universität München
Physik Department
James-Frank-Str.
D-8046 Garching

Zelazny, R.
Regional Computation Centre of the
Atomic Energy
RCCAE CYFRONET
PL-05-400 Otwock-Swierk
Poland

Ziegler, K.
Hahn-Meitner-Institut Bln GmbH
Bereich P   VICKSI
Glienicker Str. 100
D-1000 Berlin 39

Ziem, P.
Hahn-Meitner-Institut Bln GmbH
Bereich P
Glienicker Str. 100
D-1000 Berlin 39

Zwoll, K.
Kernforschungsanlage Jülich GmbH
Zentrallabor für Elektronik
Postfach 19 13
D-5170 Jülich 1

AUTHOR INDEX

Aghion, F.	351		Gontier, J.	553
Akiyama, A.	367		Goto, A.	530
Albrand, S.	188		Guan, Xia-ling	140
Atkins, V. R.	344		Guan, Xia-Ling	182
Barber, D. P.	243		Guignard, G.	237
Belmont, J. L.	188		Gurd, D. P.	332
Bogert, D.	338		Gusdal, M. I.	411
Bombi, F.	311		Halfmann, K. D.	206
Bozoki, E. S.	416		Hatton, V.	455
Bozoki, E. S.	420		He, N. W.	391
Bozsik, I.	128		Heymans, P.	300
Brand, K.	152		Hofmann, I.	128
Bremer, H. D.	243		Hofmann, I.	134
Bruckshaw, J.	411		Horrabin, C. W.	473
Buchholz, W.	557		Houtman, H.	98
Büsch, R.	391		Hukumoto, S.	361
Busse, W.	497		Igarashi, Z.	361
Cao, Qing-xi	140		Ikegami, K.	361
Cao, Qing-xi	182		Iselin, F. Ch.	146
Casstevens, B. J.	491		Ishii, K.	361
Cauvin, B.	224		Ishii, K.	367
Chabert, A.	164		Jahnke, A.	128
Chao, A. W.	59		Johnson, C.	405
Chen, Mao-bai	116		Johnstone, W. T.	473
Chen, Mao-bai	218		Joho, W.	446
Chidley, B. G.	212		Jowett, J. M.	261
Chohan, V.	405		Junior, P.	206
Crowley-Milling, M.	278		Kadokura, E.	367
Czosnyka, T.	92		Kase, M.	530
Dahl, L.	518		Katoh, T.	367
Davis, S. V.	473		Kewisch, J.	243
Deitinghoff, H.	206		Kewisch, J.	249
Dell, Jr, G. F.	176		Kikutani, E.	367
Derenchuk, V.	411		Kimura, Y.	367
Deutschman, K.	92		Kishiro, Jun-ichi	361
Diquattro, S.	351		Klotz, W.-D.	425
Divatia, A. S.	199		Klotz, W.-D.	436
Dohan, D. A.	332		Komada, I.	367
Dorenbos, T.	398		Kost, C.	98
Douglas, D. R.	122		Kost, C.	158
Dragt, A. J.	122		Kubota, T.	361
Edwards, T. R. M.	377		Kudo, K.	367
Egan-Krieger, G. von	425		Kuiper, B.	300
Egan-Krieger, G. von	436		Kulinski, S.	92
Ehrich, A.	518		Kulinski, S.	86
Eversheim, P. D.	386		Kurokawa, S.	367
Eversheim, P. D.	391		Lagniel, J. M.	553
Farrell, J. A.	267		Lagniel, J.-M.	542
Fliss, G.	524		Lancaster, J.	411
Fouan, J. P.	224		Lapostolle, P. M.	224
Fougeron, C.	553		Lee, M. J.	465
Franczak, B.	170		Lemaire, J.-L.	542
Frese, H.	275		Lewin, H. C.	243
GANIL Group,	536		Limberg, T.	243
GANIL OG and CCG,	503		Loyant, J. M.	536

Loyer, F.	536	Sheppard, J. C.	465
Ludemann, C. A.	491	Shering, G. C.	481
Maier, R.	425	Spellmeyer, B.	497
Maier, R.	436	Spurling, K.	473
Mais, H.	243	Strehl, P.	524
Marti, Y.	237	Struckmeier, J.	524
Martin, B.	497	Sullenberger, M.	465
Mattei, P.	553	Takagi, A.	361
McIlwain, A.	411	Takasaki, E. M.	361
McKee, J. S. C.	411	Takeda, S.	367
McMichael, G. E.	212	Thiessen, H. A.	255
Melen, R. E.	289	Tronc, D.	
Michaelsen, R.	497	Trowbridge, C. W.	33
Miller, M.	405	Uchino, K.	367
Miyazawa, Y.	530	Valero, S.	224
Mori, Y.	361	Warren, J. L.	255
Müller, R. W.	128	Weiland, T.	21
Nietzel, Ch.	355	Welt, H. J.	355
Odera, M.	530	Wermelskirchen, C.	355
Oh, S.	104	Wilhelm, W.	110
Oh, S.	411	Wilson, E. J. N.	11
Oide, K.	367	Woodley, M.	465
Paccalini, A.	351	Wrulich, A.	75
Panzeri, E.	351	Xu, Sen-lin	218
Pelzer, W.	497	Yano, Y.	530
Piwinski, A.	50	Yoon, M.	104
Pogson, R.	104	Zablotny, J. R.	372
Pogson, R.	411	Zaremba, S.	92
Poole, D. E.	344	Zech, E.	193
Potier, J. P.	405	Zelazny, R.	316
Promé, M.	164	Ziegler, K.	497
PS Controls Group,	300	Zimek, Z.	372
Rawlinson, W. R.	344	Zoubek, N.	206
Reeve, P. A.	158		
Ripken, G.	243		
Ripouteau, F.	188		
Rivoltella, G.	351		
Rösch, N.	193		
Rossen, P. von	386		
Rossen, P. von	391		
Rossmanith, R.	243		
Saltmarsh, Ch. G.	509		
Sauret, J.	164		
Sauret, J.	536		
Sawlewicz, L.	86		
Schaa, V.	524		
Schempp, A.	206		
Schillo, M.	355		
Schmidt, R.	243		
Schott, W.	193		
Segler, S. L.	338		
Sekutowicz, J.	86		
Sen, Wen-bin	116		
Sen, Wen-bin	218		
Sethi, R. C.	199		

Springer Series in Computational Physics

Editors: H. Cabannes, M. Holt, H. B. Keller, J. Killeen, S. A. Orszag

C. A. J. Fletcher
Computational Galerkin Methods
1984. 107 figures. XI, 309 pages.
ISBN 3-540-12633-3

R. Glowinski
Numerical Methods for Nonlinear Variational Problems
1984. 82 figures. XV, 493 pages.
ISBN 3-540-12434-9
(Originally published as "Glowinski, Lectures on Numerical Methods...", Tata Institute Lectures on Mathematics, 1980)

M. Holt
Numerical Methods in Fluid Dynamics
2nd revised edition. 1984. 114 figures. XI, 273 pages. ISBN 3-540-12799-2

O. G. Mouritsen
Computer Studies of Phase Transitions and Critical Phenomena
1984. 79 figures. XII, 200 pages.
ISBN 3-540-13397-6

O. Pironneau
Optimal Shape Design for Elliptic Systems
1984. 57 figures. XII, 168 pages.
ISBN 3-540-12069-6

M. Kubíček, M. Marek
Computational Methods in Bifurcation Theory and Dissipative Structures
1983. 91 figures. XI, 243 pages.
ISBN 3-540-12070-X

R. Peyret, T. D. Taylor
Computational Methods for Fluid Flow
1983. 125 figures. X, 358 pages.
ISBN 3-540-11147-9

Y. I. Shokin
The Method of Differential Approximation
Translated from the Russian by K. G. Roesner
1983. 75 figures, 12 tables. XIII, 296 pages.
ISBN 3-540-12225-7

Finite-Difference Techniques for Vectorized Fluid Dynamics Calculations
Editor: D. L. Book
1981. 60 figures. VIII, 226 pages.
ISBN 3-540-10482-8

D. P. Telionis
Unsteady Viscous Flows
1981. 132 figures. XXIII, 408 pages.
ISBN 3-540-10481-X

F. Thomasset
Implementation of Finite Element Methods for Navier-Stokes Equations
1981. 86 figures. VII, 161 pages.
ISBN 3-540-10771-1

F. Bauer, O. Betancourt, P. Garabedian
A Computational Method in Plasma Physics
1978. 22 figures. VIII, 144 pages.
ISBN 3-540-08833-4

Springer-Verlag
Berlin
Heidelberg
New York
Tokyo

Lecture Notes in Physics

Vol. 195: Trends and Applications of Pure Mathematics to Mechanics. Proceedings, 1983. Edited by P. G. Ciarlet and M. Roseau. V, 422 pages. 1984.

Vol. 196: WOPPLOT 83. Parallel Processing: Logic, Organization and Technology. Proceedings, 1983. Edited by J. Becker and I. Eisele. V, 189 pages. 1984.

Vol. 197: Quarks and Nuclear Structure. Proceedings, 1983. Edited by K. Bleuler. VIII, 414 pages. 1984.

Vol. 198: Recent Progress in Many-Body Theories. Proceedings, 1983. Edited by H. Kümmel and M. L. Ristig. IX, 422 pages. 1984.

Vol. 199: Recent Developments in Nonequilibrium Thermodynamics. Proceedings, 1983. Edited by J. Casas-Vázquez, D. Jou and G. Lebon. XIII, 485 pages. 1984.

Vol. 200: H. D. Zeh, Die Physik der Zeitrichtung. V, 86 Seiten. 1984.

Vol. 201: Group Theoretical Methods in Physics. Proceedings, 1983. Edited by G. Denardo, G. Ghirardi and T. Weber. XXXVII, 518 pages. 1984.

Vol. 202: Asymptotic Behavior of Mass and Spacetime Geometry. Proceedings, 1983. Edited by F. J. Flaherty. VI, 213 pages. 1984.

Vol. 203: C. Marchioro, M. Pulvirenti, Vortex Methods in Two-Dimensional Fluid Dynamics. III, 137 pages. 1984.

Vol. 204: Y. Waseda, Novel Application of Anomalous (Resonance) X-Ray Scattering for Structural Characterization of Disordered Materials. VI, 183 pages. 1984.

Vol. 205: Solutions of Einstein's Equations: Techniques and Results. Proceedings, 1983. Edited by C. Hoenselaers and W. Dietz. VI, 439 pages. 1984.

Vol. 206: Static Critical Phenomena in Inhomogeneous Systems. Edited by A. Pękalski and J. Sznajd. Proceedings, 1984. VIII, 358 pages. 1984.

Vol. 207: S. W. Koch, Dynamics of First-Order Phase Transitions in Equilibrium and Nonequilibrium Systems. III, 148 pages. 1984.

Vol. 208: Supersymmetry and Supergravity/Nonperturbative QCD. Proceedings, 1984. Edited by P. Roy and V. Singh. V, 389 pages. 1984.

Vol. 209: Mathematical and Computational Methods in Nuclear Physics. Proceedings, 1983. Edited by J. S. Dehesa, J. M. G. Gomez and A. Polls. V, 276 pages. 1984.

Vol. 210: Cellular Structures in Instabilities. Proceedings, 1983. Edited by J. E. Wesfreid and S. Zaleski. VI, 389 pages. 1984.

Vol. 211: Resonances – Models and Phenomena. Proceedings, 1984. Edited by S. Albeverio, L. S. Ferreira and L. Streit. VI, 369 pages. 1984.

Vol. 212: Gravitation, Geometry and Relativistic Physics. Proceedings, 1984. Edited by Laboratoire "Gravitation et Cosmologie Relativistes", Université Pierre et Marie Curie et C.N.R.S., Institut Henri Poincaré, Paris. VI, 336 pages. 1984.

Vol. 213: Forward Electron Ejection in Ion Collisions. Proceedings, 1984. Edited by K. O. Groeneveld, W. Meckbach and I. A. Sellin. VII, 165 pages. 1984.

Vol. 214: H. Moraal, Classical, Discrete Spin Models. VII, 251 pages. 1984.

Vol. 215: Computing in Accelerator Design and Operation. Proceedings, 1983. Edited by W. Busse and R. Zelazny. XII, 574 pages. 1984.

Selected Issues from
Lecture Notes in Mathematics

Vol. 932: Analytic Theory of Continued Fractions. Proceedings, 1981. Edited by W.B. Jones, W.J. Thron, and H. Waadeland. VI, 240 pages. 1982.

Vol. 934: M. Sakai, Quadrature Domains. IV, 133 pages. 1982.

Vol. 935: R. Sot, Simple Morphisms in Algebraic Geometry. IV, 146 pages. 1982.

Vol. 936: S.M. Khaleelulla, Counterexamples in Topological Vector Spaces. XXI, 179 pages. 1982.

Vol. 937: E. Combet, Intégrales Exponentielles. VIII, 114 pages. 1982.

Vol. 938: Number Theory. Proceedings, 1981. Edited by K. Alladi. IX, 177 pages. 1982.

Vol. 942: Theory and Applications of Singular Perturbations. Proceedings, 1981. Edited by W. Eckhaus and E.M. de Jager. V, 363 pages. 1982.

Vol. 953: Iterative Solution of Nonlinear Systems of Equations. Proceedings, 1982. Edited by R. Ansorge, Th. Meis, and W. Törnig. VII, 202 pages. 1982.

Vol. 956: Group Actions and Vector Fields. Proceedings, 1981. Edited by J.B. Carrell. V, 144 pages. 1982.

Vol. 957: Differential Equations. Proceedings, 1981. Edited by D.G. de Figueiredo. VIII, 301 pages. 1982.

Vol. 963: R. Nottrot, Optimal Processes on Manifolds. VI, 124 pages. 1982.

Vol. 964: Ordinary and Partial Differential Equations. Proceedings, 1982. Edited by W.N. Everitt and B.D. Sleeman. XVIII, 726 pages. 1982.

Vol. 968: Numerical Integration of Differential Equations and Large Linear Systems. Proceedings, 1980. Edited by J. Hinze. VI, 412 pages. 1982.

Vol. 970: Twistor Geometry and Non-Linear Systems. Proceedings, 1980. Edited by H.-D. Doebner and T.D. Palev. V, 216 pages. 1982.

Vol. 972: Nonlinear Filtering and Stochastic Control. Proceedings, 1981. Edited by S.K. Mitter and A. Moro. VIII, 297 pages. 1983.

Vol. 978: J. Ławrynowicz, J. Krzyż, Quasiconformal Mappings in the Plane. VI, 177 pages. 1983.

Vol. 979: Mathematical Theories of Optimization. Proceedings, 1981. Edited by J.P. Cecconi and T. Zolezzi. V, 268 pages. 1983.

Vol. 982: Stability Problems for Stochastic Models. Proceedings, 1982. Edited by V.V. Kalashnikov and V.M. Zolotarev. XVII, 295 pages. 1983.

Vol. 989: A.B. Mingarelli, Volterra-Stieltjes Integral Equations and Generalized Ordinary Differential Expressions. XIV, 318 pages. 1983.

Vol. 994: J.-L. Journé, Calderón-Zygmund Operators, Pseudo-Differential Operators and the Cauchy Integral of Calderón. VI, 129 pages. 1983.

Vol. 999: C. Preston, Iterates of Maps on an Interval. VII, 205 pages. 1983.

Vol. 1000: H. Hopf, Differential Geometry in the Large. VII, 184 pages. 1983.

Vol. 1003: J. Schmets, Spaces of Vector-Valued Continuous Functions. VI, 117 pages. 1983.

Vol. 1005: Numerical Methods. Proceedings, 1982. Edited by V. Pereyra and A. Reinoza. V, 296 pages. 1983.

Vol. 1007: Geometric Dynamics. Proceedings, 1981. Edited by J. Palis Jr. IX, 827 pages. 1983.

Vol. 1015: Equations différentielles et systèmes de Pfaff dans le champ complexe – II. Seminar. Edited by R. Gérard et J.P. Ramis. V, 411 pages. 1983.

Vol. 1021: Probability Theory and Mathematical Statistics. Proceedings, 1982. Edited by K. Itô and J.V. Prokhorov. VIII, 747 pages. 1983.

Vol. 1031: Dynamics and Processes. Proceedings, 1981. Edited by Ph. Blanchard and L. Streit. IX, 213 pages. 1983.

Vol. 1032: Ordinary Differential Equations and Operators. Proceedings, 1982. Edited by W.N. Everitt and R.T. Lewis. XV, 521 pages. 1983.

Vol. 1035: The Mathematics and Physics of Disordered Media. Proceedings, 1983. Edited by B.D. Hughes and B.W. Ninham. VII, 432 pages. 1983.

Vol. 1037: Non-linear Partial Differential Operators and Quantization Procedures. Proceedings, 1981. Edited by S.I. Andersson and H.-D. Doebner. VII, 334 pages. 1983.

Vol. 1041: Lie Group Representations II. Proceedings 1982–1983. Edited by R. Herb, S. Kudla, R. Lipsman and J. Rosenberg. IX, 340 pages. 1984.

Vol. 1045: Differential Geometry. Proceedings, 1982. Edited by A.M. Naveira. VIII, 194 pages. 1984.

Vol. 1047: Fluid Dynamics. Seminar, 1982. Edited by H. Beirão da Veiga. VII, 193 pages. 1984.

Vol. 1048: Kinetic Theories and the Boltzmann Equation. Seminar, 1981. Edited by C. Cercignani. VII, 248 pages. 1984.

Vol. 1049: B. Iochum, Cônes autopolaires et algèbres de Jordan. VI, 247 pages. 1984.

Vol. 1054: V. Thomée, Galerkin Finite Element Methods for Parabolic Problems. VII, 237 pages. 1984.

Vol. 1055: Quantum Probability and Applications to the Quantum Theory of Irreversible Processes. Proceedings, 1982. Edited by L. Accardi, A. Frigerio and V. Gorini. VI, 411 pages. 1984.

Vol. 1057: Bifurcation Theory and Applications. Seminar, 1983. Edited by L. Salvadori. VII, 233 pages. 1984.

Vol. 1058: B. Aulbach, Continuous and Discrete Dynamics near Manifolds of Equilibria. IX, 142 pages. 1984.

Vol. 1059: Séminaire de Probabilités XVIII, 1982/83. Proceedings. Edité par J. Azéma et M. Yor. IV, 518 pages. 1984.

Vol. 1063: Orienting Polymers. Proceedings, 1983. Edited by J.L. Ericksen. VII, 166 pages. 1984.

Vol. 1065: A. Cuyt, Padé Approximants for Operators: Theory and Applications. IX, 138 pages. 1984.

Vol. 1066: Numerical Analysis. Proceedings, 1983. Edited by D.F. Griffiths. XI, 275 pages. 1984.

Vol. 1071: Padé Approximation and its Applications, Bad Honnef 1983. Proceedings Edited by H. Werner and H.J. Bünger. VI, 264 pages. 1984.

Vol. 1072: F. Rothe, Global Solutions of Reaction-Diffusion Systems. V, 216 pages. 1984.

Vol. 1085: G.K. Immink, Asymptotics of Analytic Difference Equations. V, 134 pages. 1984.

Vol. 1086: Sensitivity of Functionals with Applications to Engineering Sciences. Proceedings, 1983. Edited by V. Komkov. V, 130 pages. 1984

Vol. 1100: V. Ivrii, The Precise Spectral Asymptotics for Elliptic Operators Acting in Fiberings over Manifolds with Boundary. V, 237 pages. 1984.